BRADLEY

A HISTORY OF AMERICAN FIGHTING AND SUPPORT VEHICLES

by
R.P. Hunnicutt

FOREWORD
by
Major General Stan R. Sheridan, USA Retired

E P B M
ECHO POINT BOOKS & MEDIA, LLC

Published by Echo Point Books & Media
Brattleboro, Vermont
www.EchoPointBooks.com

Copyright © 1999, 2015 R. P. Hunnicutt
ISBN: 978-1-62654-153-5

Cover design by Adrienne Núñez,
Echo Point Books & Media

Editorial and proofreading assistance by Christine Schultz,
Echo Point Books & Media

Printed and bound in the United States of America

CONTENTS

FOREWORD
by
Major General Stan R. Sheridan, USA Retired

CHARIOTS OF FIRE may be the title of an award winning Hollywood movie, but it is also a fitting description of the many "soldier carrying" vehicles described in this latest work on armored vehicles by Dick Hunnicutt. In fact, chariots of one kind or another have carried soldiers into battle throughout the history of warfare, dating as far back as the early Egyptians and Romans. So it is appropriate that this latest in Dick Hunnicutt's series of volumes on Fighting Vehicles should now focus on the armored carriers, both fighting and support, that inhabit today's modern battlefield. While this eminently qualified author covers the spectrum of these vehicles in this newest volume, as the title suggests, the thrust of this work is the U.S. Army's Infantry and Cavalry Fighting Vehicles and their derivatives – The BRADLEY. While I find this entire volume intriguing and a true and factual reference for the scholar of fighting vehicles, as the U.S. Army's first Program Manager for what has become known to the world as the BRADLEY Fighting Vehicle System, it appears appropriate for me to comment on the title-subject of Dick Hunnicutt's latest work.

The history of the BRADLEY was long and tortured, and today we take the design and the outstanding war fighting performance of the vehicle for granted – the two man turret, the two TOW antitank missile launcher, the highly effective 25mm cannon system, the very reliable power train, its outstanding cross country mobility, and the overall fightability of the system. This was not always so. In the beginning, in the late 1960s and early 1970s, the Army was struggling to determine and define just what it wanted as the replacement for the M113 Armored Personnel Carrier (APC) that was a workhorse during the Vietnam War and the backbone of the Army's mechanized infantry. Was the replacement to be another APC that brought fighting men to the battle in a protected "battlefield taxi" and then placed them in harms way to fight on foot; or was it to be a true fighting vehicle, giving the soldier a protected place from which to assault, fight, and kill the enemy? The result, in the early 1970s, was the latter, a fighting vehicle concept called the Mechanized Infantry Combat Vehicle or MICV, which, when translated into an all-up prototype in the mid 1970s, was unfightable: i.e., the gunner was in a one man turret; the vehicle commander was in the hull behind the driver where he could not see to command or fight the vehicle; the

crew/squad compartment was an unfightable and crowded "arms room"; and the main armament was a 20mm cannon with no armor killing capability.

In 1975 the MICV program was reoriented and combined with the Army's SCOUT and BUSH-MASTER (25mm cannon) programs into a single vehicle program, named the Infantry and Cavalry Fighting Vehicle Program, which was later renamed in 1981 for General of the Army Omar N. Bradley. With that reorientation came a reaffirmation of the Army's requirement and a redesign to today's BRADLEY Fighting Vehicle with its two man turret which places the vehicle commander up high where he can see, command, and fight the vehicle; the addition of a two TOW antitank missile launcher to give the Infantry Battalion a long range, front line, tank killing capability without increasing the Army's force structure; a re-stowed and redesigned crew compartment into a fighting compartment from which mounted infantrymen can fight; and the replacement of the less than capable 20mm cannon with the battlefield worthy 25mm cannon and its armor piercing and high explosive multipurpose ammunition. With this redesign and reorientation it became readily apparent that the technical design challenge for the developer of the new vehicle was on a par with that of designing a tank; but with the added human factors of carrying an infantry squad, allowing the vehicle to swim, and ultimately making it truly fightable for the mounted infantryman. At the same time, from a doctrine standpoint, the Mechanized Infantry found itself in much the same position as the horse mounted Cavalry when the machine gun was introduced to the battlefield. The design of the new mobile weapon system, when translated into fightable hardware, required a mounted infantry doctrine change and the development of new operational concepts and tactics in order to take full advantage of the new and added battlefield capabilities of a fighting vehicle. As an example, firing on moving targets with the 25mm cannon now required the infantry gunner to use tank gunnery techniques, totally foreign to the infantryman of the late 1970s and early 1980s whose largest automatic weapon until then was a .50 caliber machine gun. As a result of this and other operational capabilities and requirements of the new system, totally new training packages also were required for the mounted infantryman who would fight in the BRADLEY. To the Army's credit, it bridged the doctrine, training, and tactics gaps and has produced the world's most capable and finest mounted warriors.

With this new concept, design, and training direction in hand, the vehicle development program proceeded successfully through the 1970s and early 1980s fighting off the "Too Big, Too Bulky, Too High" nay-sayers, a Presidential program cancellation, and three General Officer Reviews (1976, 77, and 78) by the Army. With the program restart after the Presidential cancellation in 1977, and the reaffirmation of the requirement, the design, and the concept by the three General Officer Reviews, the program proceeded to meet its congressionally mandated first production delivery date of May 1981 without further hitches or delays. In fact, the BRADLEY was the first, and I believe the only, tracked vehicle to be approved for production by the Army and the Office of the Secretary of Defense on the first request. This was due primarily to the vehicle exceeding its overall system Reliability-Availability-Maintainability (RAM) requirements during independent government acceptance testing.

But there is more to the BRADLEY story than development, great designed-in system RAM, doctrinal changes, and the final Army acceptance and production go-ahead. The real questions facing the BRADLEY were: What do soldiers think of the vehicle? Is it really fightable? Does it meet the Army's needs? And how does it do in combat? The proof of any piece of combat equipment issued to soldiers is its performance and soldier acceptance in combat, and the BRADLEY was no exception. The real proof then of the BRADLEY was Desert Storm where it received not only its baptism of fire in combat, but complete soldier acceptance. The experience of the lead brigade of the 24th Mechanized Infantry Division's "Left Hook" operation is typical of the BRADLEY'S superb combat performance in the 100 hours of Desert Storm. The brigade's 120 BRADLEYS traveled 360 miles, fighting all the way, with no vehicle drop-outs or losses. While the 25mm armor piercing round did kill some T72 tanks from the side and rear, it was an over-kill against the Iraqi's infantry carriers (BMP), passing right through the BMPs and calling for the use of the more appropriate HEAT-MP round (High Explosive Antitank Multi-Purpose). The BRADLEY soldiers of Desert Storm, and those using the vehicle in places like Somalia and Bosnia, have resoundingly endorsed the system and put to bed the nay-sayers, the questioners, and the critics by affirming that the BRADLEY is a highly mobile and effective battlefield killing machine; it is not an APC or battlefield taxi, but it does take soldiers to the battle and lets them fight while mounted and protected; it is not a tank, nor is it heavily armored, but it does have a long range tank killing capability; and it exceeds the tank's cross country mobility and effectively compliments the tank on the battlefield. Today, with 6724 Infantry and Cavalry Fighting Vehicles in the hands of U.S. Army soldiers around the world, the BRADLEY is justly touted as the finest fighting vehicle of its kind in the world.

Looking back and forgetting the pain along the way, one can say that the BRADLEY was a success story, primarily because of the Army's belief in, and support for, a fighting vehicle and the dedicated and hand-in-hand team effort by all those directly involved in its development, production, and fielding – the Army Program Manager's Office, the Army Infantry and Cavalry users, and all of the many dedicated civilian contractors. A widely used development buzzword today is PARTNERING, or joining together of all those involved in a development program toward a common goal; but without knowing it, that is what was done with the BRADLEY in the 1970s and early 1980s, long before the word or the thought was in vogue in the Defense Department.

And what of the BRADLEY derivatives, or support vehicles, during this process? In 1975, the Army had a need for a tracked vehicle platform for the Multiple Launch Rocket System (MLRS) and the fighting vehicle chassis was chosen as the candidate platform. In reality, what the Army really wanted was a highly mobile tracked "pick-up truck" whose "truck bed" could be used for many battlefield missions; but at the time, the only money available was for the development of a MLRS carrier. Taking its lead from the very successful and reliable automotive and suspension components of the original MICV chassis, the MLRS carrier was developed, tested, accepted, and fielded with almost complete commonality with the chassis of its sister fighting vehicle. The differences between the two being in the physical rather than the mechanical aspects of the chassis. Again, the proof of this derivative was its complete success and soldier acceptance in the combat of Desert Storm. At the same time, the Army got its "pick-up truck". Today the derivative carrier's time has come for, among other uses, it is being strongly considered by the Army as a command and control vehicle, an ambulance, and a communications vehicle.

ACKNOWLEDGEMENTS

It would be difficult, if not impossible, to list all of the sources that provided information during the research for this volume. However, the most important of these must include Major General Stan Sheridan who also was kind enough to write the foreword for the book.

Once again, my thanks go to General Donn A. Starry for permission to quote from his book Mounted Combat in Vietnam.

At the Tank Automotive and Armament Command, Dr. Richard McClelland and others were a great help in obtaining material for the data sheets. Roland Asoklis provided information on the Future Scout and Cavalry System.

The Patton Museum at Fort Knox was an important source of both data and photographs. John Purdy and Charles Lemons spent a lot of time helping me sort out a number of problems.

At Aberdeen Proving Ground, Dr. William F. Atwater, Director of the Ordnance Museum, and Alan Killinger located much of the information on some of the earlier vehicles.

At United Defense (formerly FMC), William Highlander put me in touch with many of the people who worked on their vehicles. Mr. Adolf Quilici provided information on the early development of the FMC personnel carriers. John Giacomazzi and Bruce Heron were particularly helpful in locating drawings and data. Linda Johns and Pat Elliott found many photographs and technical manuals required for the project.

At the Land Systems Division of General Dynamics, Dr. Philip W. Lett obtained information on their advanced amphibian assault vehicle.

Kenneth Smith-Christmas of the U.S. Marine Corps Museum at Quantico, Virginia provided data on the LVTP-5.

My old friend Fred Crismon was the source for many very rare photographs.

Many of the field modifications to the various vehicles in Vietnam are illustrated by photographs taken by the late James Loop during his service in that country.

Michael Green, Greg Stewart, Jim Mesko, Hans Halberstadt and Scott Gourley also provided photographs of several vehicles. There would have been several more gaps in the data sheets without the help of Jacques Littlefield and his incredible collection of armored vehicles. Dean and Nancy Kleffman came up with some new photographs that were particularly helpful to the research program.

My thanks go to Jon Clemens of Armor magazine for his help with the photographs and his permission to quote from Armor.

Special thanks also go to Becky Page for the preparation of the dust jacket for this volume.

PART I

DEVELOPMENT OF SUPPORT VEHICLES DURING WORLD WAR II

During World War II, the standard armored personnel carrier was the open top, lightly armored, half-track vehicle. Above the M3A1 personnel carriers of the Third Armored Division cross the Seine River in France on 26 August 1944.

ARMORED PERSONNEL AND CARGO CARRIERS

During World War II, armored personnel carriers appeared under a variety of names. These included tractors, cargo carriers, and armored utility vehicles. When used to tow an artillery piece, they were referred to as prime movers. The armored half-track was standardized during this period as the carrier for infantry in the armored divisions of the U.S. Army. These lightly armored, open top, vehicles served in many roles including personnel carriers, mortar carriers, and self-propelled artillery. Although they were intended to combine the best features of the wheeled and full-tracked vehicles, it also was true that they combined the worst characteristics of each. They were not as efficient on roads as armored cars and they lacked the cross-country mobility of the full-tracked vehicles.

The introduction of the Sexton 25 pounder self-propelled gun resulted in the replacement of the early 105mm howitzer motor carriage M7 in the British forces. Many of these surplus vehicles were then converted to armored personnel carriers by removing the howitzer and modifying the armor and the stowage arrangement. This conversion provided a full-tracked armored personnel carrier that could easily keep up with the tanks in cross-country operations.

When the Canadian Army was equipped with Sherman tanks, a large number of the earlier Ram tanks became available for conversion to other duties. With the turret removed and the interior modified, they became personnel carriers with mobility and armor protection equal to that of the medium tank. Some old Sherman III (M4A2) medium tanks also had their turrets removed and were converted to armored personnel carriers by the British forces. All of these converted vehicles were referred to as Kangaroos.

9

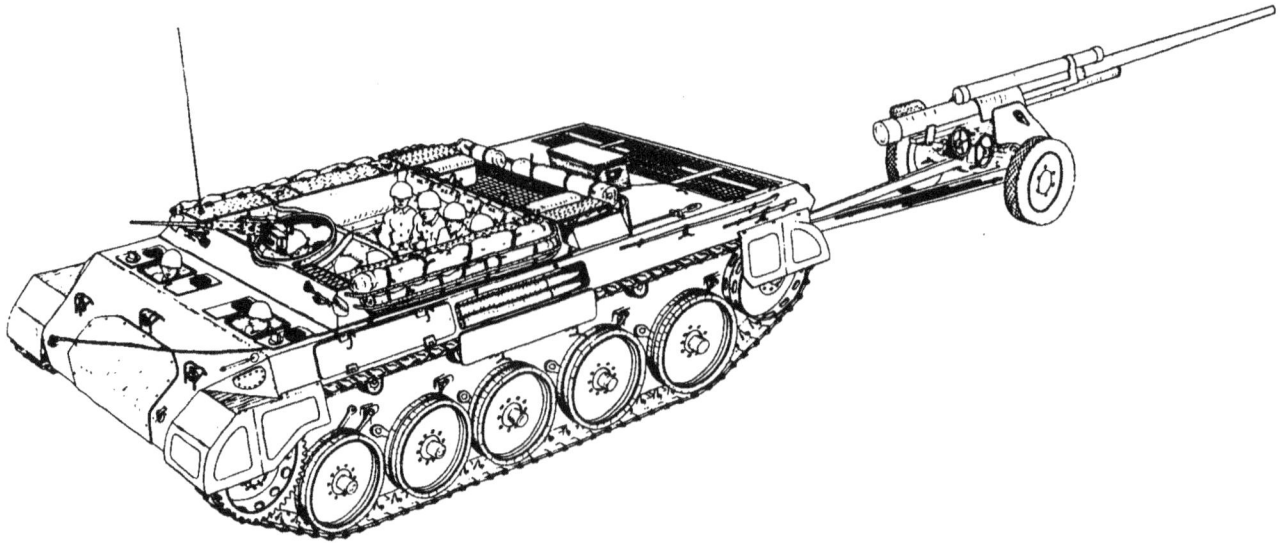

The armored utility vehicle M39 is sketched above towing an antitank gun.

The appearance of the high performance 76mm gun motor carriage M18 as a self-propelled tank destroyer revealed the need for an equivalent vehicle to serve as a personnel carrier, reconnaissance vehicle, and prime mover for the towed guns in the tank destroyer battalions. The M18 chassis was modified by removing the turret and changing the stowage to provide a suitable vehicle. Standardized as the armored utility vehicle M39, it served in the U.S. Army until the end of the war in Korea. However, its very light armor was vulnerable to rifle caliber fire at close range and the open top exposed the crew to artillery air bursts.

Numerous wartime development programs were initiated to provide full-tracked, armored, personnel and ammunition carriers as companion vehicles for the self-propelled artillery mounted on the chassis of the medium and light tanks. These companion vehicles also were conversions of the same chassis used to mount the self-propelled weapon.

The T14 cargo carrier, based upon the medium tank chassis, was modernized and standardized as the cargo carrier M30. It saw action as the companion vehicle to the 155mm gun motor carriage M12. With the introduction of the 155mm gun motor carriage M40, a new companion vehicle was designed using the same chassis. Originally designated as the cargo carrier T30, it was modified to permit the stowage of either 155mm, 8 inch, or 240mm ammunition. This modified version was designated as the cargo carrier T30E1. However, with the decision to use unarmored, high speed, tractors as the companion vehicles for the heavy self-propelled artillery, production plans for the T30E1 were canceled. A similar fate befell the cargo carrier T31. This vehicle

was designed to accompany the 8 inch howitzer motor carriage T84. The chassis of both the T84 and the T31 was based upon components of the M26 Pershing tank.

Companion vehicles for the self-propelled artillery utilizing the light tank chassis followed the same pattern. The cargo carrier T22 was designed as the companion vehicle for the 4.5 inch gun motor carriage T16 or the 155mm howitzer motor carriage T64. Like the artillery motor carriages, the T22 was based upon the light tank M5A1. The cargo carrier T23 also used the chassis of the M5A1 and was intended to accompany the 40mm gun motor carriage T65. When the artillery motor carriages were redesigned to utilize the chassis of the later light tank M24, a similar change was made in the cargo carriers. They now became the cargo carriers T22E1 and T23E1. However, neither vehicle was placed in production.

The heavy tractors T2 and T16 were early prime movers intended to serve as armored personnel carriers for the crew of the heavy artillery pieces to which they were assigned. Based upon the early medium tank, neither the T2 nor the T16 proved to be satisfactory and they never entered production. Later, in December 1943, some M31 and M32B1 tank recovery vehicles were converted into prime movers for use with the 8 inch gun and the 240mm howitzer. The turret, armament, and recovery equipment was removed and they were designated as the prime movers M33 and M34 respectively. In January 1944, 209 M10A1 tank destroyers also were converted to M35 prime movers for the heavy artillery by removing the turret. All of these served as armored personnel carriers for the artillery crews.

Above, the universal carrier T16, vehicle number 26, is at Aberdeen Proving Ground on 31 May 1943. The universal carrier T16E2 appears at the left below. Note the change in the spring arrangement on the rear bogie.

Production of the cargo carrier T16 by the Ford Motor Company began in March 1943. This vehicle, a redesigned version of the British universal carrier, was later designated as the universal carrier T16 to provide uniformity with the British nomenclature. It differed from the British vehicle in several features. The T16 suspension consisted of four road wheels in two bogies per side compared to one two wheel bogie and a single independently sprung road wheel on the earlier vehicle. The springs on the suspension bogies of the T16 were aligned in opposite directions. The T16 was powered by a Ford V-8 gasoline engine developing 102 net horsepower at 4,000 rpm. Manned by a crew of four, it was used as a prime mover for light antitank guns and as

a carrier for a variety of weapons. With a combat weight of about five tons, the T16 had a maximum road speed of 30 miles per hour and a cruising range of approximately 150 miles. The open top vehicle was protected by steel armor ranging in thickness from $^9/_{32}$ to $^7/_{32}$ inches.

A modified version of the T16 was designated as the universal carrier T16E2. An obvious difference between the two vehicles was the reversal of the rear suspension bogie on the T16E2 so that the springs were aligned in the same direction as on the front bogie. The upper front armor also was increased in thickness to $^{25}/_{64}$ inches. Production of the T16E2 began in 1945.

A total of 13,893 T16 series universal carriers were built by the Ford Motor Company ending in July 1945. An additional 5714 universal carriers were procured from War Supplies, Ltd. bringing the total to 19,607 vehicles. A total of 19,193 carriers were distributed under the Lend-Lease program with 19,079 going to Britain and 96 to the Soviet Union. None were issued to the U.S. Army.

The light tractor T18 was based upon the light airborne tank T9. Intended as the prime mover for the M3 105mm howitzer, it carried 25 rounds of 105mm ammunition in addition to the five man crew. With a combat weight of less than eight tons, the open top T18 had armor protection ranging from ½ inch on the front and sides to $^3/_8$ inches on the rear. It utilized the power train and suspension of the light tank T9. The air-cooled Lycoming engine drove the T18 at a maximum road speed of 40 miles per hour and the 55 gallon fuel tank provided a cruising range of about 200 miles. Converted by Marmon-Herrington, the second of the two pilot T18s was under test at Aberdeen Proving Ground during March 1943, but it was not released for production.

At the left is the pilot number 2 of the light tractor T18 during evaluation at Aberdeen Proving Ground. This photograph was dated 26 March 1943.

The armored utility vehicle T9 appears above on 9 July 1945. The exhaust grill can be seen in the left wall of the hull.

On 21 July 1944, Item 24520 of the Ordnance Committee Minutes (OCM), approved the procurement of two pilot T9 armored utility vehicles. Although two were originally authorized, only one was completed by the Chevrolet Division of General Motors Corporation. The objective of this program was to develop a prime mover for the 90mm gun carriage T9E1. The vehicle also was to be evaluated as an infantry carrier. In the latter respect, some features of the T9 foreshadowed those of the postwar armored personnel carriers. Although it had an open top, the box-like hull was fitted with a rear door through which the crew could dismount, even though the internal stowage made this somewhat difficult. The driver and the assistant driver were seated

in the front of the vehicle on the left and right respectively. Armored flaps on the front could be lowered and replaced by windshields when not in a combat area. When not in use, the windshields were stowed on the bulkhead behind the drivers. The engine was installed immediately behind the driving compartment and another bulkhead behind the engine separated it from the personnel in the rear of the vehicle. The air-cooled Lycoming O-435-TA engine was a slightly modified version of the O-435-T in the light tank M22. This six cylinder, horizontally opposed, power plant developed 165 net horsepower at 3,000 rpm. It drove the T9 through a Spicer synchromesh transmission, a controlled differential, the final drives,

Exhaust stacks for deep water fording have been installed on the armored utility vehicle T9 in these views. The open rear door shows the access to the rear of the vehicle.

The bottom hinged rear door of the armored utility vehicle T9 number 1 is shown above in the closed position.

and the sprockets at the front of each track. The torsion bar suspension carried the vehicle on four 25 inch diameter road wheels and a 34¼ inch diameter trailing idler on each side. The single pin, rubber bushed, steel tracks were 16 inches wide. A 10,000 pound line pull winch was installed in the left rear of the personnel compartment. With a loaded weight of ten tons, the maximum road speed of the T9 was 30 miles per hour. All hull openings were fitted with rubber seals and the fording depth was limited to 41 inches by the engine and cooling air exhaust grills on each side. Exhaust stacks could be installed on these grills to increase the fording depth.

A crew of 12 was specified for the T9 including the drivers. However, the rear compartment was extremely crowded with ten men. The vehicle was lightly armored with ¼ inch thick steel plate on the front, sides, and rear. Thus it could have been penetrated by small arms fire at close range and the open top exposed the crew to attack from above. The tests at Aberdeen revealed numerous mechanical problems, most of which could have been corrected. However, the war was over by the time the tests were completed and the project was terminated by OCM 30424 dated 11 April 1946. Already, the need for overhead protection had become obvious and new designs for armored personnel carriers were under study.

The interior arrangement of the armored utility vehicle T9 appears in the views below.

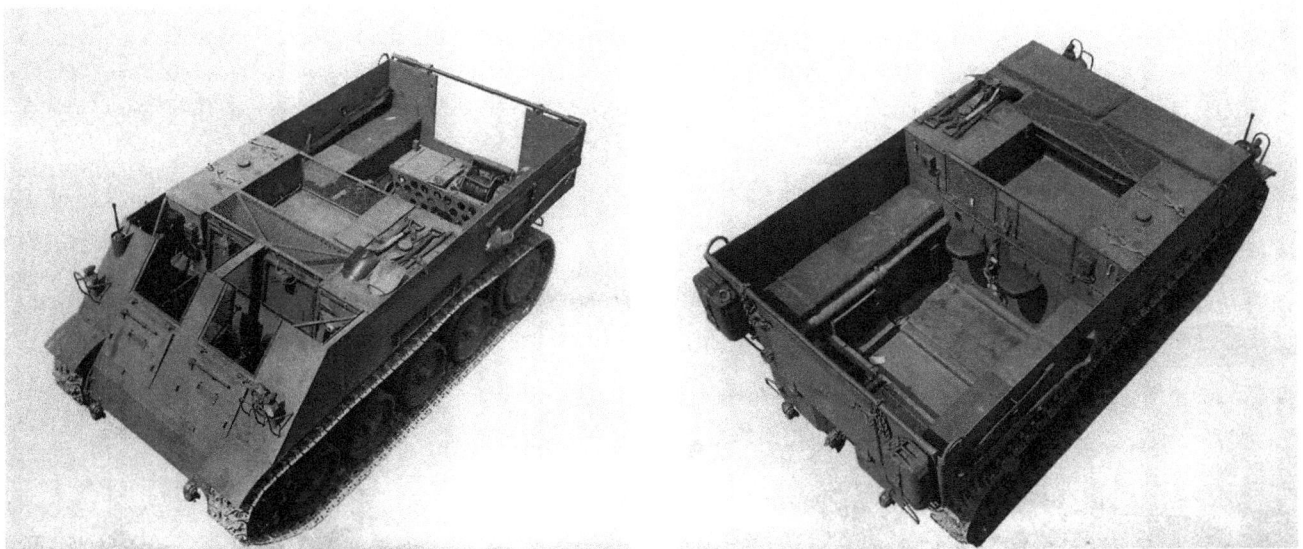

Front views of the light armored car M8, reconnaissance vehicle T8E1, reconnaissance vehicle T8, light tank M5A1, and the 75mm howitzer motor carriage M8 can be compared from left to right above.

ARMORED COMMAND AND RECONNAISSANCE VEHICLES

The use of medium and light tanks as command vehicles was standard practice in the armored units. On most of these, the modifications consisted of little more than the installation of additional radios and signaling equipment. Some other vehicles received extensive rework. For example, Brigadier General George Read, the Assistant Division Commander of the 6th Armored Division, used a highly modified M8 75mm howitzer motor carriage as his command vehicle. The turret on the M8 was removed and replaced by a fixed armor superstructure fitted with .50 caliber machine gun.

A wide variety of armored vehicles were used to perform the reconnaissance mission. These included armored cars, half-tracks, and light tanks. One experimental program attempted to develop a full-tracked armored vehicle specifically for reconnaissance duties. This development was required by a letter from the Headquarters, Army Ground Forces dated 21

December 1943. In response, OCM 22941 of 17 February 1944 directed that two M5A1 light tanks be converted to reconnaissance vehicles. These vehicles evaluated two different designs and were designated as the reconnaissance vehicles T8 and T8E1. In both cases, the turrets were removed and an M49C ring mount was installed for a .50 caliber machine gun. On the T8, this mount was high above the right rear of the open fighting compartment. The T8E1 carried the same mount lower and concentric with the turret ring opening. The T8E1 was fitted with 16 inch wide Burgess-Norton steel tracks while the T8 retained the original $11^{5}/_{8}$ inch wide rubber tracks. With a six man crew, the T8 and T8E1 had combat weights of 28,200 pounds and 30,900 pounds respectively. Both vehicles carried the SCR 506 radio in the left sponson and the SCR 528 in the right. A rack for ten antitank mines was attached to the outside of the right sponson and a 2 inch smoke mortar was located in the hull roof between the driver and the assistant driver.

Evaluated at Fort Knox, the T8E1 was considered to be a satisfactory reconnaissance vehicle and superior to the T8. However, the Armored Force Board recommended that the light tank M24 be used as the primary reconnaissance vehicle because of the greater firepower of its 75mm gun.

Brigadier General George W. Read, Jr., the assistant division commander of the 6th Armored Division, is at the left in his command vehicle converted from a 75mm howitzer motor carriage M8.

Above, the armored utility vehicle T41E1 appears at the left and the first Canadian Armored Tracked Jeep produced by Willys Overland is at the right. This photograph was dated 20 May 1944.

As mentioned previously, the armored utility vehicle T41 was standardized as the M39 and was intended to be the prime mover for the 90mm gun carriage T5E1 in the Tank Destroyer Command. Ten vehicles were designated as the armored utility vehicle T41E1 and fitted out as command and reconnaissance vehicles. These were equipped with SCR 506 and SCR 508 radios in the left and right sponsons respectively. To provide power for the additional radios, an auxiliary generator was installed in the right front sponson. Like the M39, the open top T41E1 was vulnerable to artillery air bursts and the Tank Destroyer Command proposed armor covers for both vehicles late in World War II.

Several interesting vehicles were produced for Canada by Marmon-Herrington and Willys-Overland Motors, Inc. utilizing power train components from the ¼ ton truck. Dubbed the Canadian Armored Tracked Jeep Mark I, it was a two man, open top, vehicle armored for protection against small arms ball ammunition at close range or .303 caliber AP rounds at a range of 250 yards. The proposed armament was a Bren light machine gun mounted in the center of the front armor plate. The engine and transmission were installed transversely at the rear of the hull and drove the rear mounted sprockets through a controlled differential. The suspension consisted of four road wheels in two bogies plus an independently sprung leading road wheel on each side. The vehicle was considered to be amphibious, but since it was only propelled in the water by track action, the maximum speed was 1-2 miles per hour. With all of the power train in the rear, the vehicle was down by the stern in the water. Two of the Mark I vehicles were built, but they were soon succeeded by a modified design. The weight distribution was improved in the new version with the engine and transmission relocated to the front of the vehicle and connected by a drive shaft to the differential at the rear. Six of the modified vehicles were completed.

A third design, referred to as the Canadian-American Tracked Jeep Mark II, had a larger hull which would have increased the internal volume and, no doubt, improved its flotation. Five of these were completed, but with the end of the war there was no further production of any of the tracked jeep vehicles.

Below, the later Canadian Tracked Jeep with the relocated power train is at the left and the Mark II version is at the right.

The high speed tractor M4 appears in the photographs above. Note the original narrow tracks on the horizontal volute spring suspension.

HIGH SPEED TRACTORS

Early in World War II, a large number of high speed tractors were evaluated for military use. Some of these were developed at Rock Island Arsenal and others were commercial vehicles. In February 1941, the 7 ton high speed tractor M2 was standardized. Built by the Cleveland Tractor Company to Ordnance specifications, it was driven by a Hercules WXLC3, six cylinder, gasoline engine. Its maximum speed was about 22 miles per hour on level roads. Used primarily by the Army Air Force to tow heavy aircraft, the M2 was driven by one man although seats were provided for three. A total of 8,510 M2s were produced at the Oliver Company (formerly the Cleveland Tractor Company) and the subcontractor, John Deere & Company.

During 1941, the medium tractor T9 was evaluated at Aberdeen Proving Ground. Built by the Allis-Chalmers Manufacturing Company, it had a loaded weight of 35,280 pounds. With a General Motors 6-71 diesel engine and a torque converter transmission, the T9 had a maximum road speed of 33 miles per hour. However, in 1942, the decision was made to replace the diesel engine with a 225 horsepower Waukesha gasoline engine and the designation was changed to the medium tractor T9E1. This was the vehicle standardized as the 18 ton high speed tractor M4. Designed for artillery towed loads of 18,000 to 30,000 pounds, the M4 could tow a 90mm gun at 33 miles per hour on a level road.

The high speed tractor M2 is at the right.

With a fuel capacity of 125 gallons, the cruising range was about 180 miles. The M4 was designated as Class A when carrying an ammunition box with racks for 3 inch or 90mm rounds. The Class B designation applied when carrying a cargo box with racks for 155mm howitzer, 8 inch howitzer, or 240mm howitzer ammunition. A swing crane and hoist were provided for lifting shells into the cargo box. The cab was divided into two compartments with space for the driver and two men in front and double seats for eight additional men in the rear. The vehicle was equipped with a winch having a maximum 30,000 pound pull and 300 feet of ¾ inch wire cable. An M49C ring mount for a .50 caliber machine gun was provided for antiaircraft protection. Allis-Chalmers produced 5,552 M4s in the production period lasting from March 1943 through June 1945.

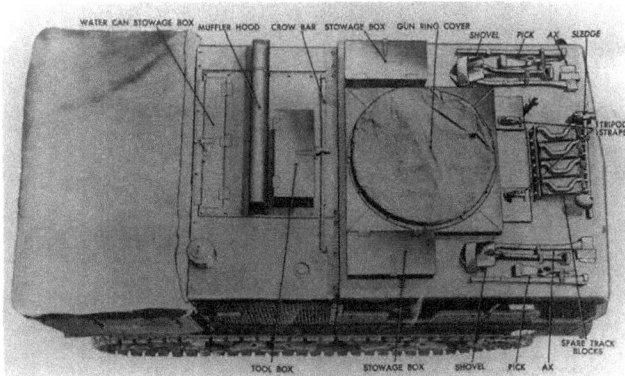

The general arrangement of the high speed tractor M4 can be seen in the top view at the left above and the driver's controls appear at the upper right. The 155mm ammunition stowage in the passenger compartment of the M4C is shown below.

Tests by the Field Artillery Board indicated that it was feasible to install a spaced out suspension and tracks with extended end connectors on the M4. This increased the track width from $16^9/_{16}$ inches to $20^1/_8$ inches with the extended end connectors on only one side of the tracks. If they were installed on both sides, the track width increased to $23^{11}/_{16}$ inches. On 28 November 1944, OCM 26329 recommended the standardization of the vehicle with the spaced out suspension as the 18 ton high speed tractor M4A1. The M4 was then reclassified as Limited Standard. This action was approved on 11 January 1945. A total of 259 M4A1s were built during the period June through August 1945.

The M4 and the M4A1 continued to serve into the postwar period. When the vehicles were modified to carry projectiles and fuzes in the crew compartment, the designations were changed to M4C and M4A1C. The number of personnel carried in these vehicles was reduced from eleven to eight. When the new M8 series cargo tractor was in short supply in 1954, Bowen-McLaughlin received a contract to refurbish and modernize some of the M4 series tractors. These vehicles were designated as the 18 ton high speed tractor M4A2.

The high speed tractors M4A2 and T13 are illustrated below and at the right respectively.

The unarmored cargo carrier T13 was a high speed tractor developed by the International Harvester Company as a companion vehicle to the 3 inch gun motor carriage M5. Development was approved by Ordnance Committee action in March 1942. The T13 used the Continental R6572 gasoline engine and a modified M3 light tank suspension. Manned by a crew of four, it had a combat weight of about 25,000 pounds. The maximum road speed was 35 miles per hour and the cruising range was approximately 150 miles. It was designed to carry 113 rounds of 3 inch ammunition. When the 3 inch gun motor carriage M5 project was canceled, the T13 program was terminated in the Spring of 1943.

These views show the high speed tractor M5 above and the high speed tractor M5A1 below. Details of the suspension are shown at the bottom right.

The experience gained on the T13 project was applied to the development of the 13 ton high speed tractor M5. This was a prime mover for artillery pieces weighing up to 16,000 pounds. These included the 105mm howitzer carriage M2, the 4.5 inch gun carriage M1, and the 155mm howitzer carriage M1. The M5 had a loaded weight of 28,300 pounds and was driven by the same Continental R6572 gasoline engine as the T13. This was a six cylinder, in-line, power plant developing 235 net horsepower at 2,900 rpm. Like the T13, the M5 used the tracks and a modified suspension from the light tank M3. The maximum road speed was 35 miles per hour and the 100 gallon fuel capacity provided a cruising range of approximately 125 miles. A folding top with side curtains was installed and the vehicle had space for nine men including the driver. A front mounted winch had a capacity of 17,000 pounds. The only armament was a .50 caliber machine gun on an M49C ring mount. The M5 was manufactured by the International Harvester Company. A total of 5,290 vehicles were produced during a run from May 1943 through May 1945.

When fitted with a new steel cab carrying eleven men, the vehicle was designated as the 13 ton high speed tractor M5A1. The overall length was increased from $191^{1}/_{8}$ inches to $196^{3}/_{8}$ inches. Production of the M5A1 totaled 589 during the period May through August 1945.

Like the 18 ton high speed tractor M4, the M5 continued to serve well into the postwar period and it was subject to additional modifications. The most important of these was the installation of a new horizontal volute spring suspension replacing the earlier vertical volute spring suspension. The new suspension was fitted with 21 inch wide, center guide, tracks replacing the $11^{5}/_{8}$ inch wide, outside guide, tracks on the earlier vehicle. After modification, the M5 and M5A1 were designated as the 13 ton high speed tractors M5A2 and M5A3 respectively.

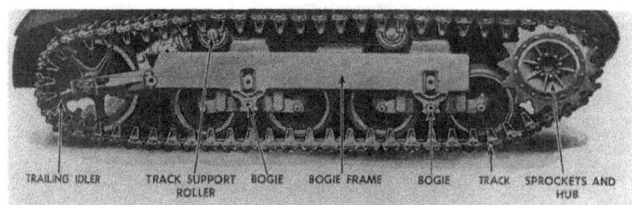

18

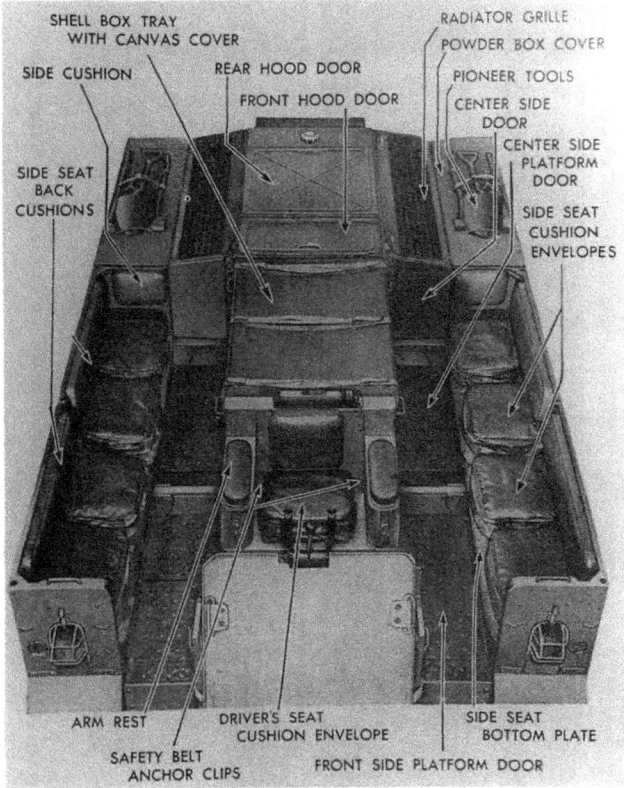

The high speed tractor M5 can be seen above with the doors and covers closed (left) and open (right). The driver's controls are at the right below.

At the left, above and below, are views of the high speed tractor M5A2. Note the new wide track horizontal volute spring suspension. The high speed tractor M5A3 is shown below at the right.

The heavy tractor T22 above is towing a 240mm howitzer tube. The early volute spring suspension on the T22 at the top right can be compared with the later design below.

In February 1942, Ordnance Committee action initiated the development of the heavy tractors T22 and T23. The T22 was designed with a fifth wheel on the rear deck for semi-trailing transport wagons of the 240mm howitzer M1and the 8 inch gun M1. The heavy tractor T23 was identical to the T22 except that the fifth wheel on the rear deck was replaced by a large cargo box for ammunition and equipment. The T23 was intended to be the prime mover for the 4.7 inch antiaircraft gun T1. The decision by the Field Artillery Board to carry the 240mm howitzer and the 8 inch gun on full-trailed transport wagons eliminated the need for a fifth wheel on the prime mover. The T22 was dropped and the T23 was standardized in June 1943 and designated as the 38 ton high speed tractor M6. It could tow artillery loads of 30,000 to 60,000 pounds and carried eleven men in two rows of seats as well as ammunition and equipment. Two Waukesha 145GZ gasoline engines supplied the power through torque converters and a constant mesh transmission. The vehicle rode on a horizontal volute spring suspension with three, two wheel, bogies per side and a large trailing idler. The double pin, center guide, tracks were $21^9/_{16}$ inches wide. Towing the tube of the 240mm howitzer M1, the M6 could reach 20 miles per hour on a level road. A 250 gallon fuel tank provided a cruising range of about 110 miles. The vehicle also was equipped with a 60,000 pound capacity winch. An M49C ring mount for a .50 caliber machine gun was on the roof.

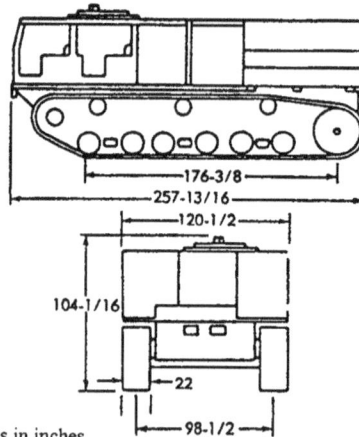

Dimensions in inches

The high speed tractor M6 is illustrated in the views below and the dimensions of the vehicle are shown in the sketch above.

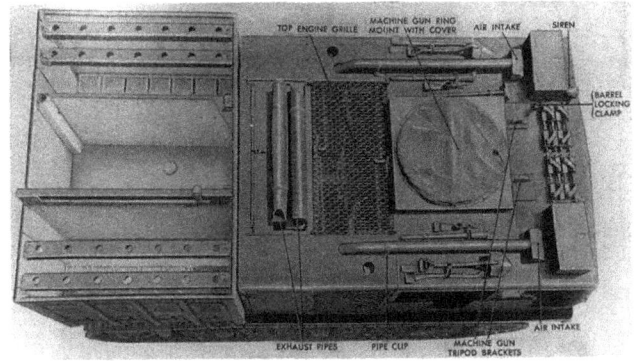

Further details of the high speed tractor M6 can be seen above. Below, a bottom view of the M6 appears at the left and the driver's controls are at the right.

The .50 caliber machine gun M2 is installed on the M49C ring mount at the right.

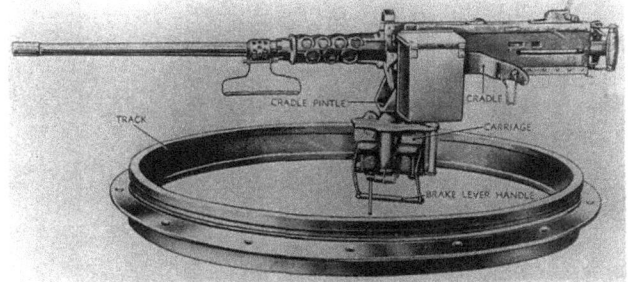

Depending upon the weapon involved, the ammunition stowage on the M6 was 24 rounds of 4.7 inch, 20 rounds of 240mm, or 24 rounds of 8 inch. Production of the M6 was at the Allis-Chalmers Manufacturing Company from February 1944 through August 1945 for a total of 1,235 vehicles.

OCM 26899, dated 8 March 1945, recommended the development of the full-tracked, high speed, crane T9 for use with the 240mm howitzer M1 and the 8 inch gun M1. The procurement of two pilots was not approved until March 1947. At that time, a contract was negotiated with the Milwaukee Excavator Company to construct the two T9s. The T9s were converted from the 38 ton high speed tractor M6 by replacing the rear cargo boxes with a 20 ton capacity crane and telescopic outriggers. The first pilot was shipped to Aberdeen Proving Ground in October 1947 for testing. The second was completed in early 1948.

The high speed crane T9, based upon the M6 tractor, is shown above.

Above, the snow tractor M7 is at the left and the snow tractor T36 is at the right.

LOW GROUND PRESSURE VEHICLES

A number of snow tractors were evaluated during World War II for use in Alaska and along the Alcan Highway. However, only one was standardized. The snow tractor M7 was a half-track vehicle with front wheels that could be replaced by skis. With a crew of two, the vehicle weighed 3,049 pounds and the ground pressure on skis was only 0.75 pounds per square inch. Using the standard engine and transmission from the ¼ ton truck, the vehicle could reach 40 miles per hour. The M7 could be used with the 1 ton snow trailer M19 which could operate on either wheels or skis. Allis-Chalmers Manufacturing Company built 291 M7s from February through August 1944.

The snow tractor T36 was approved for limited procurement in September 1944. Another two man vehicle, it was driven by the same 99 horsepower engine used in the Dodge ¾ ton truck. With a loaded weight of 7,500 pounds, it had a ground pressure of 1.73 pounds per square inch. A total of 36 of these snow tractors were built by the Iron Fireman Manufacturing Company during 1944. A few other snow tractors were procured for evaluation. These included 6 T26s, 13 T27s, 12 T28s, 6 T29s, and 6 T30s. None of the snow tractors were suitable for combat use and were mainly employed in rescue and supply operations.

The requirement for good mobility over snow covered terrain resulted in the development of an excellent low ground pressure vehicle. In May 1942, the Studebaker Corporation was requested to design and build such a vehicle in time for use by the Special Service Force in Norway during the Winter of 1942/43. In response to this request, Studebaker developed and tested the T15 cargo carrier later standardized as the M28. Although the operation in Norway was canceled, the M28, named the Weasel, was used during the occupation of Kiska in August 1943.

Below, the details of the two man cargo carrier T15 can be seen with some of the stowage required for the over the snow mission in Norway.

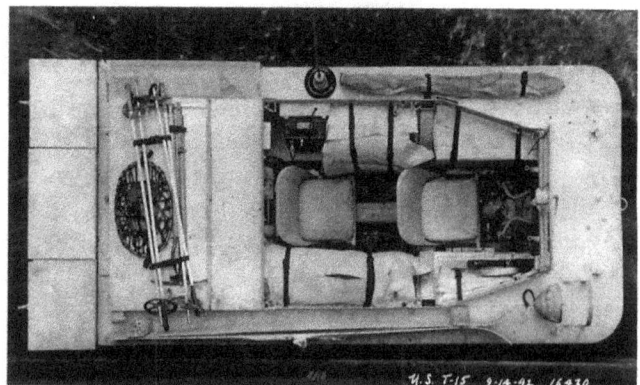

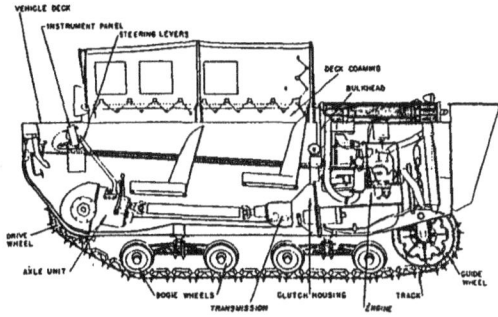

The standardized cargo carrier M28 is shown in the sectional drawing and photograph above.

Further development of the Weasel produced an improved vehicle designated as the cargo carrier T24 and later standardized as the cargo carrier M29. Although the M29 could float, it was easily swamped because of the low freeboard. To obtain satisfactory amphibious performance, flotation cells were added to the front and rear of the vehicle and the modified version was designated as the cargo carrier M29C. Both the M29 and the M29C retained the name Weasel and they remained in service long after World War II.

With a 1,200 pound load and a track width of 20 inches, the M29 and the M29C had ground pressures of 1.69 and 1.91 pounds per square inch respectively. Carrying four men including the driver, both vehicles had a cruising range on roads of 175 miles. Powered by a six cylinder, liquid-cooled, gasoline engine developing 75 net horsepower at 3,800 rpm, the Weasel had a maximum road speed of about 36 miles per hour. The M29C could reach 4 miles per hour in calm water. Both

vehicles had excellent performance in snow, mud, or swampy terrain. Production of the Weasel ended in August 1945 with totals of 766 M28s, 4,476 M29s, and 10,647 M29Cs. Both the M29 and the M29C retained the Standard A classification until July 1958 when they were reclassified as Standard B.

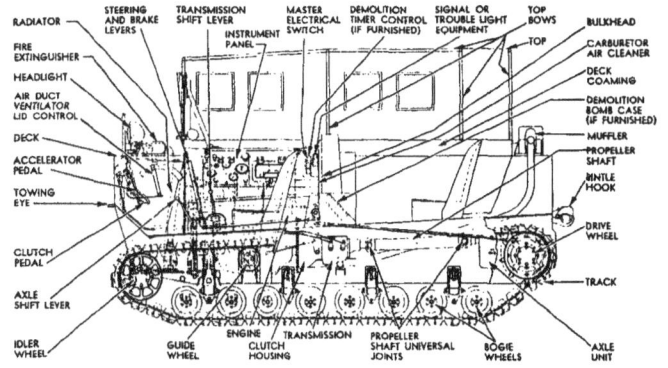

The early production cargo carrier M29 is depicted in these photographs and the sectional drawing. Note the relocation of the power train compared to the M28.

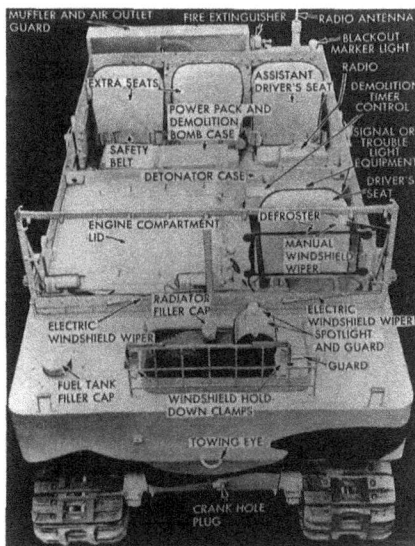

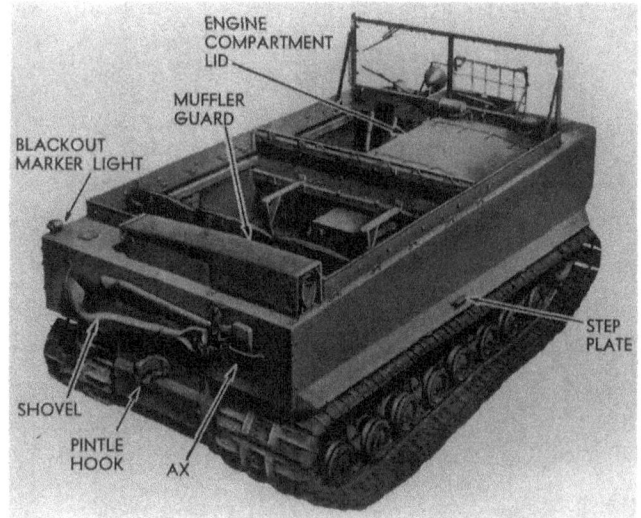

Above is the later production (after vehicle serial number 3102) cargo carrier M29.

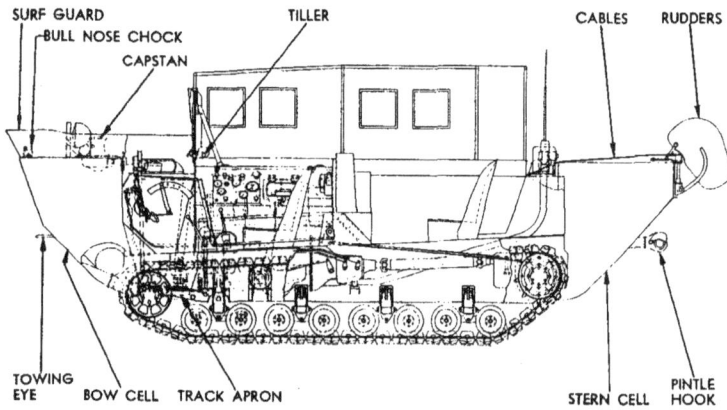

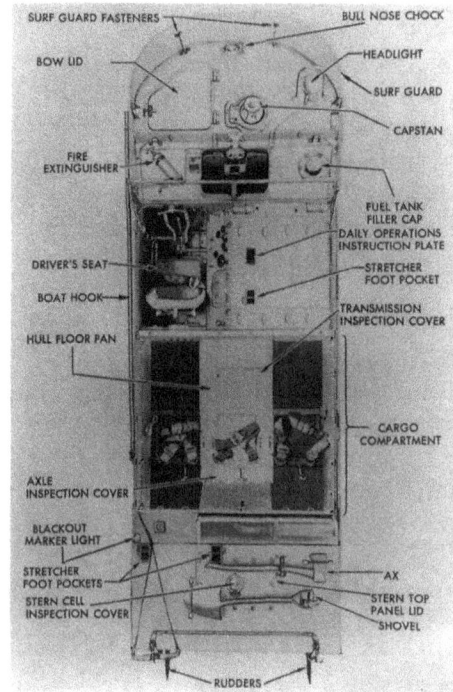

The amphibious cargo carrier M29C appears in these views and the sectional drawing. Note that it is essentially the same as the M29 with the addition of the bow and stern flotation cells.

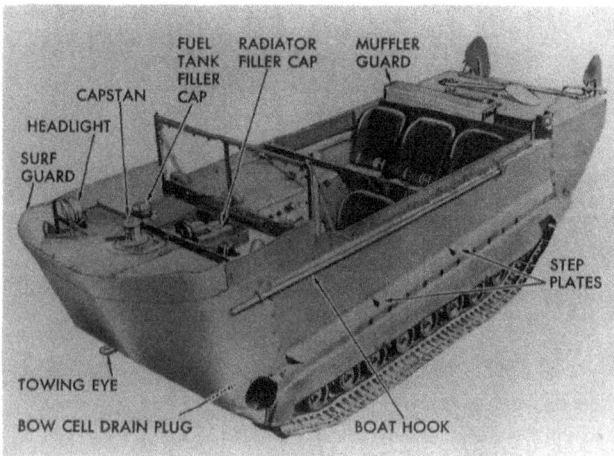

Cargo carrier M29, serial number 2581, is shown above on 17 March 1944. Note the relocation of the radio antenna compared to the photograph on page 23.

Cargo carrier M29C, serial number 4913, can be seen at the right and below. These photographs were dated 24 August 1944.

The light tractor T39 is shown above and at the right.

The war in the Pacific frequently involved operations in swampy areas and there was a requirement for a prime mover with low ground pressure to operate under these conditions. Five pilots of the light tractor T39 were built by the Lima Locomotive Works and one was shipped to the Tank Destroyer Board in February 1945 for evaluation as the prime mover for the 90mm antitank gun.

Powered by a 110 net horsepower Cadillac V8 engine with a Hydramatic transmission, the T39 had a maximum road speed of 35 miles per hour. The vehicle was supported by a flat track, torsion bar, suspension with four road wheels per side. The loaded weight of the T39 was about 9 tons and the 19½ inch wide tracks held the ground pressure to 4.4 pounds per square inch. It carried an eight man crew and was armed with a single .50 caliber machine gun. With the end of the war, the T39 did not go into production. An alternate design replaced each of the two center road wheels with a two wheel bogie and added two support rollers per side. This version was designated as the light tractor T39E1.

Another vehicle based upon the T39 light tractor was the amphibian carrier T34. It was essentially a T39 fitted with an amphibian hull using the same Cadillac V8 engine and Hydramatic transmission. The T34, also referred to as the Paddy Vehicle, had a payload of 3,000 pounds. Its ground pressure was 4.5 pounds per square inch. The maximum speed was 20 miles per hour on roads and about 4 miles per hour in water. The T34 was constructed by the Lima Locomotive Works. With the end of the war, there was no production of the T34.

The amphibian carrier T34 "Paddy Vehicle" is at the left and below. Note the track skirts added to improve performance in the water.

PART II

EARLY POSTWAR DEVELOPMENT

Above is a concept drawing of the armored utility vehicle T13.

NEW ARMORED PERSONNEL CARRIERS

As mentioned earlier, the limited mobility and protection provided by the open top half-track vehicles was recognized long before the end of World War II. In the Fall of 1944, a design study at the Office, Chief of Ordnance-Detroit proposed the development of a new vehicle to solve this problem based upon the power train and suspension components of the light tank M24. OCM 25696, dated 9 November 1944, approved the development and assigned the designation armored utility vehicle T13. It also was proposed to adapt the new vehicle to a variety of roles. These were an armored personnel carrier, an armored reconnaissance vehicle, an armored cargo carrier, and an armored prime mover.

As originally proposed, the T13 carried 18 or 22 men depending upon the internal arrangement. This compared to the 13 men in the half-track carrier which consisted of a driver and a 12 man infantry squad. Thus the capacity of the new vehicle did not correspond to the tactical units that it might be expected to carry. With an estimated combat weight of 39,000 pounds, the T13 was driven by the standard power pack from the M24 light tank consisting of two Cadillac V-8 engines with Hydramatic transmissions. The estimated maximum road speed was 35 miles per hour with a cruising range of about 250 miles. This design introduced the covered box-like hull that was to be characteristic of postwar armored personnel carriers. The driver and assistant driver were seated in the front on the left and right respectively. A .50 caliber machine gun on a ring mount over the assistant driver's hatch provided the only armament. Steel armor, ½ inch thick, was proposed for all surfaces.

The exterior components of the armored utility vehicle M44 (T16) are identified in the photographs above and below. Note the bow machine gun on this vehicle.

An unarmored cargo tractor version of the T13 designated as the T33 also was proposed. Further studies indicated that the M24 power plant was inadequate for this projected cargo tractor. As a result, the T13 project was terminated by OCM 27033 on 22 March 1945 and the development of a new vehicle was recommended utilizing the power train of the 76mm gun motor carriage M18. This action was approved by OCM 27227, dated 5 April 1945, which assigned the new designation armored utility vehicle T16. At this time, it was planned to use the T16 as the basis for the 4.2 inch mortar carrier T35. However, because of delays in the design and production program, the T35 was canceled in favor of the 4.2 inch mortar carrier T38 based upon the 105mm howitzer motor carriage M37.

The armored utility vehicle T16 was an even larger vehicle than the T13 with an overall length exceeding 20 feet. It now carried a crew of 27 consisting of the commander, driver, bow gunner, and 24 men in the personnel compartment. The welded hull was assembled from rolled steel armor ranging in thickness from $5/8$ inches on the front to $1/4$ inch on the bottom of the

sponsons. The driver and bow gunner rode in the front hull on the left and right of the engine compartment. The vehicle commander was located under a cupola on the roof just behind the engine compartment. Seats for 24 men were provided on four benches mounted in the center and along each side of the personnel compartment. A hatch with a .50 caliber machine gun on a T107 mount was installed on the rear of the hull roof. The bow gunner's .30 caliber machine gun could be relocated

Below, the M44 (T16) driver's station appears at the left and the bow gunner's compartment is at the right.

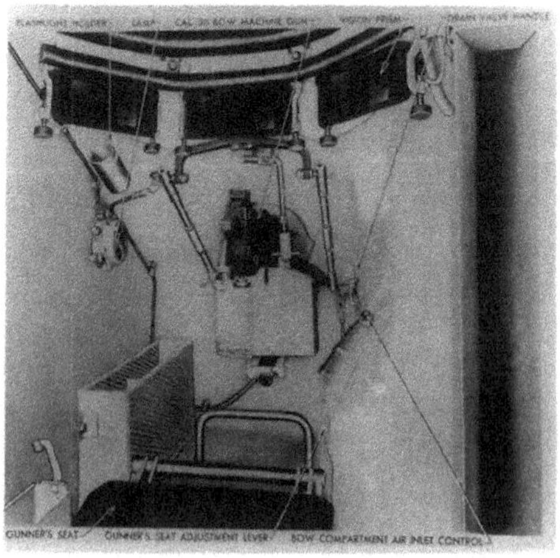

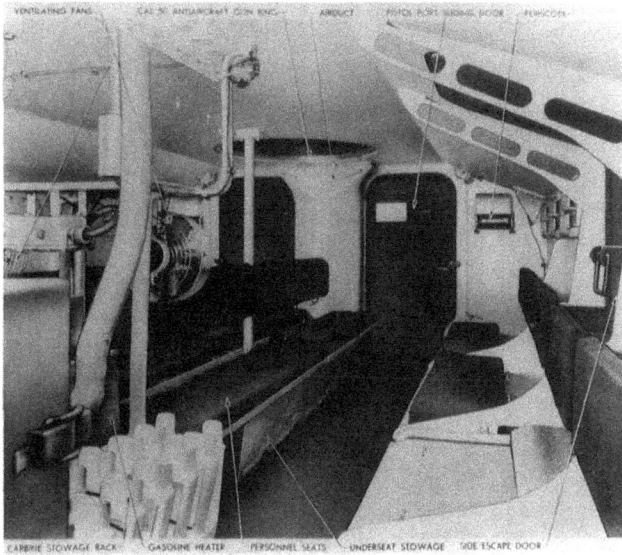

Above, the interior of the M44 (T16) troop compartment can be seen looking toward the rear (left) and toward the front (right).

to any one of four socket mounts next to the two side escape doors. The upper half of these doors opened up and the lower half opened out. Four pistol ports with sliding covers were on each side of the hull. Two large doors in the rear permitted rapid exit from the vehicle. Each door had a pistol port with a sliding cover. A fixed periscope was located in the rear wall outboard of each rear door. The driver and bow gunner each had three periscopes to the front and sides of their hatches. The commander's cupola was fitted with six periscopes for all round vision.

The air-cooled Continental R-975-D4 radial engine was mounted in the center front hull of the T16 with the crankshaft vertical. It drove the vehicle through a bevel gear, a 900AD combined torque converter and transmission, a controlled differential, and two final drives

and sprockets mounted at the front. The torsion bar suspension supported the T16 on six dual road wheels per side with 21 inch wide, single pin, tracks. An adjustable idler was at the rear of each track. Shock absorbers were fitted on the first two and last two road wheel arms on each side. The fuel tanks, batteries, oil coolers, and auxiliary generator were located in the hull beneath the personnel compartment.

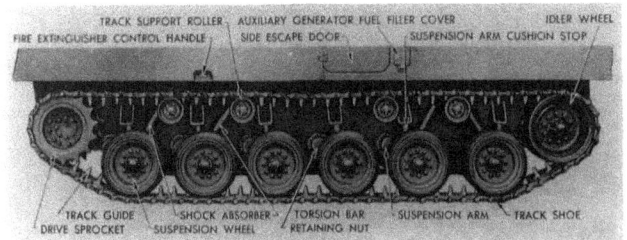

Details of the suspension and tracks on the M44 (T16) are shown above. Below, the engine and power train are installed in the vehicle at the left and right respectively.

A M44 (T16), registration number 40226288, is shown above and below during its evaluation by the Army Ground Forces Board Number 2. Note the socket mounted .30 caliber machine gun on the right side of the roof.

On 12 April 1945, OCM 27295 recommended limited procurement of the T16. Manufactured by the Cadillac Motor Car Division of General Motors Corporation, the delivery of the first six vehicles was scheduled for June 1945. With the end of the war in August, the requirement was no longer urgent and the T16s were under test at Aberdeen Proving Ground and Fort Knox for use in the postwar army. Now designated as the armored utility vehicle M44 (T16), it was evaluated as an armored infantry carrier. However, doctrine required a squad size vehicle for this application and the M44, with its capacity of 27 men, was far too large.

Below, M44 (T16), registration number 40226289, is at Fort Knox during its test program.

Additional views of M44 (T16), registration number 40226289, are shown here during the evaluation at Fort Knox. Note that the machine guns are not mounted on the vehicle.

Armored utility vehicle M44 (T16), registration number 40226287, was photographed on 29 March 1946 at Aberdeen Proving Ground. It was listed as experimental vehicle number 2 at Aberdeen.

Further details of the armored utility vehicle M44 (T16) can be seen in the photographs above and below taken at Aberdeen Proving Ground. The dimensions of the vehicle are shown in the sketch at the right.

Dimensions in inches

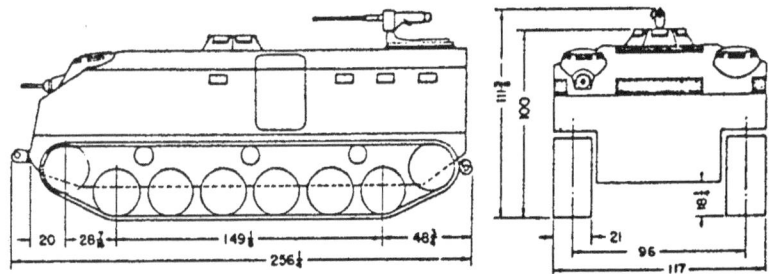

TRANSMISSION, BEVEL DRIVE AND
ACCESSORY DRIVE FILLER AND SCREEN
DIFFERENTIAL OIL FILTER
AUXILIARY GENERATOR AIR FILTER
TRACK SUPPORT ROLLERS (4 EACH SIDE)
ENGINE OIL FILLER
FINAL DRIVE FILLER (1 EACH SIDE)
TORSION ARM BEARINGS (6 EACH SIDE)
OIL COOLER BLOWER FILLER PLUG
(1 EACH SIDE)
TRACK SUSPENSION WHEELS (6 EACH SIDE)
IDLER ARM AND LINK (1 EACH SIDE)
TRACK IDLER WHEEL (1 EACH SIDE)

DIFFERENTIAL OIL FILLER
ENGINE OIL FILLER
ENGINE OIL TANK
ENGINE OIL FILTER
AIR COMPRESSOR OIL FILLER
RIGHT OIL COOLER BLOWER FILLER PLUG
DIFFERENTIAL BREATHERS
TRANSMISSION, BEVEL DRIVE AND
ACCESSORY DRIVE OIL FILLER AND SCREEN
DIFFERENTIAL OIL FILTER
LEFT OIL COOLER BLOWER
FILLER PLUG
AUXILIARY GENERATOR AIR FILTER

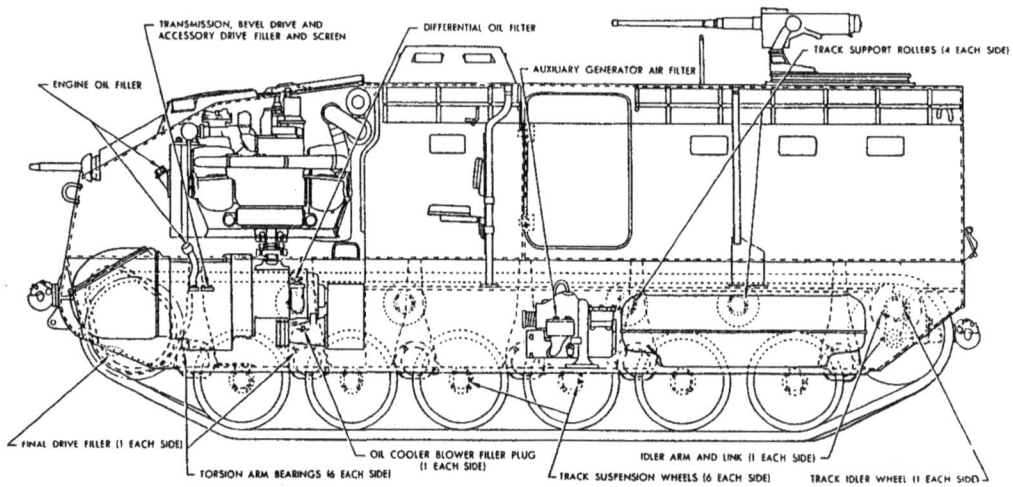

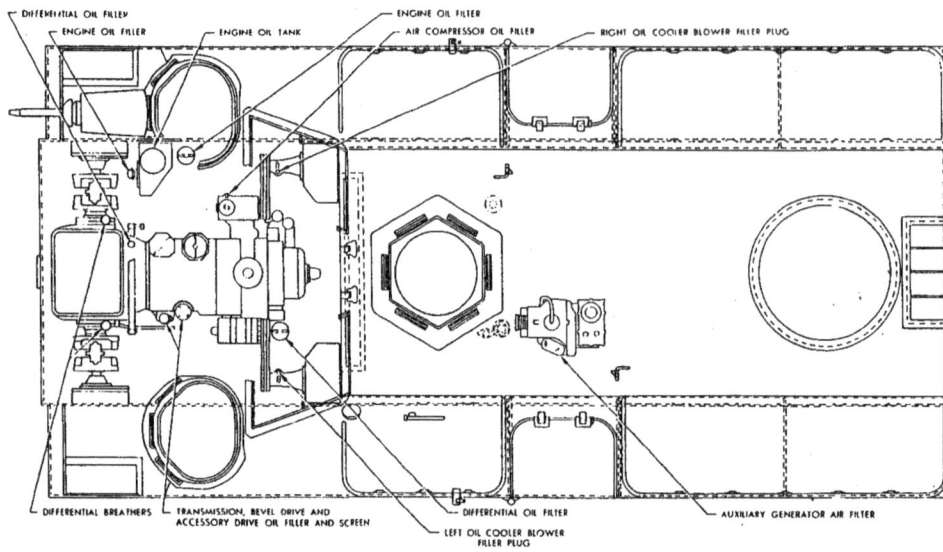

The interior arrangement of the armored utility vehicle M44 (T16) can be seen in these drawings.

Above and at the right are photographs of the armored utility vehicle M44E1. Note the new front hull and the elimination of the bow machine gun.

On 31 October 1946, OCM 31172 approved the modification of one pilot M44 (T16) to incorporate various improvements resulting from the proving ground tests and the availability of new components. Designated as the armored utility vehicle M44E1, it was powered by the 500 gross horsepower Continental AOS-895-1, air-cooled, engine with the CD-500 cross drive transmission. It was fitted with 21 inch wide, double pin, T87 tracks. Other changes included roof sections that could be opened eliminating the need for the side escape doors and pistol ports. The bow machine gun mount also was eliminated. However, it was still the same oversize vehicle and it was not considered for production.

Additional views of the armored utility vehicle M44E1 are at the right and below. The pistol ports have been welded shut.

Above, the mock-up of the armored utility vehicle T18 appears at the left and the T18 pilot number 1, registration number 9200889, is at the right. The .50 caliber machine gun remote control mounts have been added by an artist.

Even before the end of World War II, the need for a squad size armored personnel carrier was obvious. On 21 September 1945, a requirement was established for a full-tracked, 12 man, armored personnel carrier to be based upon the chassis of the cargo tractor T43. The development of the T43 had been approved by OCM 27382 on 19 April 1945. However, the T43 design utilized a volute spring suspension instead of the preferred torsion bar system. Since it was now planned to adapt the cargo tractor for a variety of vehicles, it was redesigned to use a torsion bar suspension and designated as the cargo tractor T43E1. It was the chassis of this later vehicle that provided the basis for the new armored personnel carrier. OCM 31057, dated 26 September 1946, approved the development of the armored utility vehicle T18 and the construction of four pilots. International Harvester Company (IHC) was awarded a contract to build the four pilots which differed in some details.

The original mock-up of the T18 located the driver and an assistant driver under separate hatches in the front hull on the left and right sides of the power plant. The vehicle commander was provided with a cupola on the hull roof just to the rear of the engine compartment. Two .50 caliber machine guns in separate remote control mounts were installed on the hull roof to the left and right of the commander's cupola. Three independent sights were provided to control the machine guns, any one of which could aim one or both of the .50 caliber weapons. One sight was for use by the vehicle commander. Two gunners rode in special seats installed below the other two sighting positions. The remaining personnel were seated in four rows in the rear compartment. A total crew of 14 was specified under the original design concept.

Details of the .50 caliber machine gun remote control mounts are shown at the left and the controls for the vehicle commander and the two gunners can be seen below.

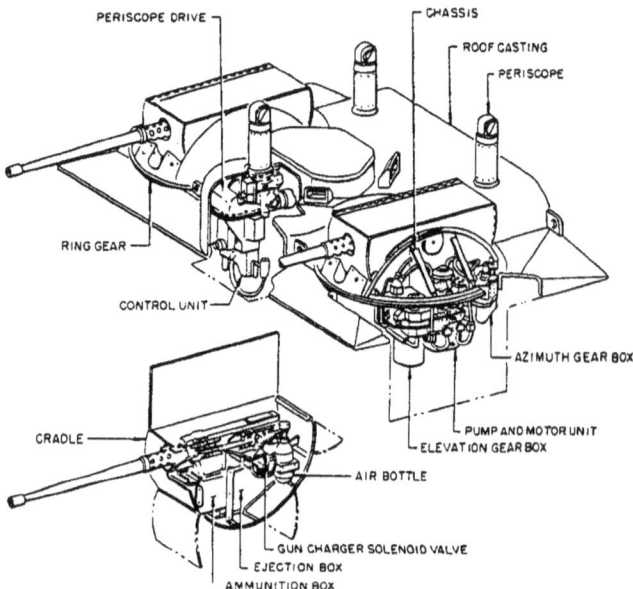

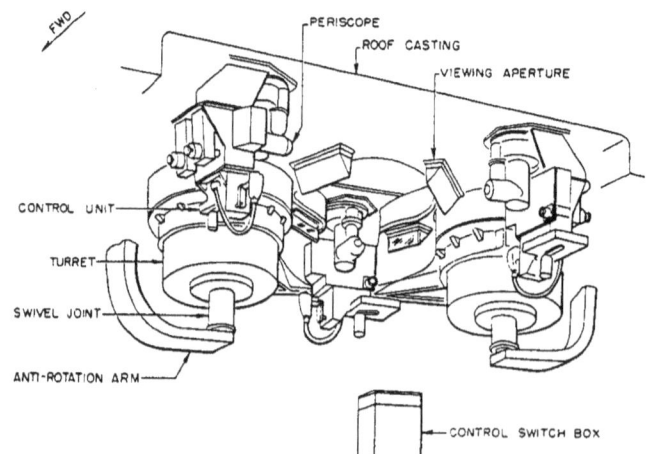

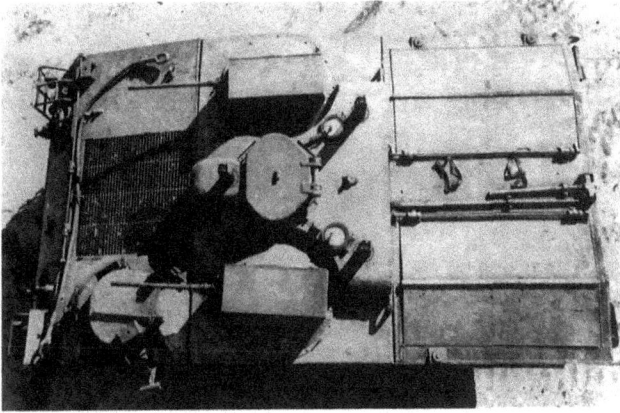

Above are two top views of the armored utility vehicle T18. The remote control .50 caliber machine guns are mounted in the left photograph. Below are two additional views of T18 pilot number 1.

As constructed, the pilot vehicles differed in several details from the mock-up. The T18 was still armed with the two remote control .50 caliber machine guns, but the assistant driver position was eliminated on this and all the other pilots.

The original T18E1 pilot was unarmed with a high cupola installed in the hull roof for the vehicle commander. The T18E2 was fitted with the T122 machine gun mount in place of the commander's cupola. This mount could be armed with either .30 caliber or .50 caliber machine guns. On both the T18E1 and the T18E2, a special seat was installed for the vehicle commander below the cupola or machine gun mount. The remaining personnel were seated in three longitudinal rows in the rear compartment. At this stage, a total crew of 13 was specified.

Armored utility vehicle T18E1, registration number 9200891, is shown below during its evaluation at Fort Knox. Note the high cupola on this vehicle. Here it is marked as pilot number 4.

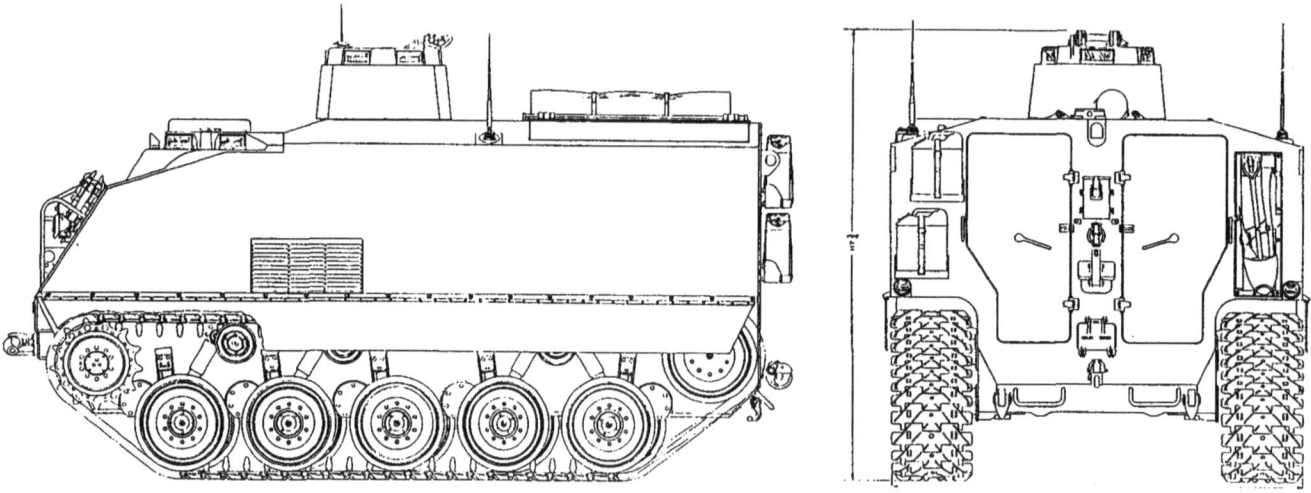

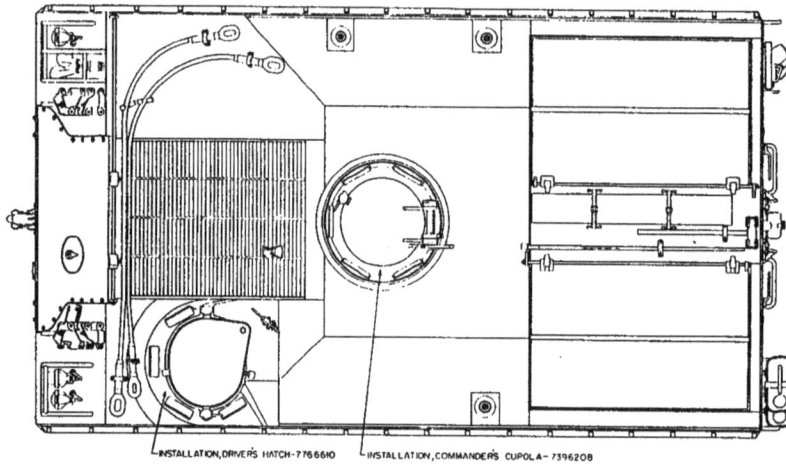

The general arrangement of the pilot armored utility vehicle T18E1 is shown in the drawing above. The internal layout of this vehicle can be seen in the drawing below.

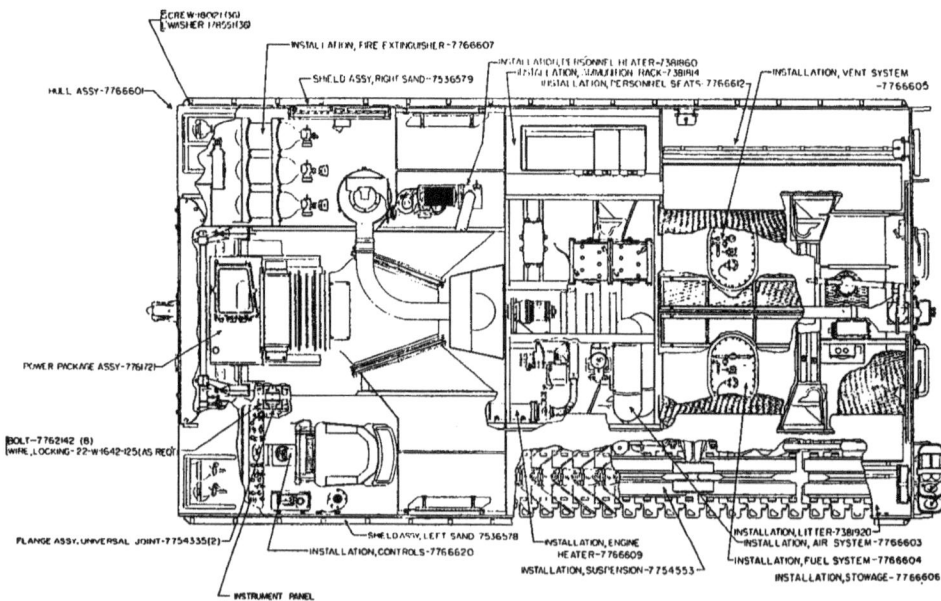

40

Above at the left, armored utility vehicle T18E1, registration number 9200891, is equipped with a dozer blade. It is now marked as pilot number 2. At the top right is armored utility vehicle T18E2, registration number 9200892, armed with two .50 caliber machine guns in the T122 mount. It is marked as T18E2 pilot number 1. Below are two additional views of the T18E2, but the .50 caliber machine guns are not installed.

Another view of the armored infantry vehicle T18E2 is at the right. Note the discoloration in the center of the side grill. The early pilot vehicles directed the engine exhaust out through this grill along with the engine cooling air. The hot exhaust discolored the center of the grill. Below are two views of the production pilot T18E1 with a separate engine exhaust pipe and the low silhouette cupola for the vehicle commander.

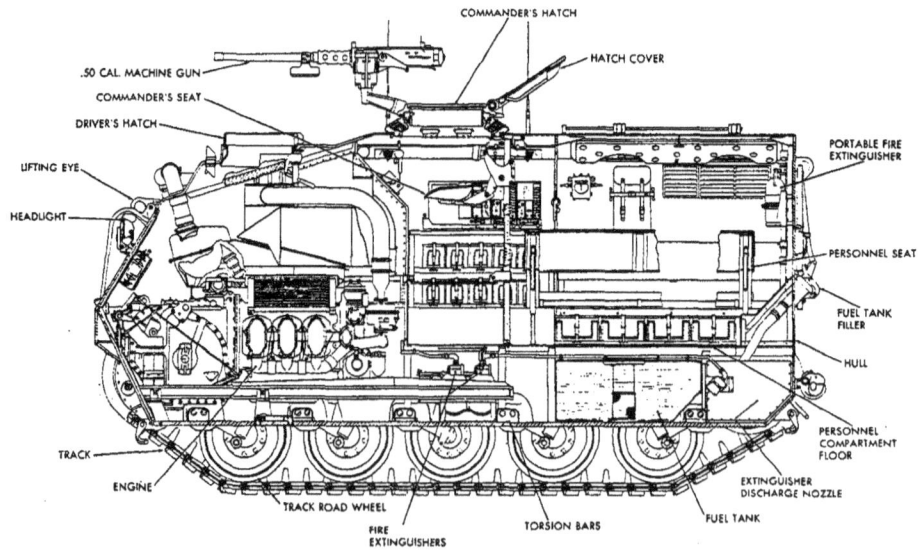

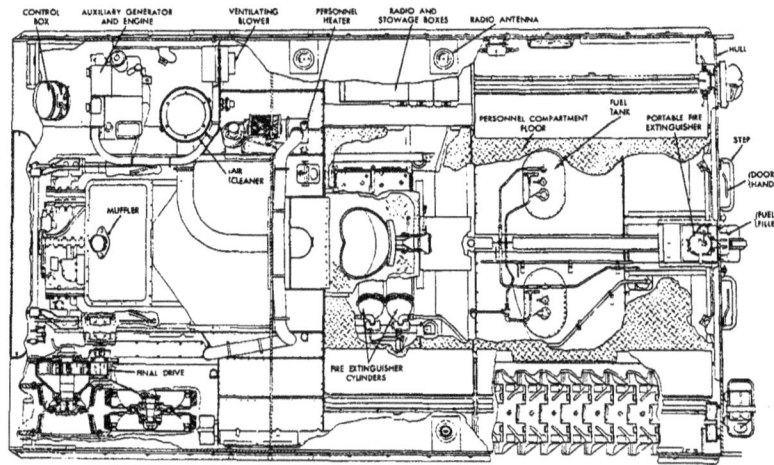

Details of the armored utility vehicle T18E1 production pilot are shown in these drawings.

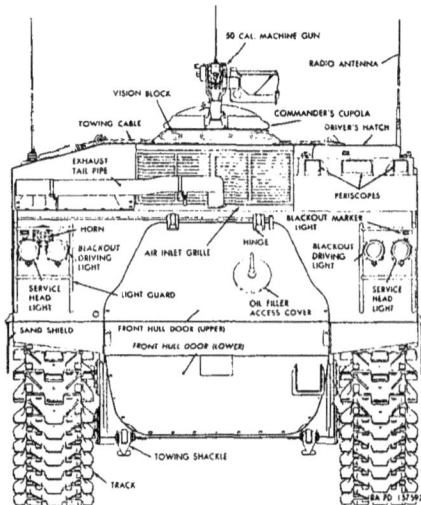

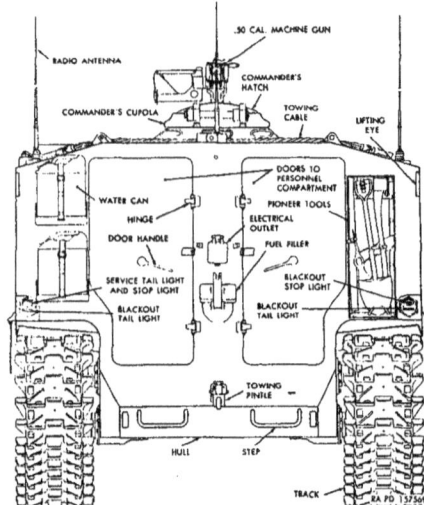

The top of the armored utility vehicle T18E1 production pilot can be seen above with the hatches open and closed. Note the new engine exhaust pipe extending from the center front to the right side of the roof. The low silhouette commander's cupola is clearly visible.

Although the original T18E1 pilot was unarmed, the commander's cupola was replaced on later vehicles by a variety of machine gun mounts. When the T18E1 was selected for production, a low silhouette commander's cupola with six vision blocks was installed. It was fitted with a .50 caliber machine gun on an external mount. The small T10 (No. 8396700) domed cupola with an internally mounted .30 caliber machine gun was installed experimentally. Later, the M13 cupola armed with a .50 caliber machine gun was evaluated on the T18E1. The pilots had a small oil filter access cover in the front upper hull door, but it was eliminated on the production vehicles. Some pilots also had a small ventilation grill at the top rear of the hull side armor. The pilots were fitted with two large folding doors in the roof of the personnel compartment. On production vehicles, they were replaced by two smaller, non-folding, doors, one on each side of the roof.

Below, the small T10 cupola armed with a .30 caliber machine gun is installed on the T18E1. At the right, the vehicle is fitted with an Aircraft Armament Model 15 cupola mounting a .50 caliber machine gun.

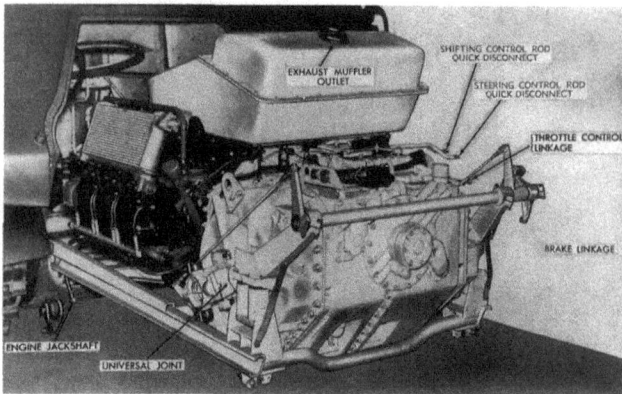
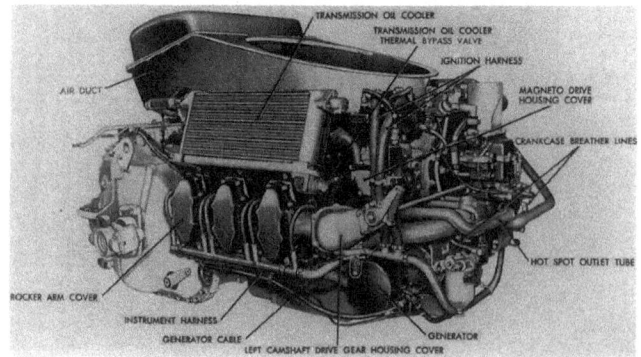

The power pack of the T18E1, consisting of the Continental AOS-895 engine and the Allison CD-500 transmission, is shown above removed from the vehicle. The disconnect points are indicated in the left photograph.

The original pilots were powered by the six cylinder, air-cooled, AO-895-2 Continental engine. The mufflers for this engine were located in the cooling air exhaust ducts behind the power plant. The engine exhaust was then discharged with the cooling air through the hull side grilles. The later AO-895-4 engine was installed in the production vehicles with a single muffler on top of the engine. A separate tail pipe carried the engine exhaust through the hull roof and to the right side of the vehicle. Later, the brush guard for the right headlight group was extended to protect the tail pipe. The AO-895 engine developed 375 gross horsepower and was coupled to a CD-500 cross drive transmission. The torsion bar suspension supported the vehicle on five dual road wheels per side with a compensating idler at the rear of each track. Three track support rollers were on each side. The 21 inch wide, single pin, tracks were driven by the front mounted sprockets. Later, a flat track suspension without support rollers was evaluated on the T18.

Production order T-24478, dated 11 January 1950, authorized procurement of five T18 type vehicles for the Field Service Division. This was reduced to two on 3 January 1952 and the remaining three were completed as T73 armored infantry vehicles.

An auxiliary generator and a personnel heater were installed in the right front hull alongside the engine compartment on the early production T18E1s. Later, they were removed and they were not included on the later production vehicles. On 18 January 1951, OCM 33541 changed the nomenclature from armored utility vehicles T18, T18E1, and T18E2 to full-tracked armored infantry vehicles T18, T18E1, and T18E2. Standardized later as the full-tracked armored infantry vehicle M75, the T18E1 entered production at the International Harvester Company (IHC) and at the Food Machinery and Chemical (FMC) Corporation. The initial contracts were for 1,000 vehicles at IHC and 730 at FMC. Later, FMC produced an additional 50 vehicles bringing the total production run to 1,780.

Exterior components on the M75 armored utility vehicle are identified in the photographs below. These views represent the earlier vehicles with serial numbers 7 through 376 (IHC) and 1007 through 1326 (FMC).

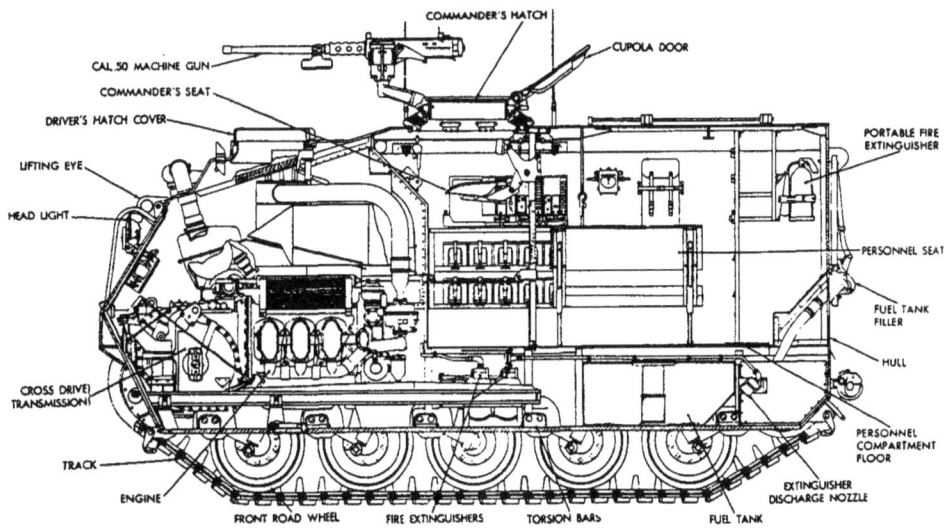

Early production armored infantry vehicle M75

Numerous changes were introduced during the production run, although the vehicle designation remained the same. Serial numbers 7 - 376 (IHC) and 1007 - 1326 (FMC) followed the early design configuration while serial numbers 377 - 1006 (IHC) and 1327 - 1736 (FMC) included the later modifications. Serial numbers 1 - 6 were assigned to the original four pilots and the two vehicles procured under production order T-24478.

Among the external changes appearing on serial numbers 377 and 1327 was the elimination of the front and side sand shields. The shock absorbers were removed from road wheel arms two and four leaving only two per side. The cushion stops for the intermediate road wheel arms on the early vehicles were replaced by fixed steel stops. The recesses for the taillights and the external fire extinguisher control were eliminated and welded steel guards were installed. A new domed fuel filler cover replaced the earlier design. When the auxiliary generator was dropped, a new flat top access cover replaced the earlier type with the outlet for the auxiliary generator engine exhaust pipe. The side access door on the early production vehicles had a semicircular recess for the door handle. Later vehicles were fitted with a flat side access door and a new design door handle.

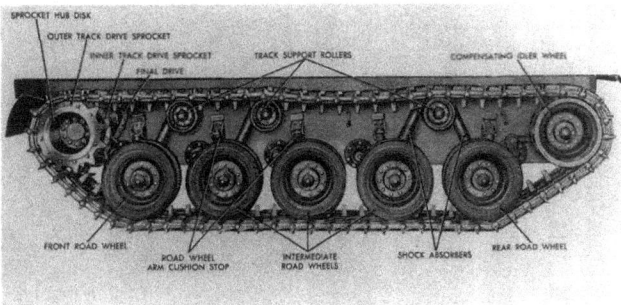

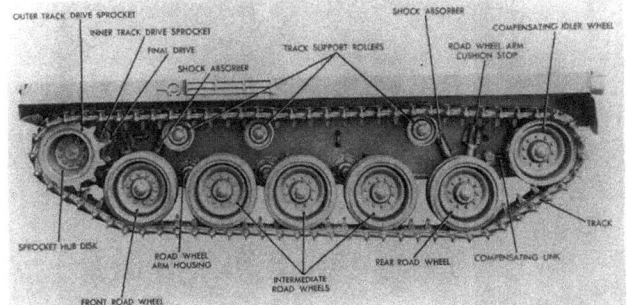

Above, the early suspension of the M75 armored infantry vehicle appears at the left and the later suspension is at the right. Note the reduced number of shock absorbers on the later version. Below are photographs of the later M75, serial numbers 377 through 1006 (IHC) and 1327 through 1736 (FMC).

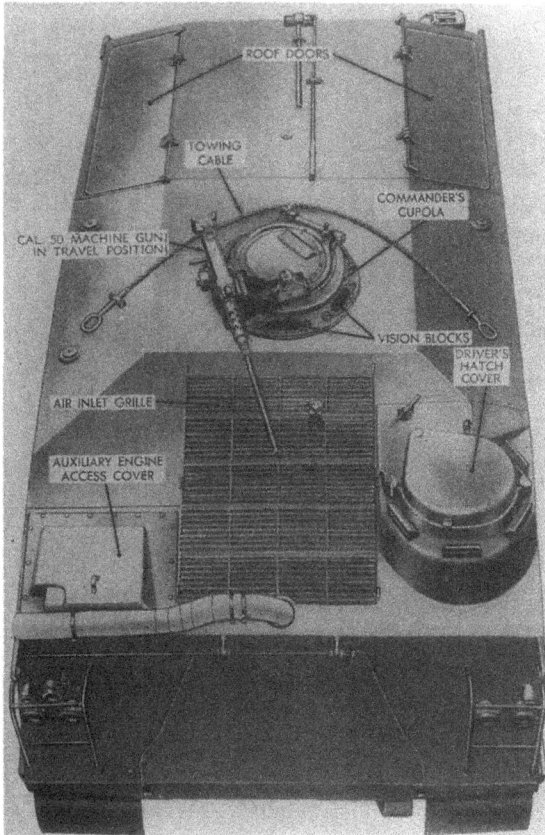

Above are top views of the armored infantry vehicle M75 with serial numbers 7 through 376 (IHC) and 1007 through 1326 (FMC) at the left. Serial numbers 377 through 1006 (IHC) and 1327 through 1736 (FMC) are at the right. Note the elimination of the auxiliary engine on the later vehicles.

Other modifications included a welded internal hull structure replacing the earlier riveted version. The hull drain valves were eliminated and drain plugs were installed. The two 75 gallon rubber fuel tanks were replaced by a single 150 gallon metal tank in the newer vehicles. New controls were fitted for the rear and roof doors and a redesigned fire extinguisher system was installed. The new instrument panel included a tachometer.

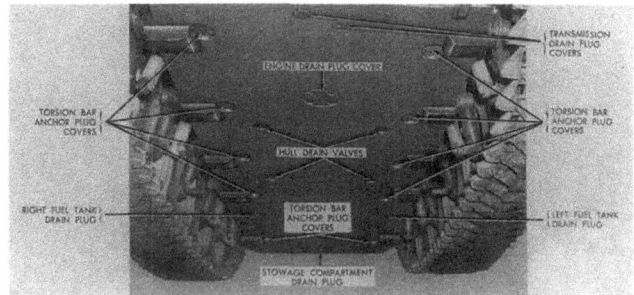

The bottom of the armored infantry vehicle M75 is above at the right, looking from the front to the rear. Below, at the left, the rear doors of the M75 are open. The driver's controls on the M75 are at the bottom right.

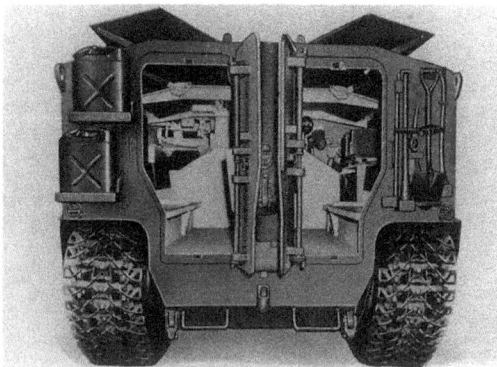

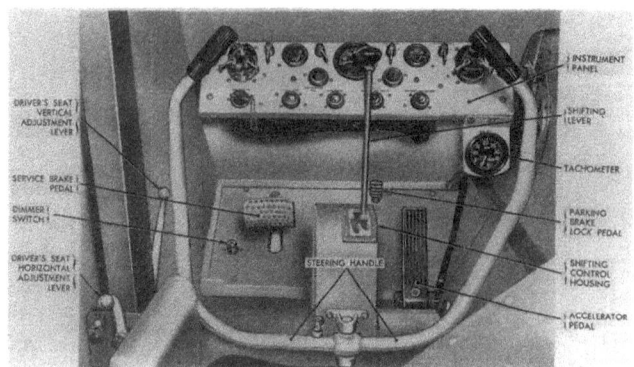

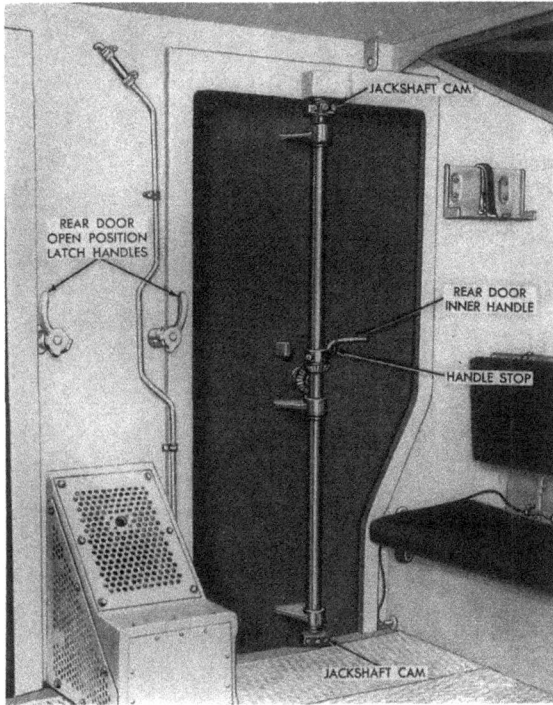

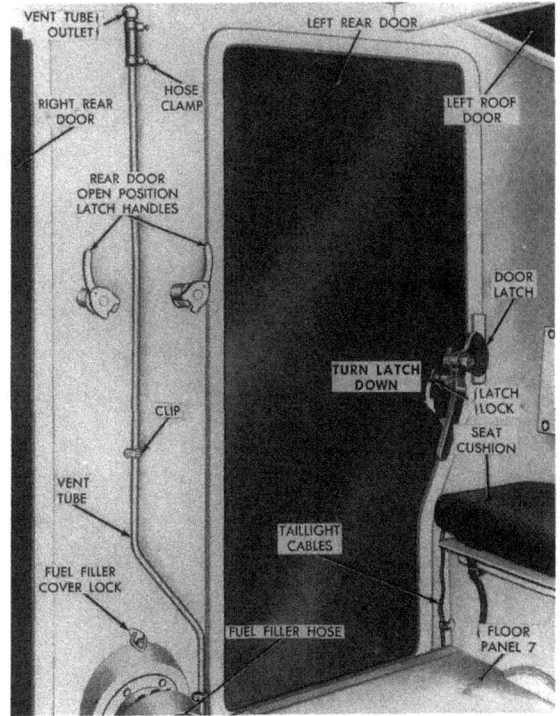

Above are interior views of the rear door controls on the armored infantry vehicle M75 with serial numbers 7 through 376 (IHC) and 1007 through 1326 (FMC) at the left. Serial numbers 377 through 1006 (IHC) and 1327 through 1736 (FMC) are at the right. The radio installation in the M75 can be seen below at the right.

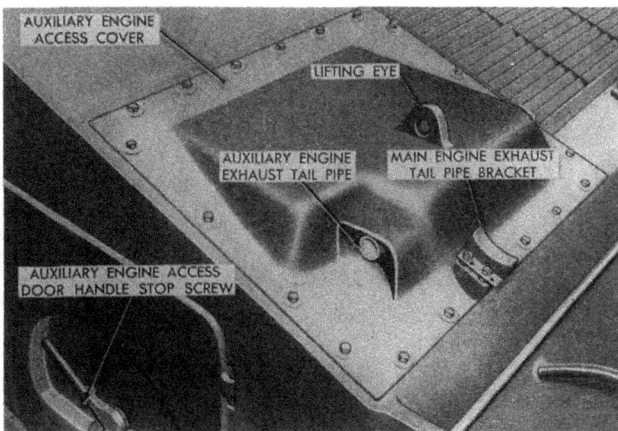

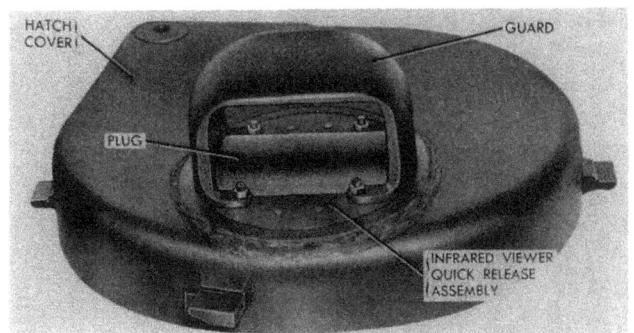

Above is the auxiliary engine cover on the early M75 and the driver's hatch is at the right. Below, the driver's instrument panel for vehicles 7 through 376 (IHC) and 1007 through 1326 (FMC) is at the left. The later instrument panel for vehicles 377 through 1006 (IHC) and 1327 through 1736 (FMC) is at the right.

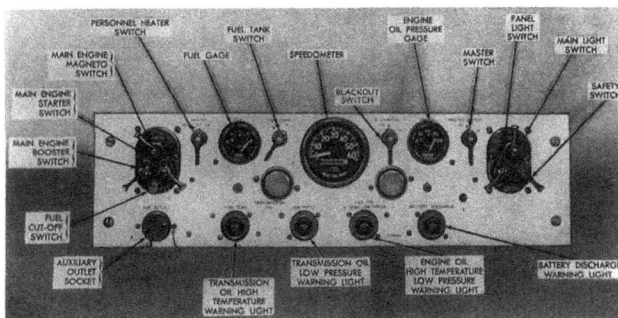

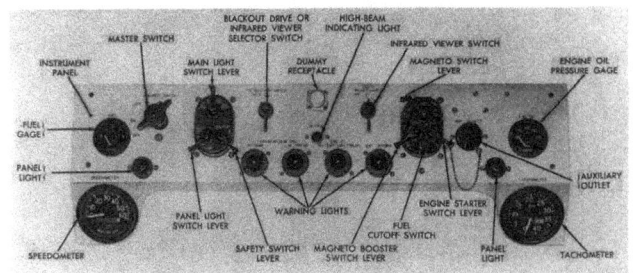

The 4.2 inch tracked mortar carrier T64 appears in the photographs above and below. Note that a canvas cover replaced the armor roof on the rear of the vehicle.

Studies at the Infantry School and the Armor School indicated a requirement for a full-tracked, armored, carrier for the 81mm, 105mm, and 4.2 inch mortars. Since it was desirable to use the M75 chassis as the basis for a family of vehicles, it was selected for this application. The designations 81mm mortar tracked carrier T62, 105mm mortar tracked carrier T63, and 4.2 inch mortar tracked carrier T64 were assigned for these vehicles. The mount was designed to handle the firing loads of the 4.2 inch mortar and it could be used with the 81mm and 105mm mortars by means of an adapter. The armor on the top and sides of the M75 was cut back at the rear of the vehicle to provide an open area for the mortar. A single pilot T64 was converted from an M75 and evaluated at Aberdeen and Fort Knox. By the time the tests were complete, no further production of the M75 was planned and interest shifted to developing a mortar carrier based upon the M59 personnel carrier. Although the design work was completed on the T62 and T63, no pilots were converted.

The M75 armored infantry vehicle was deployed to Korea in the Summer of 1953 and was effectively used during the final stages of the fighting. It was particularly valuable in resupplying outposts and in the evacuation of troops and wounded from isolated front line positions. Troop tests in the United States also confirmed the value of this type of armored personnel carrier. However, it was extremely expensive with unit costs of approximately $72,000. In view of this high cost, a program was initiated to develop a less expensive armored personnel carrier.

At the right, the 4.2 inch mortar is installed in the T64 carrier.

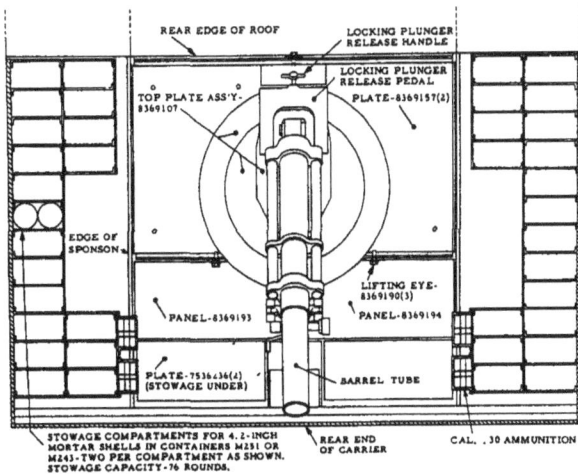

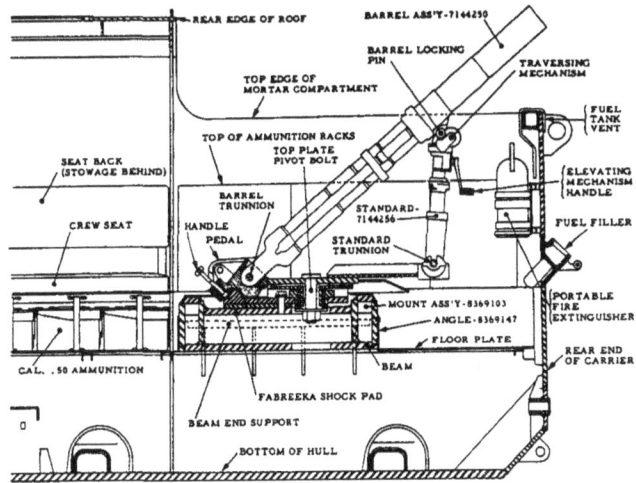

Details of the 4.2 inch mortar installation in the T64 carrier are shown in the drawings above.

These photographs show the 4.2 inch tracked mortar carrier T64, registration number 9211417, during its evaluation at Fort Knox.

Above are the pilot armored infantry vehicles T59 (left) and T59E1 (right). The T59E1 is fitted with the large, hydraulically operated, trim vane.

A LOWER COST ARMORED PERSONNEL CARRIER

Although it was generally satisfactory, the armored infantry vehicle T18E1 (M75) was too costly for procurement in large numbers. On 8 November 1951, OCM 33981 initiated a program to develop a less expensive vehicle for the role of the T18E1 (M75). The OCM authorized the procurement of six pilot vehicles (F1 - F6) which were proposed and manufactured by FMC Corporation. The first four were powered by the six cylinder GMC Model 302 engine then used in the 6 x 6 truck M135. The last two utilized the Cadillac, Model 331, V8 engine. All pilots were fitted with a modified version of the truck Hydramatic transmission and a controlled differential. The first four pilots were designated as the armored infantry vehicle T59 and the last two became the armored infantry vehicle T59E1. In both cases, each vehicle was powered by two engines which were installed with their transmissions in the left and right sponsons of the box shaped hull. Power passed through each transmission to a right angle drive at the front of each sponson and then to the controlled differential located in the center front of the hull. Propeller shafts from the differential drove the final drives and sprockets at the front of the tracks. The vehicle was steered using the brake levers operating the controlled differential. The torsion bar suspension supported the vehicle with five dual road wheels per side on the 21 inch wide, single pin, tracks. These were the same T91E3 tracks used on the T18E1 (M75) and the 76mm gun tank M41.

The driver and vehicle commander rode in the front hull on the left and right respectively in both the T59 and T59E1. Three periscopes were located in front of

the driver's hatch and the commander had a cupola with six vision blocks providing a 360 degree view. This cupola was fitted with an external mount for a .50 caliber machine gun. A five man folding seat on each side of the personnel compartment provided space for a ten man infantry squad. An hydraulically actuated ramp with a small vision port formed the rear wall of the hull. When opened, it provided rapid, unobstructed, access to the interior of the vehicle. With the ramp opened and the personnel seats folded down, a jeep could be driven inside. Roof hatches also were installed in the personnel compartment, however, their configuration varied on the six pilots. The vehicles were assembled with welded homogeneous steel armor $5/8$ inches thick on the front, sides, and rear. It was $3/8$ inches thick on the roof and 1 inch thick on the floor providing good mine protection.

Both the T59 and T59E1 were designed to be amphibious and the rear ramp was fitted with a rubber seal. The vehicles were propelled by track action in the water and a folding trim vane was mounted on the front. The short, manually operated, trim vane on pilots 1 through 4 was replaced by a larger, hydraulically operated, design on pilots 5 and 6. A number of other features varied among the pilots as indicated in the table.

Another photograph of the armored infantry vehicle T59E1 appears at the right

The armored infantry vehicle T59 and T59E1 pilots appear at the left and right respectively above and below.

At the right is an additional photograph of the armored infantry vehicle T59E1 during the test program at Fort Knox.

The differences between the six pilot T59 and T59E1 armored infantry vehicles are listed in the table below.

ITEM	PILOT MODEL NO. 1 (T59)	PILOT MODEL NO. 2 (T59)	PILOT MODEL NO. 3 (T59)	PILOT MODEL NO. 4 (T59)	PILOT MODEL NO. 5 (T59E1)	PILOT MODEL NO. 6 (T59E1)
Air Compressor	None	None	None	Midland Stl. Prod. Co. - #N-4025 Ser. #XN4025E8068	None	None
Bumper Spring	Volute Springs on 1 & 5. Solid stops on 2, 3 & 4.	Volute Springs	Volute Springs	Volute Springs	Same as #1	Same as #1
Cargo Access Cover	Single Hatch Dwg. #1013113	Double Hatch Dwg. #1013454	Double Hatch Dwg. #1013454	Double Hatch Back Half Hinged Dwg. #1013455	Same as #4	Same as #4
Choke Installation	Hand Choke	Hand Choke	Hand Choke	Hand Choke	Automatic Choke	Automatic Choke
Commander's Cupola	T18 #7952500	T18 #7952500	T18 #7952500	T18 #7952500	T18 #7952500	Pac. Car & Fdry. #7993247
Engine	GMC	GMC	GMC	GMC	Cadillac	Cadillac
Engine Access Panels	Has 4	Has 4	Has 4	6 (2 Cover Fan Compartment)	Has 4	Has 4
Front & Rear Bogie Arm	Splined to Trunnion & Welded	Splined to Trunnion & Welded	Splined to Trunnion & Welded	Shrink Fit to Trunnion & Welded	Splined to Trunnion & Welded	Shrink Fit to Trunnion - No Weld
Generator & Regulator	Delco 25 Amp.	Delco 25 Amp.	Delco 25 Amp.	Delco 25 Amp.	70 Amp.	70 Amp.
Hull Fastenings for Suspension	Nuts Welded to Hull	Hull Tapped	Hull Tapped	Hull Tapped	Hull Tapped	Hull Tapped
Rubber Wheels Drive Sprocket	Has	None	None	None	None	Has
Sand Shield	Long-1010228	Short-1014049	Short-1014049	Short-1014049	Long-1010228	Long-1010228
Torsion Bar Anchors	1010197 Right 1010196 Left	1010197	1010197	1010197	1010197	1010197
Trim Vane	Manual Short	Manual Short	Manual Short	Manual Short	Hydraulic Long	Hydraulic Long
Welding Rod Used	Austenitic	Ferretic	Ferretic	Ferretic	Ferretic	Austenitic

51

BOX-RADIO TERMINAL
INSTALLATION D-1013470

CUPOLA-COMMANDER'S
VISION K-7993247

COMMANDER'S SEAT
INSTALLATION E-1013162

FAN AND RADIATOR
INSTALLATION D-1013710

POWER TRAIN
INSTALLATION E-1013130

HULL ASSEMBLY
E-1010191

ELECTRICAL INSTALLATION
E-1013351

TRIM VANE INSTALLATION
E-1013229 SHOWN
(OR E-1013902)
MANUALLY OPERATED

DRIVE ASSEMBLY
RIGHT ANGLE
E-1011774

DIFFERENTIAL
ASSEMBLY E-1011700

FIRE EXTINGUISHER
INSTALLATION E-1013342

SUSPENSION INSTALLATION
E-1011736

FUEL SYSTEM
INSTALLATION
C-1013153

SEAT-PERSONNEL
INSTALLATION
E-1013567

BILGE PUMP
INSTALLATION
D-1013192

RAMP INSTALLATION
E-1013164

inches

HORN INSTALLATION
D-1013435

DOME LIGHT INSTALLATION
D-1013419

ENGINE ACCESS PANEL

COVER-FUEL FILLER
C-1013178

INSTALLATION-POWER PLANT
(GMC) E-1013143 (T59)
SHOWN

INSTALLATION
CADILLAC ENGINE
E-1013659 (T59E1)

CARGO COMPARTMENT DOOR
INSTALLATION E-1013454

AIR INTAKE GRILLE

COVER RADIATOR FILLER
C-1013299

AIR EXHAUST GRILLE

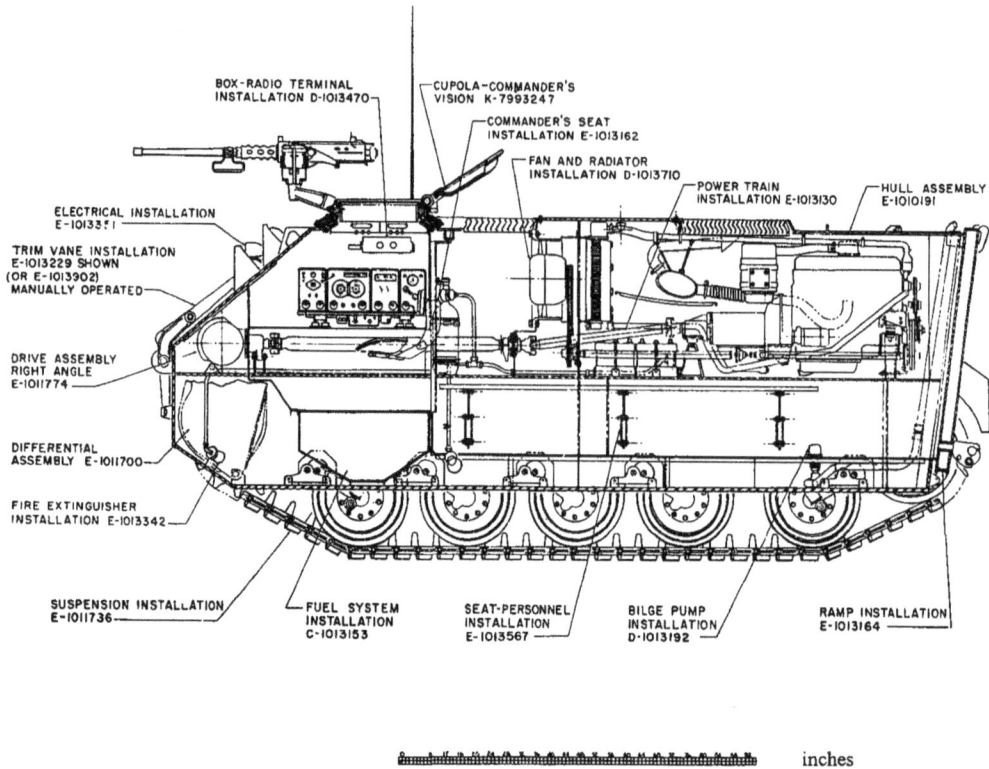

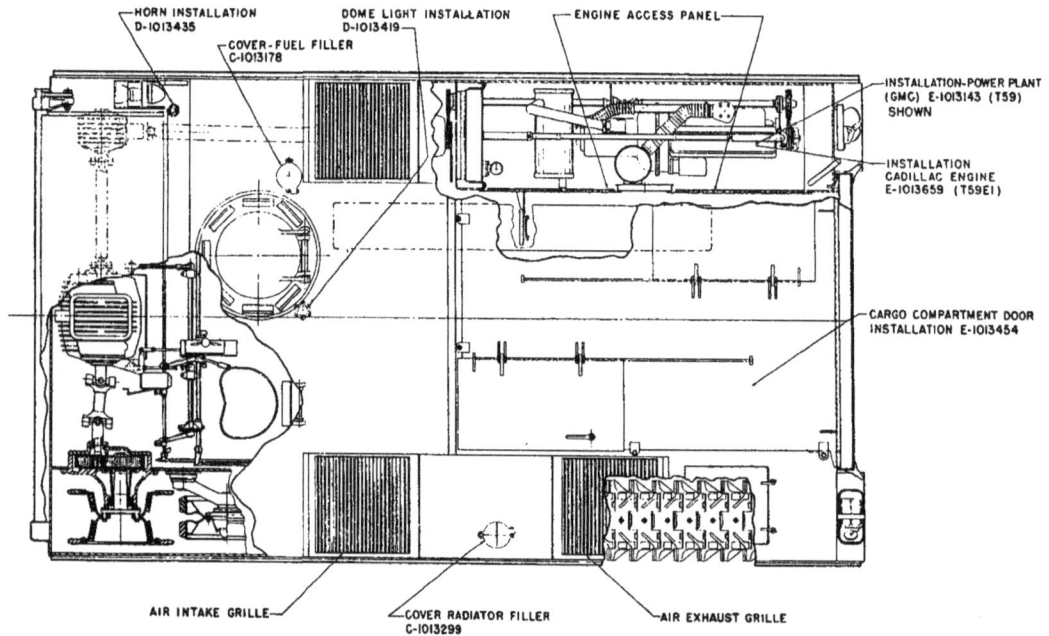

The general arrangement of the armored infantry vehicle T59 is shown in these drawings.

52

The armored infantry vehicle T59 number 1 is shown above and at the right at Aberdeen Proving Ground on 8 September 1952.

As completed, the pilots had a loaded weight of about 21 tons. The GMC Model 302 in the T59 developed 146 gross horsepower compared to 192 gross horsepower for each of the Cadillac Model 331 engines in the T59E1. Thus, the T59 had a gross power to weight ratio of about 13.7 horsepower per ton compared to 18.4 horsepower per ton for the T59E1. Both had a maximum design road speed of 32 miles per hour. However, in water, the track propulsion gave the T59 a maximum speed of 4.3 miles per hour compared to 5.0 miles per hour for the T59E1.

The rear of T59 pilot number 1 can be seen above open and closed. Note that there is a vision port but no door in the rear ramp. Below are top views of T59 number 1 showing the hatches open and closed.

The armored infantry vehicle T73 above is at Fort Knox during the test program

It also was desirable to consider a less expensive version of the armored infantry vehicle T18E1 (M75) then in production. In response to a verbal request from the Assistant Chief of Staff, G4, a work directive, dated 3 January 1952, was issued to complete the three remaining T18E1s under production order T-24478 as lower cost vehicles. Built by the International Harvester Company, they were subsequently designated as the armored infantry vehicle T73. The cost reduction was achieved by using a less expensive version of the Continental AO-895-4 engine, the Allison XT-500 transmission, and a modified hull structure. The T73 retained the T18E1's homogeneous steel armor $5/8$ inches thick on the front, sides, and rear. The top and floor were still ½ and 1 inch thick respectively. The driver rode in the left front above the sponson alongside the engine compartment. Four periscopes were installed around his hatch. The vehicle commander was in the left center under a cupola with six vision blocks and an external mount for a .50 caliber machine gun. The personnel compartment seated ten infantrymen on two

Additional photographs of the armored infantry vehicle T73 at Fort Knox are at the right and below. Note the low silhouette of this vehicle.

folding benches along the side walls. A single large door in the rear with a pistol port provided access to the vehicle. Two roof doors could be opened for loading cargo or to permit the troops to fire from the vehicle. With the roof doors locked in the open position, a six inch firing space remained between the bottom of the doors and the hull roof.

The T73, with its 375 gross horsepower AO-895-4 engine had a gross power to weight ratio of 18.5 horsepower per ton compared to the 13.7 horsepower per ton for the T59 and 18.4 horsepower per ton for the T59E1. However, it was not amphibious and access from the rear was limited to the single door unlike the ramp which opened the entire rear of the T59 and T59E1. The latter vehicles also were much quieter in operation.

Exterior components of the armored infantry vehicle T73 are identified in the views above and at the left below. At the right is the driver's station in the T73.

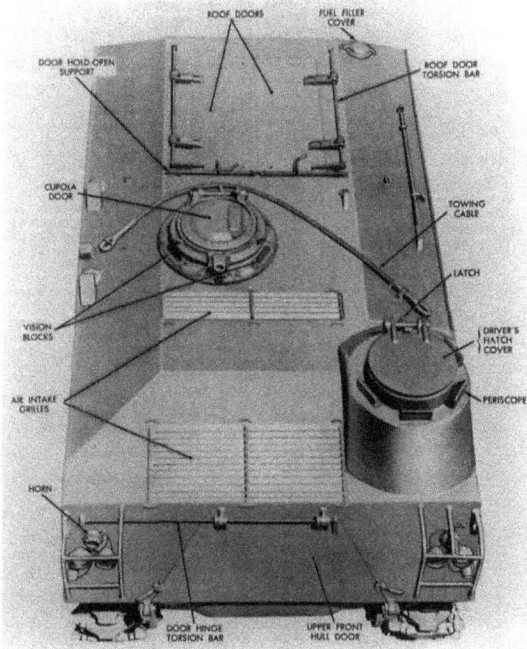

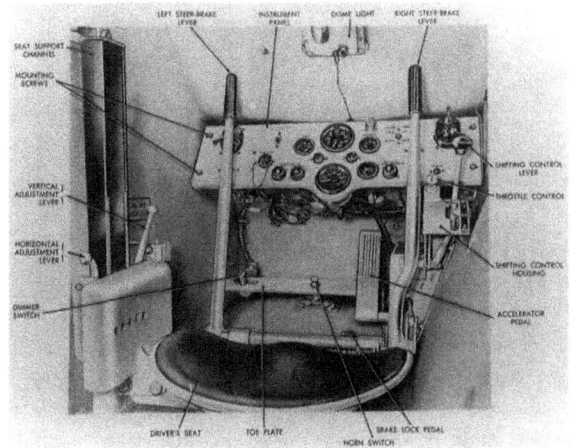

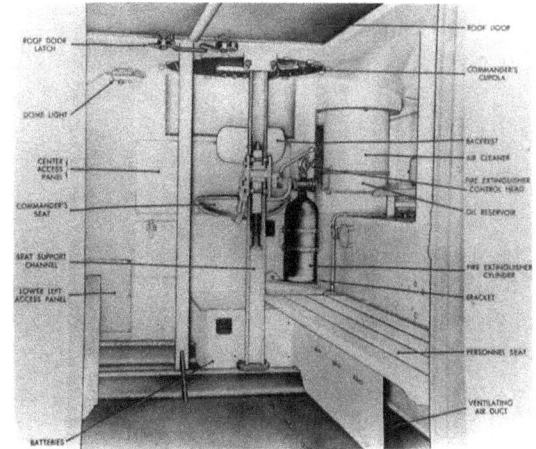

The commander's station in the T73 appears at the right and the front and rear of the vehicle can be seen below.

55

Above, from left to right, are the armored infantry vehicles T73, T59, and T18E1 at Aberdeen Proving Ground on 3 October 1952. Note the low silhouette of the T73 compared to the other vehicles.

After service tests, the Army selected the T59 as the new armored infantry vehicle and further development of the T73 was canceled. On 24 May 1953, the Assistant Chief of Staff, G4, directed that the armored infantry vehicle M59 (T59) be classified as standard and the armored infantry vehicle M75 (T18E1) also be retained as standard. A contract was awarded to FMC Corporation for production of the M59 which sub-sequently was designated as the armored personnel carrier M59. The first production configuration vehicle was delivered in August 1953. The initial contract was for 2,385 M59s, but over 6,300 were produced by the time production ended in 1960.

The production M59 differed from the pilots by the installation of an escape hatch door in the rear ramp. The small vision port in the ramp was shifted to one side of the door. This door was installed in vehicles F7 - F31 and F41 up. M59s F32 - F40 replaced the escape hatch door with a small rectangular escape hatch. The vision port on these vehicles was moved back to the top center of the ramp. This vision port was eliminated on the later production vehicles. The roof of the personnel compartment consisted of two hatches (one front and one

rear) bolted to the hull. Since they were hinged to each other, either hatch could be unbolted and folded on top of the other to permit loading of cargo. The front hatch had a door on the left side and the rear hatch had one on the right.

The production vehicles were fitted with a re-designed, manually operated, trim vane. The control handle for the trim vane was on the outside upper front armor within reach of either the driver or the vehicle commander. The GMC Model 302 engines were coupled to Model 300MG Hydramatic transmissions in vehicles F7 - F590. In M59s F591 up, Hydramatic transmissions Model 301MG were installed. These transmissions had four automatic forward and one reverse speeds which were transmitted through two differential speed ranges,

The early production armored infantry vehicle M59 is shown at the right and below. Note the vision port in the rear ramp to the left of the door.

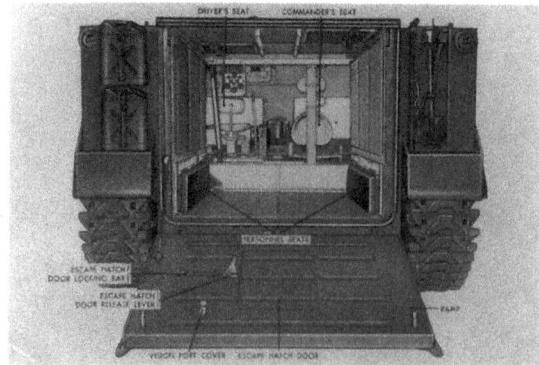

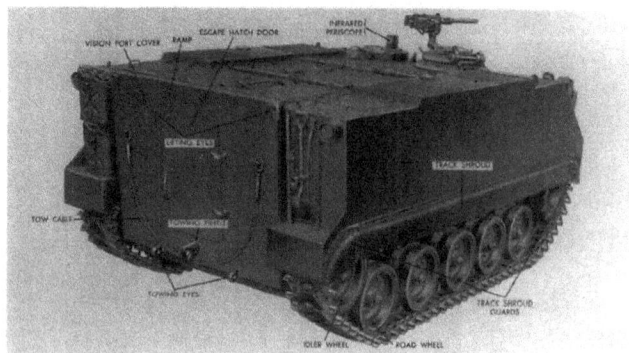

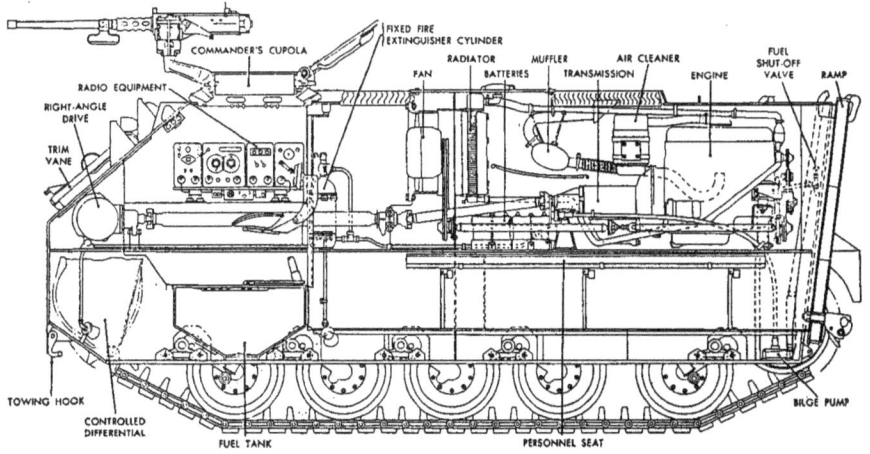

Armored Infantry Vehicle M59

Labels on sectional drawing: COMMANDER'S CUPOLA, FIXED FIRE EXTINGUISHER CYLINDER, RADIATOR, FAN, BATTERIES, MUFFLER, AIR CLEANER, TRANSMISSION, ENGINE, FUEL SHUT-OFF VALVE, RAMP, RADIO EQUIPMENT, RIGHT-ANGLE DRIVE, TRIM VANE, TOWING HOOK, CONTROLLED DIFFERENTIAL, FUEL TANK, PERSONNEL SEAT, BILGE PUMP

Labels: VISION PORT COVER, ESCAPE HATCH COVER, COVER RELEASE LEVER, RAMP

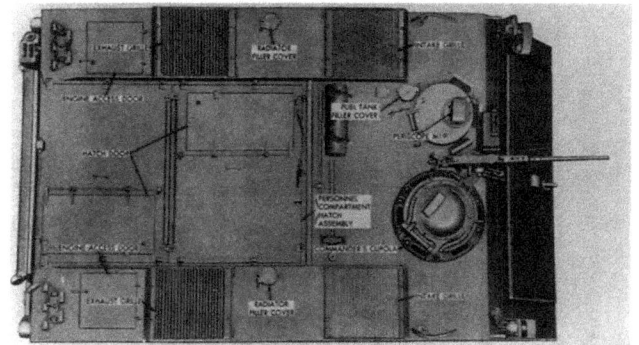

Labels: ESCAPE HATCH DOOR, DOOR HANDLE, RELEASE HANDLE LATCHING DETENT, RELEASE HANDLE SAFETY DETENT, RAMP SAFETY LOCK, DOOR RELEASE INNER HANDLE, RELEASE HANDLE STOP, RAMP

A sectional drawing of the armored infantry vehicle M59 appears at the top of the page. Above, the small escape hatch on M59s F32 through F40 is compared to the escape hatch door on the later vehicles. Details of the roof hatches can be seen at the right and below

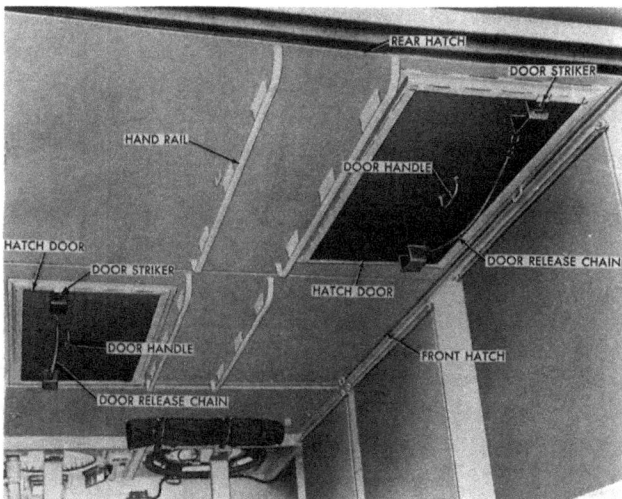

Labels: REAR HATCH, DOOR STRIKER, HAND RAIL, DOOR HANDLE, DOOR RELEASE CHAIN, HATCH DOOR, DOOR STRIKER, HATCH DOOR, DOOR HANDLE, FRONT HATCH, DOOR RELEASE CHAIN

Labels: HATCH DOOR, DOOR STRIKER, DOOR RELEASE CHAIN, TORSION SPRING, HATCH DOOR, TORSION SPRING, DOOR RELEASE CHAIN, REAR HATCH, FRONT HATCH

Above is one of the two folding troop seats inside the M59. When both were folded up, a jeep could be driven inside the vehicle. At the right is the M59 driver's station. Below are exterior and interior views of the driver's hatch.

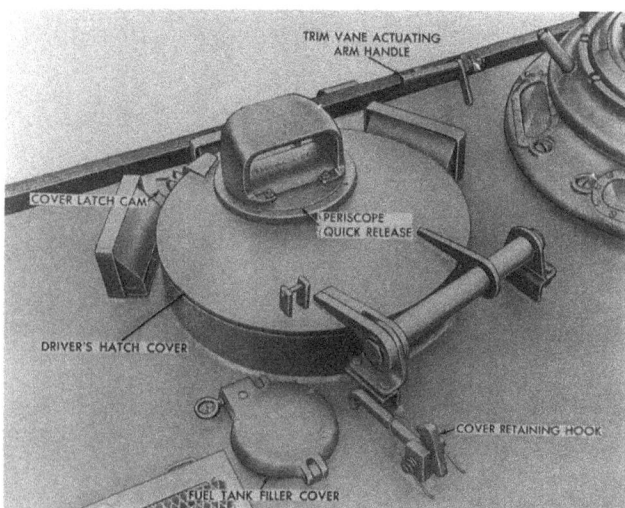

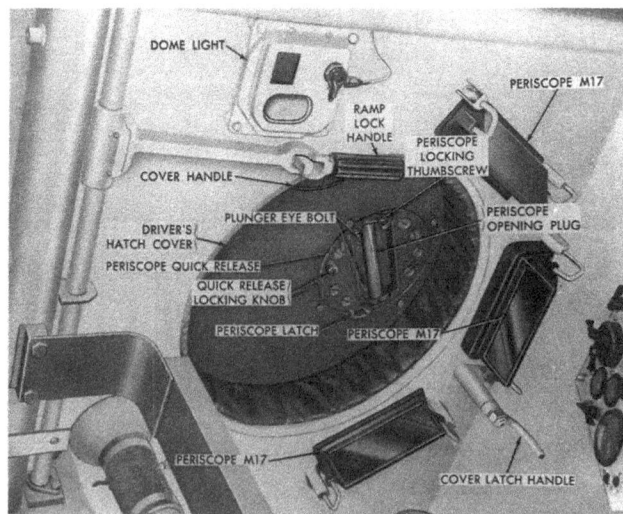

a high range and a low range. The driver's shift lever was used to control both the transmission and the differential. In addition to neutral, there were three transmission ranges in both the high and low differential ranges. These were DRIVE, HILLY, and reverse (REV). The transmission shifted automatically to the appropriate speed in each range. Early vehicles also had a neutral position for only the transmission in each differential range. The left and right auxiliary shifting levers were used to engage the transmission without engaging the controlled differential. This permitted the drive shafts to rotate without moving the vehicle, providing power for the ramp hydraulic system. The M59 could be operated on one engine in the low differential range. The drive shaft to the dead engine

was removed to avoid damage to the Hydramatic transmission. However, in an extreme emergency, the vehicle could be operated by removing the cross shaft locking pin and lowering the auxiliary shifting lever to the disengaged position. The ignition for the dead engine should be turned off in all cases and the fuel shut-off valves should be closed on vehicles F7 - F786. Later vehicles did not have fuel shut-off valves.

The commander's platform was adjustable in height and was attached, along with a folding seat, to the seat post under the commander's cupola on vehicles F7 - F2941. On F7 - F1312, the cupola was the same as on the pilots with six vision blocks and an external rotating mount for a .50 caliber machine gun. On F1313 - F2941, a new cupola was installed with four M17 periscopes

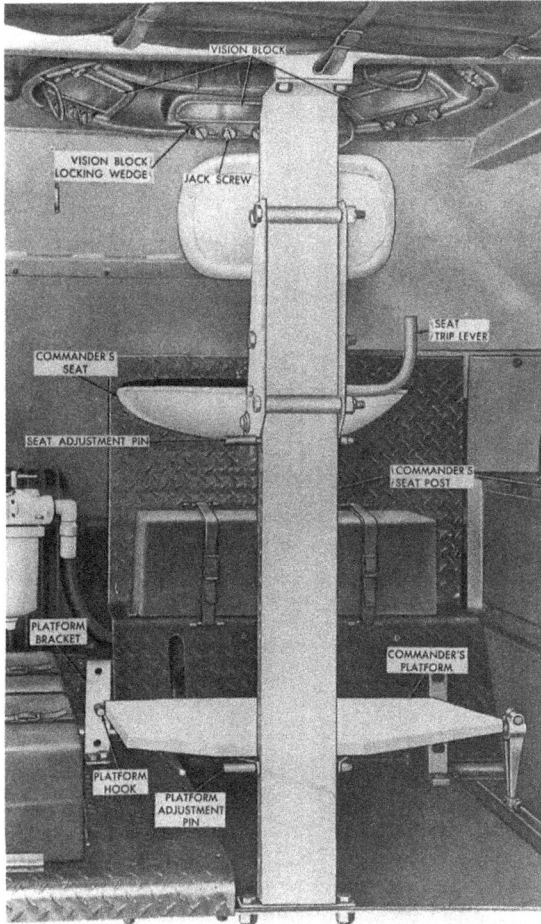

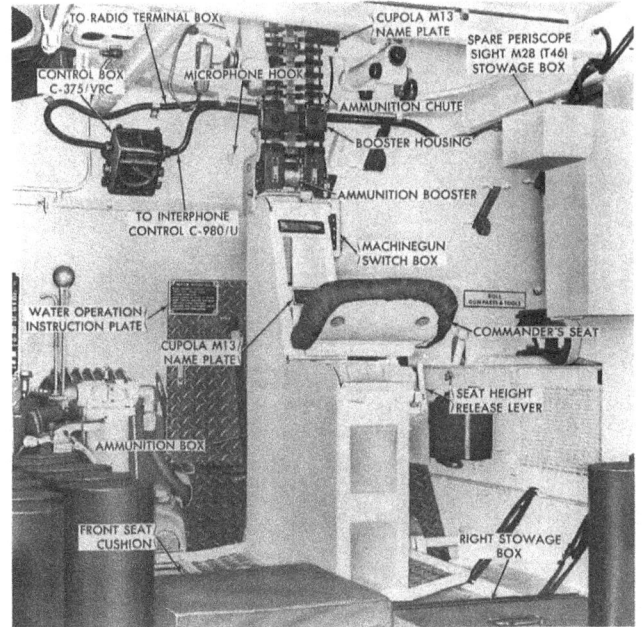

At the left is the commander's seat and platform on the early M59s. Above is the commander's seat with the M13 cupola and ammunition system on the late vehicles.

replacing the vision blocks. This cupola also carried a .50 caliber machine gun on an external mount with 360 degree rotation. Beginning with vehicle F2942, the M13 cupola was provided for the vehicle commander. This cupola was armed with an internally mounted, turret type, .50 caliber machine gun. The weapon was elevated and the cupola rotated manually. The ammunition system consisted of a 735 round ammunition box with the commander's seat attached. It was installed on a slip ring on the hull floor and connected to the cupola through a torque mount so that it could rotate with the cupola. An electric system fired the machine gun and powered the ammunition booster. The cupola contained four vision blocks with two in the body and two in the door. An M28 periscope sight in the cupola roof was used to aim the .50 caliber machine gun.

At the right are exterior and interior views of the commander's cupola on the intermediate M59 armored infantry vehicles, F1313 through F2941.

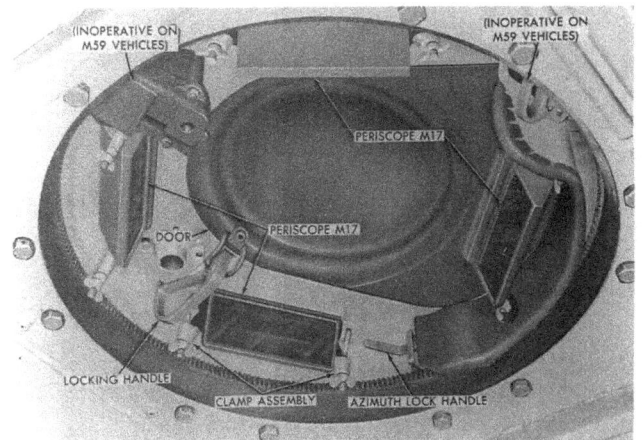

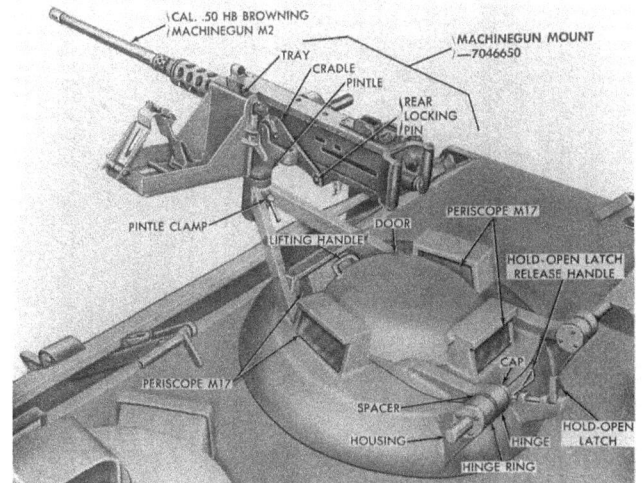

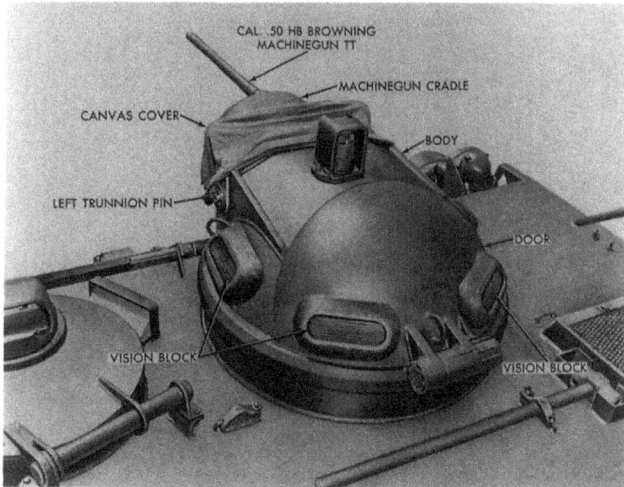

CAL. .50 HB BROWNING MACHINEGUN TT

MACHINEGUN CRADLE

CANVAS COVER

BODY

LEFT TRUNNION PIN

DOOR

VISION BLOCK

VISION BLOCK

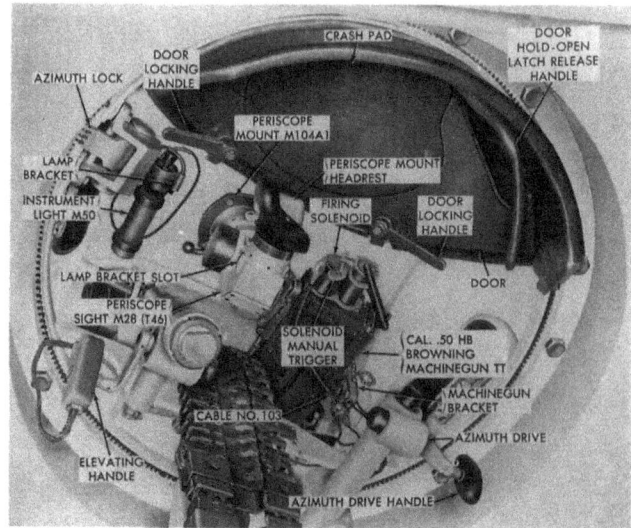

CRASH PAD

DOOR HOLD-OPEN LATCH RELEASE HANDLE

DOOR LOCKING HANDLE

AZIMUTH LOCK

PERISCOPE MOUNT M104A1

PERISCOPE MOUNT HEADREST

LAMP BRACKET

FIRING SOLENOID

DOOR LOCKING HANDLE

INSTRUMENT LIGHT M50

LAMP BRACKET SLOT

DOOR

PERISCOPE SIGHT M28 (T46)

SOLENOID MANUAL TRIGGER

CAL. .50 HB BROWNING MACHINEGUN TT

MACHINEGUN BRACKET

CABLE NO. 103

AZIMUTH DRIVE

ELEVATING HANDLE

AZIMUTH DRIVE HANDLE

Above are exterior and interior views of the commander's M13 cupola installed on F2942 and later M59s. At the lower right, a late M59 is swimming.

The M59 was used for a number of experimental installations. The French SS-10 and SS-11 antitank missile systems were mounted on the vehicle for evaluation. A remote controlled .50 caliber machine gun also was installed experimentally over the commander's station on another M59. Other test programs included a flat track suspension eliminating the track support rollers.

CAL. .50 HB BROWNING MACHINEGUN TT

CUPOLA M13

TRIM VANE

The photographs at the left show the M59 with missile armament. The upper view shows the installation of the SS-10 missile system during its tests at Fort Knox. In the lower photograph, the M59 is armed with SS-11 missiles. Below is an experimental installation of a .50 caliber machine gun in a remote control mount on the M59.

The 4.2 inch self-propelled mortar M84, based upon the armored infantry vehicle M59, is shown above and at the left.

The requirement still existed for a full tracked mortar carrier. In March 1953, a project was initiated to convert the armored infantry vehicle T59 (M59) into carriers for the 81mm, 105mm, and 4.2 inch mortars. These vehicles were designated as the T82, T83, and T84 respectively. The initial effort was concentrated on the T84 and it was completed before work began on the T82 and T83. By this time, the requirement for the 105mm mortar had been dropped and the T83 project was terminated. Work continued on the 81mm self-propelled mortar T82, but it also was canceled in favor

of a self-propelled mortar based upon the new M113 armored personnel carrier.

The 4.2 inch self-propelled mortar T84 consisted of the 4.2 inch M30 mortar installed on a modified M59. The roof hatches on the personnel compartment were redesigned to include a front hatch, a center hatch, and a rear hatch. The front hatch was bolted to the hull and connected to the center hatch by a torsion hinge. The center and rear hatches also were hinged together and they could be folded up on top of the front hatch providing an open area for the mortar. Hatch doors were located in the front and rear roof hatches. The escape door in the rear ramp was shifted to the right as viewed from the outside. This provided stowage space on the left rear for the base plate and rotator assembly from the M24A1 ground mount for the 4.2 inch mortar. OCM 36027, dated 25 November 1955, standardized the new vehicle as the 4.2 inch, full-tracked, self-propelled mortar M84.

Below, the 4.2 inch mortar installation can be seen through the open rear ramp at the left and from the open top at the right. Note the folding roof hatch which could be closed to provide armor protection when the mortar was not in action.

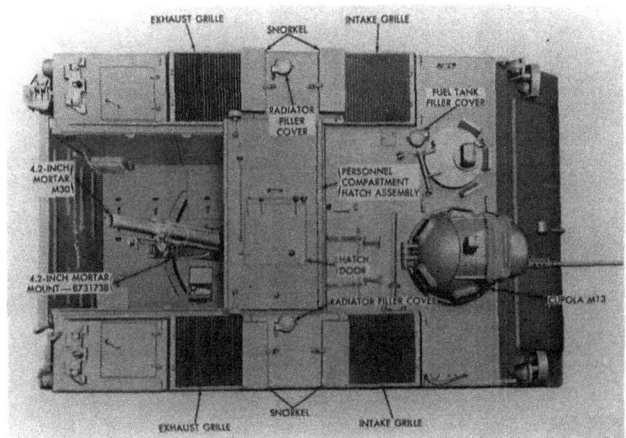

The two photographs on this page show the early 4.2 inch self-propelled mortar T84 during the test program at Fort Knox.

The M84 was manned by a crew of six consisting of the commander, driver, gunner, loader, and two ammunition handlers. It carried 88 rounds of 4.2 inch ammunition. The combat weight of the M84 was about 4500 pounds heavier than the M59. As a result, the freeboard was reduced when the vehicle was afloat. To keep water out of the cooling air intake and exhaust grilles, folding snorkels were installed. These were to be erected around the grilles prior to entering the water. Other features of the M84 were the same as the late production M59 including the M13 cupola for the vehicle commander. Production of the M84 at FMC Corporation began in January 1957.

With a cost approximately $1/3$ that of the M75, the M59 was popular with the Army. However, operations with the troops revealed some shortcomings. With its inexpensive commercial engines, it was somewhat underpowered. The low power was particularly obvious with the heavier M84 which had its maximum speed reduced to 27 miles per hour on land and 3.5 miles per hour in water. The cruising range also was considered to be inadequate. FMC proposed modifications to correct these problems, but they were rejected by the army since interest had shifted to a new vehicle then under development. It would eventually appear as the armored personnel carrier M113.

The late production armored infantry vehicle M59 with the M13 cupola machine gun mount above can be compared with the early vehicle below armed with external .50 caliber machine gun.

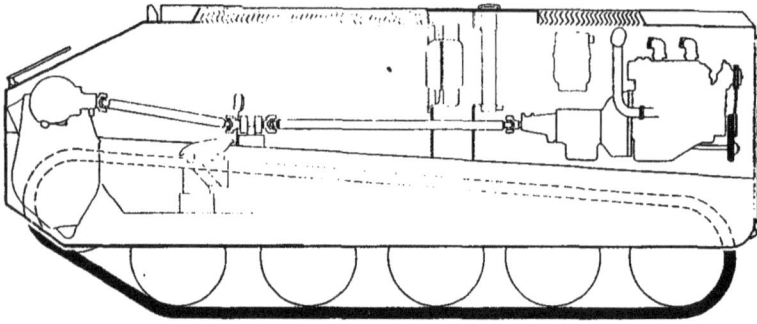

The sectional drawing and model above depict the TS 28 concept for a lightweight version of the M59 armored infantry vehicle proposed at the Questionmark III conference.

It is interesting to note that a concept for a lightweight version of the M59 was proposed by the Ordnance Tank Automotive Command during the Questionmark III conference in June 1954. With an estimated combat weight of 34,260 pounds, it was to utilize the power train originally installed in the T59E1. These more powerful engines would have improved the performance and the cruising range was estimated to be 177 miles.

During the early 1950s, the Tractor Division of the Allis-Chalmers manufacturing Company produced two small personnel carriers. Designated as the infantry tracked utility vehicles T55 and T56, they differed only in length. The T55 with an overall length of 148 inches was intended to carry six men including the driver. The T56, increased in length to 184½ inches, was designed for ten men including the driver. Driven by a General Motors type 302, liquid-cooled, gasoline engine and an Allison XT-90 transmission, the T55 had a maximum road speed of about 30 miles per hour and a cruising range of approximately 150 miles. The space inside these vehicles was extremely limited and they were not amphibious. Although they did not prove suitable as personnel carriers, the T55 provided the basis for the Ontos antitank vehicle armed with the 106mm recoilless rifle.

The drawings below show the infantry tracked utility vehicles T55 (left) and T56 (right). Note the extremely limited space inside these vehicles.

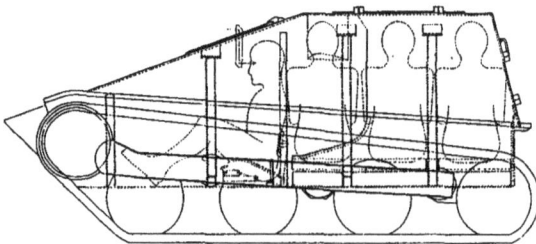

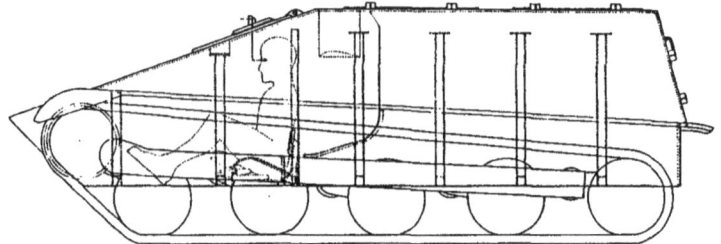

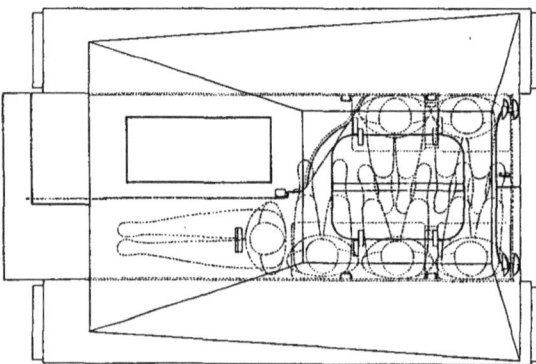

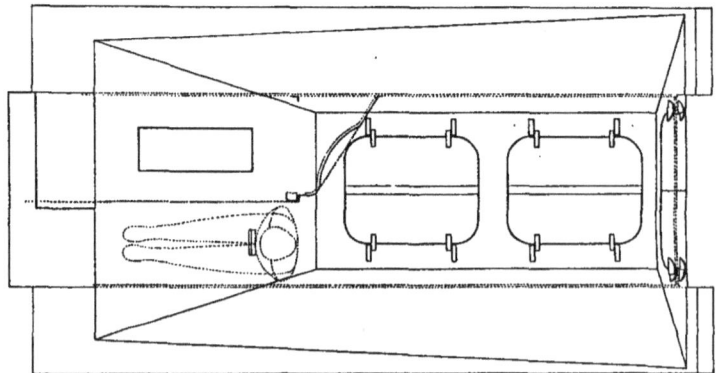

The ten man infantry tracked utility vehicle T56 above can be compared with the smaller six man T55 below.

At the left, the top armor has been removed from the infantry tracked utility vehicle T55 revealing the engine and power train. Again, note the limited space inside the vehicle. Below, the driver's controls can be seen in his station alongside the engine.

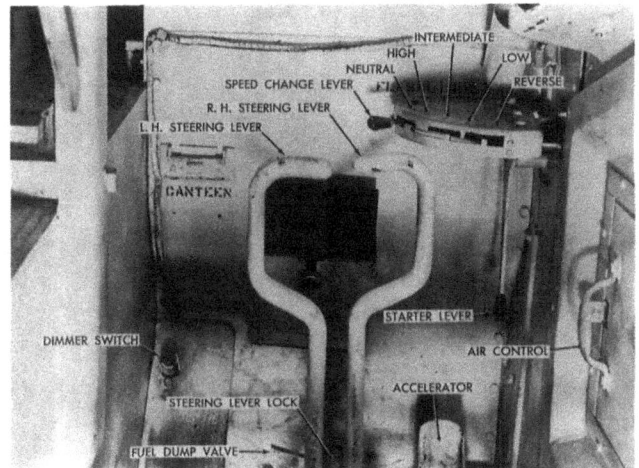

Above is the cargo tractor M8 (T42).

HIGH SPEED TRACTORS AND CARGO CARRIERS

To provide improved prime movers for the artillery, three new high speed tractors were authorized early in 1945. Designated as the cargo tractors T42, T43, and T44, their respective payload capacities were specified as 15,000, 6,000, and 30,000 pounds. The T42 paralleled the development of the armored utility vehicle T16 and used the same power train and suspension components. Like the T16, the T42 replaced an earlier design concept, in its case designated as the T33, which had been based upon the components of the light tank M24. However, the light tank engine did not have sufficient power and it was replaced in the T16 and the T42 by the power train from the 76mm gun motor carriage M18. As an artillery prime mover, the T42 was intended for use with towed loads of 18,000 to 32,000 pounds. Six pilot T42s were built and the vehicle was standardized as the cargo tractor M8 in December 1945. Three of the pilots were rebuilt replacing the air-cooled R-975-D4 radial engine, the 900AD combined torque converter-transmission, and the controlled differential with the power train from the light tank T41. This consisted of the air-cooled AOS-895 engine with the CD-500 transmission. The new version was designated as the cargo tractor M8E1. Further modification included a new version of the engine and transmission then being installed in the latest model of the light tank. These were the AOS-895-3 and the CD-500-3. The vehicle also was redesigned to insure maximum interchangeability of parts between the light tank and the cargo tractor. With the new designation cargo tractor M8E2, 480 vehicles were authorized for production at the Allis-Chalmers Manufacturing Company on 31 July 1950. Later, the M8E2 was standardized as the cargo tractor M8A1. When the M8A1 was fitted with the AOSI-895-5 fuel injection engine, it was reclassified as the full-tracked, high speed tractor M8A2.

At the left is the cargo tractor T42.

66

Above and below, the cargo tractor M8E1 appears at the left and the M8E2 (M8A1) is at the right. Both vehicles are equipped with a dozer blade in the upper photographs.

The high speed wrecker T4 is shown above and in the drawing at the right.

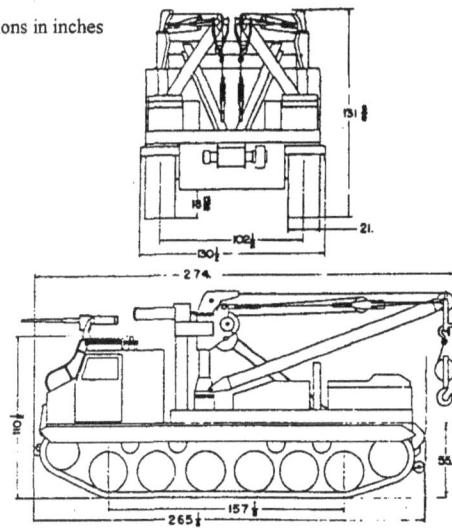

Dimensions in inches

In April 1946, OCM 30587 approved the development of a wrecker suitable for use by artillery, cavalry, armor, and tank destroyer units. Designated as the full-tracked, high speed, wrecker T4, the wrecker equipment was installed on the chassis of the cargo tractor M8E1. Hydraulic failures during the test program resulted in modifications and the new wrecker equipment was mounted on two M8E2 cargo tractors. The new wrecker was designated as the T4E1.

The high speed tractor T4E1 appears at the right.

The photographs above and below show the cargo tractor T43E1 during its test program at the Army Field Forces Board Number 2.

The cargo tractor T43 was designed as a prime mover for the 155mm howitzer and lighter artillery loads. As originally designed, it utilized a bogie type suspension with volute springs. Since it was desirable to use the latest type suspension on the entire new family of vehicles, it was redesigned to use the same type of torsion bar suspension employed on the light tank T41. It also was powered by the AOS-895-2 engine with the CD-500-2 transmission. Pilots of the original T43 design were never built and the new version was designated as the cargo tractor T43E1. This same chassis serve as the basis for the armored utility vehicle T18. The first of three T43E1 pilots was completed at the International Harvester Company in January !949. Engineering and service tests revealed the need for numerous modifications. The CD-500-2 transmission was replaced by the Allison XT-500-3. The designation was now changed to the cargo tractor T43E2. Twelve of the T43E2s were ordered. Five of these were additional test

Below is a sectional drawing of the high speed tractor T43E1

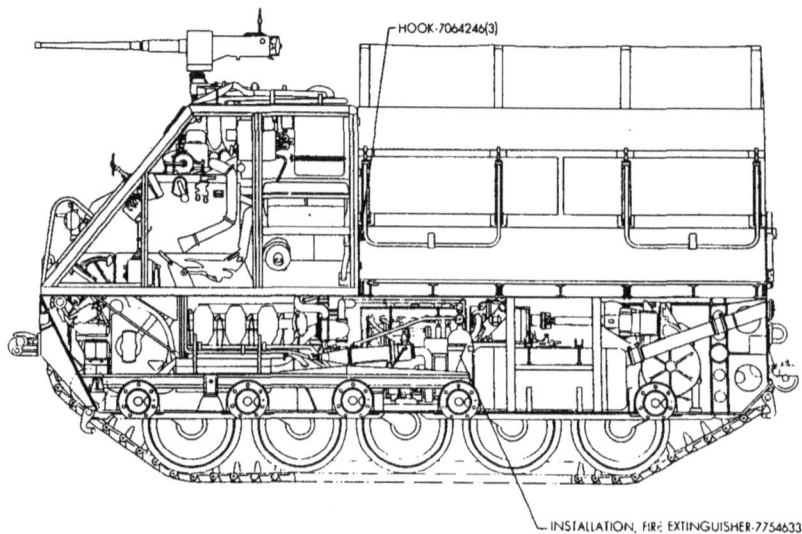

68

The cargo tractor T43E2 appears in the views above and below. The dimensions of the T43E2 are shown in the sketch below.

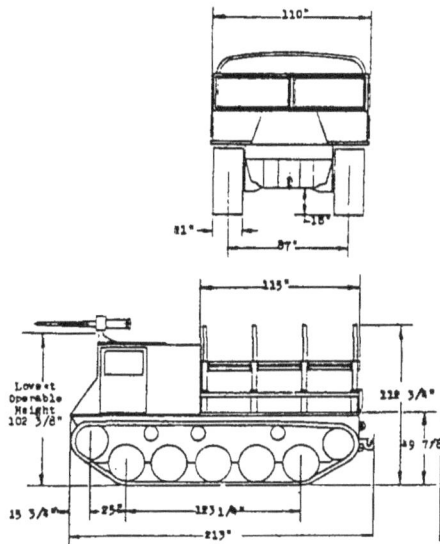

Dimensions in inches

pilots and the remaining seven were allocated to the U. S. Marine Corps.

OCM 31160, dated 24 October 1946, approved the procurement of the full-tracked, high speed, wrecker T5 based upon the T43E1 cargo tractor. The hydraulic failures during the test of the T4 wrecker resulted in the adaptation of commercial wrecker equipment which was now installed on the chassis of the cargo tractor T43E2. Designated as the full-tracked, high speed, wrecker T5E1, two pilots were procured for evaluation.

Below are a dimensional sketch and a photograph of the high speed wrecker T5E1.

Dimensions in inches

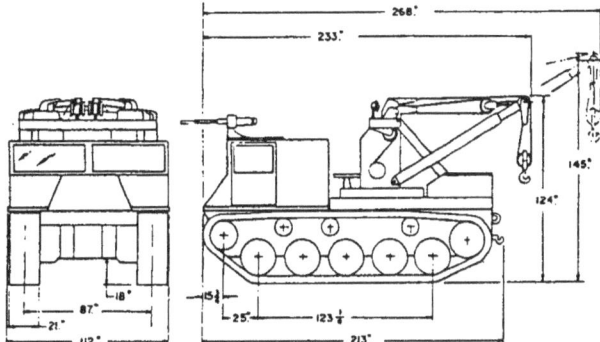

The cargo tractor T44 is shown in the photographs above and below as well as in the dimensional sketch.

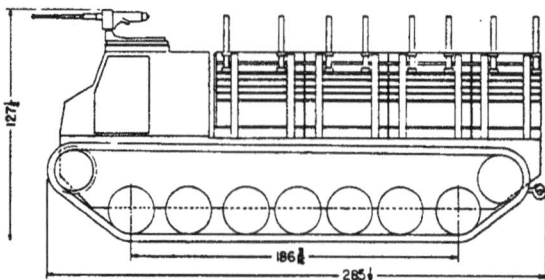

Dimensions in inches

Below is a photograph of the high speed wrecker T6E1.

The cargo tractor T44 was developed to provide a high speed prime mover for artillery towed loads of 40,000 to 65,000 pounds. Although six pilots were authorized originally, the number was reduced to two in March 1946. The design was modified to use the new Continental AV-1790-3 air-cooled engine with the CD-850-1 cross-drive transmission. The engine developed 810 gross horsepower at 2,800 rpm. A new vertical drum winch also was included. With a crew of two, the combat weight of the T44 was 99,075 pounds. Later, the number of pilots was increased to four, all of which were built at Detroit Arsenal.

After the outbreak of war in Korea, the T44 was redesigned by the Pacific Car and Foundry Company to use the latest components such as the AV-1790-5B engine and the CD-850-4A transmission. The redesigned vehicle was designated as the cargo tractor T44E1, but no pilots were built. However, two T6 high speed wreckers were built based upon the T44E1 design. The T6 was fitted with a crane and was intended for units that would be equipped with the T44E1 cargo tractors. The crane also could be used to emplace the 240mm howitzer and the 8 inch gun.

Above, the cargo tractor T85 is at the left and the cargo tractor T86 is at the right.

Although the tests of the cargo tractors were not complete, a study indicated that it would be more economical to replace the T43, M8, and the T44 series of vehicles with a new line of two tractors as prime movers for the medium, heavy, and very heavy artillery. Two separate approaches were followed to achieve this objective. The first involved modification of the tractors already under development. This produced two vehicles designated as the cargo tractors T85 and T86 which were based upon the cargo tractors T43E2 and the M8E2 respectively. International Harvester Company assembled four pilots of the T85 using many components from the T43E2. Combat loaded, it had a weight of 41,000 pounds including a payload of 12,000 pounds. OTCM 36445, dated 14 February 1957, standardized the T85 as the cargo tractor M85.

Allis-Chalmers Manufacturing Company modified two M8E2 cargo tractors to meet the T86 specification and they were submitted for test. Intended for use with the heavy or very heavy artillery having towed loads of 33,000 to 62,500 pounds, the T86 had a payload capacity of 12,000 pounds.

The second approach to obtaining two tractors that could meet all of the requirements involved the development of two new designs. Both of these were intended to provide lower cost prime movers for the artillery.

For medium and heavy artillery loads, the cargo tractor T93 was proposed based upon major components from the armored infantry vehicle M59. With an estimated combat weight of 42,750 pounds, the T93 was powered by a Chrysler V8, liquid-cooled, engine producing 400 gross horsepower at 4,200 rpm. Built by FMC, the T93 utilized the suspension and tracks from the M59. Four pilots were authorized and two of these were fitted with solid sides and hatch seals to permit amphibious operation. The amphibious vehicles were designated as the cargo tractor T93E1.

The proposed low cost prime mover for the very heavy artillery was designated as the cargo tractor T94. A new design by the Allis-Chalmers Manufacturing Company, it had an estimated combat weight of 72,000 pounds with a 12,000 payload. The T94 was powered by the AOSI-895, air-cooled, engine developing 525 gross horsepower at 2,800 rpm. It used the XT-500 transmission. Four pilots were authorized for construction.

The cargo tractor T93E1 is above and the cargo tractor T94 is below.

As early as April 1945, the Army Ground Forces had requested the development of an amphibious cargo carrier with a payload capacity of 1½ tons. The T34 Paddy Vehicle had been a partial response to that requirement. On 16 August 1945, OCM 28777 approved the development of the 1½ ton amphibian carriers T37 and T37E1. The requirement also was included in the War Department Equipment Board Report. On 30 October 1947, OCM 31792 recommended that two pilots of the amphibian cargo carrier T46 be built in lieu of development of the 1½ ton amphibian carriers T37 and T37E1. The objective of this program was to provide a light amphibian carrier and prime mover, convertible to a winterized vehicle, suitable for cargo or personnel and capable of being transported by air. General Motors Corporation was awarded the contract to design and build the two pilot vehicles. Later, the number of pilots was reduced to one when it became necessary to redesign the suspension and transmission. The initial configuration used a trailing idler, flat track, torsion bar suspension with five dual road wheels per side. Each road wheel was fitted with a 6:00 x 12 pneumatic tire. The sectional, band type, tracks were 28½ inches wide. The boat-like hull had the engine compartment in the front, the crew compartment for the driver and assistant driver in the center, and the cargo compartment in the rear. A door in the rear of the cargo compartment was hinged at the bottom and could be swung down to serve as a loading platform. The pilot vehicle was powered by the Continental AO-268-2, air-cooled, engine with the CD-150-1 transmission. A wobble stick control was used with this transmission. The T46 was driven in the water by two shrouded propellers at the rear which could be folded up for land operation.

The first modification to the new carrier was to the suspension. The trailing idler was eliminated and replaced by a small raised idler. The remaining four dual road wheels per side retained the 6:00 x 12 pneumatic tires and rode on the same 28½ inch wide tracks. At this stage, an access ladder was attached to the left side of the vehicle and a spare wheel was mounted on the left side at the rear. The engine exhaust pipe came out of the

The early configuration of the cargo carrier T46 is illustrated in the photograph above and the dimensional sketch below. Note the trailing idler suspension.

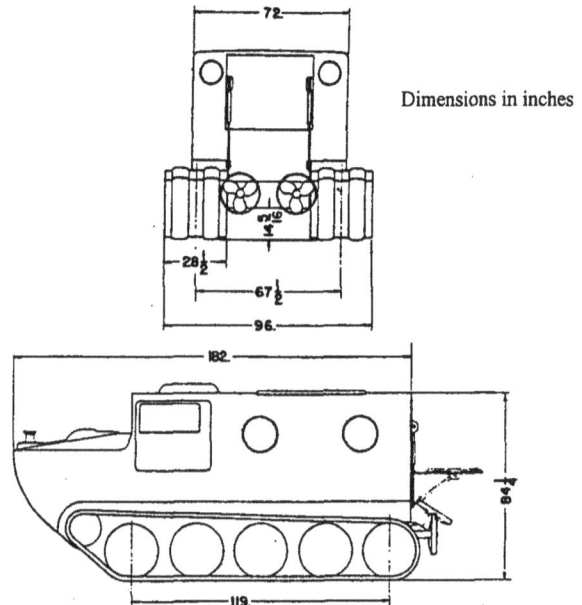

Dimensions in inches

engine compartment, extended toward the rear along the left side and then up to the top.

Additional changes replaced the two propellers with a single propeller and a late model CD-150 transmission was installed which was controlled by a

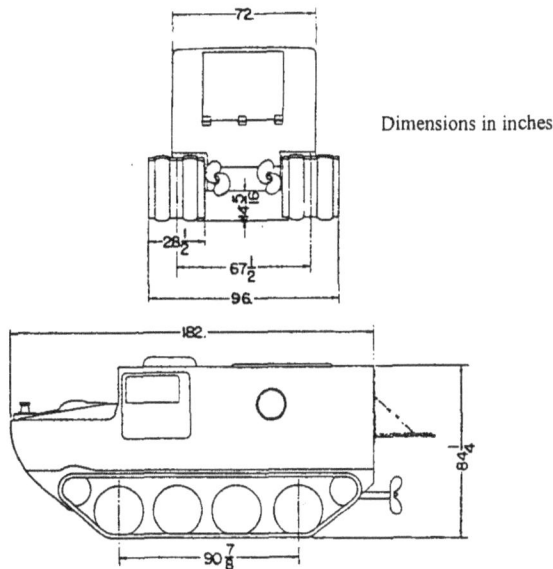

The modified cargo carrier T46 appears above and in the dimensional sketch at the left. Note the new suspension. Below is a photograph of the cargo carrier T46E1.

Dimensions in inches

handle bar. The earlier engine was replaced by the AO-268-3 and the fuel cells were mounted on the outside with one on each side of the hull. The engine exhaust pipe was relocated now extending inside the hull up through the roof. The suspension rode on 6.60 x 15 pneumatic tires and the track width was increased to 30 inches. The rebuilt vehicle was designated as the amphibious cargo carrier T46E1 and it was shipped to Aberdeen Proving Ground for tests in November 1949. After a successful test program, production of the T46E1 began at the Pontiac Division of General Motors in January 1951 and continued until June 1954. Standardization was approved in March 1953 as the amphibious cargo carrier M76 and it was named the Otter.

Further modifications were made during the production run. The external fuel cells installed near the center of each side were shifted toward the rear on the later vehicles. This required the relocation of the fuel and water cans, spare wheel, pioneer kit, and shutter stowage cover. A folding propeller guard was provided for water operation. The steps on the rear were moved slightly and a swivel type towing pintle was installed on the late vehicles. The hull was assembled by riveting aluminum alloy sheet and the production vehicles were equipped with the AOI-268-3A fuel injection engine. A 5,000 pound capacity winch was located under the rear seat in the cargo compartment.

The cargo carrier T46E1 is shown above and below with a dimensional sketch at the bottom left. Note the single propeller.

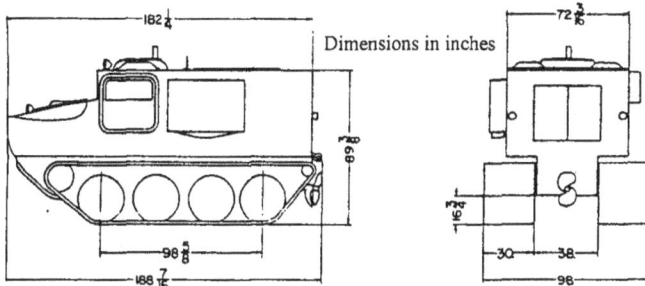

Dimensions in inches

The early and late production versions of the cargo carrier M76 (T46E1) are shown above and below at the left and right respectively. Note the relocation of the fuel cells and the external stowage.

The Otter had a combat weight of 12,045 pounds and a rated payload of 3,000 pounds resulting in a ground pressure of 2.0 psi. It had space for ten men including the two man crew. The maximum speed was 28 miles per hour on roads and 4½ miles per hour in water. A fuel capacity of 70 gallons provided a cruising range on roads of about 200 miles.

Details of the M76 suspension can be seen above at the right and top views of the early (left) and late (right) production M76 (T46E1) vehicles are shown below.

PART III

A NEW FAMILY OF TRACKED CARRIERS

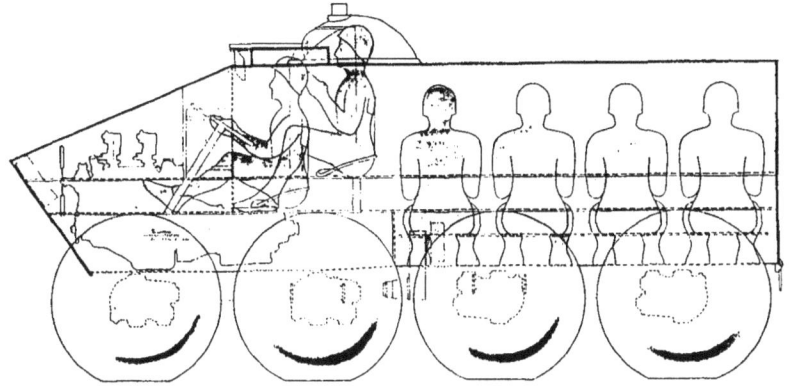

A model and a sectional drawing of the proposed 10 man armored, wheeled, carrier are shown above.

NEW LIGHTWEIGHT DESIGNS

In June 1954, Detroit Arsenal initiated concept studies for a series of lightweight combat vehicles utilizing the same universal chassis. These vehicles were to be not only suitable for air transport, but also capable of being dropped by parachute during phase 1 of airborne operations. On 30 September 1954, the Army Field Forces issued military characteristics for a family of two armored carriers with weights of 16,000 pounds and 8,000 pounds. The heavier vehicle was intended to carry a ten man infantry unit and to provide a suitable chassis for self-propelled weapons, a cargo carrier, and an ambulance. The lighter vehicle was to carry a four man crew and be used as a command and reconnaissance vehicle or as a mount for the battalion antitank rifle. Both vehicles could be wheeled or tracked and were to be amphibious with all round light armor protection. The 16,000 pound tracked vehicle corresponded to the universal chassis design concept already under study at Detroit Arsenal.

Above is a model of the proposed ten man armored, tracked, carrier. Below, a model of a proposed four man armored, wheeled, carrier is at the left armed with a recoilless rifle. At the bottom right is a model of a proposed four man armored, tracked, vehicle intended for use as a command post.

At the right is a mock-up of the armored, wheeled, personnel carrier T115.

Preliminary mock-ups of the various vehicle configurations were completed at the Arsenal and presented to the Continental Army Command (CONARC) in June 1955. At that time, the wheeled version of the ten man vehicle was dropped from further consideration because of cross-country mobility requirements. The original design of the larger tracked vehicle also was modified to carry 12 men in addition to the driver. On 5 January 1956, Item 36049 of the Ordnance Technical Committee Minutes (OTCM) officially initiated the development and defined the military characteristics of the air transportable, armored, multi-purpose vehicle family. These vehicles were designated as the T113 armored, full-tracked, personnel carrier (13 men), the T114 armored, full-tracked, personnel carrier (4 men), and the T115 armored, wheeled, personnel carrier (4 men). However, the T115 was abandoned without further development, although a mock-up was completed.

Following a meeting at the Pentagon in May 1956, a contract was awarded to FMC Corporation for the development of the new vehicle and it authorized the construction of 16 pilots. They were to consist of ten armored personnel carriers, two 81mm self-propelled mortars (later designated as the T257), three carriers for the Dart antitank missile (later designated as the T149), and one basic chassis for experimental use. To obtain the maximum amount of information, CONARC requested that eight of the pilots be fabricated using aluminum alloy armor and be powered by a standard Ordnance air-cooled engine. The remaining eight vehicles were to be assembled with steel armor and a commercial liquid-cooled engine. FMC completed the mock-up of the T113 which was inspected in October

1956 and approval was received for the construction of the pilot personnel carriers and self-propelled mortars. Two versions of the armored personnel carrier were completed during 1957. One of these was the original T113 with aluminum alloy armor powered by the AOSI-314-2, air-cooled, engine with a geared steering, X-drive, transmission. The other utilized steel armor and was driven by a Ford Model 368-UC, liquid-cooled, V8 engine using the same transmission. To avoid confusion with the T113, this vehicle was designated as the armored personnel carrier T117. Five T113s and five T117s were manufactured by FMC. In addition, one aluminum T113 hull and one steel T117 hull were delivered for ballistic tests. Consideration also was given to the use of 2 inch thick magnesium alloy armor on the T113. However, this project was canceled.

With gross weights of 17,600 pounds and 19,530 pounds respectively, the T113 and T117 were very similar in appearance. External stiffeners were fitted on the thinner roof of the T117 and the towing eyes welded to the front hull also were thinner on the steel vehicle. Both the T113 and the T117 could be dropped by parachute. Engineering and service tests of both types began in the Fall of 1957.

Below, the armored personnel carrier T113, pilot number 2, is at the left with aluminum alloy armor. The steel armored personnel carrier T117, pilot number 2 is below at the right. Note the heavier towing eyes on the aluminum vehicle.

Additional views of the armored personnel carrier T113, pilot number 2, and the armored personnel carrier T117, pilot number 2, are shown at the left and right respectively above and below.

Top views of the T113 and T117 number 2 pilots are shown above at the left and right respectively. Note the stiffener on the roof of the T117 steel vehicle. The interior of the two vehicles can be seen below through the open ramps. The T113 is on the left and the T117 is on the right.

The photograph above shows the launching of a Dart antitank missile. The concept drawing is a proposed installation of the Dart missile on the armored personnel carrier. Below at the right, the armored personnel carrier T113, pilot number 3, is swimming.

Above, the armored personnel carrier T113, pilot number 3, is under test by the Infantry Board. At the right is another T113 pilot. Below, a T117 pilot is at the left. The roof stiffener and the smaller towing eyes are clearly visible on the T117.

The armored personnel carrier T113E1 is shown above. Note the change in the front hull configuration. These early vehicles were all armed with a .30 caliber machine gun on the commander's hatch.

M113 SERIES OF ARMORED PERSONNEL CARRIERS

By late 1957, CONARC requirements for the new armored personnel carrier had changed. A lower cost power package was preferred and improved armor protection was desired, even if there was a weight penalty. To meet these revised requirements, two new vehicles were proposed, both with aluminum alloy armor. The first, with an air drop weight of less than 17,500 pounds, was intended for use by the airborne forces and the second, with heavier armor and a maximum combat weight of 24,000 pounds, was to equip the armored divisions. Its armor was to be equal or superior to that on the armored personnel carrier M59. On 9 October 1958, OTCM 36890 assigned the designations armored personnel carriers T113E1 and T113E2 to the lighter and heavier vehicles respectively. The project for the T149 Dart missile carrier was canceled and its resources transferred to the new T113E1/T113E2 program. Eight pilots were authorized consisting of four T113E1s and four T113E2s. Delivery of the new vehicles was complete in November 1958.

The T113E1 and T113E2 differed from the original T113 pilots in several ways. The most obvious external changes were the new configuration of the front armor and the suspension system. The blunt vertical bow with the narrow trim vane on the T113 was replaced by a sloping upper front plate extending down to the top of the tracks and a much larger trim vane was installed. A

modified suspension with a separate idler replaced the trailing idler suspension on the T113. Tests on the T113 had shown that the aluminum alloy towing eyes frequently failed so they were replaced with cast steel towing eyes bolted to the armor plate. The new vehicles were powered by the Chrysler Model 361B, liquid-cooled, gasoline engine with the Allison TX200-2 transmission. Later, the Chrysler engine was designated as the 75M.

Another view of the armored personnel carrier T113E1 is at the right.

81

These photographs show the armored personnel carrier T113E2. It is noted that the T113E1 and T113E2 have steel towing eyes bolted to the front hull.

Service tests by the Armor and Infantry Boards were completed in January 1959. After evaluating these tests, CONARC concluded that the T113E2 could meet both requirements if its weight could be reduced by 400 pounds. This weight reduction was achieved by reducing the armor thickness on the floor and rear of the vehicle as well as on the bottom of the sponsons. The modified T113E2 was classified as the full-tracked armored personnel carrier M113, Standard A, by OTCM 37037 dated 2 April 1959. The M59 armored personnel carrier was reclassified as Standard B. After completion of two preproduction pilots, full production of the M113 began at FMC Corporation in January 1960.

The external configuration of the production M113 vehicles followed that of the T113E2. The driver's station was in the left front hull alongside the power plant. His hatch in the hull roof was surrounded by four M17 periscopes and the hatch cover was fitted for the installation of an M19 infrared periscope. The commander was located in the center of the vehicle behind the driver and the power plant compartment. The commander's cupola in the hull roof had five M17 periscopes and was armed with a .50 caliber machine gun on a rotating mount replacing the .30 caliber weapon on the early pilots. A platform for the commander was adjustable in height and the seats for

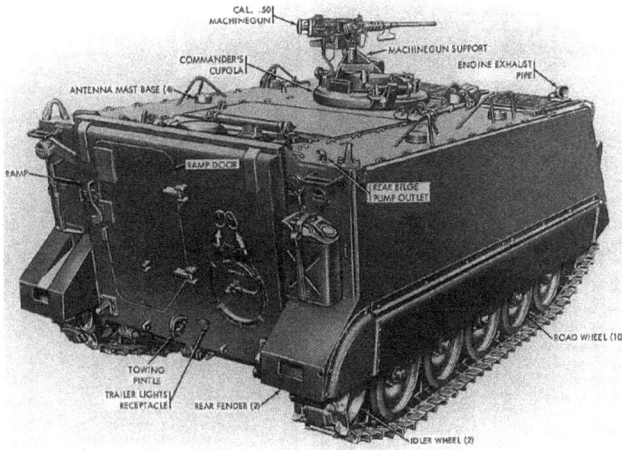

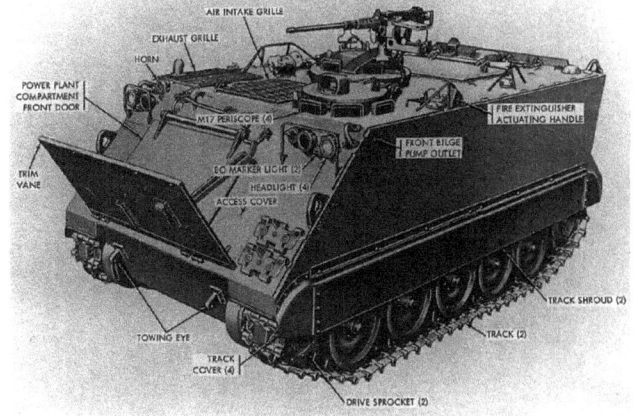

Exterior components on the armored personnel carrier M113 are identified in the photographs above. The operation of the trim vane controls is illustrated below at the right.

both the driver and commander also were adjustable in height to permit operation with their heads exposed in the open hatches. The remaining 11 crew members were seated on folding benches on each side of the troop compartment and on a jump seat in the center behind the commander's seat.

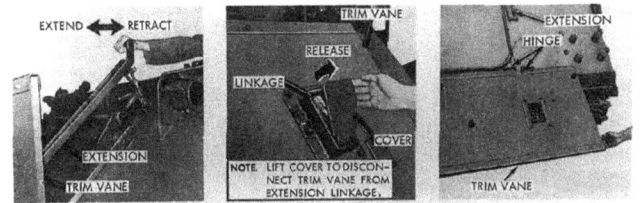

Armored Personnel Carrier M113

The crew seating arrangement in the M113 can be seen in the top view at the right.

Dimensions in inches

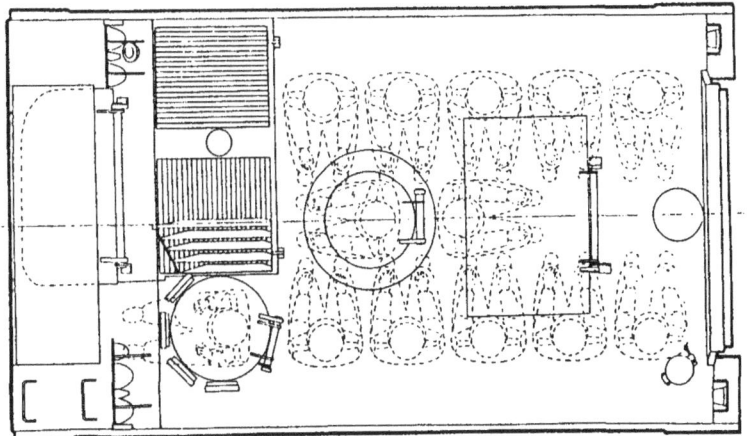

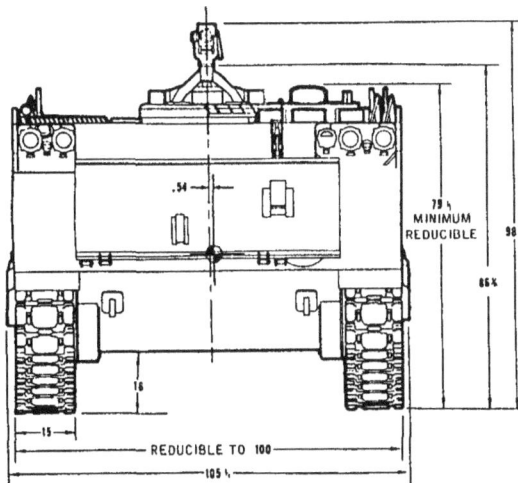

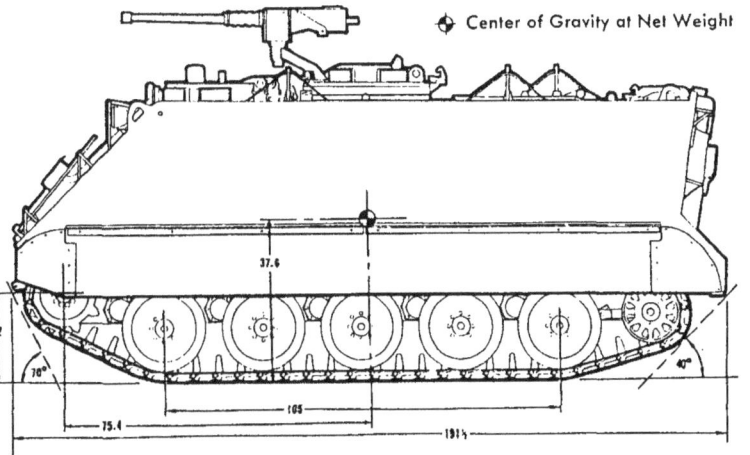

83

Components on the top and inside the armored personnel carrier M113 are identified in these photographs.

The box-like hull was assembled by welding 5083 aluminum alloy rolled plate 1½ inches thick on the roof, front, and rear. The sides and floor were 1³/₄ and 1¹/₈ inches thick respectively. The ramp at the rear was hydraulically operated and also contained a door for use when the ramp was closed. All openings had watertight seals to permit amphibious operation. The power pack was in the right front hull consisting of the Chrysler Model 75M, liquid-cooled, gasoline engine with the Allison TX200-2 transmission. This V8 engine developed 215 gross horsepower at 4,000 rpm. It was connected to the transmission through a transfer case with a 1:1 gear ratio. The transmission had six forward speeds and one reverse and it shifted automatically between the speeds in each shift range. These ranges were 1-2, 3-4, 3-5, 3-6, reverse, and neutral. To provide steering, the power was transmitted through a controlled differential at the front of the power plant compartment and then to the final drives and sprockets at the front of each track. The vehicle was supported on a flat track torsion bar suspension with five dual road wheels per side. Track tension was maintained by an hydraulically adjusted idler at the rear of each track. The single pin, center guide, steel tracks were 15 inches wide and were fitted with detachable rubber pads. The air drop weight of the M113 was 18,600 pounds and its combat weight was estimated to be 22,900 pounds. With a fuel capacity of 80 gallons, the cruising range was about 200 miles. It

The internal dimensions of the M113 are shown below at the left. At the bottom right, the power flow is illustrated from the engine through the transmission, differential, and final drives to the tracks.

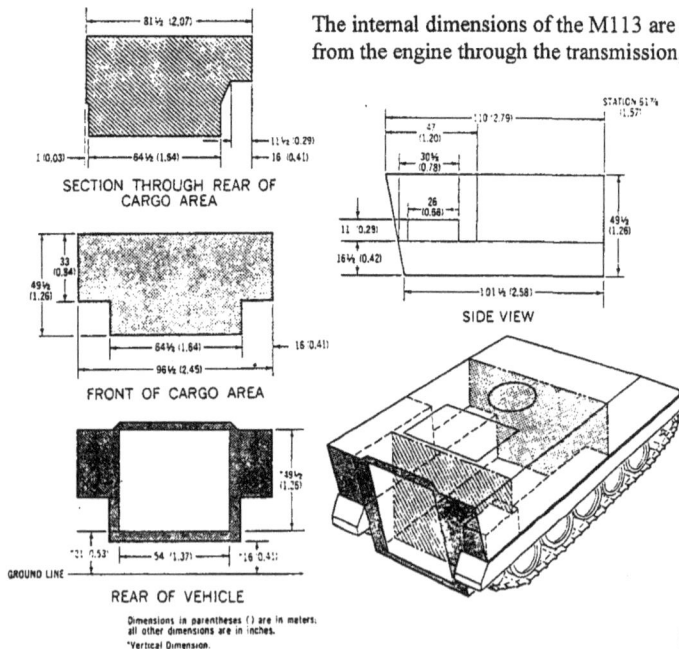

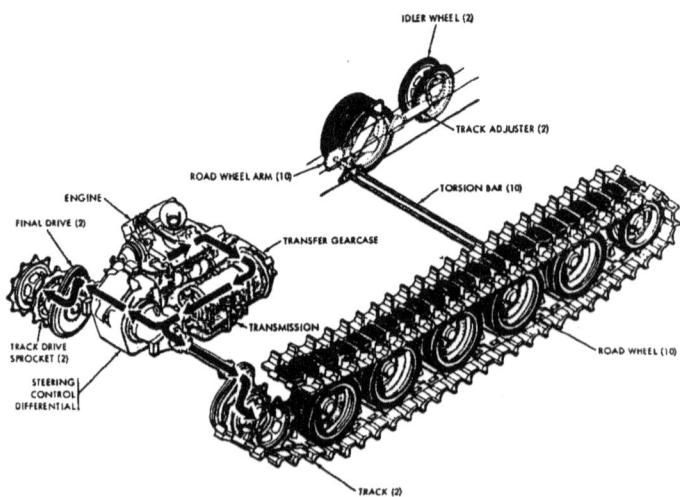

The Chrysler 75M liquid cooled engine is shown above removed from the M113. Below is a view of the power plant compartment with the engine and other power train components installed.

could reach a maximum speed of 40 miles per hour on a level road. In water, the freeboard was about 14 inches when combat loaded. Propelled by track action in the water, the M113 could reach a forward speed of approximately 3½ miles per hour. Needless to say, amphibious operations were restricted to calm water as the vehicle was not intended for use in surf or rough water.

For cold weather operation, a winterization kit was provided. This consisted of a personnel heater installed in the right front of the troop compartment. Ducts distributed heat to the troop compartment and to the battery box at the right rear. Hot air could be diverted from the troop compartment to the power plant compartment to aid starting in cold weather.

To use the carrier as an ambulance, a kit was available to install four litters in the troop compartment. These litters were supported by chains attached to the roof and floor.

A total of 14,813 M113s were built by the time production ended in 1968. United States forces received 4,974 vehicles and 9,839 were allocated to foreign military sales.

Details of the suspension on the M113 armored personnel carrier can be seen below. At the right is a view through the open ramp of the M113 with the litter kit installed.

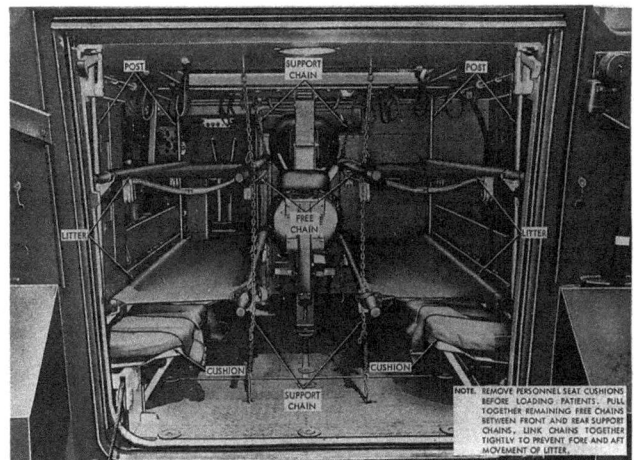

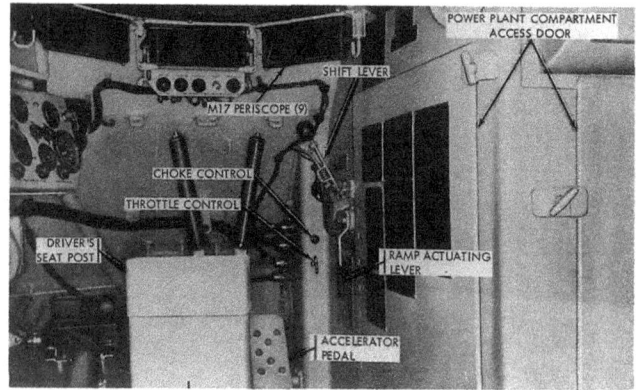

Above are the left and right sides of the M113 driver's station. Below at the left is the radio installation.

Above at the right, the M113 is fitted with windshields for the driver and vehicle commander. Below, note that this M113 has two additional periscopes installed, one on each side of the troop compartment.

The internal stowage on each side of the M113 can be seen in the two photographs above. A view of the personnel heater is at the left.

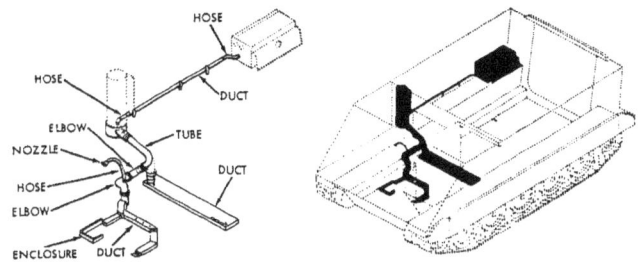

The drawing above shows the ducts used to distribute the warm air from the personnel heater. Note the duct leading to the batteries in the right rear sponson. Below is a photograph of the M113 armored personnel carrier swimming.

These two photographs show the M113 armored personnel carrier during test operations.

At the right is the power pack consisting of the 6V53 diesel engine and the XTG-90-2 transmission used in the M113E1 armored personnel carrier.

In June 1959, even before production began on the M113, a contract was awarded to evaluate the use of a commercial, compression ignition, engine in the new armored personnel carrier. This was in line with the policy to develop a line of diesel powered combat vehicles. This study concluded that the M113 could easily be modified to use the General Motors 6V53, liquid-cooled, diesel engine. Various data sheets rated this V6, two stroke cycle, power plant at 210-215 gross horsepower at 2,800 rpm. A 212 horsepower figure was used for the data sheets in this volume. The engine was coupled to the Allison XTG-90-2 transmission which incorporated a torque converter with four manually selected speeds forward and one reverse. It included geared steering in the 2nd, 3rd, and 4th forward speed ranges and clutch brake steering in 1st and reverse. This eliminated the need for the controlled differential required in the M113. It was expected that the elimination of the controlled differential would compensate for the heavier weight of the diesel engine.

FMC Corporation built three engineering pilot vehicles using the new power train and they were designated as the armored personnel carrier M113E1. Tests at Aberdeen Proving Ground, Fort Knox, and Fort Greely revealed numerous transmission failures resulting in its modification and a new designation as the XTG-90-2A. The tests also revealed that the performance with the new power train was superior in all respects to that of the M113. Fuel economy of the new diesel engine increased the cruising range of the M113E1 to about 250 miles compared to the 200 miles of the M113 using the same 80 gallon fuel tank. The use of diesel fuel also reduced the fire danger.

During this same period, FMC Corporation proposed the construction of additional pilots powered by the 6V53 engine with an Allison TX-100 automatic transmission. Since steering was not incorporated in the TX-100, the use of the DS-200 controlled differential was required as in the M113. It was expected that this power train would reduce manufacturing costs. The proposal was approved and FMC built three pilot vehicles designated as the armored personnel carrier M113E2. The fuel tank capacity in the M113E2 also was increased to 95 gallons providing a cruising range of about 300 miles. Tests of the M113E2 were highly successful and it was classified as the full-tracked armored personnel carrier M113A1, Standard A, by Army Materiel Command Technical Committee Minutes (AMCTCM) Item 950 on 16 May 1963. After completion of two preproduction pilots, full production of the M113A1 began at FMC in late 1964.

The armored personnel carrier M113E2 appears in the two photographs below.

These views show details of the armored personnel carrier M113E2. The new power pack is installed in the power plant compartment at the bottom right.

The M113A1 armored personnel carrier is above. The new power train is shown below along with the tracks and suspension.

Except for the power train and fuel tank, the configuration of the M113A1 was identical to the M113. Since there was no ignition switch, a fuel cutoff control was provided to stop the diesel engine. The TX-100 transmission consisted of an hydraulic torque converter with a lock-up clutch and a basic three speed, constant mesh, planetary gear train. The drive ranges were 1, 1-2, 1-3, neutral, and reverse. It shifted automatically to the appropriate speed in ranges 1-2 and 1-3. Steering was provided by the brake levers on the controlled differential as in the M113.

The left and right sides of the driver's station in the M113A1 can be seen below.

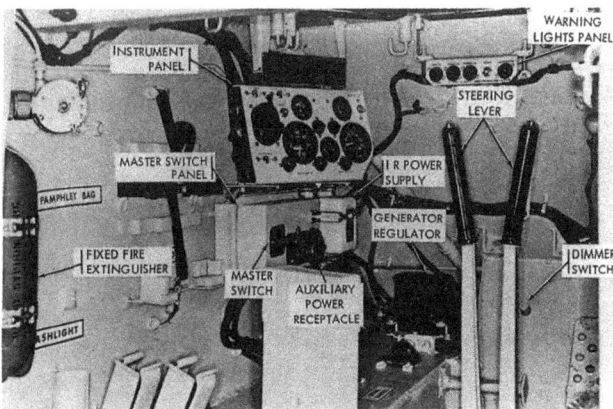

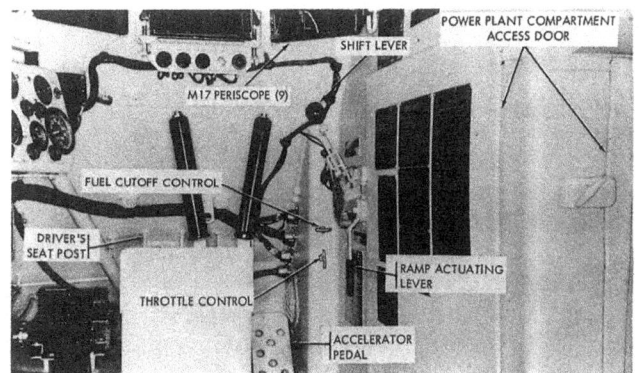

The M113A1 armored personnel carrier appears above and the drawing below depicts the hospital hood protectors connected to the M8A3 gas particulate filter unit used when the litter kit was installed.

The personnel heater kit was retained on the M113A1 for cold weather operation, but it no longer supplied hot air for the battery box or the power plant compartment. An engine coolant heater kit warmed and circulated coolant through the engine and the battery box heat exchanger. A windshield kit was available to protect the driver from cold wind when driving with his head exposed in the open hatch. An M8A3 gas particulate filter unit could be installed to provide purified air to four M14A1 gas masks or four hospital hood protectors when the litter kit was installed.

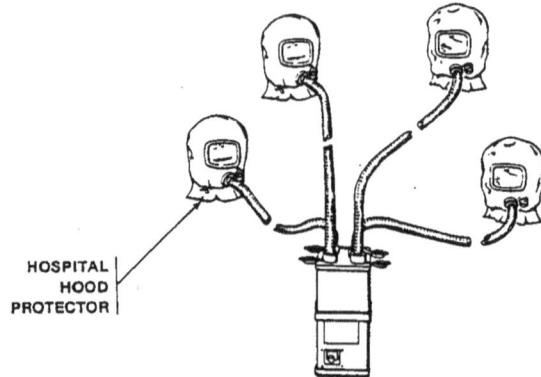

HOSPITAL
HOOD
PROTECTOR

The two photographs below show the driver's windshield kit installed. The hatch could be opened and closed without interfering with the windshield.

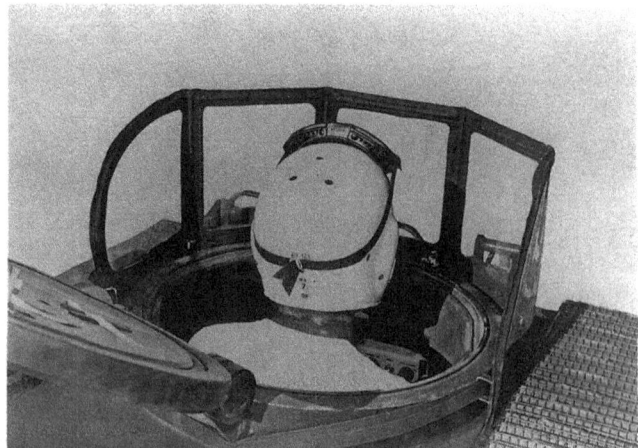

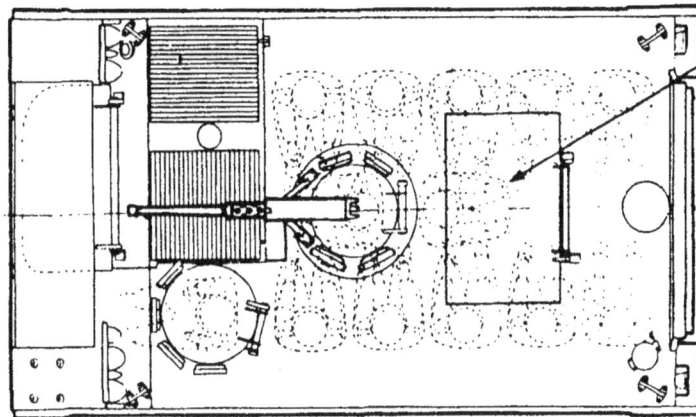

CARGO HATCH OPENING
47.5 (1,21) BY 30.75 (0,78)

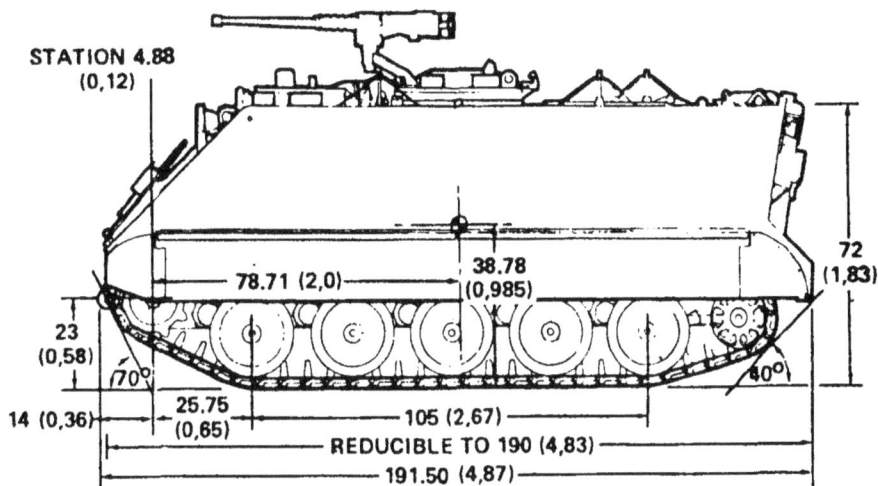

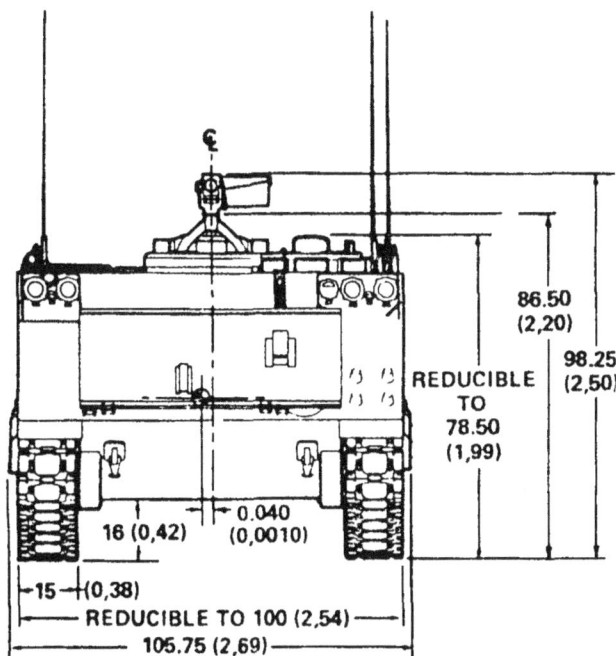

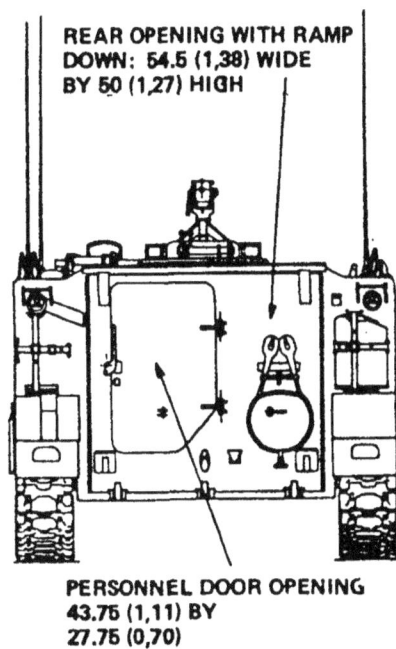

STATION 4.88
(0,12)

72
(1,83)

78.71 (2,0)   38.78
                     (0,985)

23
(0,58)
      70°   60°

14 (0,36)   25.75
              (0,65)   105 (2,67)
              REDUCIBLE TO 190 (4,83)
              191.50 (4,87)

REAR OPENING WITH RAMP
DOWN: 54.5 (1,38) WIDE
BY 50 (1,27) HIGH

86.50
(2,20)

REDUCIBLE
TO
78.50
(1,99)

98.25
(2,50)

0.040
(0,0010)

16 (0,42)

15 (0,38)
REDUCIBLE TO 100 (2,54)
105.75 (2,69)

PERSONNEL DOOR OPENING
43.75 (1,11) BY
27.75 (0,70)

Dimensions in inches (meters)

Armored Personnel Carrier M113A1

93

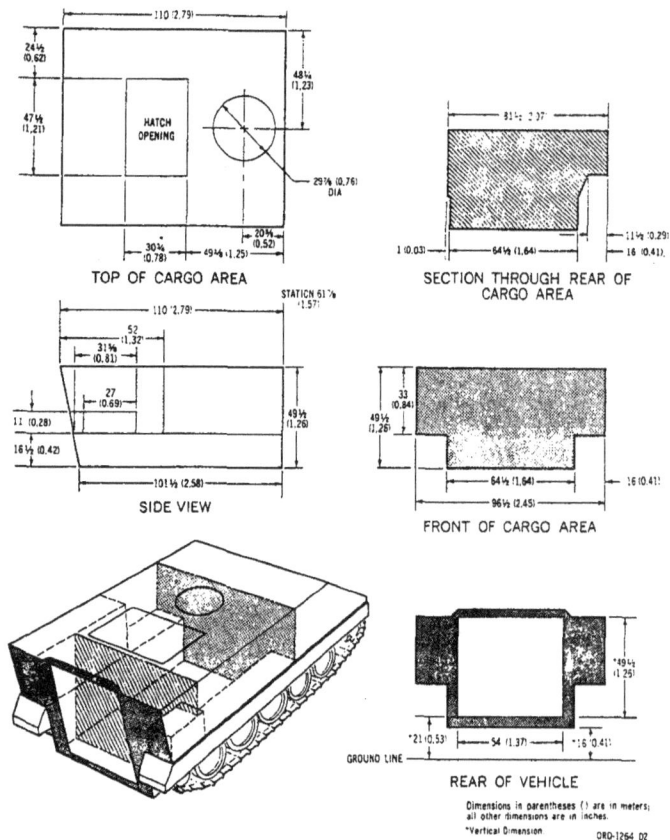

TOP OF CARGO AREA

SECTION THROUGH REAR OF CARGO AREA

SIDE VIEW

FRONT OF CARGO AREA

REAR OF VEHICLE

Dimensions in parentheses ( ) are in meters; all other dimensions are in inches.
*Vertical Dimension

ORD-1264 02

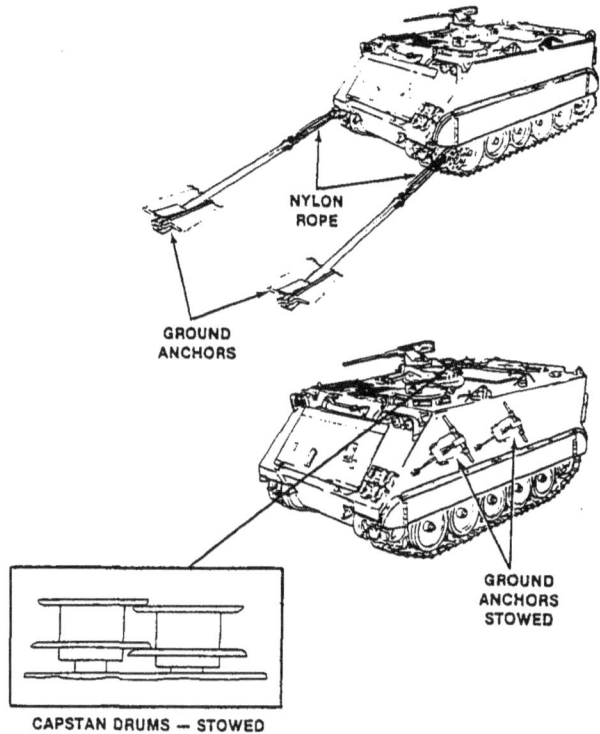

NYLON ROPE

GROUND ANCHORS

GROUND ANCHORS STOWED

CAPSTAN DRUMS — STOWED

The interior dimensions of the M113A1 armored personnel carrier can be seen above at the left. The use of the capstan drums to extricate the vehicle from boggy areas is shown in the sketch above.

Experience in Vietnam had shown the value of sprocket mounted capstans on the carriers. Use of these capstans with anchors allowed a bogged vehicle to pull itself out of swampy areas without assistance. In line with this experience, a capstan and marine recovery kit was provided for the M113A1. It consisted of two ground anchors with nylon ropes which wound onto the two capstan drums attached to the final drives.

A machine gun stowage kit allowed the .50 caliber machine gun on the M113A1 to be removed from the commander's cupola and stowed on the top deck. When the machine gun was stowed, a ground laser locator designator (GLLD) was mounted in its place on the cupola. When the GLLD was installed, the cupola was locked to the top deck by the cupola lock kit to prevent any rotation.

The combat weight of the M113A1 increased to over 12 tons with the heavier diesel engine and the 95 gallon fuel capacity. However, the maximum speed was now over 40 miles per hour on level roads. The maximum water speed remained at about 3½ miles per hour.

Production of the M113A1 continued at FMC into 1979 for a total of 23,576 vehicles for the United States and foreign military sales. Additional vehicles were produced for direct sales by FMC.

The stowage of the .50 caliber machine gun when using the GLLD is illustrated below.

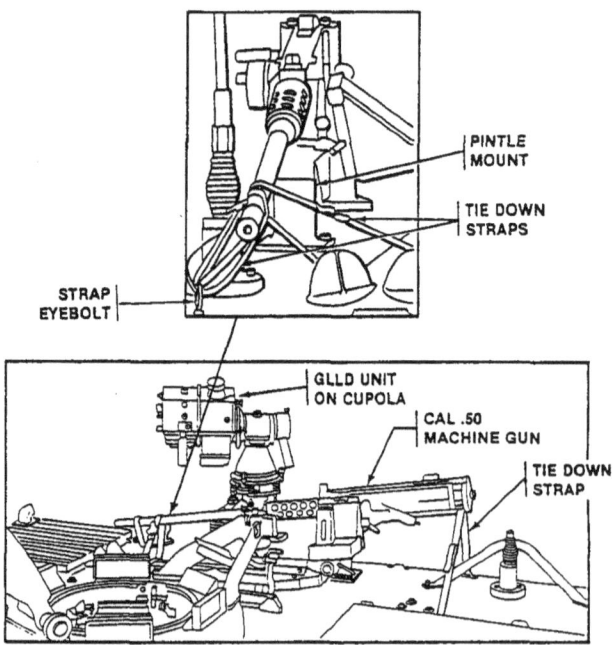

PINTLE MOUNT

TIE DOWN STRAPS

STRAP EYEBOLT

GLLD UNIT ON CUPOLA

CAL .50 MACHINE GUN

TIE DOWN STRAP

Components of the vulnerability reduction kit are shown alongside the vehicle at the right.

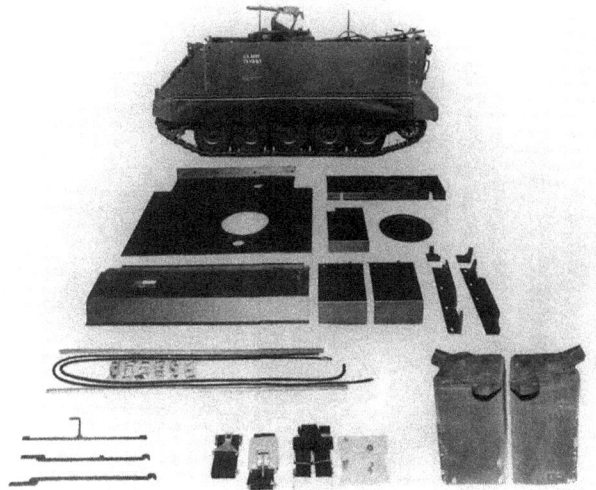

As a result of experience with mines in Vietnam, a vulnerability reduction kit was developed for the armored personnel carriers. This kit included a steel belly armor plate installed under the front of the vehicle and armored exterior fuel tanks mounted on the rear. The fuel lines also were rerouted away from the floor. Buoyancy cells were attached to the trim vane and on each side of the hull front to maintain the vehicle trim when afloat. An emergency ramp release also was provided as well as a shoulder harness for the driver.

The photographs above and below show the vulnerability reduction kit installed on the armored personnel carrier. Note the scales painted on the side of the vehicle to determine the freeboard when swimming with the extra weight of the vulnerability reduction kit.

The mine resistant M113A1 with the double bottom can be seen in the two top photographs. Note that the suspension is fitted with track return rollers. At the left, the tall double bottom vehicle can be compared with the standard M113A1.

Another effort to minimize mine damage resulted in the development of a version of the M113A1 with a double floor during 1968. This increased the height of the vehicle requiring longer road wheel arms and adjustment of the torsion bars to maintain the ground clearance. However, this version of the vehicle did not enter production.

Low speed and limited maneuverability when afloat were always a problem for the armored personnel carriers. One attempt to improve the swimming performance resulted in the installation of two water-jet propulsion units on the rear of the personnel carrier. The water-jets were driven by propeller shafts on top of the vehicle. Steering vanes on the water-jets improved the maneuverability in the water, but the new propulsion system was not installed on the production vehicles.

Below is the armored personnel carrier with the water-jets installed to improve the performance when swimming. The propeller shafts for the water-jets can be seen along the top at the right side and the rear of the vehicle.

At the right is a photograph of the M113A1E1 armored personnel carrier.

Although the M113A1 was extremely reliable, work continued to improve both the reliability and the performance. Four major areas were under consideration. These were the engine cooling system, the suspension, the fuel system, and the power train. Five prototype vehicles incorporating improvements in all four areas were constructed and designated as the armored personnel carrier M113A1E1.

The engine cooling system was modified by reversing the location of the fan and the radiator. The installation included a new fan and a radiator with increased capacity and a separate expansion tank. The new arrangement drew fresh ambient air through the radiator before it was warmed by passing over the engine and increased the cooling efficiency for both the engine and transmission. The new location also reduced the buildup of dust and oil on the radiator core and provided an approximate ten degree Fahrenheit reduction in the coolant temperature during normal operating conditions.

A new suspension system was introduced with high strength torsion bars which increased the road wheel

travel from six to nine inches. A stronger rear idler assembly was installed and raised two inches to reduce the chance of ground impact. The overall ground clearance also was increased from 16 to 17 inches. New shock absorbers were located on road wheels one, two, and five compared to the old shock absorbers on road wheels one and five on the M113A1. These changes greatly improved the cross-country performance resulting in a three to ten mile per hour increase in speed depending upon the terrain.

External fuel tanks similar to those developed for the reduced vulnerability kit replaced the internal fuel tank in the M113A1E1. These two tanks had the same 95 gallon total capacity as the internal tank and they were armored for equivalent ballistic protection. The new tanks were installed on the rear of the vehicle with one on each side of the ramp opening. In addition to improved safety, the use of the external tanks increased the space inside the vehicle by 16 cubic feet. The tanks were interchangeable and were bolted in place to permit rapid replacement in the field. The installation of the external fuel tanks increased the weight of the M113A1E1 by about 900 pounds and the length by 17 inches.

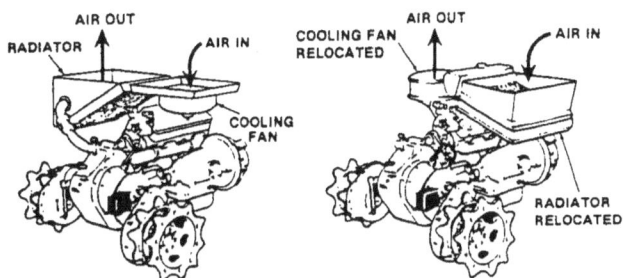

Above, the cooling system on the M113A1E1 on the right is compared to that of the M113A1 on the left. The new suspension appears below and the external fuel tanks are at the bottom right.

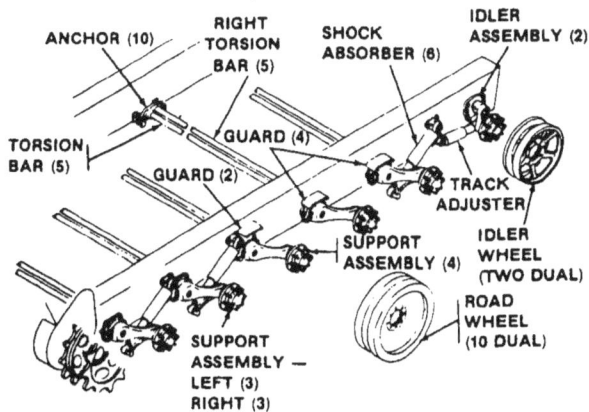

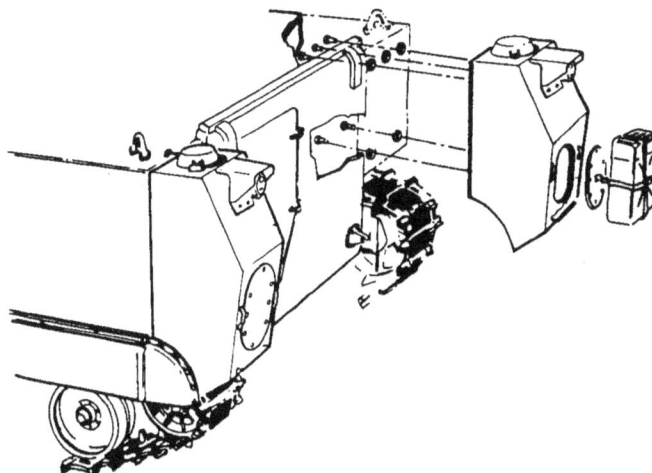

97

The new power pack consisting of the 6V53T diesel engine and the X200-3 transmission is shown above. The driver's controls in the M113A1E1 are at the right.

The power train in the M113A1 was replaced in the M113A1E1 by the turbocharged 6V53T diesel engine with the Allison X200-3 cross drive transmission. The new engine developed 275 gross horsepower at 2,800 rpm and had an improved cooling system. The cross drive transmission had four forward speeds compared to three in the Allison TX-100. It also included hydrostatic steering, eliminating the need for the controlled differential in the earlier vehicle. The steering brake levers in the M113A1 were replaced by a steering wheel and a brake pedal. The vehicle now had neutral steer capability allowing it to pivot in its own length. The five

prototype vehicles with the new components, completed development tests in October 1978 followed by operational tests ending in May 1979.

Although further development was required for the new power train, the revised engine cooling system, improved suspension, and optional external fuel tanks were introduced into production during the latter half of 1979. With these improvements, the vehicle was now designated as the full tracked armored personnel carrier M113A2.

A smoke grenade launcher kit was provided for the M113A2. This kit consisted of two, four tube, smoke

Components of the M113A2 without the optional external fuel tanks are identified below.

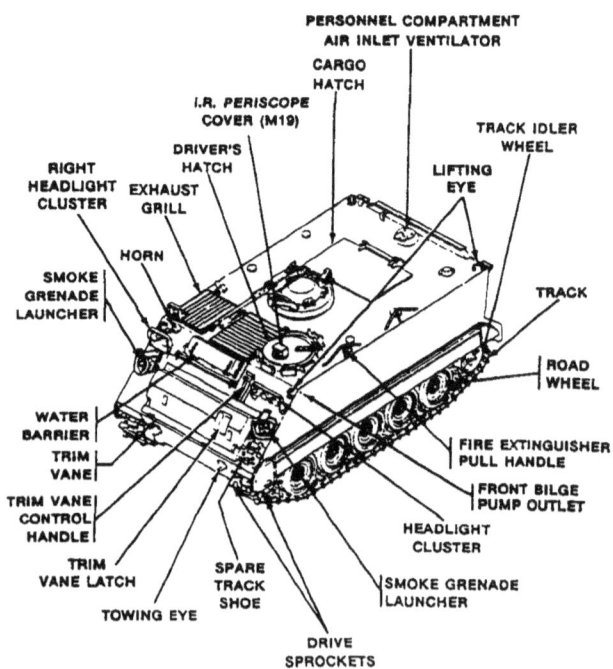

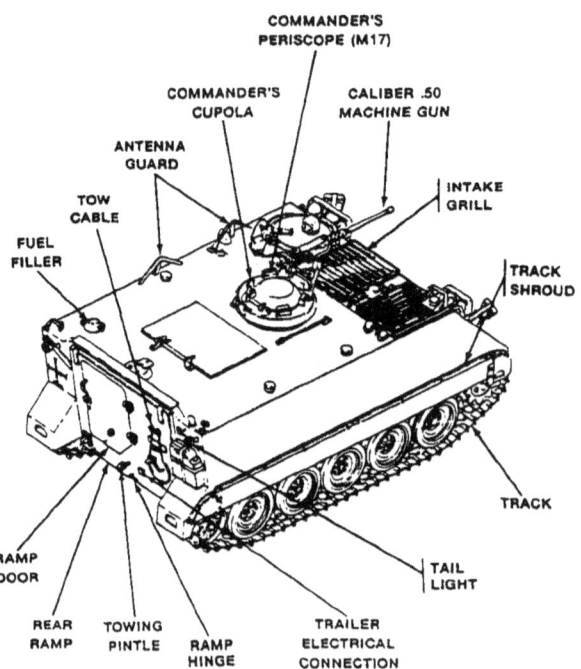

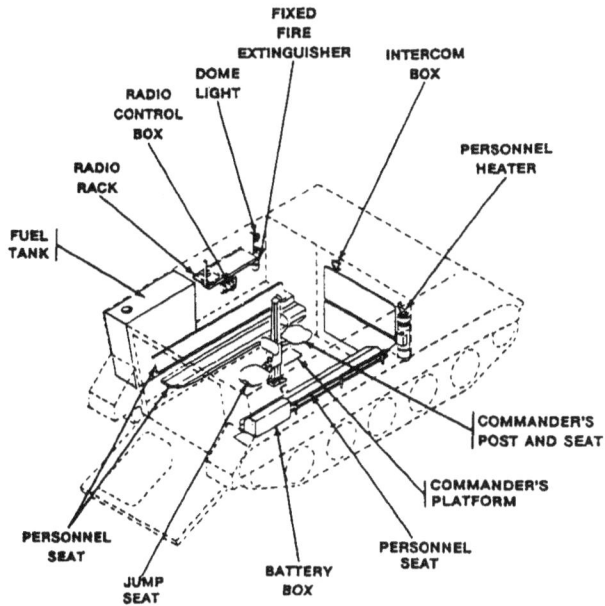

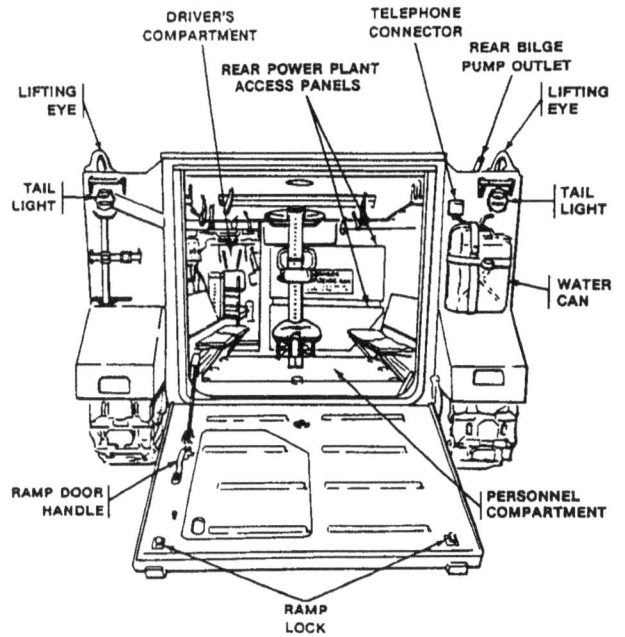

grenade launchers. One launcher was installed on each side of the front hull below the headlights. The control box for the launchers was mounted on the fire wall in the troop compartment.

Without the external fuel tanks, the combat weight of the M113A2 had now increased to about 12½ tons. This, of course, reduced the freeboard of the vehicle when it was afloat. As a result, the Training and Doctrine Command (TRADOC) dropped the swimming requirement for these personnel carriers and banned swimming the vehicle during training operations. The marginal swimming capability was retained for emergency use during wartime. Such use required extreme care and attention to weight distribution when in the water. Inflatable flotation cells also were evaluated for use with these vehicles.

The various components inside the M113A2 armored personnel carrier are identified in the drawings above. A photograph of the M113A2 with the external fuel tanks is below.

The dimensions of the M113A2 with and without the external fuel tanks can be seen below.

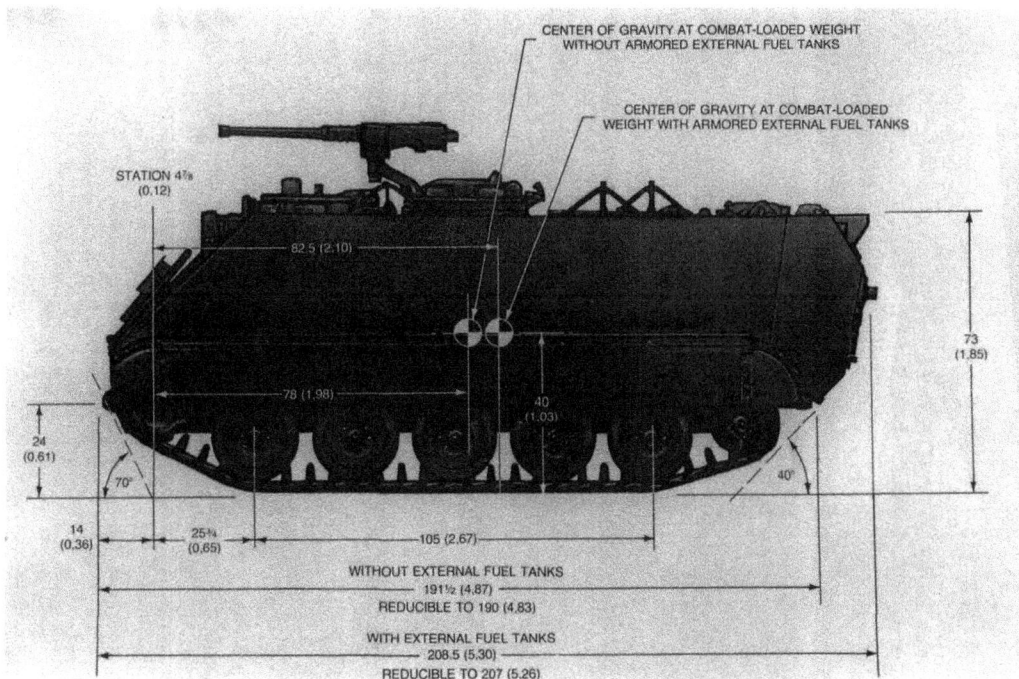

Above are photographs of the armored personnel carrier M113A3. The external fuel tanks are now standard equipment. Also, note the high buoyancy trim vane.

The development program continued and additional modifications replaced the X200-3 transmission with the X200-4 and later the X200-4A. The latter would permit future engine upgrades to 350 horsepower. Sliding Kevlar spall liners were installed on both sides in the troop compartment to reduce the effect of any armor penetration. Evaluation of the vehicle, now designated as the M113A2E1, indicated that the performance was much improved over the M113A2. Despite a combat weight of over 13½ tons, the new turbocharged engine provided a power to weight ratio of over 20 gross horsepower per ton.. After some minor modifications, the M113A2E1 was standardized as the full tracked armored personnel carrier M113A3. The new power train components were referred to as the RISE (Reliability Improvements for Selected Equipment) package. FMC also provided a hydrostatic steer differential as a kit to replace the DS-200 controlled differential in the M113A1 and M113A2. Used with the TX-100 transmission, the brake levers were replaced by a steering wheel and the earlier vehicles were provided with a neutral steer capability.

Except for the external fuel tanks on the rear, the outward appearance of the M113A3 was about the same as the original M113. The next modification was to drastically change that appearance. To improve the survivability of the M113A3 on the battlefield, FMC offered the P-900 applique armor kit as an option.

At the left are views of the M113A3 with the trim vane raised and lowered. Note the attachment points on the side for applique armor. The drawing below shows the hydrostatic steer differential provided as a kit for installation in the M113A1 and M113A2.

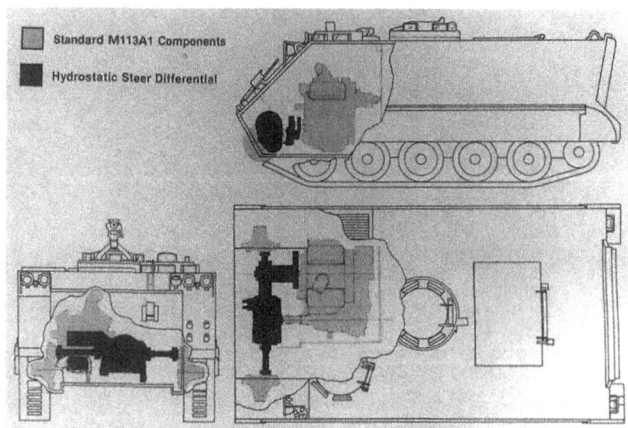

Standard M113A1 Components

Hydrostatic Steer Differential

The armored personnel carrier M113A3 is fitted here with the P900 applique armor kit and the armor shields to protect the vehicle commander when firing the .50 caliber machine gun.

Bolted to the aluminum armor on the front, sides, and rear, this kit greatly increased the protection level on the M113A3. Mine protection also was enhanced by steel armor in the floor. However, the extra armor did add two tons to the weight of the vehicle. Thus the performance with the new 275 horsepower engine was about the same as the M113A2 when the armor kit was installed. The armor gun shield kit, originally supplied for the armored cavalry assault vehicle (ACAV), was installed around the commander's hatch.

Additional views of the M113A3 show the rear of the armored personnel carrier and the applique armor added to the ramp.

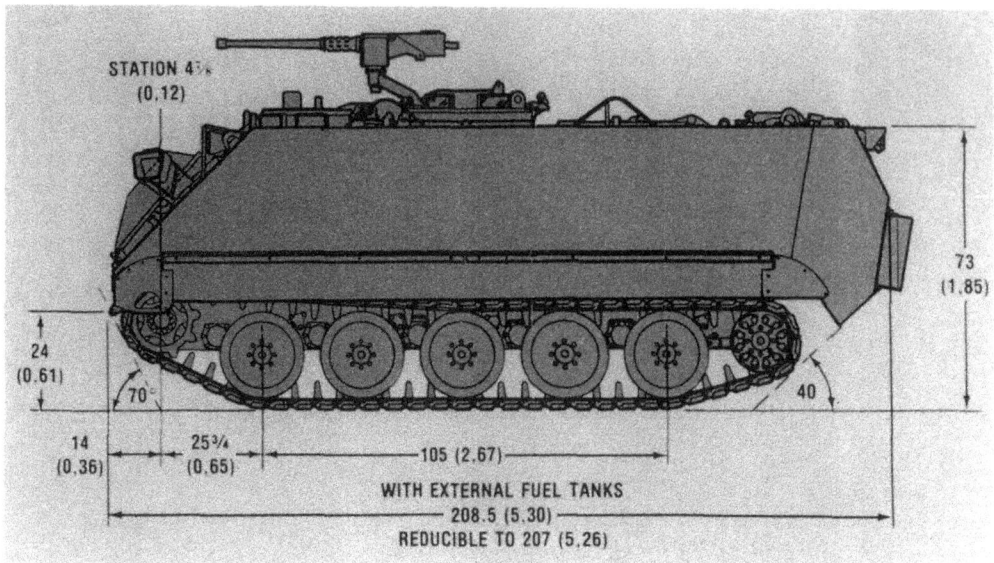

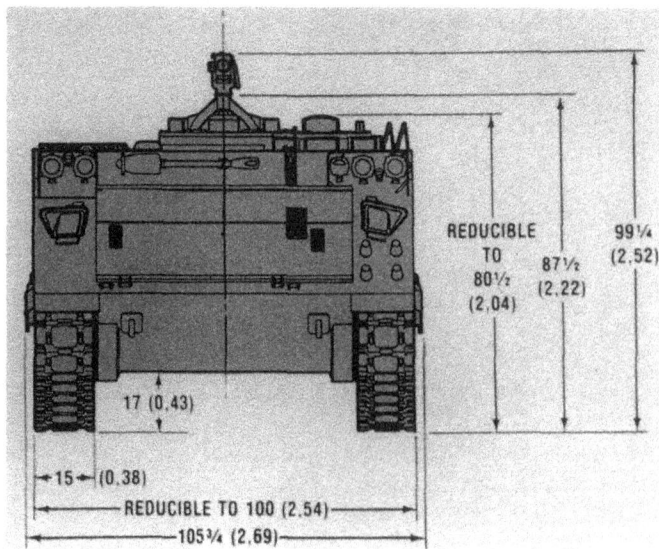

The drawings above show the dimensions of the M113A3 armored personnel carrier without the applique armor or the commander's armor shields. Below, the M113A3 driver's station is at the left and the troop compartment is at the right. Note the spall liners installed in the latter.

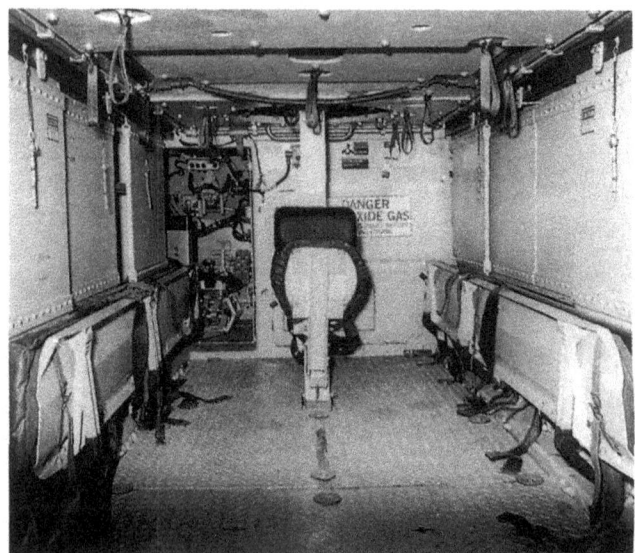

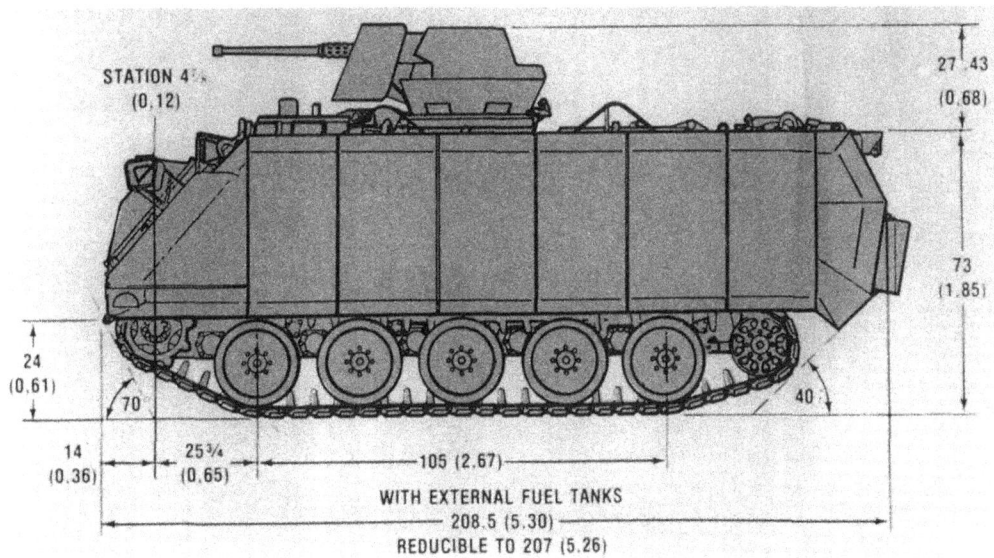

STATION 4½
(0.12)

27.43
(0.68)

73
(1.85)

24
(0.61)

70

40

14
(0.36)

25¾
(0.65)

105 (2.67)

WITH EXTERNAL FUEL TANKS
208.5 (5.30)
REDUCIBLE TO 207 (5.26)

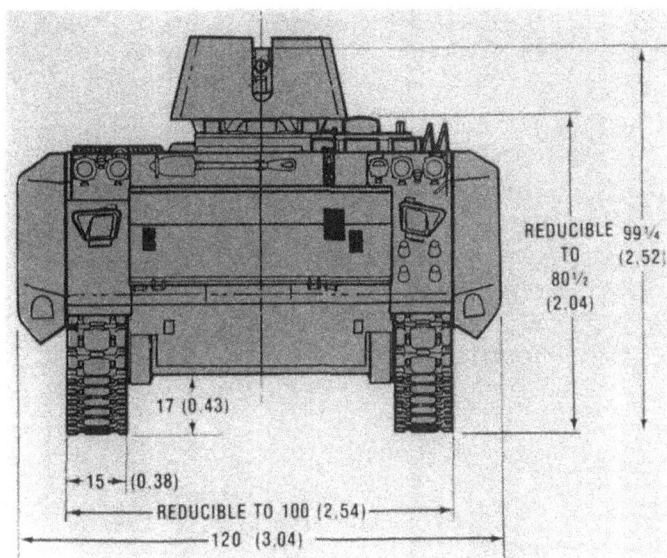

REDUCIBLE
TO
80½
(2.04)

99¼
(2.52)

17 (0.43)

15 (0.38)

REDUCIBLE TO 100 (2.54)
120 (3.04)

The drawings above show the dimensions of the M113A3 armored personnel carrier with the applique armor and the commander's armor shields installed. Below, the rear applique armor can be seen at the left and a close view of the commander's armor shields is at the right. Needless to say, the open hatch cover protected his back.

Above, the U.S. Marine Corps armored personnel carrier at the left has been fitted with ceramic tile armor. The vehicle at the right has Israeli EAAK applique armor on the sides.

Efforts to improve the level of protection on the M113 family included numerous applique armor kits as well as field modifications. In addition to the P-900 kit described previously, other passive armor upgrades utilized titanium and composite armor to provide protection against 14.5mm projectiles and in some cases against 30mm rounds. Reactive armor was designed to defeat the shaped charge warheads of the rocket propelled grenades (RPG). Other improvements under consideration included the battlefield combat identification system (BCIS), the AN/VVS-2(V)1A night viewer for the driver, the enhanced position locating reporting system (EPLRS), the M13 gas particulate filter unit, the improved cold start system, laser protection for the M17 unity optical devices, the TM 10-4 quick erect antenna mast (QEAM), the M113A3 pontoon swim system kit, brackets for the precision lightweight GPS receiver (PLGR), the new solid state vehicle inter-communication system (VIS), a water/ration heater, and a new tie down design to improve transportability.

On several occasions, completely enclosed turrets were proposed for the vehicle commander permitting him to fire the .50 caliber machine gun without exposure to enemy fire. Other armament combinations also were considered for the commander including twin 7.62mm machine guns. In Vietnam, a United States Navy turret armed with twin 7.62mm machine guns was installed as a field modification.

Here are two more examples of Israeli applique armor installed on the personnel carrier, both by Rafael.

Two enclosed turrets for the vehicle commander of the armored personnel carrier are shown below. At the left, the turret is armed with a single .50 caliber machine gun. At the right, the armament consists of two 7.62mm machine guns.

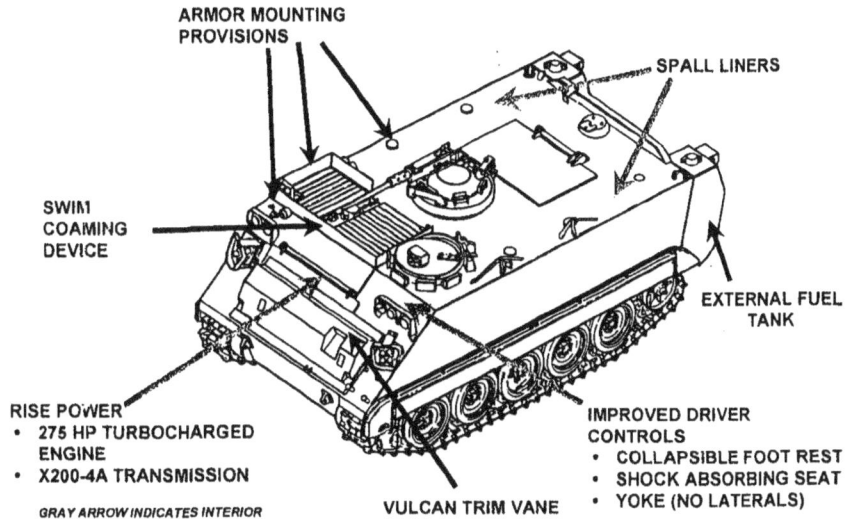

ARMOR MOUNTING
PROVISIONS

SPALL LINERS

SWIM
COAMING
DEVICE

EXTERNAL FUEL
TANK

RISE POWER
• 275 HP TURBOCHARGED
  ENGINE
• X200-4A TRANSMISSION

GRAY ARROW INDICATES INTERIOR

IMPROVED DRIVER
CONTROLS
• COLLAPSIBLE FOOT REST
• SHOCK ABSORBING SEAT
• YOKE (NO LATERALS)

VULCAN TRIM VANE

Improvements incorporated into the M113A3 armored personnel carrier are illustrated above. Below, the attachment of a reactive armor kit is shown in the drawing at the left. At the right below, an armored personnel carrier has the Urdan Industries Ltd. applique armor installed.

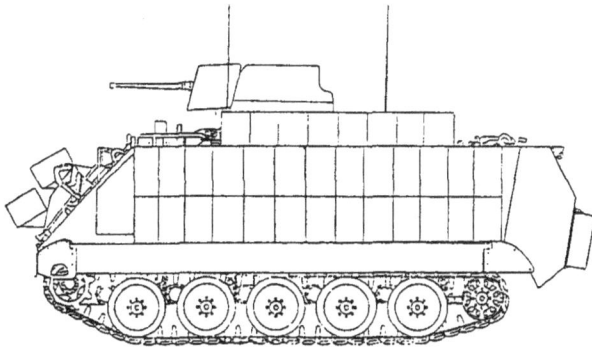

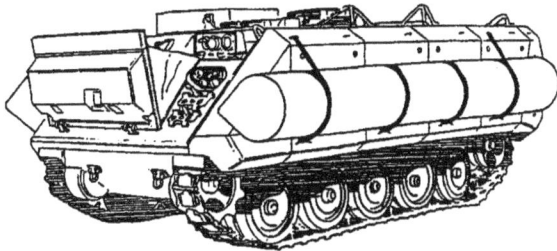

The drawing above at the left shows the pontoon swim system kit for the M113A3. Below at the left is a sketch of the quick erect antenna mast. At the right another applique armor kit is illustrated in the photograph (above) and the drawing (below).

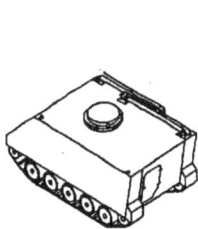

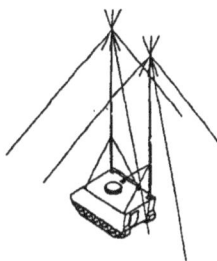

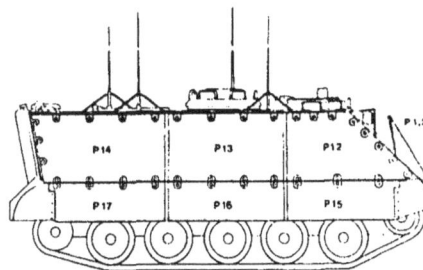

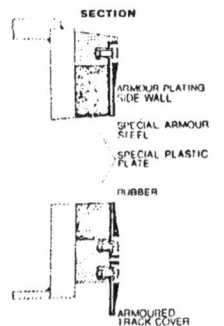

The photographs on this page show the stretched M113A1 using the components from the M113A1E1. Note that the vehicle is fitted with the high flotation trim vane.

The M113 series of armored personnel carriers provided the basic chassis for a wide variety of vehicles. A lengthened version of the M113A1E1 was referred to as the stretched M113A1. It utilized the same high strength torsion bar suspension and power train as the original M113A1E1. The stretched vehicle had six road wheels per side and the length of the hull was extended by 26½ inches. The payload was doubled compared to the M113A1 allowing it to carry over 6,700 pounds of cargo in an additional 70 cubic feet of cargo space. Two of the lengthened M113A1E1s were produced and one was evaluated at Fort Knox and the other at Aberdeen Proving Ground. The stretched vehicle concept also was

The stretched M113A1 is shown below with the rear ramp and the troop compartment roof hatch in the open position.

The stretched M113A1 above is at Fort Knox during its evaluation. At the right is a concept drawing of an even longer version with seven road wheels per side and wider tracks.

applied to other variants of the M113 family. Design studies were completed for even longer vehicles with seven road wheels per side.

Although the stretched M113A1 did not go into production, it provided valuable information for future designs. One of these, also constructed by FMC, was the

The dimensions of the stretched M113A1 appear on the drawing below.

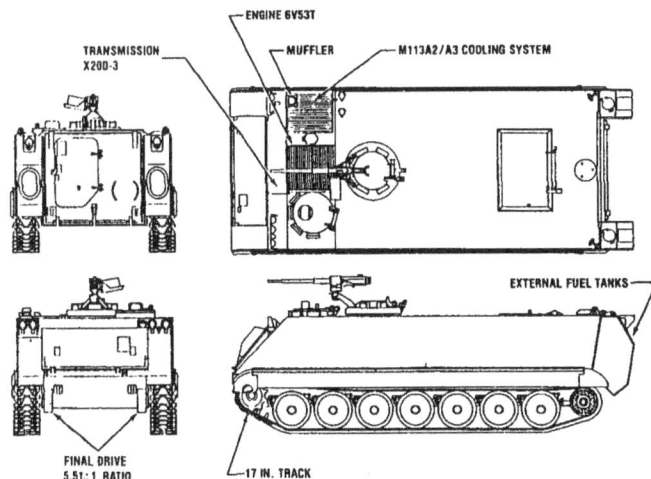

ENGINE 6V53T
TRANSMISSION X200-3
MUFFLER
M113A2/A3 COOLING SYSTEM
EXTERNAL FUEL TANKS
FINAL DRIVE 5.51:1 RATIO
17 IN. TRACK

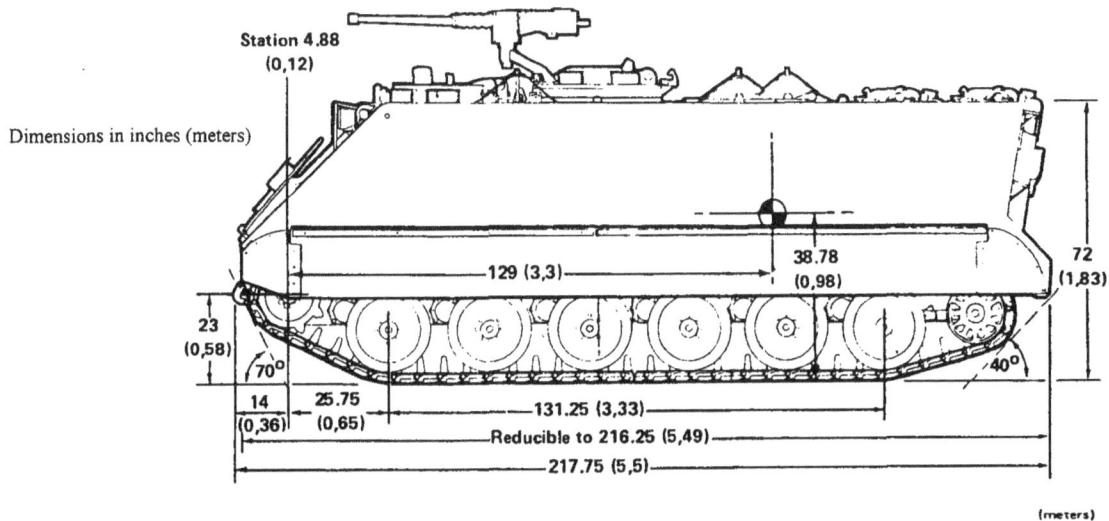

Dimensions in inches (meters)

Station 4.88 (0,12)

129 (3,3)

38.78 (0,98)

72 (1,83)

23 (0,58)

70°

40°

14 (0,36)

25.75 (0,65)

131.25 (3,33)

Reducible to 216.25 (5,49)

217.75 (5,5)

(meters)

The mobile tactical vehicle light appears in the photographs above. Note that the vehicle was not fitted with a trim vane as it was not intended to swim.

mobile tactical vehicle light (MTVL) with six road wheels per side. The suspension was improved providing more than 15 inches of wheel travel. Powered by an upgraded version of the 6V53T diesel engine that developed 350 horsepower at 2,800 rpm, it drove the vehicle through the X200-4A transmission. The gross vehicle weight was 36,000 pounds, but the maximum road speed was 41 miles per hour with the new power train. A fuel capacity of 120 gallons provided a cruising range of about 300 miles. As on the standard M113s, the hull was constructed of 5083 aluminum alloy plate. However, the protection could be improved by the use of titanium applique armor on the upper front and sides. Spaced, expanded steel, armor could be added to the rear and the bottom could be reinforced by steel laminate armor.

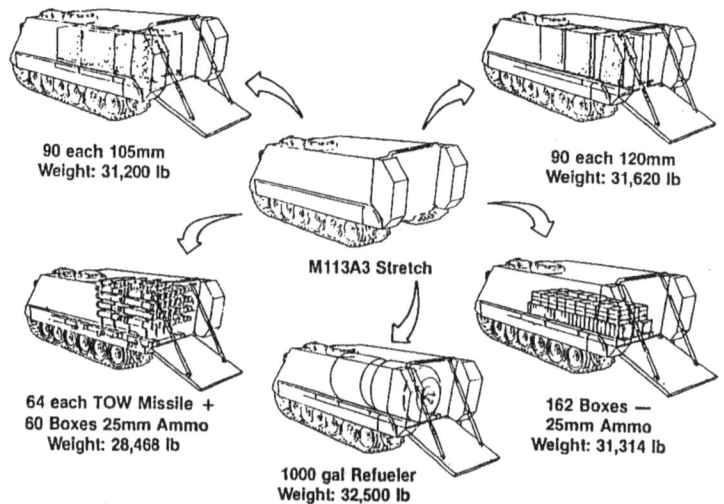

90 each 105mm
Weight: 31,200 lb

90 each 120mm
Weight: 31,620 lb

M113A3 Stretch

64 each TOW Missile +
60 Boxes 25mm Ammo
Weight: 28,468 lb

162 Boxes —
25mm Ammo
Weight: 31,314 lb

1000 gal Refueler
Weight: 32,500 lb

In the drawings above are a number of proposed applications for the stretched M113A3. Below, the drawing shows the length of the mobile tactical vehicle light.

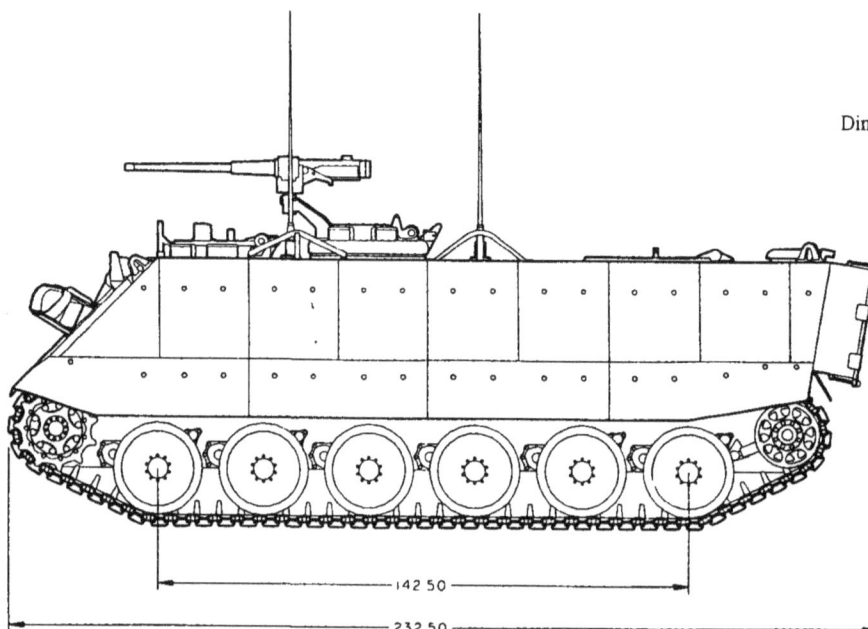

Dimensions in inches

142 50

232 50

108

Although new members of the M113 family continue to appear, one variant makes use of the older vehicles of the series. These were modified to resemble the BMP-2 infantry fighting vehicle used in many parts of the world. Designated as the Opposing Forces (OPFOR) Surrogate Vehicle/BMP-2, they were intended for use at the National Training Center. Converted at the Red River Army Depot, the OSV (BMP-2) will replace the M551 Sheridans modified to resemble the BMP-1. Equipped with the RISE power package, the new OSVs used 50 per cent less fuel than the M551s and had improved reliability. Although converted from M113 series vehicles. the OSVs utilized components from the Bradley Fighting Vehicle for the turret. In January 1996, the requirement was projected to be 190 vehicles, all for use at the NTC.

The Opposing Forces Surrogate Vehicle is shown in the photographs above and below.

Below is the M113 referred to as the "Hot Rod". It was powered by two 440 cubic inch Chrysler gasoline engines driving through two modified 727 transmissions. During tests at Fort Knox in September 1979, the "Hot Rod" reached an average speed of 75.76 miles per hour over a 500 foot gravel track.

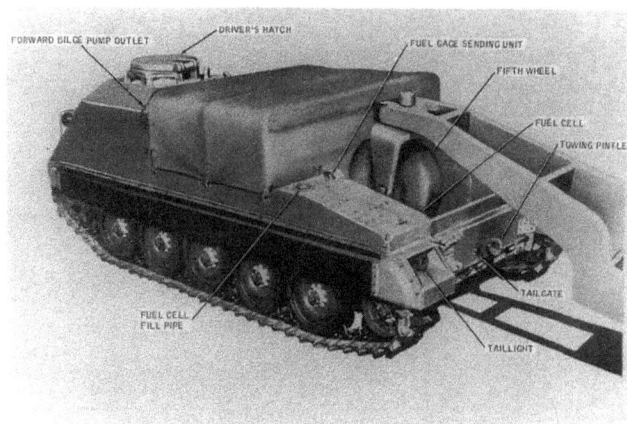

The high speed tractor T122 is shown in the two photographs above.

A UNIVERSAL CHASSIS BASED UPON THE M113 SERIES

Early in the development program, it was obvious that a successful armored personnel carrier would provide the basis for an entire family of vehicles. Such a family would utilize many of the same components greatly simplifying logistical support. Although the T149 was dropped along with the Dart missile, another antiaircraft weapon was under development which would make use of the T113 type chassis. This was the 37mm gun T250, a six barrel Gatling type gun being developed both as a towed weapon, Vigilante A, and as a self-propelled gun, Vigilante B. A modified version of the T113E1/E2 armored personnel carrier was proposed as the prime mover for the Vigilante A. Designated as the high speed, full-tracked, tractor T122, it retained the suspension, tracks, and power train of the personnel carrier. The driver and the engine compartment in the front hull were protected by aluminum alloy armor as were the two fuel cells located in the open top rear compartment. These cells were installed in the rear corners of the compartment on each side of the fifth wheel support. The contents of the rear compartment could be protected by the installation of an armor blanket on three aluminum bows. This blanket consisted of 12 ply, unbonded nylon, bar laminate, armor cloth.

The high speed tractor T122 is at the left below with the cargo area open. The driver's compartment in the T122 is at the right below.

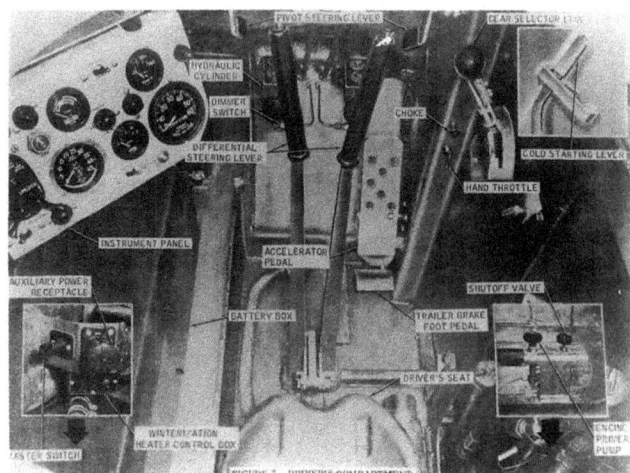

110

High Speed Tractor T122

Dimensions in inches

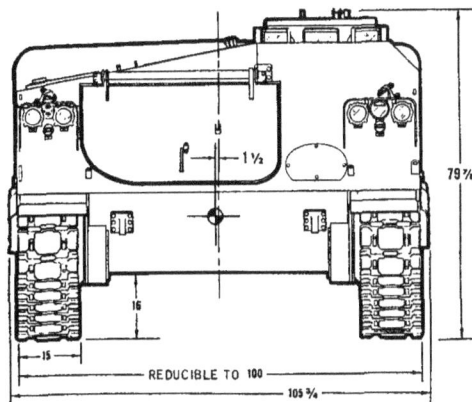

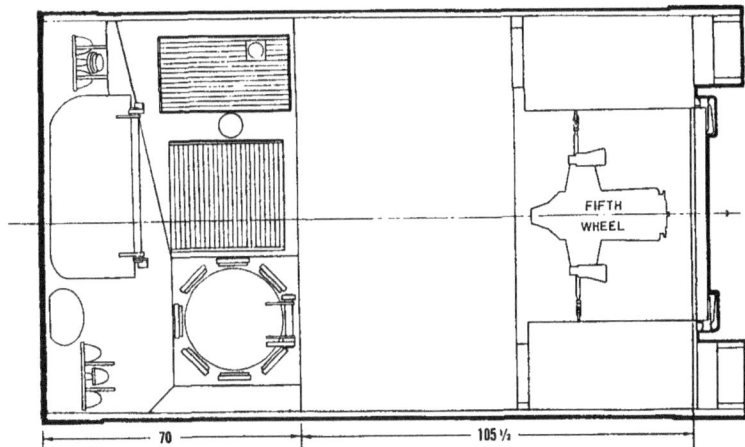

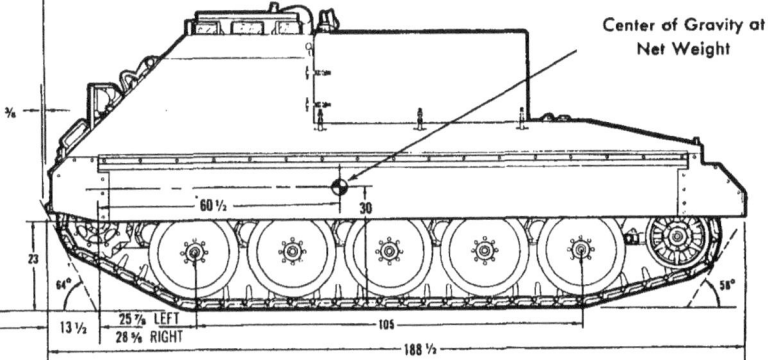

FMC received a contract on 9 June 1958 to build a prototype and two pilot tractors. The prototype was accepted for test in July 1959 and the first pilot was scheduled for completion in March 1960. However, funds were not available for the second pilot. With the termination of the Vigilante weapon system, the T122 project also was canceled.

The self-propelled Vigilante B was installed on a low silhouette armored vehicle based upon the same components as the personnel carrier. Designated as the self-propelled gun chassis T249, it was slightly longer than the other vehicles with the ground contact length increased from 105 inches to 110¾ inches. Additional details of the T249 appear in the section on antiaircraft vehicles.

The engine compartment of the high speed tractor T122 can be seen above. Below are two views of the self-propelled gun chassis T249.

111

Two views of the missile equipment carrier XM474 are at the left. Above, the vehicle is fitted with the amphibious kit and is emerging from the water after swimming.

During the initial development of the Pershing guided missile system, it was intended to use two types of tracked vehicles to transport the missile and its support equipment. The missile equipment carrier was designated as the XM474 and was based upon components from the T113 series of vehicles developed by FMC. The servicing equipment carrier was designated as the XM476 and it was based upon the T116 amphibious cargo carrier. This was the vehicle standardized later as the M116 Husky. It was developed and

Missile Equipment Carrier XM474

Dimensions in inches

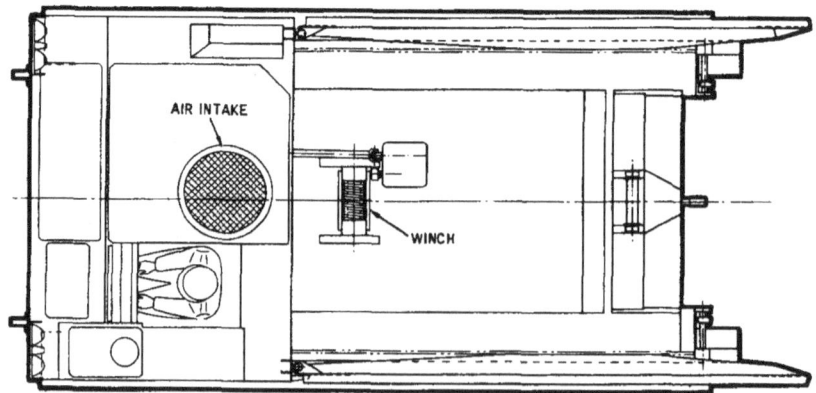

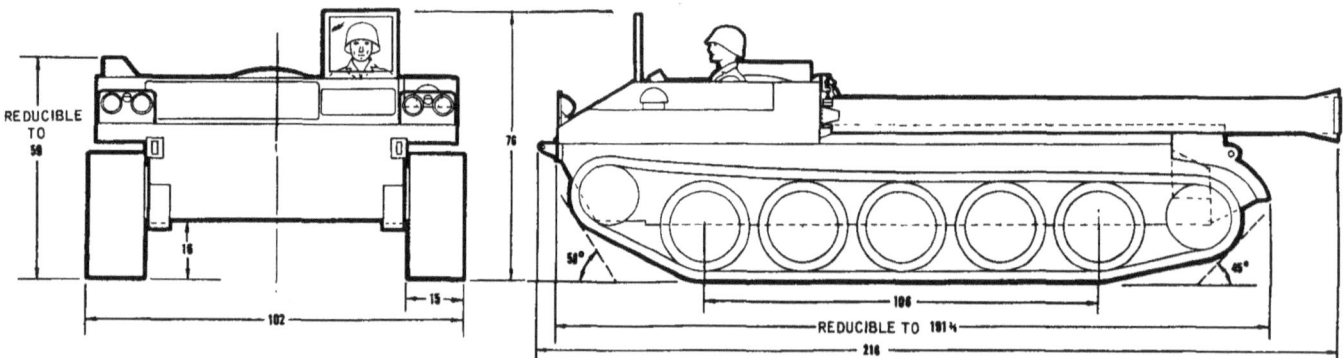

Above, the XM476 carrier is at the left and the basic XM474 vehicle is at the right.

manufactured by the Pacific Car and Foundry Company. Prototypes were constructed of both the XM474 and the XM476, but further analysis indicated that the XM474 could meet all of the ground transportation requirements of the Pershing system. As a result, the XM476 program was terminated. Further development resulted in changes to the XM474. To meet air transport requirements, the vehicle was reduced in width from 102 inches to 100 inches and it was lengthened, increasing the ground contact length to 111 inches. The first modified design concept was designated as the XM474E1, but the final, lengthened, version became the XM474E2.

Four of the XM474E2s, equipped with bolt-in kits were required to transport the complete Pershing system. This included the missile itself on the transporter-erector-launcher, the communications hut, the fire control hut, the reentry unit with the warhead, and the personnel and miscellaneous hardware required to launch the missile.

The XM474E2 was a low silhouette, unarmored, vehicle assembled by welding 5083 aluminum alloy with

These three photographs show the XM474E2 carrier.

the same suspension, tracks, and power train components as the M113 armored personnel carrier. As on the M113, the driver rode in the left front with the engine and transmission on the right. Without the missile system components, the XM474E2 weighed about 11,900 pounds. The vehicle had a maximum fording depth of 42 inches, but it could be made amphibious by the installation of a kit.

The Pershing missile system was deployed initially using the XM474E2. However its primary theater of operation was in Europe. Here, the excellent road system did not require the use of transporters with good cross-country performance. As a result by 1967, the Pershing system was being deployed using five 5 ton 8 x 8 trucks.

Above, the XM474E2 is carrying its Pershing missile. At the left below, the missile is being launched. The basic XM474 carrier could swim if fitted with the flotation kit as seen at the right below.

Missile System Carrier XM474E2

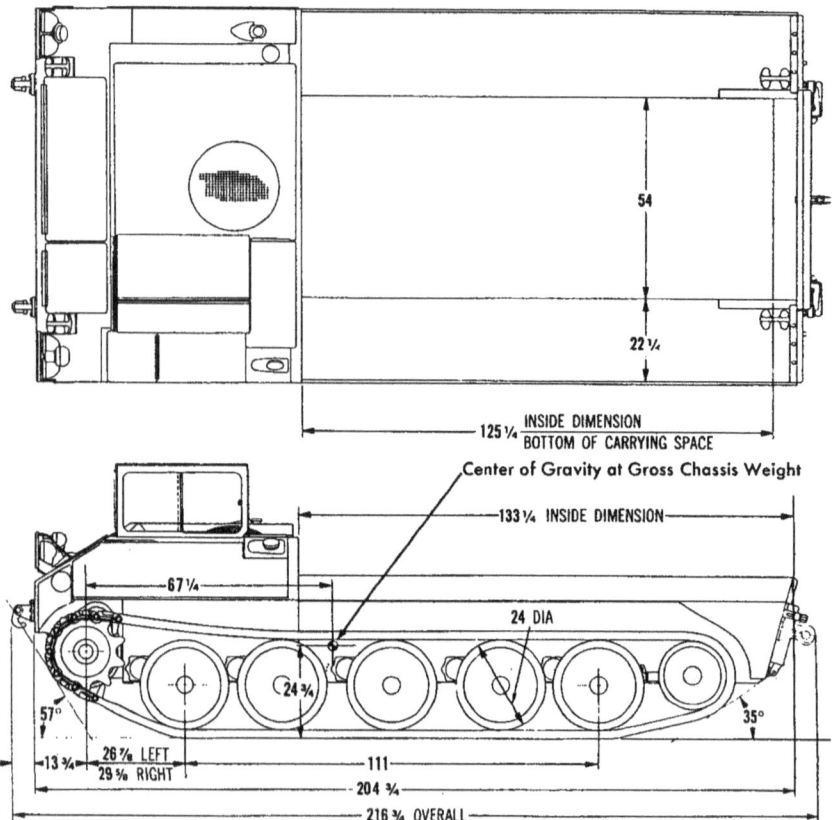

Dimensions in inches

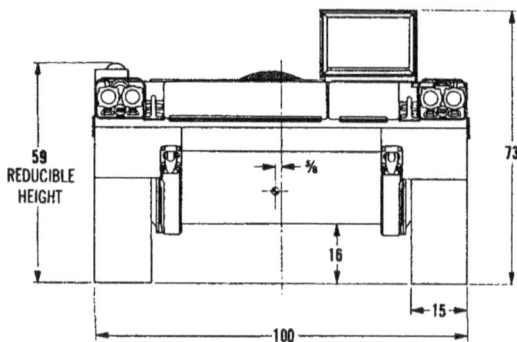

54

22 1/4

125 1/4 INSIDE DIMENSION
BOTTOM OF CARRYING SPACE

Center of Gravity at Gross Chassis Weight

133 1/4 INSIDE DIMENSION

67 1/4

24 DIA

59 REDUCIBLE HEIGHT

73

5/8

16

24 3/4

57°

35°

15

6 1/2

13 3/4

26 7/8 LEFT
29 5/8 RIGHT

111

100

204 3/4

216 3/4 OVERALL

114

At the left above, the XM474, the T122, and the M113 are lined up left to right. At the top right, the XM474E2 is loaded with the communications pack number 7.

Following the T122 and the XM474 series development, numerous proposals were presented for a variety of carriers using the components of the M113. One concept combined the armored front section of the T122 with an unarmored rear section based upon the XM474E1 design. Later, a similar approach combined the armored front section from an M113 with an unarmored rear cargo body from the XM474E2.

In 1960, a Signal Corps requirement resulted in the development by FMC of what would be a truly universal carrier based upon the earlier work on the XM474E2. Designated as the XM548, it was intended to carry the control center and crew of the AN/MPQ-32 counter-battery radar system. When the control center was re-moved from the vehicle, it would then serve as a cargo carrier. However, weight increases during the development of the radar system soon exceeded the capacity of the M113 suspension on the XM548 and it was no longer suitable for that application. Although six pilot vehicles had been ordered, only two were delivered. One of these was under test by the Electronics Command and the other was used as a cargo carrier during Exercise Swamp Fox II. The latter vehicle, fitted with roll bars instead of a cab, was dubbed "Catamount" and it showed excellent performance during the exercise. This pilot was then evaluated at Fort Sill to provide information for further development. It was eventually returned to FMC and converted into a 5 ton wrecker.

The pilot XM548 "Catamount" is shown at the left above. Views of the XM548 with the cab are above and at the left.

The diesel powered XM548E1 cargo carrier is shown above during its test operations.

Item 2182 of the Army Materiel Command Technical Committee Minutes (AMCTCM), dated 25 June 1964, initiated the development of a new version of the vehicle designated as the 6 ton tracked cargo carrier XM548E1. This vehicle utilized the power train and running gear of the M113A1 armored personnel carrier. With a fuel capacity of 105 gallons, the new diesel powered vehicle had a cruising range of about 300 miles. Standardized as the cargo carrier M548, the unarmored vehicle had a combat weight of slightly over 14 tons including its 6 ton payload.

The hull of the M548 was a watertight welded assembly of 5083 aluminum alloy and the open top bed was $130^5/_8$ inches in length and 63¾ inches wide between the sponsons. The bed could be raised to the sponson level providing a full width of 96½ inches. In the lower position, the bed was only 26½ inches above ground level. Access to the rear of the cargo compartment was provided by two watertight tailgates. The lower tailgate was hinged at the bottom and the

upper tailgate could be swung open from either end. With the tailgates and track shrouds in place, the vehicle was amphibious. In water, the freeboard ranged from 24 inches unloaded to 12 inches with a full 6 ton load. The M548 was equipped with six aluminum bows and a vinyl coated cover for the cargo area. Tie down eyes were provided to secure the cargo. Four men could be seated in the cab under a vinyl coated fabric cab cover. An M49A1 or an M66 ring mount for a machine gun could be installed over the center of the cab. An opening in the cab roof provided access to the machine gun mount. A winch with a 20,000 pound line pull was installed in the center front hull below the driving compartment. The maximum speed of the M548 was 38 miles per hour on hard roads and about 3½ miles per hour in water.

At the right is another view of the XM548E1 cargo carrier.

An additional photograph of the XM548E1 is above. The standard
M548 cargo carrier can be seen at the right and below.

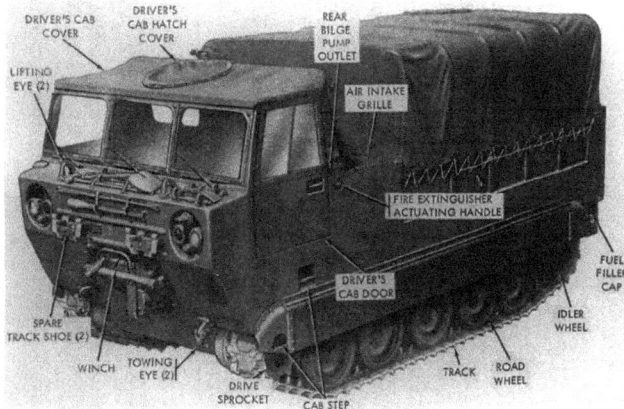

DRIVER'S CAB COVER
DRIVER'S CAB HATCH COVER
REAR BILGE PUMP OUTLET
LIFTING EYE (2)
AIR INTAKE GRILLE
FIRE EXTINGUISHER ACTUATING HANDLE
DRIVER'S CAB DOOR
FUEL FILLER CAP
SPARE TRACK SHOE (2)
WINCH
TOWING EYE (2)
DRIVE SPROCKET
CAB STEP
TRACK
ROAD WHEEL
IDLER WHEEL

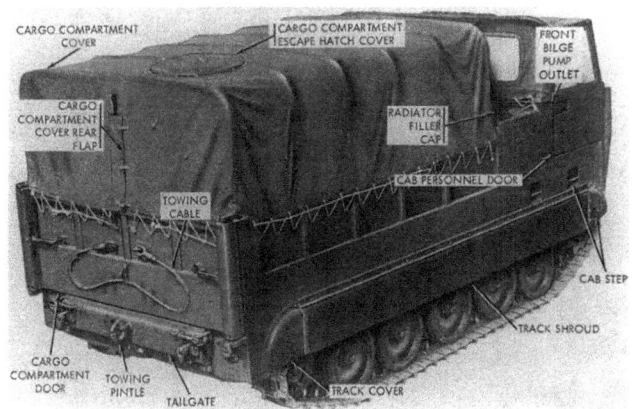

CARGO COMPARTMENT COVER
CARGO COMPARTMENT ESCAPE HATCH COVER
FRONT BILGE PUMP OUTLET
CARGO COMPARTMENT COVER REAR FLAP
RADIATOR FILLER CAP
TOWING CABLE
CAB PERSONNEL DOOR
CAB STEP
CARGO COMPARTMENT DOOR
TOWING PINTLE
TAILGATE
TRACK COVER
TRACK SHROUD

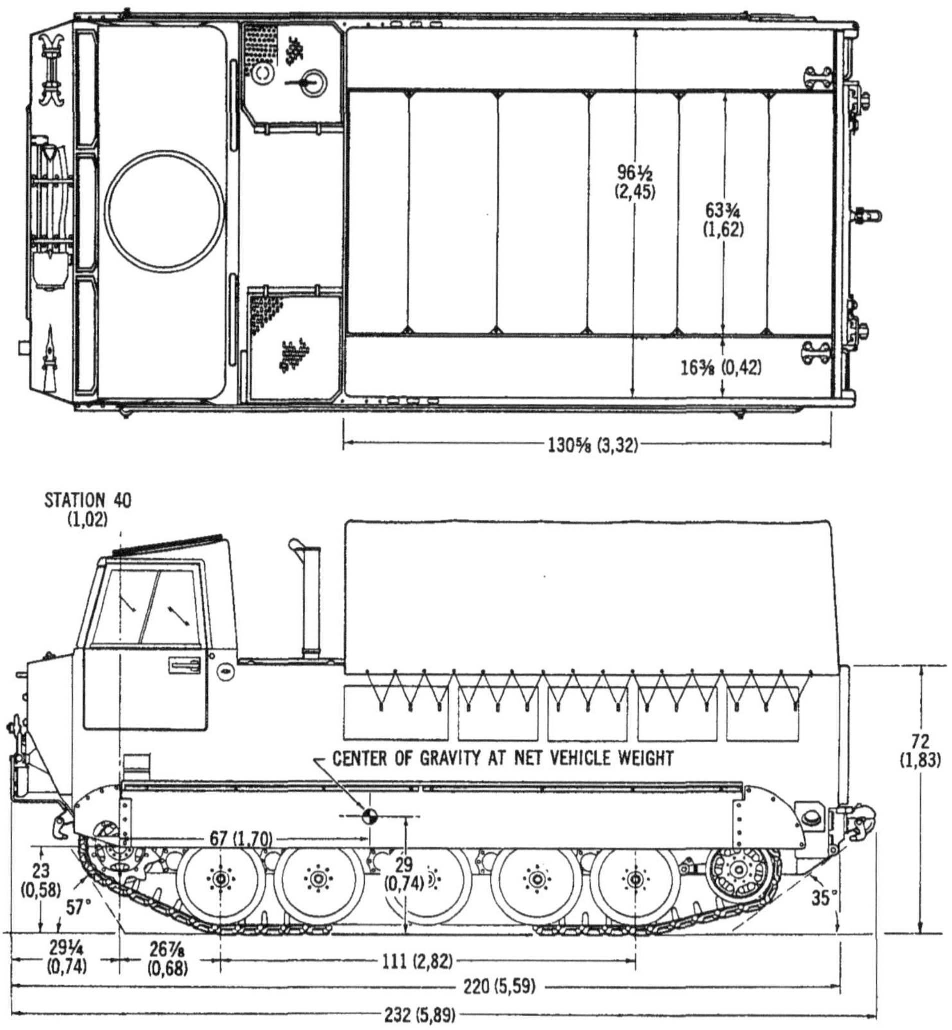
96½
(2,45)
63¾
(1,62)
16⅝ (0,42)
130⅝ (3,32)
STATION 40
(1,02)
CENTER OF GRAVITY AT NET VEHICLE WEIGHT
72
(1,83)
67 (1,70)
29
(0,74)
23
(0,58)
57°
35°
29¼
(0,74)
26⅞
(0,68)
111 (2,82)
220 (5,59)
232 (5,89)

Cargo Carrier M548

Dimensions in inches (meters)

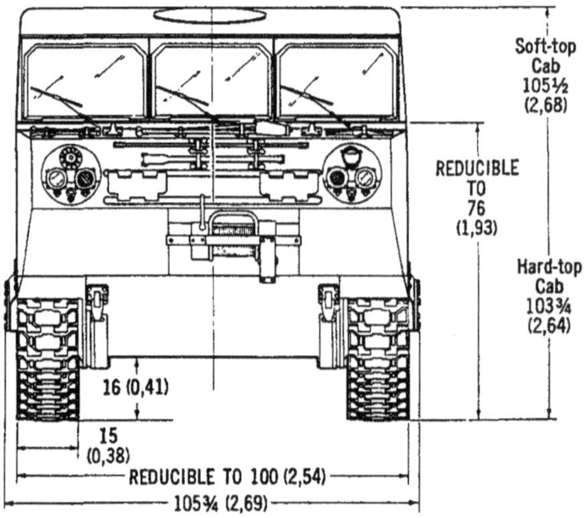
Soft-top
Cab
105½
(2,68)
REDUCIBLE
TO
76
(1,93)
Hard-top
Cab
103¾
(2,64)
16 (0,41)
15
(0,38)
REDUCIBLE TO 100 (2,54)
105¾ (2,69)

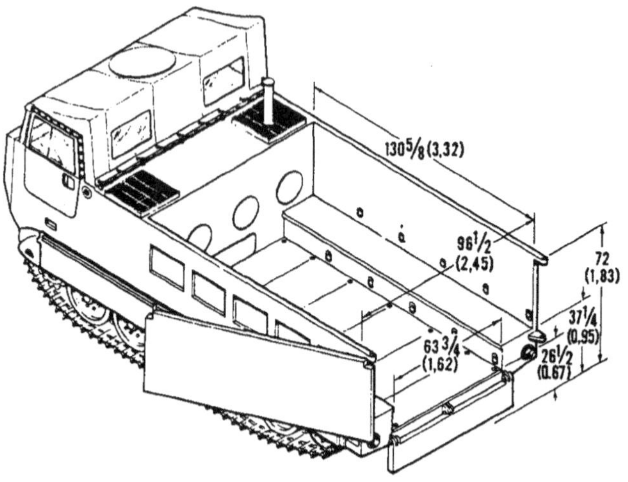
130⅝ (3,32)
96½
(2,45)
72
(1,83)
37¼
(0,95)
63¾
(1,62)
26½
(0,67)

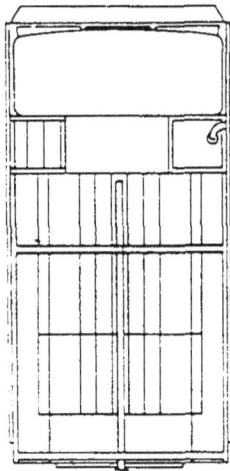

105-MM AMMUNITION LOAD
200-ROUNDS CAPACITY

105-MM LOAD TABULATION	WEIGHT (LB)
100 2-rd standard boxes (120 lb each)	12,000
1 man with equipment	250
Trolley hoist	380
TOTAL LOAD	12,630

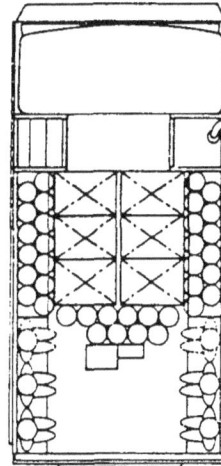

175-MM AMMUNITION LOAD
36-ROUNDS CAPACITY

175-MM LOAD TABULATION	WEIGHT (LB)
6 6-rd projectile pallets (948 lb each)	5,688
36 propellant charge cans (99 lb each)	3,564
1 8-rd fuze box	23
2 15-rd fuze boxes (55.8 lb each)	112
8 men with equipment (250 lb each)	2,000
Trolley hoist	380
TOTAL LOAD	11,767

155-MM AMMUNITION LOAD
80-ROUNDS CAPACITY

155-MM LOAD TABULATION	WEIGHT (LB)
10 8-rd projectile pallets (gas, 862 lb each)	8,620
40 2-rd propellant charge boxes (58.5 lb each)	2,340
5 16-rd fuze boxes (55.8 lb each)	279
4 men with equipment (250 lb each)	1,000
Trolley hoist	380
TOTAL LOAD	12,619

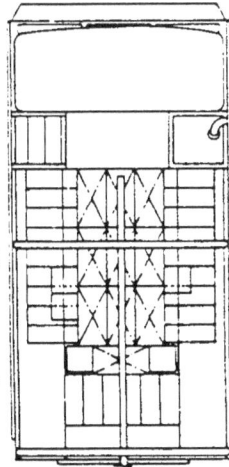

8-IN AMMUNITION LOAD
36-ROUNDS CAPACITY

8-IN LOAD TABULATION	WEIGHT (LB)
6 6-rd projectile pallets (1,253 lb each)	7,518
36 propellant charge boxes (50 lb each)	1,800
1 8-rd fuze box	23
1 30-rd fuze box	91
8 men with equipment (250 lb each)	2,000
Trolley hoist	380
TOTAL LOAD	11,812

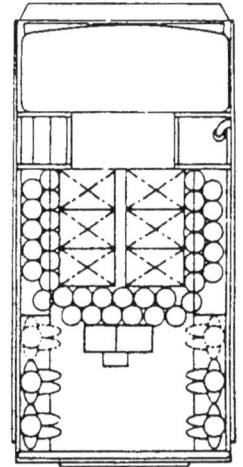

The ammunition loads for the M548 in its role as an accompanying vehicle for the self-propelled artillery are shown above.

The initial application of the M548 was as an accompanying vehicle for the self-propelled artillery. Specifically, it was intended to carry ammunition and the extra crew members for the 175mm self-propelled gun M107, the 105mm self-propelled howitzer M108, the 155mm self-propelled howitzer M109, and the 8 inch self-propelled howitzer M110. For the M107 and the M110, it carried 36 complete rounds of 175mm or 8 inch ammunition as well as eight men including the driver. Operating with the M108, the M548 carried 200 105mm rounds and the driver. For the M109, the load was 80 rounds of 155mm ammunition and four men including the driver.

The role of the ammunition carrier was only the beginning for the M548. It was proposed for a wide range of applications and many of them were adopted. The M548 chassis was modified for a variety of special tasks, many of which are described in later sections. Numerous kits also were developed to extend the capability of the vehicle.

Two views of the cargo carrier M548 are shown below.

119

Labels (front view): ANTENNA MOUNTING BASE, HORN, SERVICE HEADLIGHT, BLACKOUT MARKER LIGHT, INFRARED HEADLIGHT, WINDSHIELD WIPER (3), BLACKOUT HEADLIGHT, SERVICE HEADLIGHT, BLACKOUT MARKER LIGHT, INFRARED HEADLIGHT, WINCH CHAIN

Labels (rear view): CARGO COMPARTMENT DOOR LATCH (2), STOP LIGHT TAILLIGHT, TOWING HOOK (2), TRAILER LIGHTS RECEPTACLE, FUEL COMPARTMENT DRAIN PLUG, CARGO COMPARTMENT COVER FLAP STRAP (3), BLACKOUT STOP LIGHT, TAILGATE LATCH (2)

The front and rear of the cargo carrier M548 are shown above. Below is the driver's station and at the right is the .50 caliber machine gun mount.

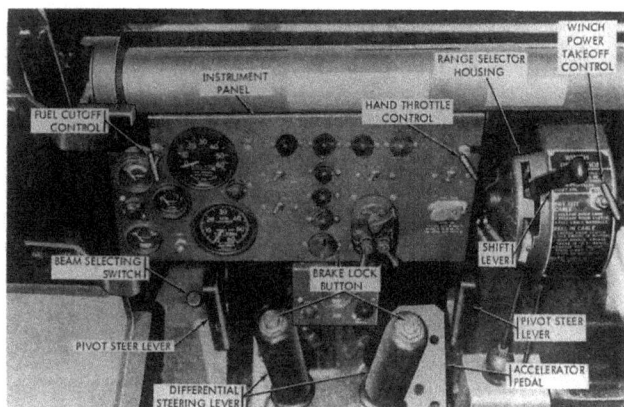

Labels (driver's station): FUEL CUTOFF CONTROL, INSTRUMENT PANEL, HAND THROTTLE CONTROL, RANGE SELECTOR HOUSING, WINCH POWER TAKEOFF CONTROL, BEAM SELECTING SWITCH, BRAKE LOCK BUTTON, SHIFT LEVER, PIVOT STEER LEVER, PIVOT STEER LEVER, DIFFERENTIAL STEERING LEVER, ACCELERATOR PEDAL

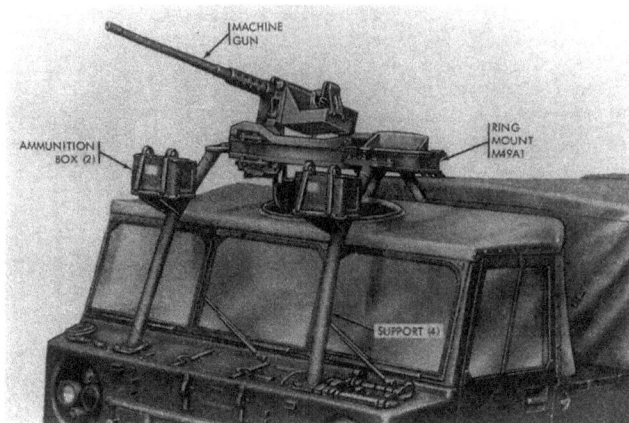

Labels: MACHINE GUN, AMMUNITION BOX (2), RING MOUNT M49A1, SUPPORT (4)

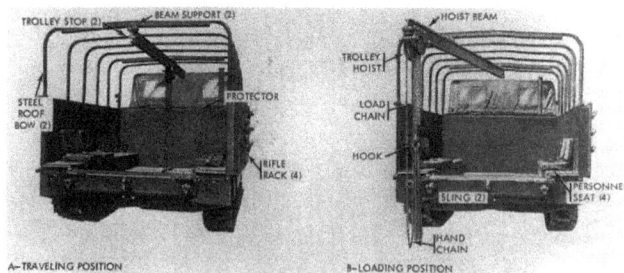

Labels (hoist): TROLLEY STOP (2), BEAM SUPPORT (2), STEEL ROOF BOW (2), PROTECTOR, RIFLE RACK (4), A—TRAVELING POSITION, HOIST BEAM, TROLLEY HOIST, LOAD CHAIN, HOOK, SLING (2), HAND CHAIN, PERSONNEL SEAT (4), B—LOADING POSITION

Above is the hoist installation on the M548. At the upper right, an M60 machine gun has been installed. Below, the early (left) and late (right) power plant compartment access covers are shown.

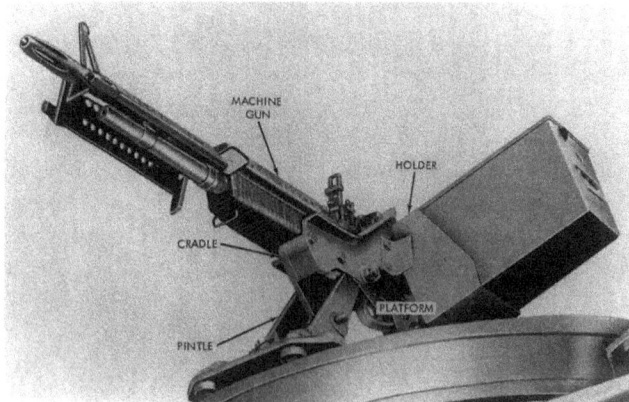

Labels: MACHINE GUN, HOLDER, CRADLE, PLATFORM, PINTLE

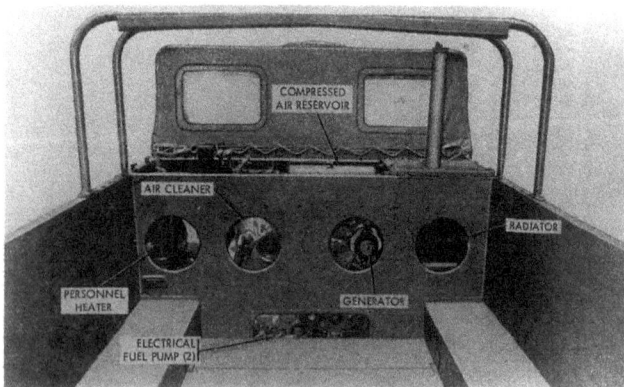

Labels (early cover): COMPRESSED AIR RESERVOIR, AIR CLEANER, RADIATOR, PERSONNEL HEATER, GENERATOR, ELECTRICAL FUEL PUMP (2)

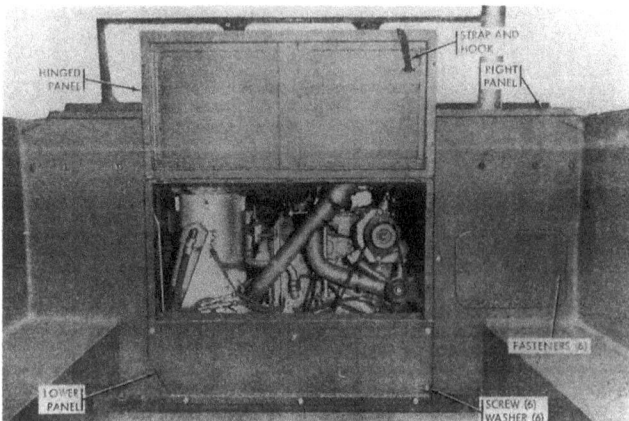

Labels (late cover): HINGED PANEL, STRAP AND HOOK, RIGHT PANEL, FASTENERS (6), LOWER PANEL, SCREW (6) WASHER (6)

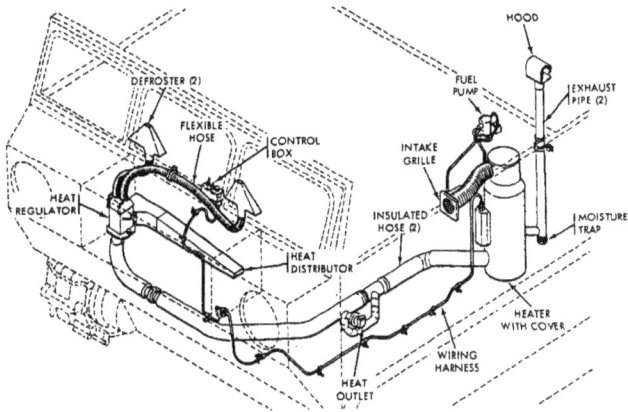

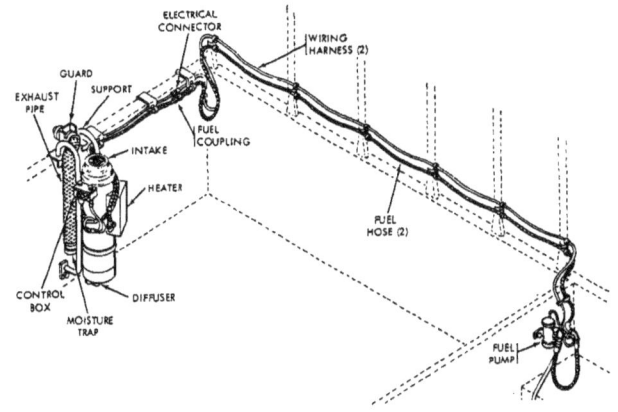

The installation of the personnel heater kits for the cab (left) and the cargo area (right) can be seen in the sketches above. Below, the insulated cab is at the left and the insulated cargo cover is at the right.

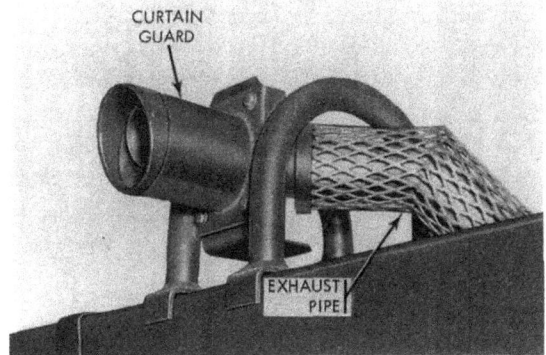

At the right is the curtain guard on the cargo area heater exhaust attached at the center of the cargo area door. The insulation provided in the cab (left) and the cargo area (right) can be seen in the photographs below.

121

The early and late engine compartment covers are installed at the left and right respectively. The winch and its drive are shown below.

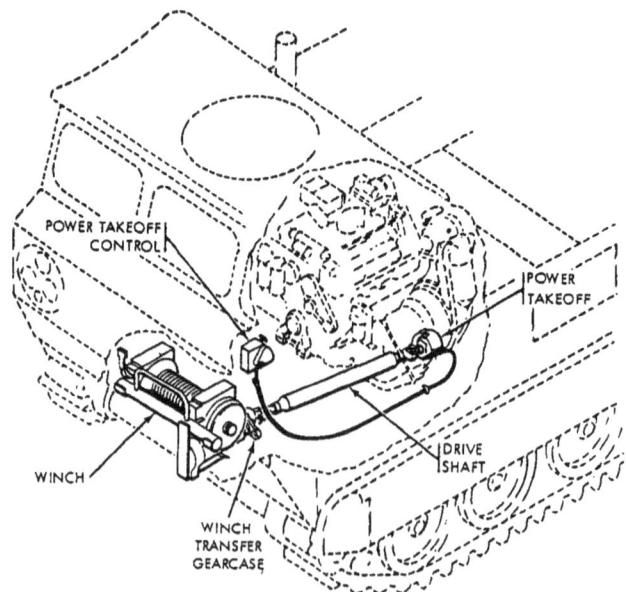

The cargo carrier M548A1 is at the left below.

With the introduction of the improved engine cooling system and suspension on the M113A2 armored personnel carrier, these modifications also were applied to the M548 and it was designated as the M548A1. These features were introduced into new production and the depot rebuilt vehicles in 1982. A 1,500 pound capacity chain hoist was mounted on an I-beam attached to the second and rear cargo cover bows. This hoist could lift loads within three feet of the tailgate and position them anywhere within the cargo area.

Installation of the RISE power pack was initiated on the depot rebuild line in 1994 converting the M548A1 to the M548A3. This power pack, consisting of the 275 horsepower 6V53T engine and the Allison X200-4 transmission, increased the vehicle performance and simplified driver training. These changes provided the M548A3 mobility comparable to the M1 tank and the M2/M3 Bradley fighting vehicle.

Like the armored personnel carrier, the M548 was lengthened with the addition of a sixth road wheel on each side. In 1978, a stretched version was designed with either an open cargo compartment or a closed, armor protected, body. The bed of the unarmored version was increased in length to $150^7/_8$ inches. The width dimensions remained the same as on the standard M548. The General Motors 6V53T diesel engine was installed developing 275 gross horsepower. It drove the vehicle through an Allison X200-3 transmission. The payload for the unarmored stretched vehicle was increased to 8 tons and the speed and cruising range remained the same as on the standard M548. The combat weight of the lengthened vehicle was 33,300 pounds.

The stretched M548 is shown in these photographs. The armored version is at the right. Dimensions of both types can be seen in the drawings below.

The armored version of the stretched M548 also had a combat weight of 33,300 pounds, but the payload remained at the 6 tons of the standard vehicle. The overall bed length for the armored, stretched, M548 was $132^7/_8$ inches with a bed width of 100 inches. It had the same power train as the unarmored vehicle and its performance was the same. One of the armored stretched vehicles, sometimes referred to as the M548-S, was completed and shipped to Fort Sill for evaluation.

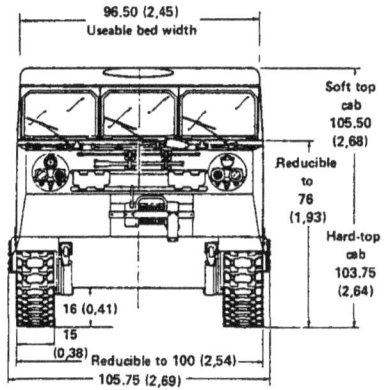

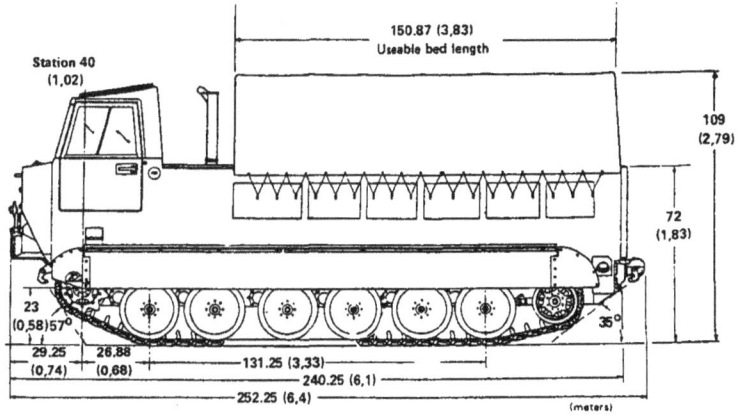

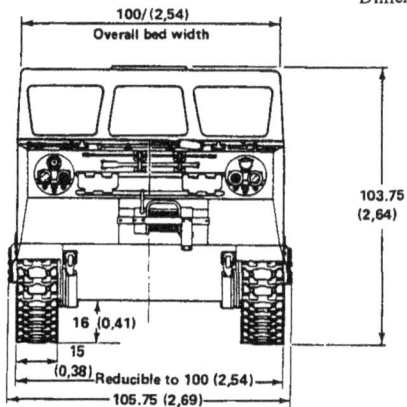

Dimensions in inches (meters)

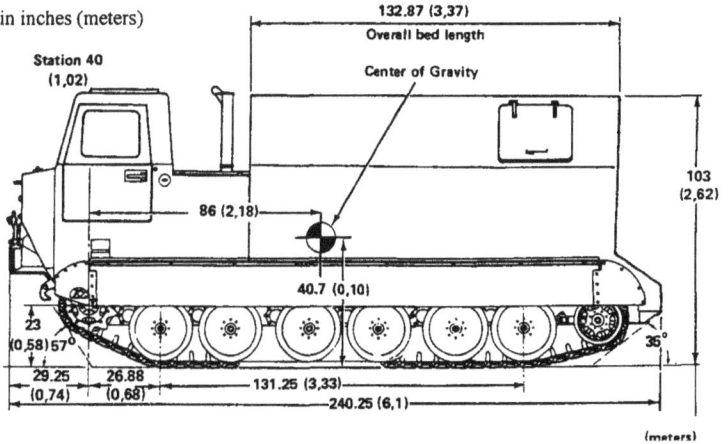

123

The guided missile equipment carrier M667 appears in the two photographs above. The various components of the carrier can be seen in the drawings below.

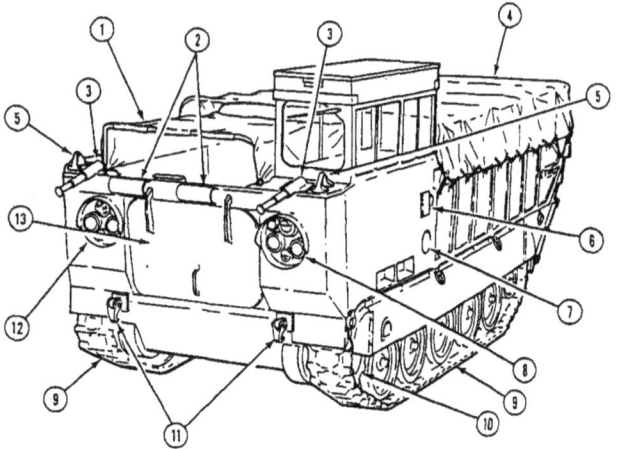

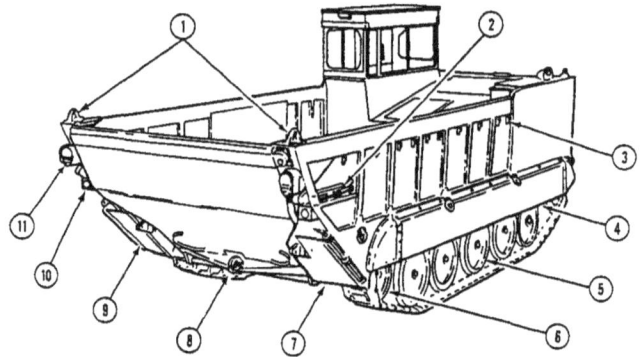

1 Front cover
2 Bumper
3 Cover and bow stowage arm
4 Cargo compartment cover
5 Front lifting eye
6 Fire extinguisher outside handle and guard
7 Fuel filler
8 Left headlight cluster
9 Track
10 Drive sprocket
11 Towing hook
12 Right headlight cluster
13 Front power plant access door

1 Rear lifting eye
2 Ramp lock handle
3 Tiedown hook
4 Track shroud
5 Road wheel
6 Idler wheel
7 Right swim vane
8 Tow pintle
9 Left swim vane
10 Reflector (late models)
11 Stop light – taillight – turn signal light (late models)

A modified version of the M548 was proposed as a carrier for the Little John missile system. However, a 2½ ton 6 x 6 truck and wheeled trailers were used to handle the 318mm M51 rocket.

The introduction of the Lance missile required a lightweight, unarmored, tracked vehicle to provide adequate battlefield mobility. Three carriers were used to support the missile system. Using the power train and suspension components from the M548, the basic vehicle was designated as the M667. It had a counterbalanced rear ramp to provide easy access to the interior of the carrier. A suspension lockout system stabilized the vehicle for missile launching or loading operations. The basic M667 was easily adapted to meet the requirements of the self-propelled launcher M752 or the loader transporter M688.

Production of the M667 at FMC totaled 168 vehicles for the United States and 163 for allied nations. After the Lance missile was declared obsolete, the Army directed the disposal of the missile and carrier hardware in January 1992.

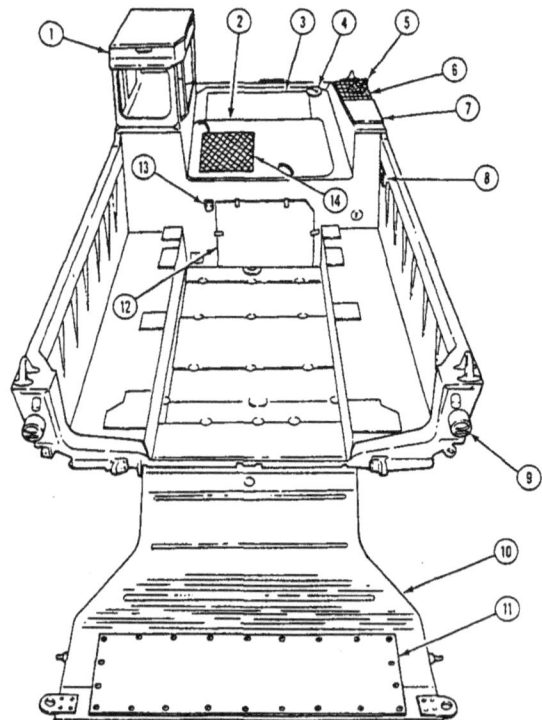

1 Driver's cab
2 Engine access panel
3 Differential access panel
4 Coolant filler cover
5 Exhaust guard
6 Exhaust grille
7 Battery box
8 Dome light
9 Stop light – taillight (early models)
10 Ramp
11 Ramp blast plate
12 Rear power plant access cover
13 Auxiliary power receptacle
14 Air inlet grille

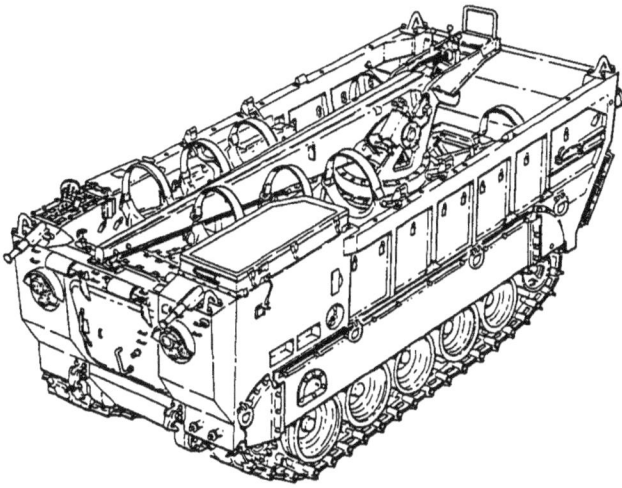

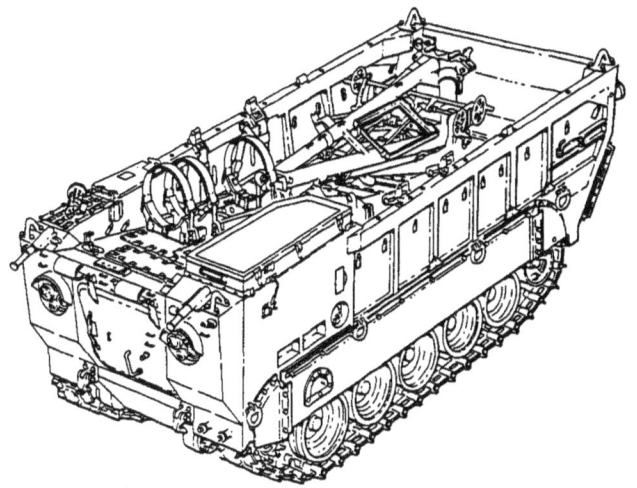

Above, the loader transporter M688 for the Lance guided missile system is at the left and the self-propelled launcher M752 is at the right. Below, the left swim vane is shown stowed (left) and extended (right).

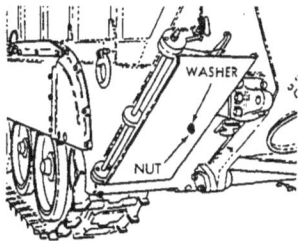

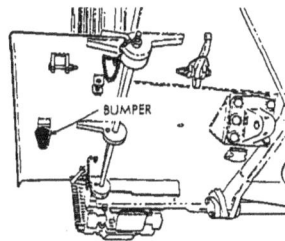

The Lance missile is shown on the self-propelled launcher below and at the right.

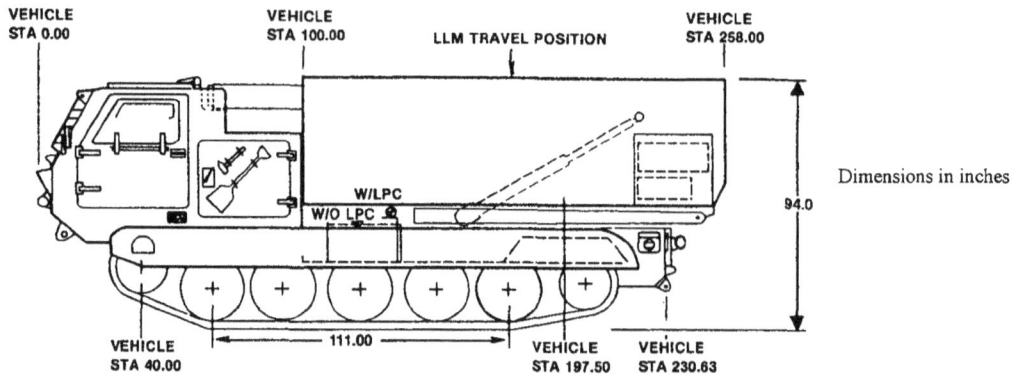

VEHICLE STA 0.00 VEHICLE STA 100.00 LLM TRAVEL POSITION VEHICLE STA 258.00

W/LPC
W/O LPC

Dimensions in inches

94.0

111.00

VEHICLE STA 40.00 VEHICLE STA 197.50 VEHICLE STA 230.63

The lightweight multiple launch rocket system is depicted in the drawing above and by the model at the right.

The successful development of the multiple launch rocket system (MLRS) emphasized the need for a lighter weight system that could be transported in C130 aircraft and lifted by the CH47D helicopter. In May 1983, the Vought Corporation evaluated several concepts to meet this requirement. They included tracked and wheeled carriers as well as a trailer mounted launcher. After the study was complete, the concept proposed was based upon the M113 family of vehicles. It utilized the suspension and power train from the M113A2 armored personnel carrier. A fire-from-cab capability was attained by the installation of a cab similar to that on the full size MLRS. The single launcher module with six rockets was mounted on the rear deck of the vehicle. With a combat loaded weight of 23,200 pounds, the lightweight MLRS had an estimated maximum road speed of 38 miles per hour. Its cruising range was about 300 miles. Without the rocket pod, the weight was 18,190 pounds and it could be carried by the CH47D helicopter. It also could be transported in the C130 aircraft and landed using the low altitude parachute extraction system (LAPES).

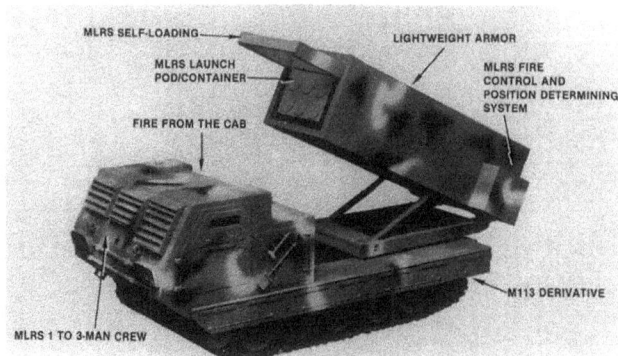

MLRS SELF-LOADING LIGHTWEIGHT ARMOR
MLRS LAUNCH POD/CONTAINER MLRS FIRE CONTROL AND POSITION DETERMINING SYSTEM
FIRE FROM THE CAB
M113 DERIVATIVE
MLRS 1 TO 3-MAN CREW

A further development of the concept vehicle above was the XM1108 universal carrier. It utilized many components from the M113 family as well as those from other development programs. The lower part of a stretched, late model, M113 series hull was combined with the crew compartment from the M993 multiple launch rocket system (MLRS). The cab provided space for three men with the same protection as on the M113A3 armored personnel carrier. With a payload capacity of 15,700 pounds, the XM1108 was proposed as a carrier for several missile systems or as an upgraded cargo carrier.

The dimensions of the XM1108 universal carrier are shown in the drawing below.

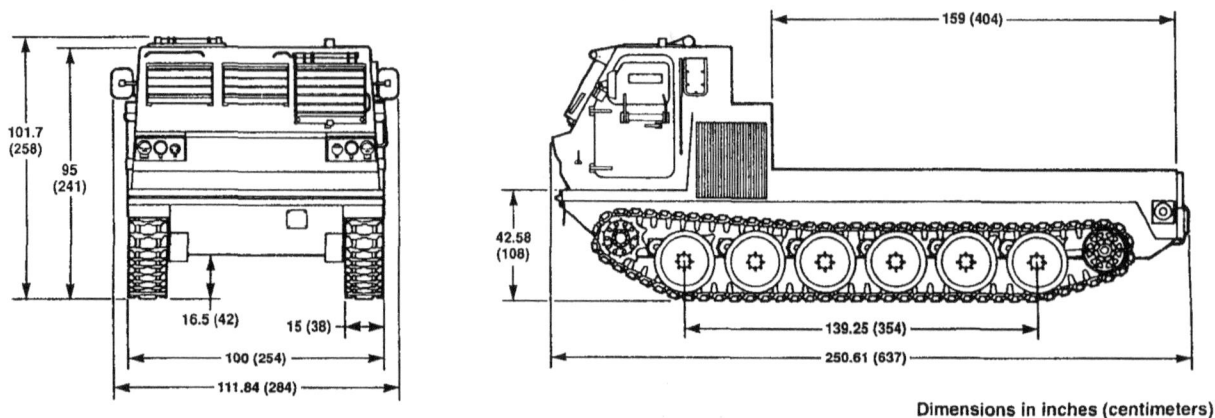

101.7 (258)
95 (241)
16.5 (42) 15 (38)
100 (254)
111.84 (284)

159 (404)
42.58 (108)
139.25 (354)
250.61 (637)

Dimensions in inches (centimeters)

126

The universal carrier XM1108 appears in these photographs. Below at the left, the cab is tilted forward exposing the engine and at the right is a view of the driver's controls.

Above are two views of a pilot amphibian cargo carrier T107.

LOW GROUND PRESSURE VEHICLES

The Army Equipment Development Guide in 1950 listed the requirement for three new amphibious cargo carriers. They were to have payload capacities of ½, 1½, and 2½ tons, but the ground pressure was not to exceed 2 pounds per square inch. Although its ground pressure was slightly above the specified 2 pounds, the T46E1 then under development generally met the requirement for the 1½ ton payload vehicle. To develop a successor to the Weasel, Studebaker Corporation was awarded a contract to build a new ½ ton amphibious cargo carrier. Designated as the T107, two pilots were built and tested at Aberdeen Proving Ground and the Arctic Test Branch of CONARC. These engineering tests lasted from June 1954 to August 1956. Although the T107 had a more powerful engine and a larger fuel capacity than the M29C, it was heavier and its speed and cruising range were not much better. The ground pressure also was 2.4 pounds per square inch. The tests concluded that the T107 was less durable and reliable than either the M29C or the M76.

Even before the test program was complete a meeting was held in Washington between representatives of the Chief of Staff, the Chief of Ordnance, and CONARC to review the status of the Weasel replacement program. Three solutions to the problem were

under consideration. The first was to redesign the T107 to improve its performance and the second was to continue the development of the T60 cargo carrier for this application. The T60 was a light vehicle evaluating the spaced link track concept. The third option was to adapt the Canadian Beaver, ¾ ton, cargo carrier for American manufacture. However, none of these alternatives were accepted. As a result, the T107 project was terminated and on 15 November 1956, a new development program was initiated for the amphibious cargo carrier T116. On 22 May 1957, a contract was awarded to the Pacific Car and Foundry Company for the design and construction of four pilot vehicles. The first pilot T116 was shipped to Aberdeen Proving Ground on 27 November 1958 and the second went to the CONARC Arctic Test Board on 22 December 1958. These vehicles were powered by the Continental 8AO-198, air-cooled engine with the Hydramatic 198-M transmission. Because of production costs and problems during the service tests, it was directed that the power pack be replaced by a Chevrolet V8, liquid-cooled, engine with a

The spaced-link track test rig for the amphibian cargo carrier T60 is at the right for tests at Aberdeen Proving Ground on 7 October 1953.

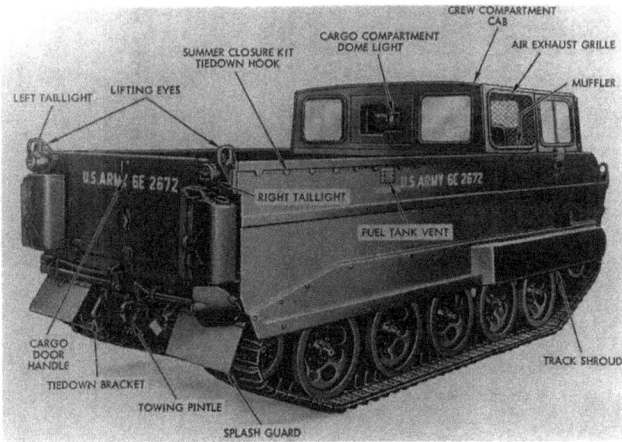

The amphibian cargo carrier M116 is shown above.

military standard Hydramatic transmission. The modified pilots were designated as the amphibious cargo carrier T116E1. After further tests, the T116E1 was type classified as the amphibious cargo carrier M116, Standard A, on 15 December 1960. The rated payload capacity of the vehicle also was increased from ½ to 1½ tons. The M116 was named Husky and a contract was awarded to the Pacific Car and Foundry Company to build three preproduction pilots. However, Blaw-Knox Company received the initial production contract on 15 December 1961 for a run of 197 vehicles.

Empty, the M116 weighed 6,700 pounds. With a driver and fuel, this weight increased to about 7,600 pounds resulting in a ground pressure of approximately 1.9 pounds per square inch. Adding a 3,000 pound payload increased the weight to 10,600 pounds and the ground pressure to 2.6 pounds per square inch. The

liquid-cooled Chevrolet engine developed 160 gross horsepower at 4,600 rpm and it drove the vehicle through a Hydramatic 305-MC transmission. The Husky had a maximum speed of 37 miles per hour on roads and 4 miles per hour in water. Propulsion in water was by the action of the shrouded tracks. The cruising range was about 300 miles on land and approximately 22 miles in water. Space was available for 10 men in winter gear or 13 men in summer gear in addition to the driver. The M116 was easily transported by air or dropped by parachute.

The cargo cover is installed on the M116 at the right. The cargo door is shown open and closed below.

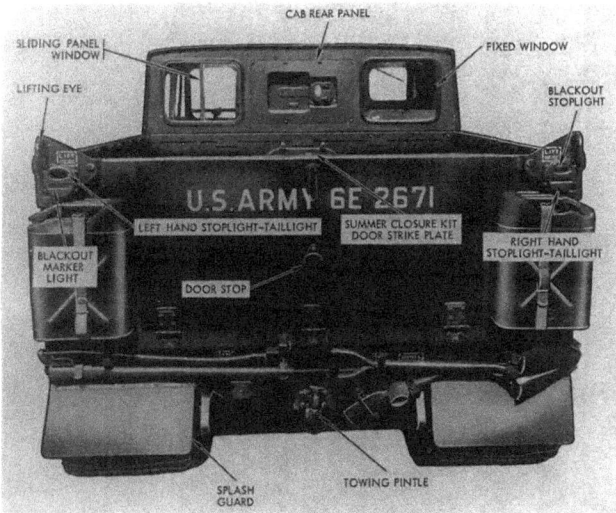

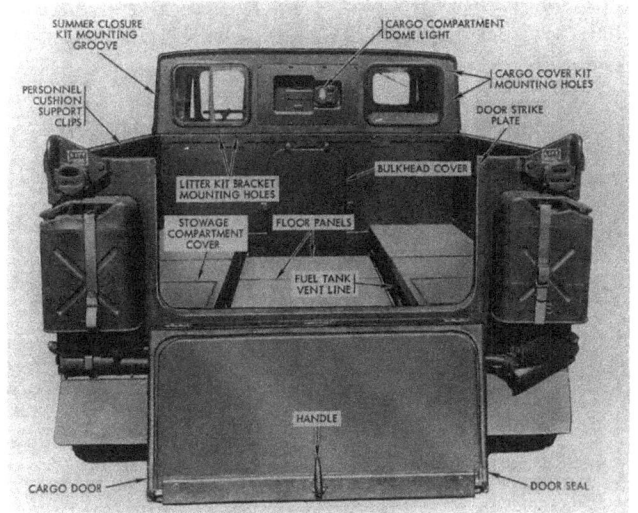

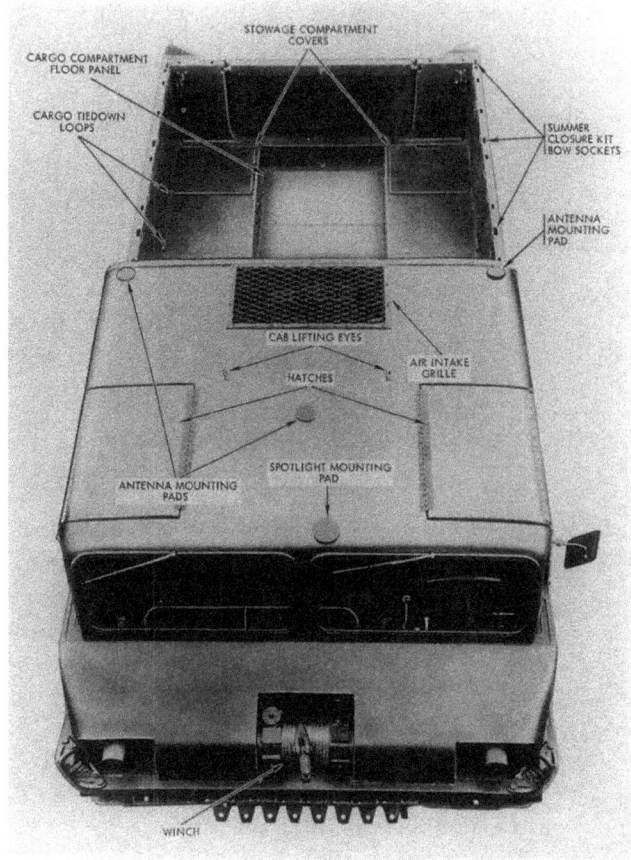

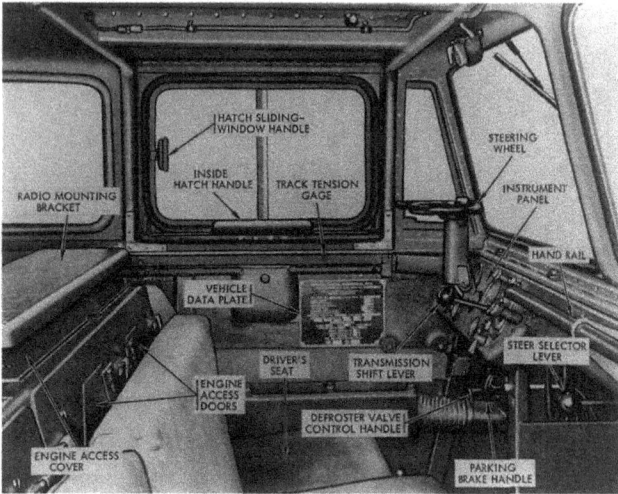

Above are the driver's controls in the amphibian cargo carrier M116 and at the right is a top view of the vehicle.

The watertight aluminum alloy hull was assembled by welding. The driver was located in the left front of the fiberglass cab. Access to the driving compartment was through two hatches in the cab roof. The engine was just behind the driver in the power plant well that extended across the full width of the cab. The engine air intake grille was in the center of the cab roof. Both the cooling air and the engine exhaust exited the vehicle through a grille on the right side of the cab. Power passed from the transmission through the geared steer unit to the front drive sprockets on each side of the vehicle. The 20 inch wide, band type, tracks were carried on a flat track, torsion bar, suspension with five dual road wheels per side. Track tension was maintained by an adjusting idler at the rear of each track. A winch with a bare drum capacity of 5,000 to 6,000 pounds was installed in the front of the vehicle. The cargo com-

partment in the rear had a movable center deck that could be raised to provide a flat cargo deck across the full width of the hull. Entrance or exit from the cargo compartment was through a hinged door at the rear. The 65 gallon fuel tank was under the cargo compartment with the filler cap on the left side of the hull. Cargo compartment closure kits were available for both winter and summer operation. The winter kit could be fitted with ski racks on the top. The summer kit was a canvas cover installed over three bows. A personnel cushion kit was available for installation in the cargo compartment. A litter kit allowed four litters to be carried in the cargo compartment.

Below, details of the suspension on the M116 can be seen at the left and the vehicle is swimming at the right.

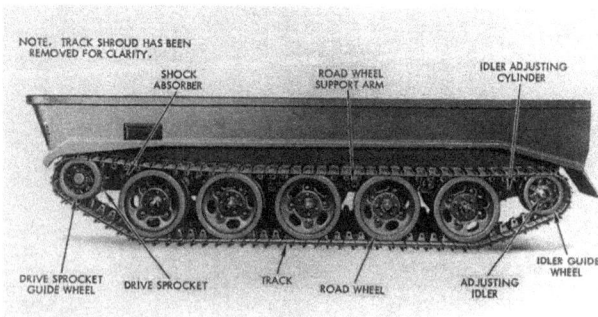

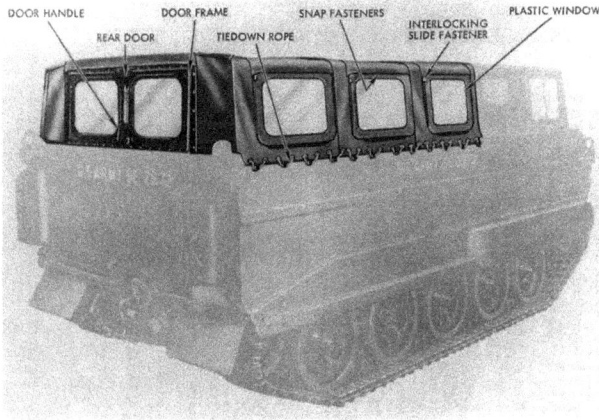

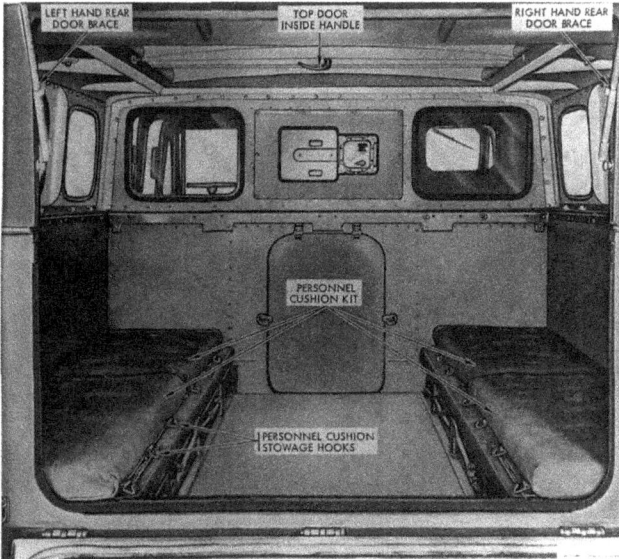

The summer closure kit on the M116 is at the top left and the winter closure kit is above.

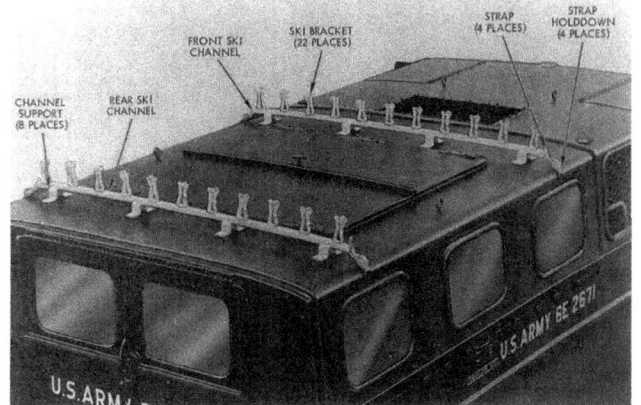

Above, the interior of the winter closure kit is at the left and the roof ski rack is at the right. Below is a view of the M116 with the winter closure kit during test operations.

The M116 amphibian cargo carrier is fitted with the summer closure kit in the top photographs and the cargo compartment is open at the right and below. The "Davy Crockett" nuclear weapon system has been installed in the vehicle in the bottom photograph.

Above, the aluminum armor test rig is at the left and the armored assault vehicle test rig is at the right. Both were assembled from welded aluminum alloy armor.

The M116 provided the basic chassis for a number of experimental programs. One of these was the aluminum armor test rig fabricated at the Tank Automotive Center in 1962. This was a lightweight vehicle armed with two machine guns and manned by a crew of two. Later, it was dubbed the armored assault vehicle test rig and evaluated at Aberdeen Proving Ground during July 1965 to determine its transportability by the CH-47 helicopter. The test results indicated that with additional changes, it would be an effective combat vehicle for the Air Mobile Division in Southeast Asia.

These additional views of the armored assault vehicle test rig were taken at Aberdeen Proving Ground on 2 August 1965.

The armored assault vehicle XM729 is shown above. Like the test rigs, it was protected by welded aluminum alloy armor. The dimensions of the XM729 can be seen in the drawing at the bottom of the page.

After modification, the little vehicle was fitted with a new twin gun cupola and a commander's station. This station had a 360 degree view through a ring of vision blocks. The cupola was armed with a 7.62mm M73E1 machine gun and a 40mm M75 grenade launcher. An additional 7.62mm M73E1 machine gun was installed in a ball mount in the center of the front hull for use by the driver although some drawings show it offset to the left. Designated as the armored assault vehicle XM729, it was intended for counterinsurgency operations in Southeast Asia. With its two man crew, The XM729 had a combat weight of 10,500 pounds. Its performance was similar to that of the M116 with a maximum speed of 37 miles per hour on roads and 3½ miles per hour in water.

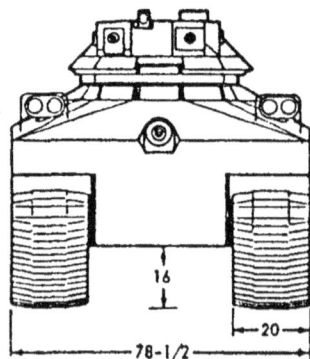

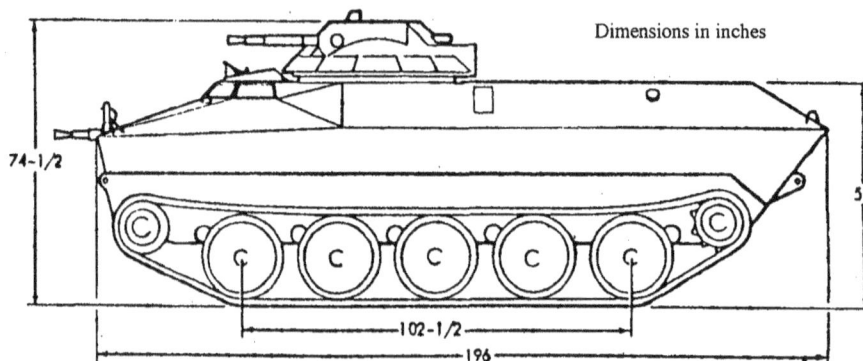

Dimensions in inches

134

Above, this armored M116 cargo carrier is armed with an M60 machine gun and a 40mm automatic grenade launcher. The armored M116 at the right and below is provided with two M60 machine guns, one of which is mounted in an open top turret.

In February 1965, the Army initiated the development of the XM733 assault vehicle as part of the Remote Area Mobility Study (RAMS). This was the same study that resulted in the XM729. The XM733 was essentially an armored M116. It was an open top vehicle armed with 7.62mm and .50 caliber machine guns. A 40mm grenade launcher also could be installed. On the XM733E1, a cupola was provided to protect the gunner.

Below, the XM733 is at the left and the XM733E1 is at the right. Needless to say, the troops installed a wide variety of armament in these vehicles.

One of the many combinations of armament applied to the XM733 is shown in the top view at the left above. Above at the right is another of the many lightweight armored vehicles proposed during this period. This one was the Thiokol Spryte 1301.

These five photographs show the XM733 during its evaluation. Here it is apparently fitted with a single M60 machine gun in addition to the 40mm automatic grenade launcher.

Here is another photograph of XM733, registration number 13B084. It is now armed with two M60 machine guns. Below are two views of XM733, registration number 13B049. Note the five machine gun armament.

Below is another variant designated as the XM755 armed with an 81mm mortar. At the bottom right is another lightweight armored assault vehicle. This one is armed with a cupola mounted .50 caliber machine gun and a 7.62mm weapon in the bow.

When the XM733 reached the troops, they installed their own armament. On the U.S. Marine Corps vehicles above and below, note the .50 caliber machine guns in the top photos and the rear firing mortar below.

The Army canceled the XM733 program in October 1966, but the Pacific Car and Foundry Company received a contract to manufacture 93 of the light armored vehicles for the U. S. Marine Corps. On 22 December 1966, Pacific Car and Foundry was awarded another contract to build 111 unarmored XM733s for the U.S. Navy and Marine Corps. Since an unarmored XM733 was essentially an M116, these vehicles were designated as the M116A1 to avoid confusion. The first production M116A1 was delivered to Aberdeen Proving Ground in December 1967.

The M116 also was proposed as a carrier for the M29 Davy Crockett Battle Group Atomic Weapon System.

The "Davy Crockett" nuclear weapon system is mounted on the M116 carrier below.

Above is the pilot 81mm self-propelled mortar T257 based upon the early T113 armored personnel carrier.

FIRE SUPPORT VEHICLES

As mentioned earlier, construction was authorized in May 1955 for two experimental self-propelled mortars along with the original T113 pilots. In October 1956, the preliminary design was reviewed and approved for development by FMC. The new vehicle, based upon the T113 armored personnel carrier, was designated as the 81mm, full-tracked, self-propelled mortar T257 in August 1957. At that time, further development of the T82 self-propelled mortar was canceled. The T257 pilots constructed at FMC were delivered in March 1958. These self-propelled mortars used the basic chassis of the T113 with aluminum alloy armor and the trailing idler suspension. Like the T113, they were powered by the AOSI-314-2 air-cooled engine with the X-drive transmission. The vehicles were armed with the 81mm mortar M29 on an M23A1 mount in the

troop compartment. A circular folding roof hatch opened to provide a clear field of fire for the mortar. A beam across the bottom of the troop compartment supported the mortar mount to absorb the shock loads from firing. At first this was a steel beam, but it was replaced later by an aluminum beam welded into the hull structure. A circular base for the rotator was welded on top of the beam. The rotator itself was secured by a retaining ring bolted to the circular base. This ring had 56 teeth on the

The mortar can be seen through the open ramp of the 81mm self-propelled mortar T257 at the right.

The two additional views above show the mortar installation in the 81mm self-propelled mortar T257. The internal arrangement of the T257 can be seen in the drawing below.

inside surface to lock the rotator in any one of 56 positions in the 360 degrees of traverse. The traversing mechanism in the mortar mount permitted further adjustment of the mortar in traverse. The feet of the mortar mount bipod were secured by locking pins in a supporting bracket. This bracket held the mortar in firing position or the weapon could be lowered to a stowed or travel position. A separate base plate was carried so that the mortar could be removed from the vehicle and fired from the ground.

Tests during 1958 revealed some slight deformation in the rotator base ring after firing and modifications were recommended. After evaluation by CONARC, the T257 was considered suitable for troop use. However, the T113 upon which it was based had been superseded by the modified T113E2 now standardized as the M113 armored personnel carrier. A new directive also required

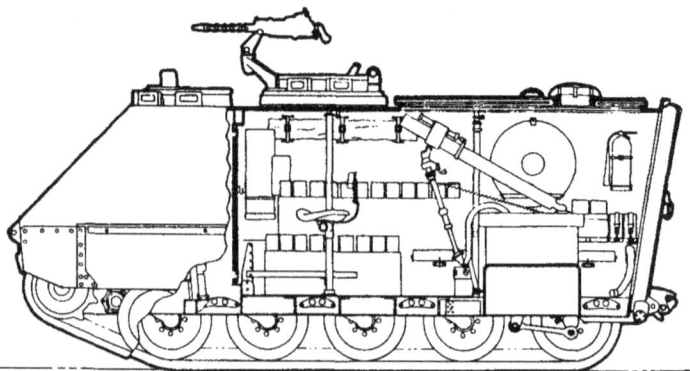

that the 81mm and 4.2 inch mortars be interchangeable in the new carrier. FMC constructed a mock-up of the modified vehicle now designated as the 81mm/4.2 inch self-propelled mortar T257E1. Based upon the M113 armored personnel carrier, it was powered by the same Chrysler 75M engine with the Allison TX200-2

The mock-up of the 81mm/4.2 inch self-propelled mortar T257E1 is shown below with the 4.2 inch mortar installation on the left and the 81mm mortar installed on the right.

Although the vehicle above is still marked T257E1, it is armed with the 4.2 inch mortar and should be correctly designated as the XM106. Note the new front hull configuration based upon the M113 armored personnel carrier.

transmission. The split circular hatch in the troop compartment roof consisted of a single section opening to the left and a folding double section opening to the right. As on the T257, the 81mm mortar had a 360 degree traverse. When the 4.2 inch mortar was installed, it fired toward the rear with a maximum 90 degree traverse. About this time, the 4.2 inch mortar M30 was designated as the 107mm mortar M30 in line with the policy to convert all weapon calibers from inches to millimeters.

To avoid confusion when armed with the two different weapons, separate designations were assigned to the carrier. When fitted with the 81mm mortar M29, it remained the 81mm self-propelled mortar T257E1.

Using the 107mm mortar M30, it became the 107mm self-propelled mortar XM106. The new system eliminating the T designations had now gone into effect. FMC built three pilots of the T257E1/XM106 and the first was delivered to Aberdeen Proving Ground in May 1961 for engineering tests. The remaining two were shipped to Fort Benning and Fort Knox in July 1961 for service tests.

Below and at the right is the 107mm (4.2 inch) self-propelled mortar XM106. The perforated metal attached to the ramp surface was to provide better footing during wet or freezing weather.

Above are two photographs of the 81mm self-propelled mortar M125A1.

The introduction of the diesel engine in the M113A1 armored personnel carrier resulted in similar modifications to the mortar carriers. When powered by the power pack from the M113A1, they were designated as the 81mm self-propelled mortar T257E2 and the 107mm self-propelled mortar XM106E1. In October 1964, the T257E1 was type classified as the 81mm self-propelled mortar M125, Standard B and the T257E2 became the 81mm self-propelled mortar M125A1,

Standard A. In a similar manner, the XM106 became the 107mm self-propelled mortar M106, Standard B and the XM106E1 was type classified as the 107mm self-propelled mortar M106A1, Standard A.

The M125 and the M125A1 carried 114 rounds of 81mm ammunition. A total of 88 rounds of 107mm ammunition was stowed in the M106 and the M106A1. Both types of mortar carriers were manned by a crew of six including the driver.

81mm Self-propelled Mortar M125A1

Dimensions in parentheses ( ) are in meters; all other dimensions are in inches.

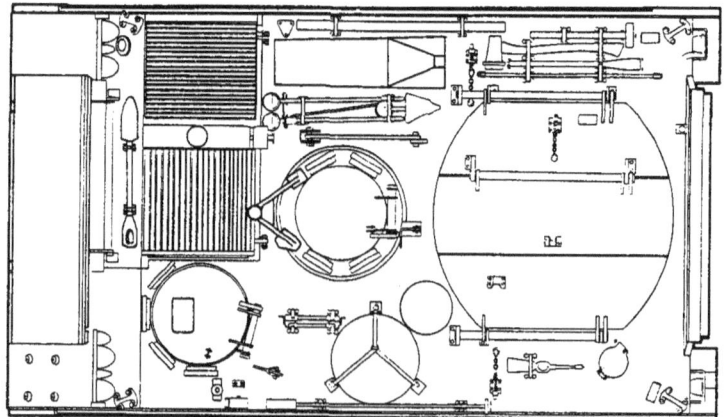

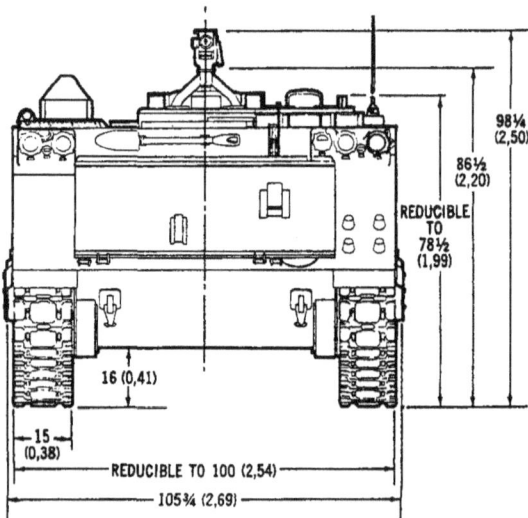

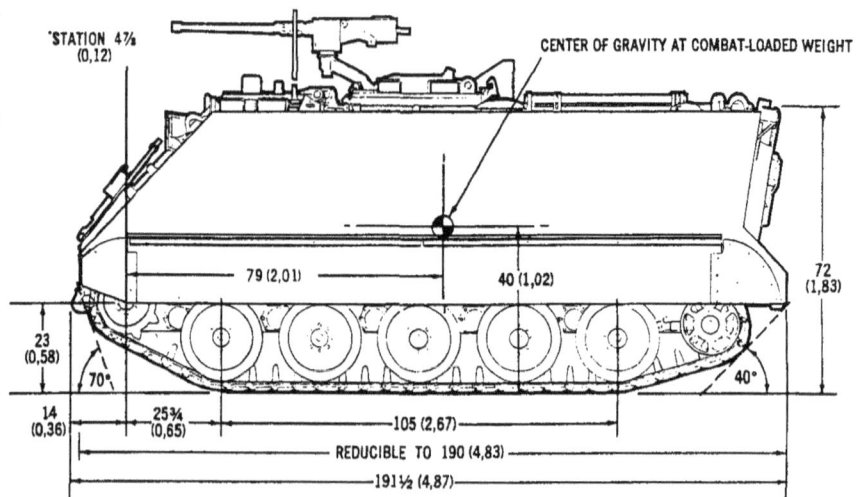

The 107mm (4.2 inch) self-propelled mortar M106A1 is shown above. The drawing at the left indicates the dimensions of the space available inside the vehicle.

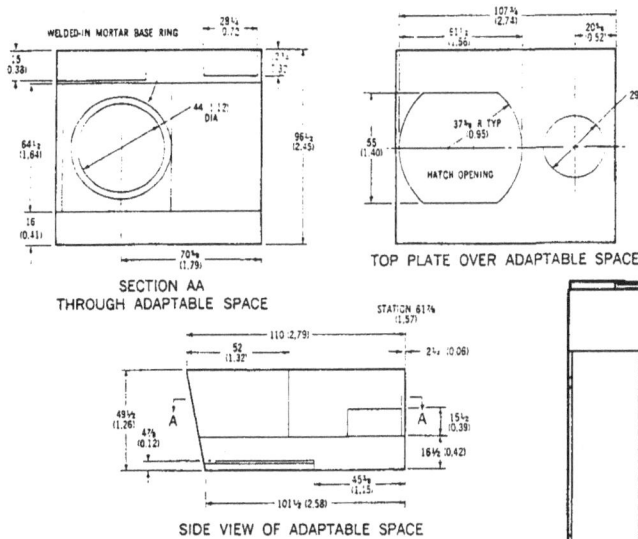

SECTION AA
THROUGH ADAPTABLE SPACE

TOP PLATE OVER ADAPTABLE SPACE

SIDE VIEW OF ADAPTABLE SPACE

107mm Self-propelled Mortar M106A1

Dimensions in parentheses ( ) are in meters;
all other dimensions are in inches.

Limited production of the XM106 was authorized prior to standardization. A total of 860 M106 vehicles were produced including 589 for United States forces. They were followed by 1,316 M106A1s of which 982 were allocated to United States forces. The M125 was not released for production, but a total of 2,252 M125A1s were completed including 460 for United States forces.

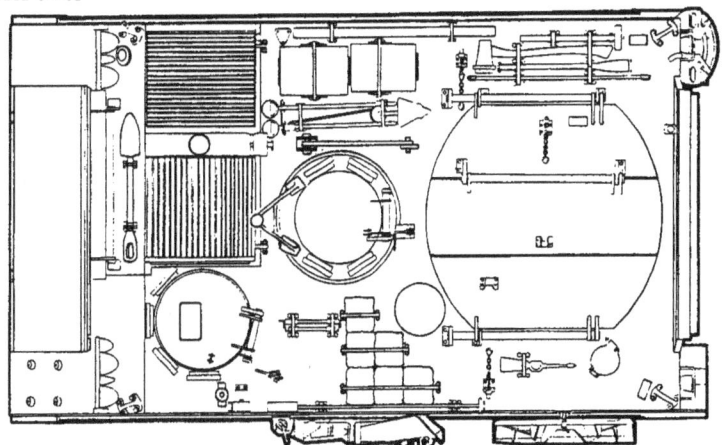

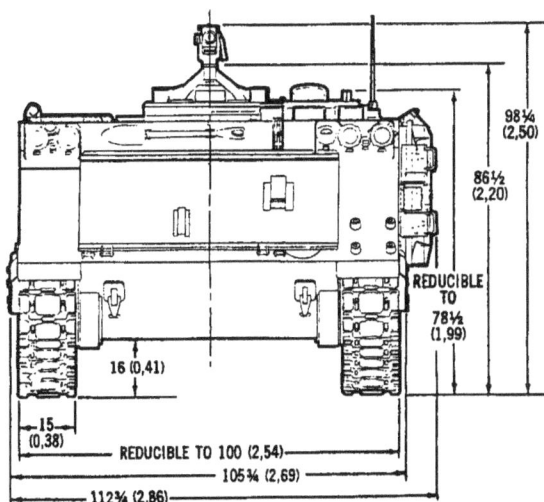

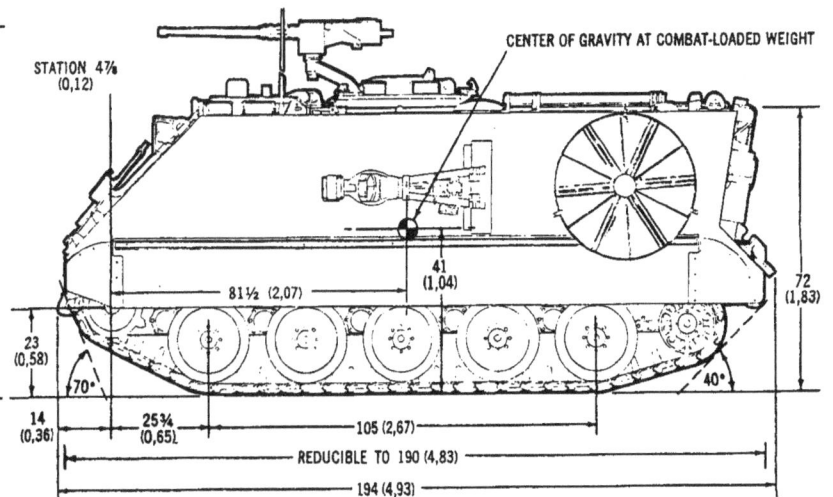

143

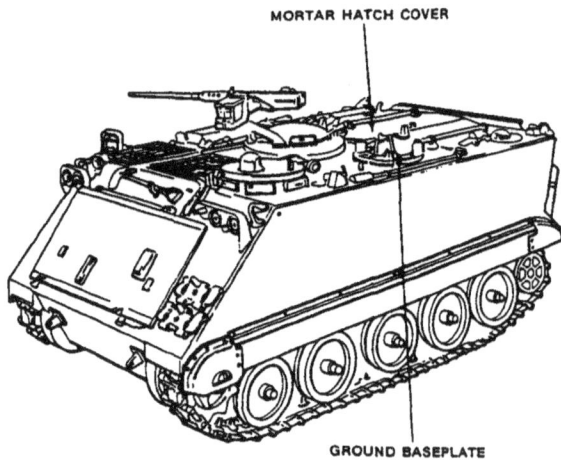

MORTAR HATCH COVER

GROUND BASEPLATE

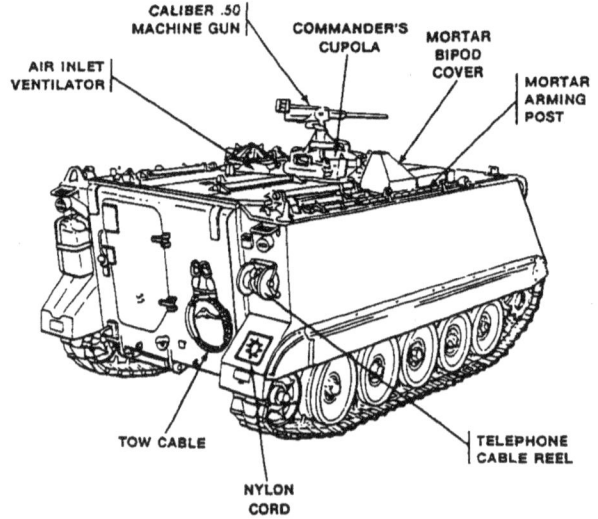

CALIBER .50 MACHINE GUN

COMMANDER'S CUPOLA

MORTAR BIPOD COVER

AIR INLET VENTILATOR

MORTAR ARMING POST

TOW CABLE

NYLON CORD

TELEPHONE CABLE REEL

Details of the 81mm self-propelled mortar M125A2 can be seen in the drawings above and the photograph at the left.

When the M113A2 armored personnel carrier appeared with the improved suspension and engine cooling system, these features also were applied to the mortar carriers and they were designated as the 81mm self-propelled mortar M125A2 and the 107mm self-propelled mortar M106A2. Other features remained the same as on the earlier models.

The drawings below show the internal arrangement and the mortar mount in the 81mm self-propelled mortar M125A2.

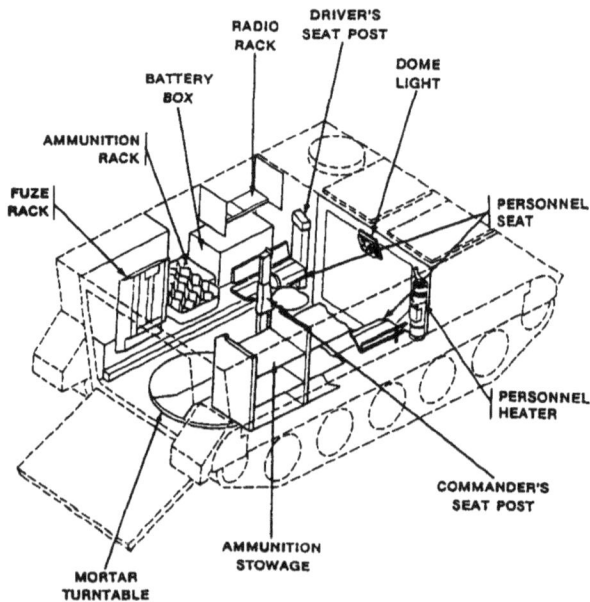

RADIO RACK

DRIVER'S SEAT POST

DOME LIGHT

BATTERY BOX

AMMUNITION RACK

FUZE RACK

PERSONNEL SEAT

PERSONNEL HEATER

COMMANDER'S SEAT POST

MORTAR TURNTABLE

AMMUNITION STOWAGE

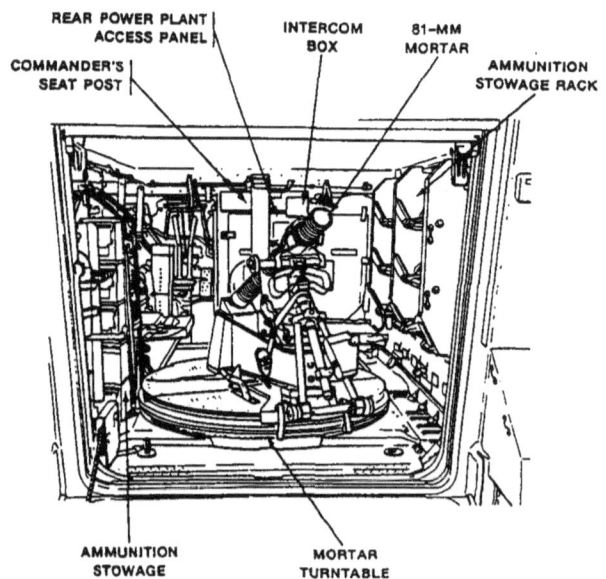

REAR POWER PLANT ACCESS PANEL

INTERCOM BOX

81-MM MORTAR

COMMANDER'S SEAT POST

AMMUNITION STOWAGE RACK

AMMUNITION STOWAGE

MORTAR TURNTABLE

144

The internal stowage and the mortar mount are visible in these views of the 81mm self-propelled mortar M125A2.

I.R. PERISCOPE COVER (M19)
DRIVER'S HATCH
DRIVER'S PERISCOPE (M17)
MORTAR HATCH
EXHAUST GRILL
GROUND BASEPLATE (STOWED)
RIGHT HEADLIGHT CLUSTER
TRIM VANE
TRIM VANE LATCH
TOWING EYE
LEFT HEADLIGHT CLUSTER
LIFTING EYE
MORTAR BRIDGE (STOWED)

COMMANDER'S CUPOLA
AIR INLET VENTILATOR
MACHINE GUN TRIPOD
MORTAR HATCH COVER
AIMING POST
WATER CAN
TRAILER ELECTRICAL CONNECTION
TOW CABLE
ROTATOR ASSEMBLY
TELEPHONE CABLE REEL

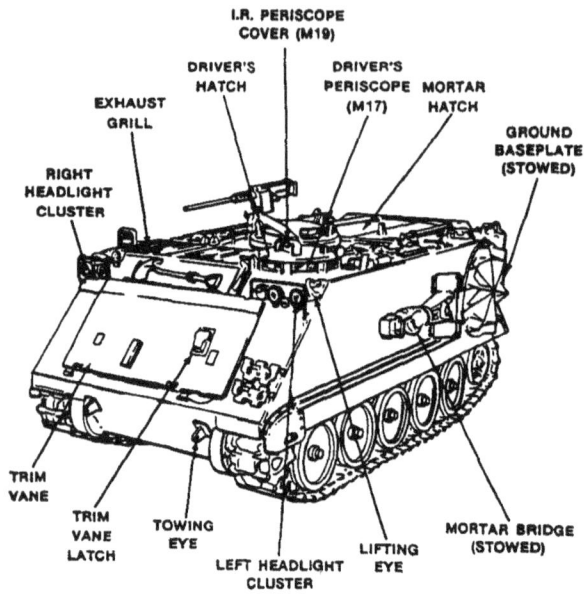

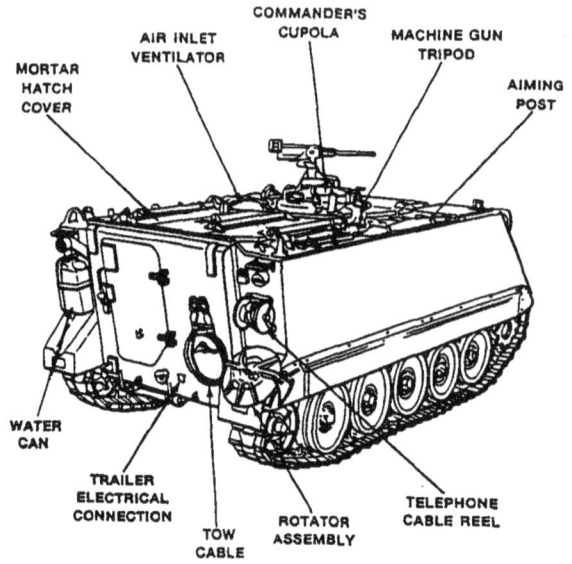

The drawing above and the photograph at the left show the 107mm (4.2 inch) self-propelled mortar M106A2. The drawings at the bottom show the internal stowage and the mortar installation in the vehicle.

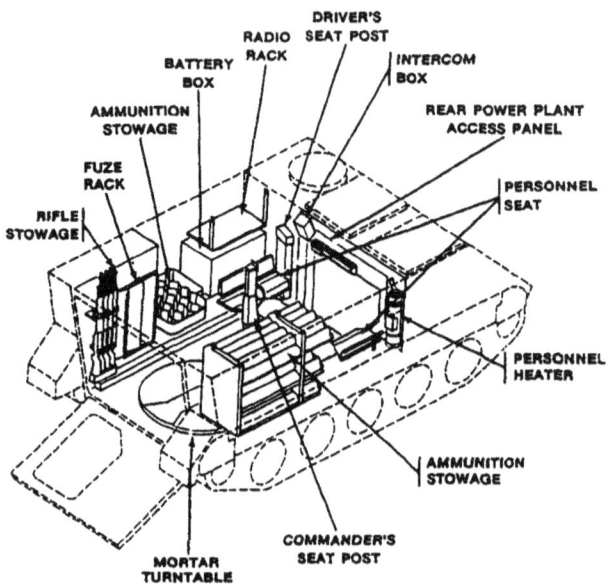

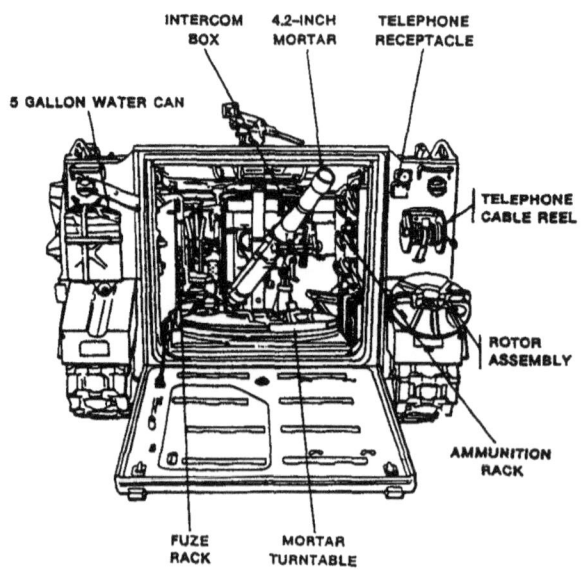

RADIO RACK
DRIVER'S SEAT POST
BATTERY BOX
INTERCOM BOX
AMMUNITION STOWAGE
REAR POWER PLANT ACCESS PANEL
FUZE RACK
PERSONNEL SEAT
RIFLE STOWAGE
PERSONNEL HEATER
AMMUNITION STOWAGE
MORTAR TURNTABLE
COMMANDER'S SEAT POST

INTERCOM BOX
4.2-INCH MORTAR
TELEPHONE RECEPTACLE
5 GALLON WATER CAN
TELEPHONE CABLE REEL
ROTOR ASSEMBLY
AMMUNITION RACK
FUZE RACK
MORTAR TURNTABLE

146

The 107mm self-propelled mortar M106A2 is shown in these photographs.

I.R. PERISCOPE COVER (M19)
DRIVER'S HATCH
DRIVER'S PERISCOPE (M17)
EXHAUST GRILL
MORTAR HATCH
RIGHT HEADLIGHT CLUSTER
GROUND BASEPLATE (STOWED)
TRIM VANE
TRIM VANE LATCH
TOWING EYE
LEFT HEADLIGHT CLUSTER
LIFTING EYE

MORTAR HATCH COVER
AIR INLET VENTILATOR
COMMANDER'S CUPOLA
MACHINE GUN TRIPOD
FUEL TANK
AIMING POST
WATER CAN
TOW CABLE
TRAILER ELECTRICAL CONNECTION
TELEPHONE CABLE REEL
FUEL TANK

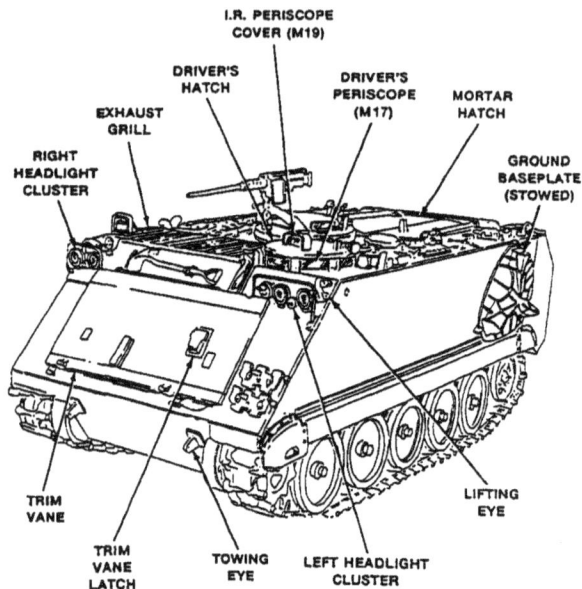

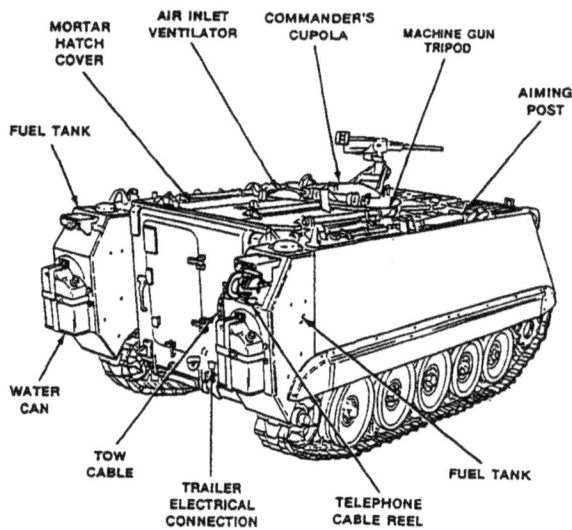

The 120mm self-propelled mortar M1064 is shown in the photograph and drawings on this page.

Later. some of the 107mm mortar carriers were upgraded by modifying the turntable and ammunition racks and installing the 120mm mortar M121. As on the earlier vehicles, a separate base plate was carried to permit firing the mortar from the ground. Designated as the 120mm self-propelled mortar carrier M1064, it was manned by a crew of four including the driver. Like on the 107mm mortar carrier, the 120mm weapon fired toward the rear with a maximum traverse of 90 degrees. The M1064 also was fitted with the external fuel tanks. When the RISE power train was installed in the vehicles, the designation was changed to the 120mm self-propelled mortar carrier M1064A3.

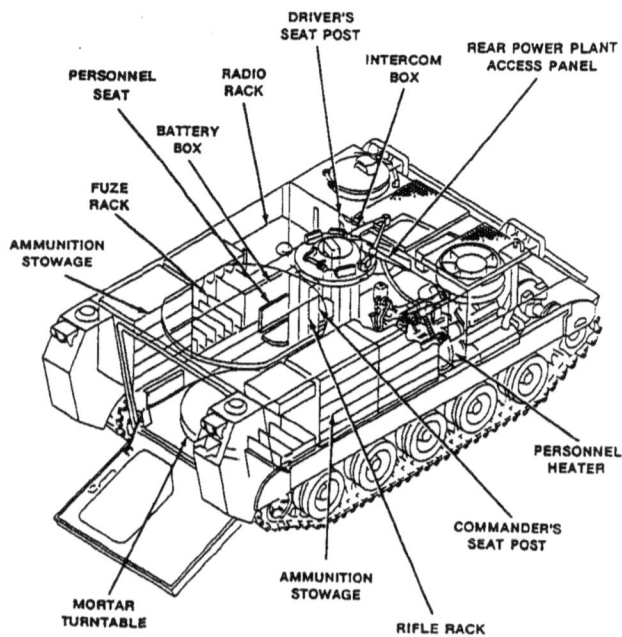

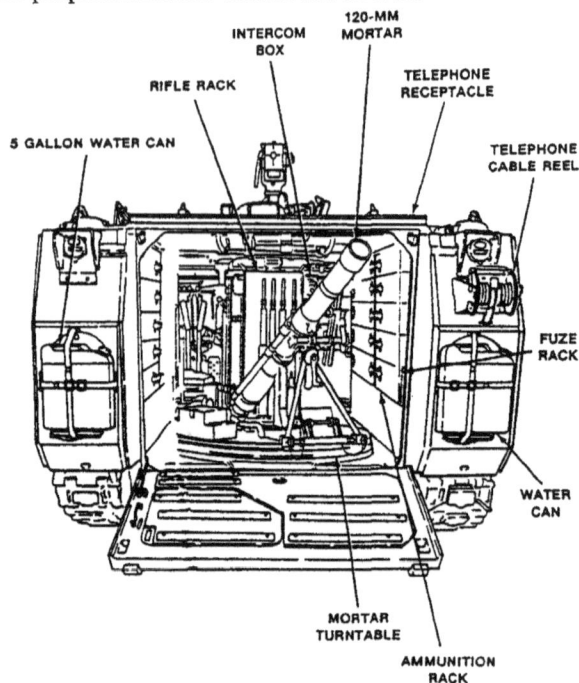

DRIVER'S SEAT POST
INTERCOM BOX
REAR POWER PLANT ACCESS PANEL
PERSONNEL SEAT
RADIO RACK
BATTERY BOX
FUZE RACK
AMMUNITION STOWAGE
PERSONNEL HEATER
MORTAR TURNTABLE
AMMUNITION STOWAGE
COMMANDER'S SEAT POST
RIFLE RACK

INTERCOM BOX
120-MM MORTAR
RIFLE RACK
TELEPHONE RECEPTACLE
TELEPHONE CABLE REEL
5 GALLON WATER CAN
FUZE RACK
WATER CAN
MORTAR TURNTABLE
AMMUNITION RACK

148

A turret armed with the Royal Ordnance 120mm breech loading mortar was installed experimentally on an M113A2 in 1987. This carrier was fitted with flotation cells on the sides and a new high displacement trim vane to compensate for the increased weight. The external fuel tanks also were installed.

These views show the turret mounted Royal Ordnance 120mm mortar installed on the armored personnel carrier M113A2.

149

The army's nuclear weapon, referred to as the "Davy Crockett" heavy weapon system, was stowed aboard an M113 armored personnel carrier.

The M113 and the M113A1 also were adapted as the carrier for the Davy Crockett nuclear weapon. Designated as the Battle Group Atomic Weapon System M29 (XM29) heavy, vehicle mounted, it consisted of the 155mm recoilless gun M64 (XM64E2) with the spotting gun XM77E1 and the recoilless gun tripod mount M121 (XM121). The 155mm recoilless gun was an open breech, single shot, smooth bore, muzzle loaded weapon. The 37mm smooth bore spotting gun was mounted coaxially under the M64 gun. The XM388 Davy Crockett projectile was 30 inches long, 11 inches in diameter, and weighed 76 pounds. The yield could be selected up to 250 tons. Obviously too large to fit in the 155mm tube, the projectile was propelled by a launching piston in the recoilless gun. The maximum range was 4,000 meters.

Note that the "Davy Crockett" was removed from the vehicle and fired from the tripod. The internal stowage of the vehicle can be seen above and at the bottom right.

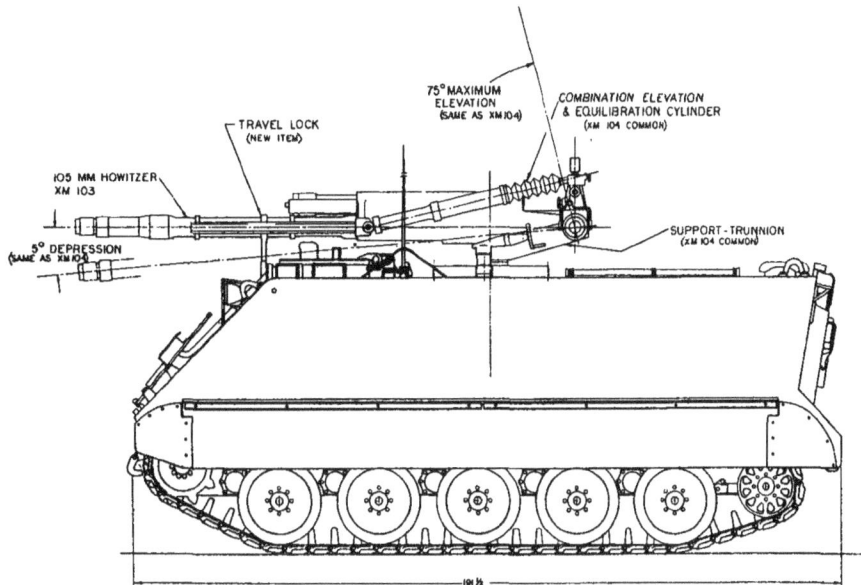

75° MAXIMUM
ELEVATION
(SAME AS XM104)

COMBINATION ELEVATION
& EQUILIBRATION CYLINDER
(XM 104 COMMON)

TRAVEL LOCK
(NEW ITEM)

105 MM HOWITZER
XM 103

SUPPORT-TRUNNION
(XM 104 COMMON)

5° DEPRESSION
(SAME AS XM104)

Dimensions in inches

Above, the 105mm howitzer XM103 is mounted on top of an M113 series armored personnel carrier.

The M113 series was an extremely popular carrier for self-propelled weapons. It provided the basic chassis for a wide variety of fire support vehicles, some of which could have been classified as self-propelled artillery, tank destroyers, or light tanks. One proposal mounted the standard 105mm howitzer M2A1 in the open top of the crew compartment on an M113. A later concept study placed the lightweight XM103 105mm howitzer on top of an M113 using components from the XM104 self-propelled howitzer. Another fire support combat vehicle concept installed a howitzer in the front armor of a modified M113. Firing ports were located in the side walls for the crew and a 7.62mm machine gun was carried on an external mount at the commander's hatch.

The fire support vehicle concept is illustrated by the photograph above and the drawing below.

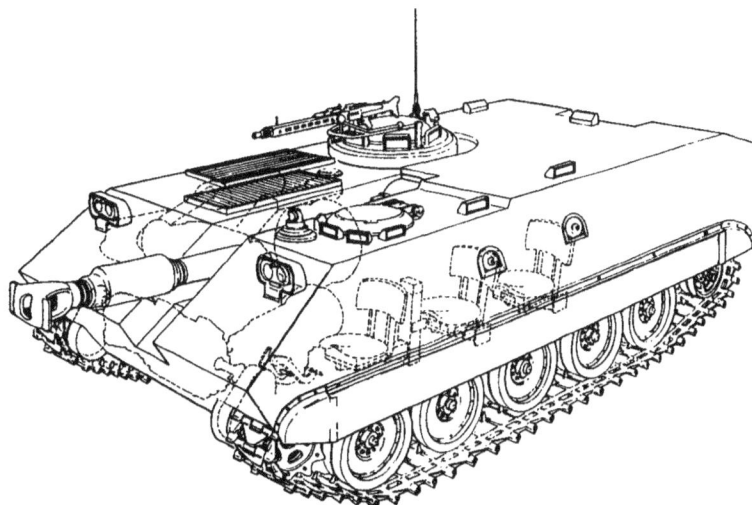

151

Above is the Australian modification of the M113A1 armored personnel carrier to mount the turret from the British Saladin armored car.

A large number of turret mounted weapons were installed on the M113 series both in the United States and abroad. In fact, it seems that anyone with a turret mounted weapon proposed its installation on the M113, no doubt influenced by the large numbers of these vehicles in service worldwide. In Australia, the turret from the Saladin armored car, armed with a 76mm gun, was adapted to the M113A1. Later, a similar installation used the turret from the British Scorpion reconnaissance vehicle which was armed with a later model of the 76mm gun. This conversion also utilized the M113A1 chassis and the vehicle retained its swimming capacity with the addition of side flotation cells and a high displacement trim vane.

The LP90 Cadillac Gage turret from the V-300 armored car was mounted on the roof of the M113A1. Armed with the Cockerill Mark III 90mm gun and a coaxial 7.62mm machine gun, it was described as the fire support vehicle M113 with 90mm cannon. This 90mm gun had a short 12 inch recoil and weighed only about 1,000 pounds. The combination of the turret and mount weighed approximately 4,900 pounds including ammunition and a two man crew. The cannon was fitted with a muzzle brake to reduce the recoil force. The total 90mm ammunition stowage was 42 rounds. In addition to the coaxial machine gun, another 7.62mm weapon was mounted on the turret roof.

An agreement between GIAT in France and FMC resulted in the installation of the GIAT TS90 turret on the M113A2. This turret carried the GIAT CS90 90mm gun. An initial concept drawing showed this vehicle with side flotation cells and a high displacement trim vane. However, both were eliminated from the test vehicle.

Below is the M113A1 armored personnel carrier modified by the installation of the turret from the British Scorpion reconnaissance vehicle. Note the flotation cells on the sides and the high displacement trim vane.

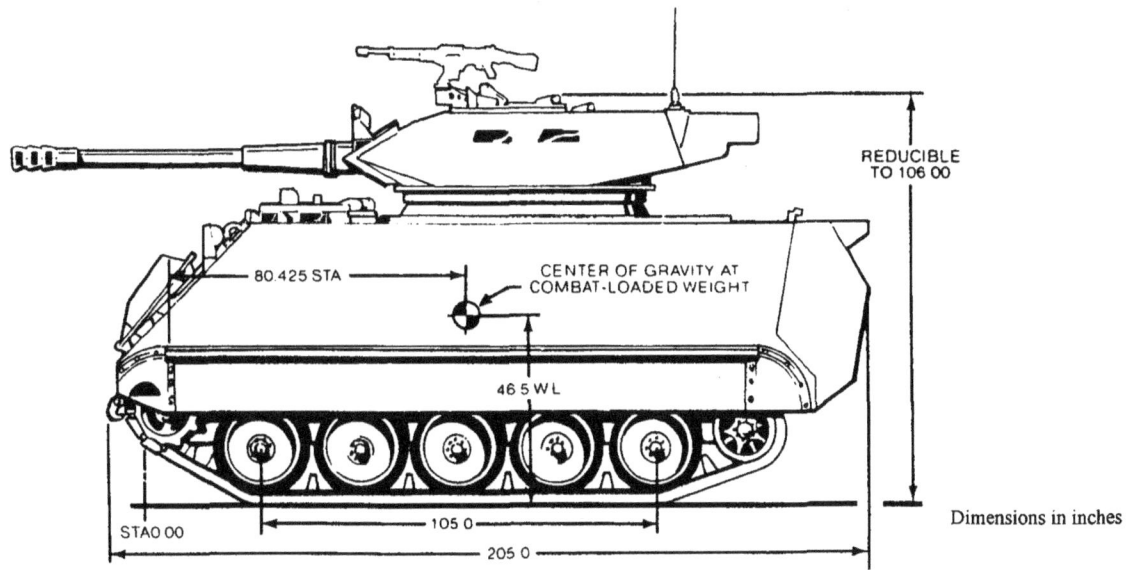

REDUCIBLE TO 106 00

80.425 STA

CENTER OF GRAVITY AT COMBAT-LOADED WEIGHT

46 5 WL

STA0 00

105 0

205 0

Dimensions in inches

The drawing above and the photographs below show the M113 series vehicle armed with a turret mounted 90mm Cockerill cannon. The vehicle is fitted with the external fuel tanks and the high displacement trim vane.

Below, the M113A2 has been fitted with the GIAT TS 90 turret.

The photographs on this page show the installation of the turret mounted Israeli Military Industries 60mm high velocity gun on the M113A1. The drawings at the bottom show the alternate turret arrangements.

Development of the 60mm automatic gun by Israeli Military Industries (IMI), resulted in another variant of the M113A1. Designated as the Hyper-Velocity Medium Support (HVMS) weapon system, it was installed in a turret mount on the M113A1. Firing an armor piercing, fin stabilized, discarding sabot (APFSDS) round at a muzzle velocity of 1,620 meters per second (5,315 feet per second), it could penetrate the front armor of the Soviet T62 tank at a range of 2,000 meters. With a rifled barrel 70 calibers in length, the IMI 60mm gun could fire a three round burst automatically in 1.5 seconds or semiautomatically fire five to six rounds per minute. Two designs for the automatic loader were proposed. The first mounted two magazines holding three rounds each on the gun just back of the breech ring. The second design was more complex with four seven round magazines permitting the use of four types of ammunition. With both designs, the commander was located on the left side of the cannon and the gunner was on the right.

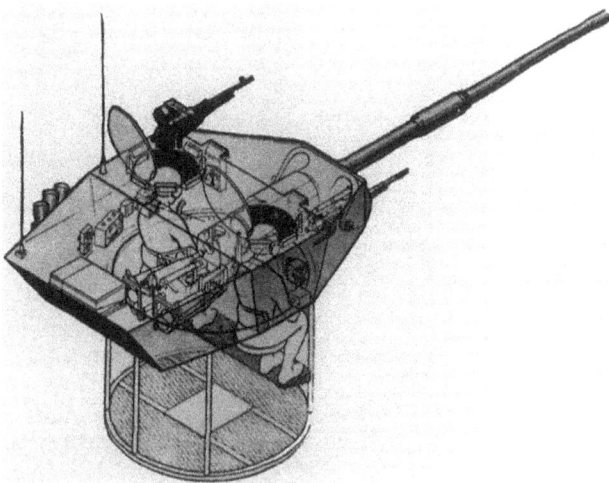

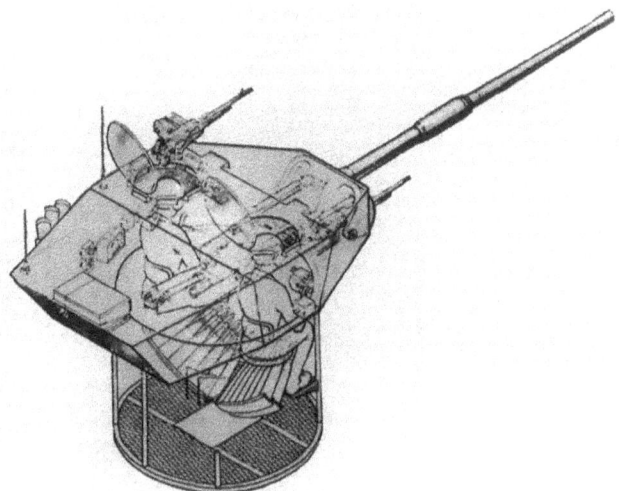

154

Above is a turret mounted 20mm cannon on an M113 series armored personnel carrier. The drawing at the right shows the installation of the 20mm Hispano Suiza cannon in the M113.

Numerous efforts were made to increase the firepower of the M113 family by replacing the commander's .50 caliber machine gun with a heavier automatic weapon. Some of these were on open mounts and others were installed in turrets. The Hispano-Suiza HS-820 20mm gun was evaluated by the United States Army in a variety of mounts on the M113 for use against both ground and air targets. Other 20mm weapons were proposed for installation on the M113 family both in the United States and abroad. The 40mm automatic grenade launcher was fitted experimentally in an open top turret. A later turret was armed with the grenade launcher and a .50 caliber machine gun in a coaxial mount.

Below are photographs of the 20mm Hispano Suiza cannon on the type 502 ring mount installed in the M113.

A wide variety of weapons were proposed for the M113. Above the turret mounted type 763 Hispano Suiza 20mm gun is installed. Below the General Electric version of the Oerlikon 20mm gun has been mounted on this photograph dated 21 August 1975.

In these four photographs, the 40mm automatic grenade launcher XM182 is turret mounted on an M113. Note the counterweight on the rear of the turret for balance. These photographs were dated 26 September 1968.

The armored personnel carrier M113A3, above and below, is armed with a turret mounted 40mm automatic grenade launcher and a .50 caliber machine gun. Two 4 tube smoke grenade launchers are installed on the turret.

Above, the mock-up at the left shows the British Fox turret armed with the Rarden gun installed on the M113 and at the right, the M113 is modified to accept the Rh 202 turret.

A one man electric drive turret was installed by FMC. This turret was armed with an M242 25mm gun and a coaxial M240 7.62mm machine gun. The M242 was an externally powered weapon known as the Bushmaster or Chain Gun (A trademark of the McDonnell Douglas Helicopter Company). It fired single shots or automatically at rates of 100, 200, or 500 shots per minute. Combat loaded, this turret weighed 3,100 pounds without the gunner. A total of 165 ready rounds of 25mm ammunition were provided.

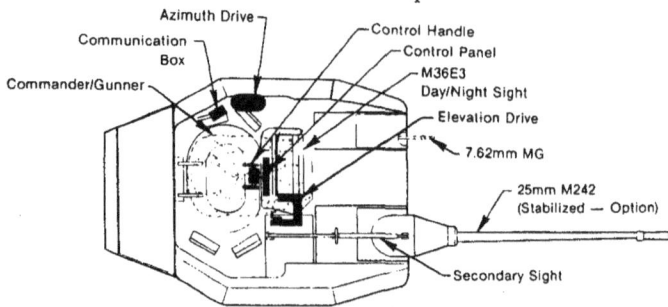

The drawing at the left and the three photographs above and below show the one man electric drive turret armed with the 25mm M242 cannon installed on the M113 series vehicle.

The M113A2 in these photographs is fitted with the two man turret armed with the 25mm M242 cannon and a 7.62 coaxial machine gun. Note the similarity to the Bradley turret minus the TOW missile launcher.

A two man turret armed with the same 25mm gun and 7.62mm coaxial machine gun also was under test. This turret was essentially the same as on the Bradley fighting vehicle without the launcher for the TOW missile, although the latter was offered as an option. The turret was assembled from 5083 and 7039 aluminum alloy plate with steel applique armor. Mounted on the M113A2 armored personnel carrier, the vehicle weight, without the crew, was 26,840 pounds. The weapon was provided with 300 25mm ready rounds and an additional 600 rounds were stowed.

The dimensions of the M113A2 with the two man turret can be seen in the drawing below.

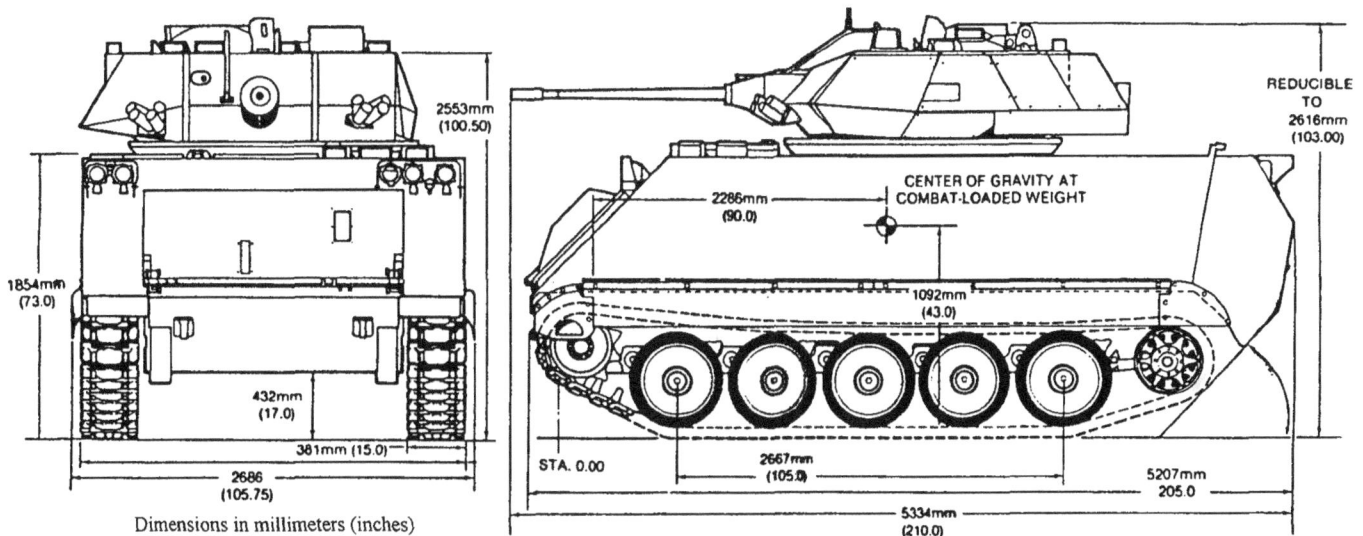

Dimensions in millimeters (inches)

The M113 appears here with the TAT 252 turret (above), the LAV 25 turret (top right), and the ASP-30 on an open mount (lower right).

The Emerson Electric Company developed a mount for the M242 cannon utilizing the M27 mount previously used with the 20mm gun on the M114A1E1 tracked carrier. This mount, referred to as the TAT 251, could be installed on the M113 series of armored personnel carriers. A later turret installation of the Bushmaster by Emerson Electric was referred to as the TAT 252. It also could be mounted on the M113 series.

The ASP-30 automatic, self-powered, cannon produced by the McDonnell Douglas Helicopter Company was installed for test as a replacement for the commander's .50 caliber machine gun on the M113 series vehicles. This 30mm weapon weighed 115 pounds and had a muzzle velocity of 2,700 feet per second. It could fire single shots or automatically at a rate of 400 to 450 rounds per minute.

The CVAST (Combat Vehicle Armament Systems Technolgy) turret developed for the Bradley fighting vehicle was installed experimentally on the M113. This was a cleft turret mounting a 35mm Ares automatic gun and it is described in the section on the Bradley.

Here the M113 is fitted with the CVAST turret (below) and the ASP-30 in a protected mount (upper and lower right).

Above are two views of the command post carrier XM577. Note the raised top on the rear of the vehicle to provide headroom.

COMMAND POST, CONTROL, AND COMMUNICATION VEHICLES

In September 1959, CONARC transmitted the military characteristics of four command post vehicles to the Chief of Research and Development. They consisted of a command post trailer, a light tracked command post carrier, a light wheeled command post carrier, and a medium wheeled command post carrier. The Chief of Research and Development approved only the light tracked command post carrier rejecting the wheeled carriers and the trailer. Subsequently, CONARC reaffirmed the requirement for the command post trailer and the medium wheeled command post carrier, but they are outside the scope of this history. OTCM 37465, dated 7 July 1960, recorded the approved characteristics of a light, tracked, command post carrier. Since such a vehicle was urgently required by the U. S. Army in Europe, CONARC recommended its immediate type classification without service tests and requested that the four prototype vehicles under construction be delivered at the earliest possible date. On 14 December 1961, OTCM 37932 type classified the light tracked command post carrier XM577 for limited production. It was intended to replace the M59 armored personnel carriers modified in the field as command post vehicles.

The XM577 command post carrier can be seen at the right. Note the auxiliary generator on the top front.

161

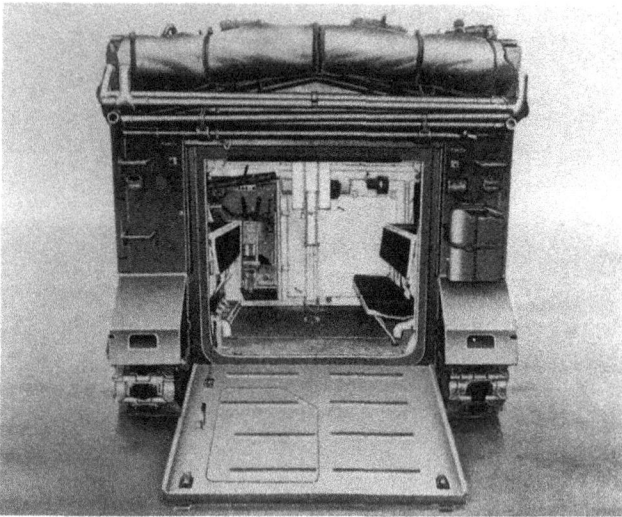

The interior of the command post carrier M577 can be seen through the open ramp in the photographs above. Note the tent stowed on the top rear.

The four pilot vehicles, modified from M113s, were completed at Detroit Arsenal on 1 February 1962. The first was shipped to Aberdeen Proving Ground in March for engineering tests. The remaining three went to the Armor, Infantry, and Artillery Boards in February for a two month service evaluation. Acceptance of the first production lot of 270 XM577s began in December 1962 and was complete in May 1963. In March 1963, AMCTCM Item 640 classified the M577 command post carrier as Standard A. Acceptance of a second production run of 674 M577s began in November 1963 and was completed in mid 1964.

Compared to the M113, the most obvious difference of the M577 was the high silhouette necessary to provide adequate headroom in the personnel compartment. The front of the vehicle extending to the rear of the driving and engine compartments was identical to the M113. In the personnel compartment, the height of the hull was increased by 25¼ inches providing headroom inside the vehicle of 74¾ inches. The single 80 gallon fuel tank of the M113 was replaced by two 60 gallon tanks installed with one on each side of the personnel compartment. Each tank served as the support for a 90 inch folding table. The commander's seat and cupola with the .50 caliber machine gun on the M113 were eliminated. They were replaced by a folding platform and a circular hatch at the commander's station without any vision devices. The vehicle was equipped with a crew compartment heater, rifle racks, eight interior lights, and two blue blackout lights. A five man troop seat could be attached to either side of the personnel compartment and provision was made on the

At the right, two M577s have their tents connected to form a large command post.

right side for a large map board. Radios were installed on a shelf along the left wall and on the right front wall. To provide adequate electric power, a 28 volt, 150 ampere, auxiliary generator was carried on the outside front wall of the personnel compartment. It could be operated in that location or dismounted for use on the ground. A lifting davit and a 50 foot generator cable were provided for dismounted operation.

To extend the working area, a tent could be attached to the rear of the vehicle providing an additional 120 square feet of floor space. Fitted with a blackout entrance, it could be used to connect two or more vehicles for command post operation. When not in use, the tent was stowed on the top rear of the vehicle.

Manned by a crew of five including the driver, the M577 was initially issued on a basis of seven vehicles to tank or mechanized infantry battalions, six per brigade headquarters, and two per division headquarters. At the battalion level, they replaced the M113s used by the S-3, S-3 air, S-2, S-1/4, communications officer, mortar platoon fire direction center, and the medical aid station.

With the introduction of the diesel engine in the M113A1 armored personnel carrier, similar changes were made in the command post carrier. Powered by the 6V53 diesel engine, it retained the original 120 gallon fuel capacity resulting in a cruising range increase from approximately 270 miles to about 370 miles. The new vehicle retained the gasoline driven auxiliary generator.

STATION 4.88
(0,12)

97.25
(2,47)

81.38 (2,07)

43.38
(1,10)

23
(0,58)

70°

40°

14 (0,36)

25.75
(0,65)

105 (2,67)

REDUCIBLE TO 190 (4,83)

191.50 (4,87)

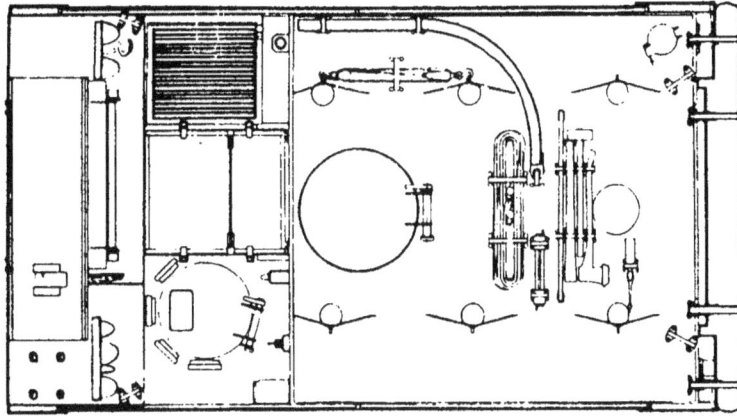

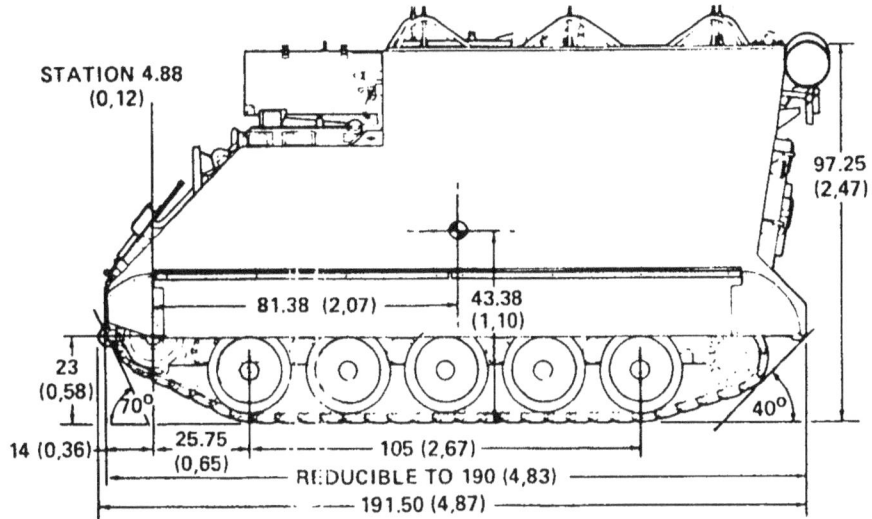

Command Post Carrier M577A1

Dimensions in inches (meters)

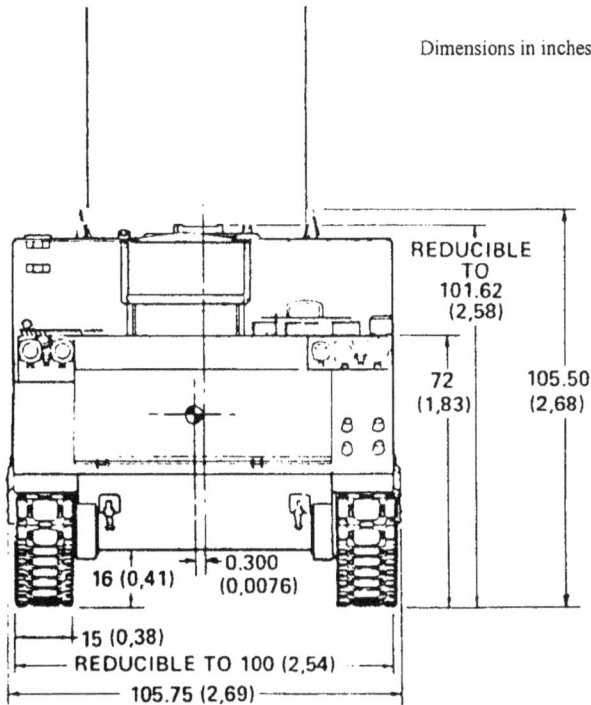

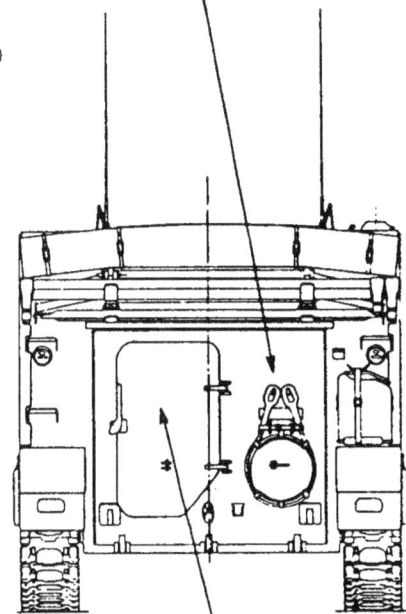

REAR OPENING WITH
RAMP DOWN:
54.5 (1,38) WIDE BY
50 (1,27) HIGH

REDUCIBLE
TO
101.62
(2,58)

72
(1,83)

105.50
(2,68)

16 (0,41)

0.300
(0,0076)

15 (0,38)

REDUCIBLE TO 100 (2,54)

105.75 (2,69)

PERSONNEL DOOR OPENING
43.75 (1,11) BY 27.75 (0,70)

The dimensions of the interior space in the command post carrier M577 can be seen in the drawing at the right.

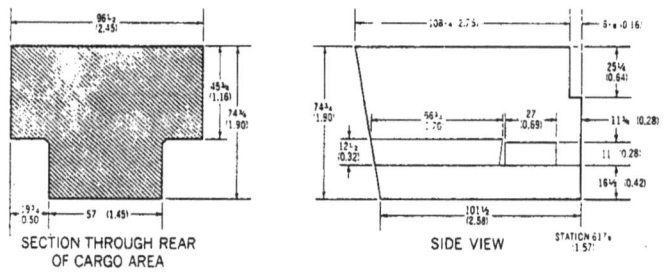

SECTION THROUGH REAR OF CARGO AREA

SIDE VIEW

FRONT OF CARGO AREA

On 16 December 1963, AMCTCM Item 1807 type classified the diesel powered vehicle as the M577A1 command post carrier, Standard A and reclassified the M577 as Standard B. Production of the M577A1 began in September 1964 with an initial run of 1,225 vehicles.

The XM15 collective protection equipment was intended for use with the M577 and M577A1. It provided protection for the personnel inside the command post carriers without the necessity of wearing individual masks or protective clothing. Operating off of the engine generator, the XM15 consisted of a gas particulate filter unit, a pressure sensing control network, and a protective entrance. The latter was an airlock that allowed contamination free entrance and exit from the vehicle.

When the improved engine cooling system and the new suspension were installed on the M113A2 armored personnel carrier, these changes also were applied to the command post carrier and it was designated as the M577A2. Production of new M577A2s began in July 1979 and the modifications were introduced during depot overhaul of the earlier vehicles in August 1979. Like the other models of the command post carrier, the M577A2 was air transportable, but it was not air droppable. Like the M113A2 and later vehicles, swimming the M577A2 was prohibited during peace-time operations.

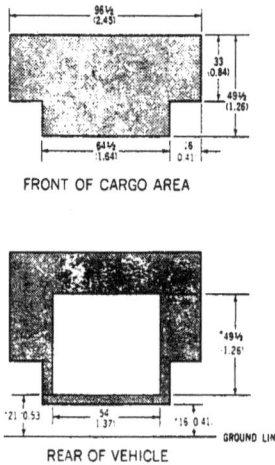

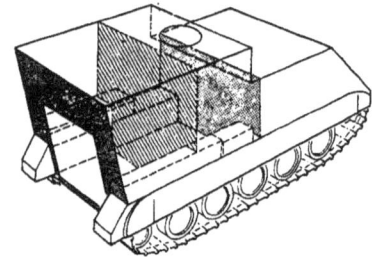

REAR OF VEHICLE

(The vehicle is equipped with a tent extension at the rear, allowing the working area of the Command Post to be more than doubled.)

Dimensions in parentheses ( ) are in meters; all other dimensions are in inches.
*Vertical Dimension.

Incorporation of the RISE power pack and new driver's controls resulted in another change of designation. The vehicle now became the M577A3 command post carrier. In addition, the gasoline powered 4.2 kW auxiliary generator could be replaced by a 5 kW diesel driven auxiliary generator. Other improvements could include armor enhancements, the BCIS, contact spall liners, the driver's night viewer, and other modifications proposed for the armored personnel carrier. The M577A3 conversion program began in 1994.

Below, the command post carrier M577 at the left is fitted with the NBC collective protector. At the bottom right is the command post carrier M577A2.

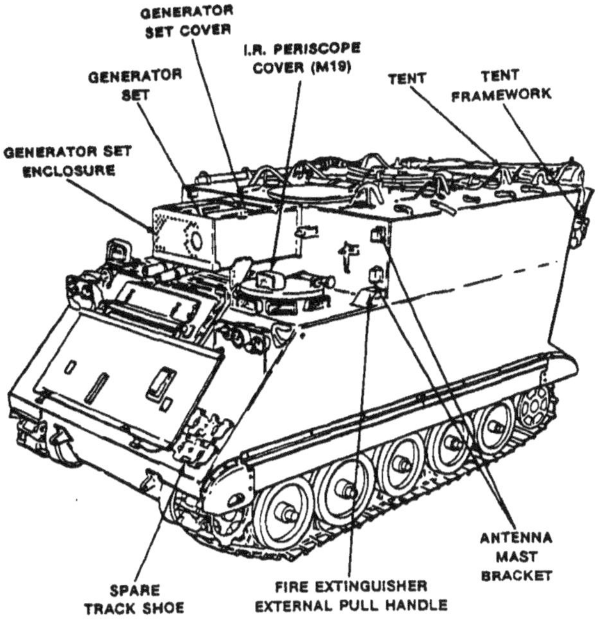
GENERATOR SET COVER
I.R. PERISCOPE COVER (M19)
GENERATOR SET
TENT
TENT FRAMEWORK
GENERATOR SET ENCLOSURE
ANTENNA MAST BRACKET
SPARE TRACK SHOE
FIRE EXTINGUISHER EXTERNAL PULL HANDLE

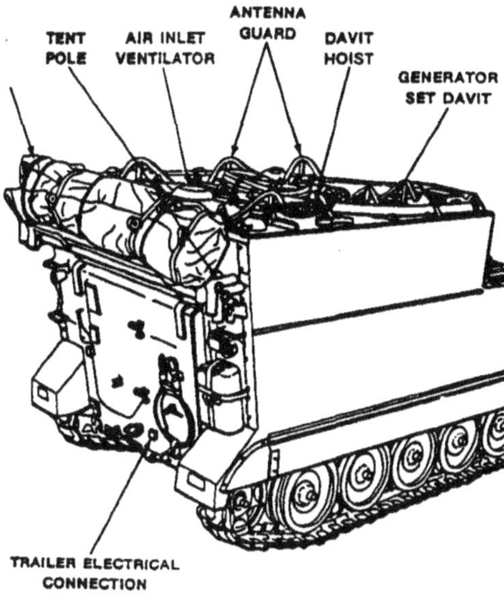
ANTENNA GUARD
TENT POLE
AIR INLET VENTILATOR
DAVIT HOIST
GENERATOR SET DAVIT
TRAILER ELECTRICAL CONNECTION

Details of the command post carrier M577A2 can be seen in the drawings above and below.

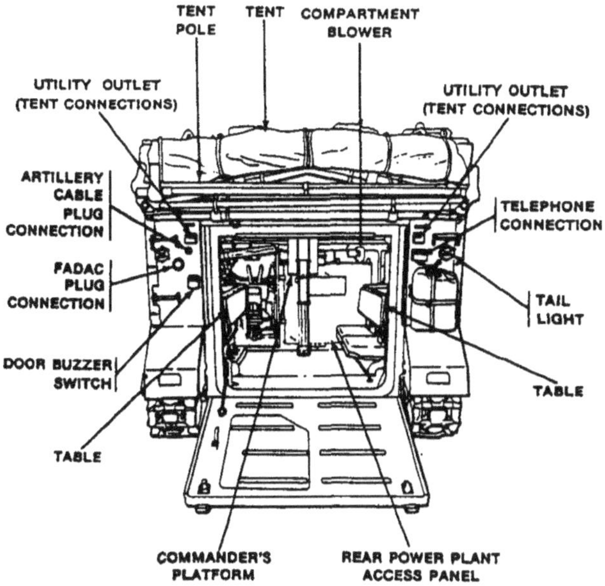
TENT POLE
TENT
COMPARTMENT BLOWER
UTILITY OUTLET (TENT CONNECTIONS)
UTILITY OUTLET (TENT CONNECTIONS)
ARTILLERY CABLE PLUG CONNECTION
TELEPHONE CONNECTION
FADAC PLUG CONNECTION
TAIL LIGHT
DOOR BUZZER SWITCH
TABLE
TABLE
COMMANDER'S PLATFORM
REAR POWER PLANT ACCESS PANEL

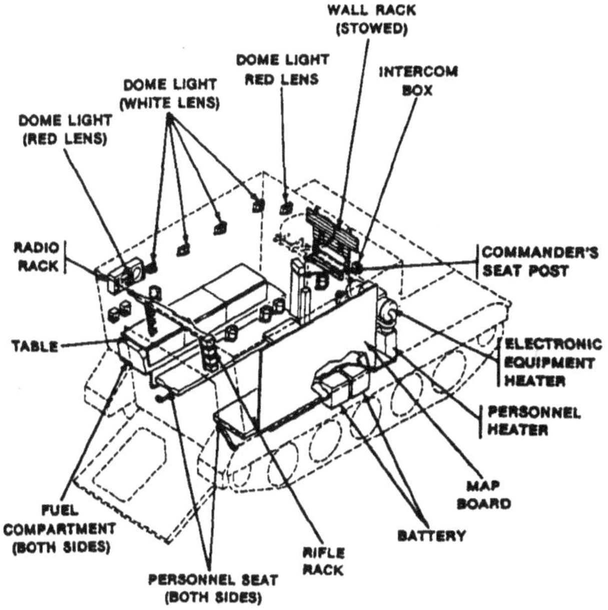
WALL RACK (STOWED)
DOME LIGHT RED LENS
DOME LIGHT (WHITE LENS)
INTERCOM BOX
DOME LIGHT (RED LENS)
RADIO RACK
COMMANDER'S SEAT POST
TABLE
ELECTRONIC EQUIPMENT HEATER
PERSONNEL HEATER
MAP BOARD
FUEL COMPARTMENT (BOTH SIDES)
BATTERY
RIFLE RACK
PERSONNEL SEAT (BOTH SIDES)

Below, the water barrier curtain is in the stowed position at the left and details of the map board and table are shown at the right.

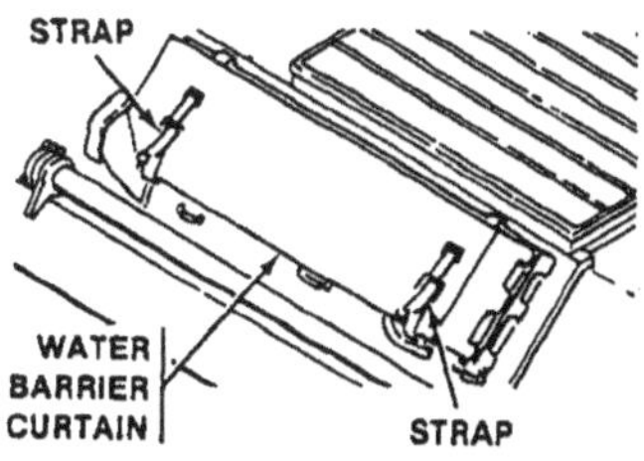
STRAP
WATER BARRIER CURTAIN
STRAP

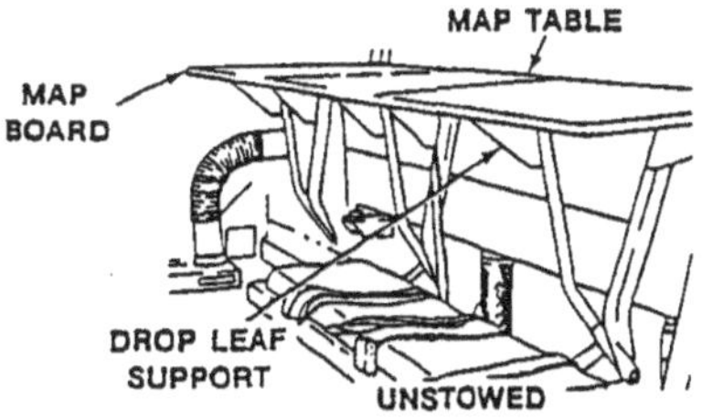
MAP TABLE
MAP BOARD
DROP LEAF SUPPORT
UNSTOWED

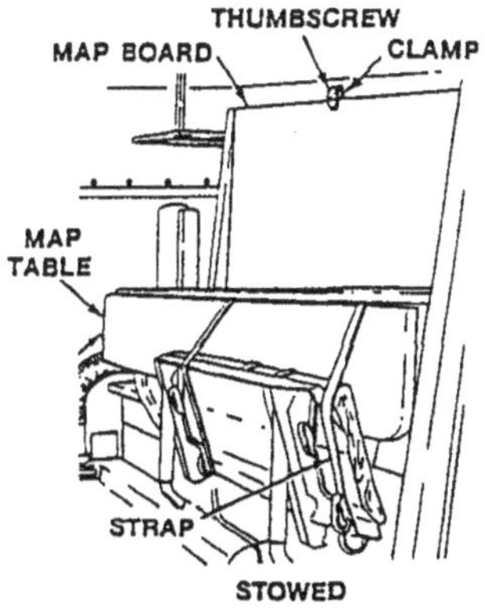
THUMBSCREW
MAP BOARD
CLAMP
MAP TABLE
STRAP
STOWED

The command post carrier M577A3 is shown in these two photographs.

The M1068 standard integrated command post system (SICPS) was a modified version of the M577A2. Intended to accommodate the army tactical command and control system (ATCCS), it was manned by a crew of four consisting of a commander, driver, and two command post operators. In addition to the ATCCS equipment, the M1068 carried the 5 kW diesel auxiliary generator, a power/data distribution system, and a ten meter antenna mast. A new extension tent with lighting was stowed on top of the vehicle. It was intended that roughly two thirds of the M577A2 fleet would be converted to the M1068 configuration. When the M1068 was fitted with the RISE power pack, it was designated as the M1068A3 standard integrated command post system.

The interior of the M1068 standard integrated command post system can be seen above. The drawings below show the details of the M1068.

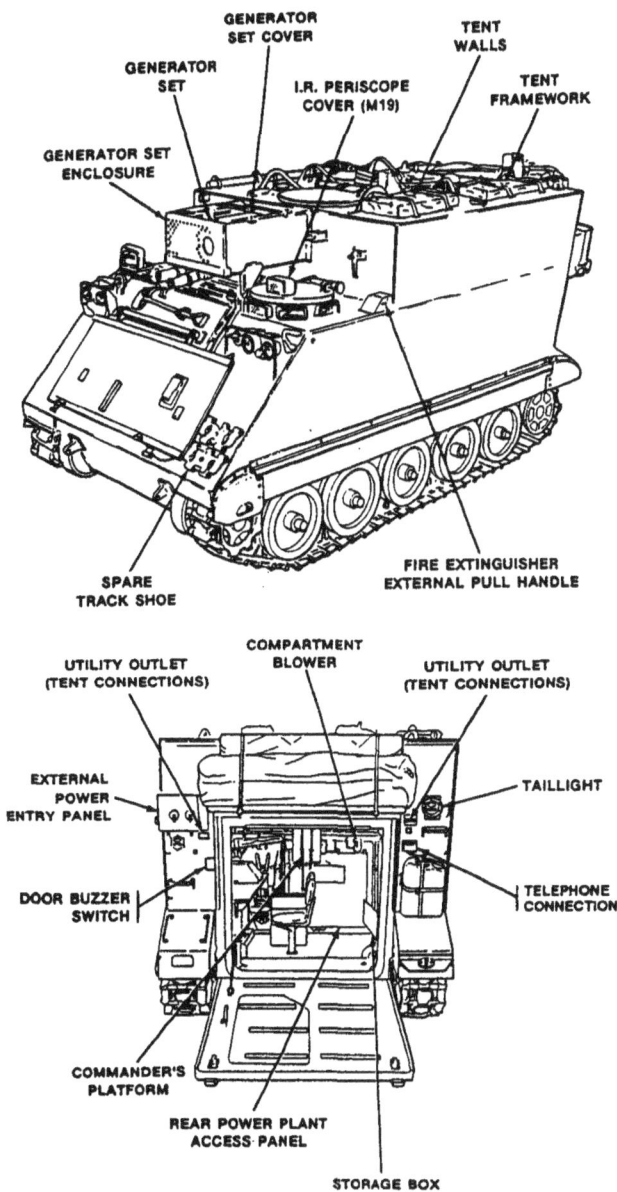

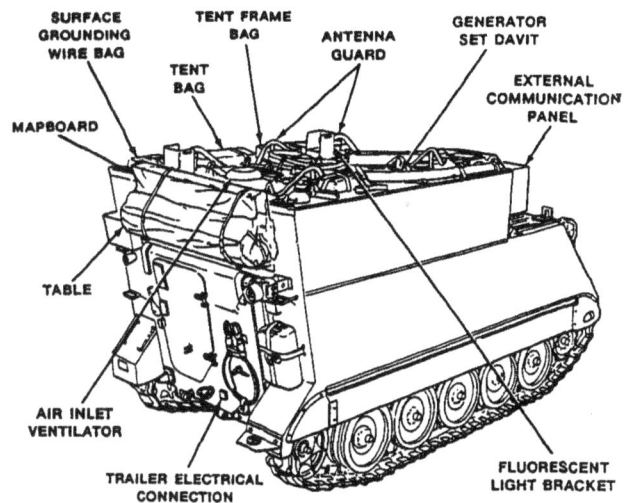

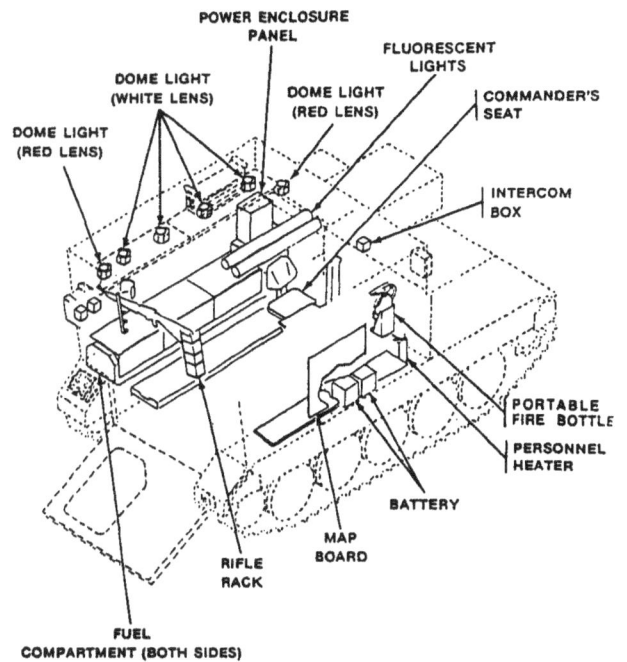

The XM577A4 stretch command and control vehicle appears in the two photographs above. The interior of the vehicle is shown below.

A stretched version of the M577A3 was developed by United Defense L. P. (formerly FMC) as an independent research and development program. Referred to as the XM577A4 stretch armored tactical command and control system, it had six road wheels per side and a longer hull for increased interior space and payload capacity. It also could be fitted with bolt-on armor as well as spall liners for additional side and roof protection. The XM577A4 stretch had an interior volume of 502 cubic feet and a payload capacity of over 5,000 pounds. It also was considered as a platform for a hazardous materials recovery vehicle and an armored medical treatment vehicle.

The photograph at the left shows the hazardous material response vehicle based upon the XM577A4. In the drawing at the left bottom, the vehicle is equipped with a dozer blade and a recovery arm. Below is a sketch of the proposed armored medical treatment vehicle.

Dimensions in inches (centimeters)

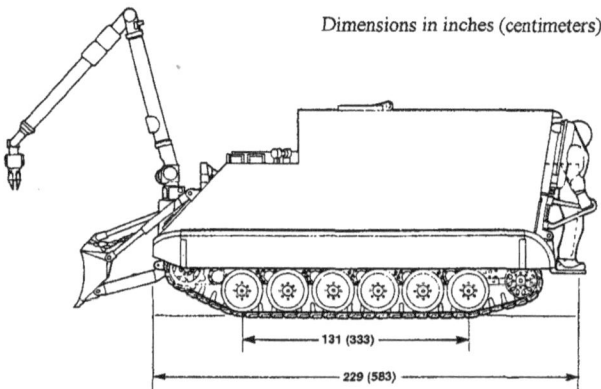

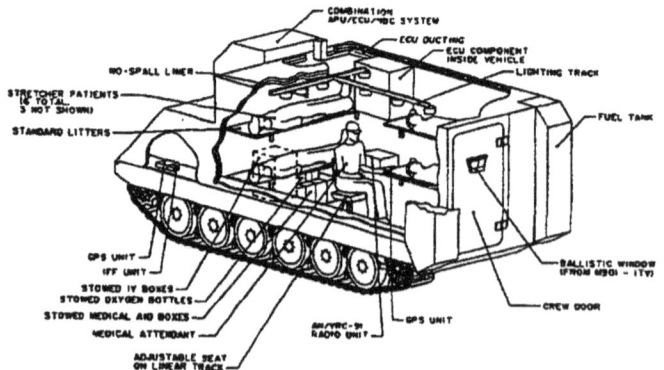

The views on this page show the proposed tactical operations center vehicle based upon the late M113 series. In the bottom photographs, the vehicle is open and the map board extended for use.

A modified version of the late model M113 series was proposed as a tactical operations center vehicle. With the normal silhouette of an M113, it could not be easily distinguished from the standard armored personnel carrier on the battlefield. Manned by a crew of six, it provided stowage space for radios and map boards. The latter could be used standing in the open cargo hatch. The vehicle was fitted with external stowage racks on each side and was equipped with the external fuel tanks.

Above, a FIST-V pilot appears at the left and an early FIST-V at Emerson Electric Company is at the right. The major FIST-V assemblies can be seen below.

The 1975 recommendations of the Close Support Study Group resulted in the development of the fire support team vehicle (FIST-V) to provide field artillery support to armor/cavalry and mechanized infantry units. Developed by the Emerson Electric Company, it utilized many components from M901 improved TOW vehicle then in production. The chassis was that of the M113A2 armored personnel carrier equipped with the external fuel tanks. The hammer head TOW launcher was used to mount the ground/vehicle laser locator designator (G/VLLD) as well as the AN/TAS-4 night sight and the north seeking gyrocompass. Originally designated as the XM981, it was standardized as the M981 fire support team vehicle. In appearance, it was almost identical to the M901 improved TOW vehicle. The early pilot used a different design for the hammerhead, but the production vehicles differed only slightly from the M901. When the RISE power pack was installed in the vehicle, it was designated as the M981A3.

The FIST-V appears below with the hammer head in the stowed position (left) and erected (right).

Fire Support Team Vehicle M981

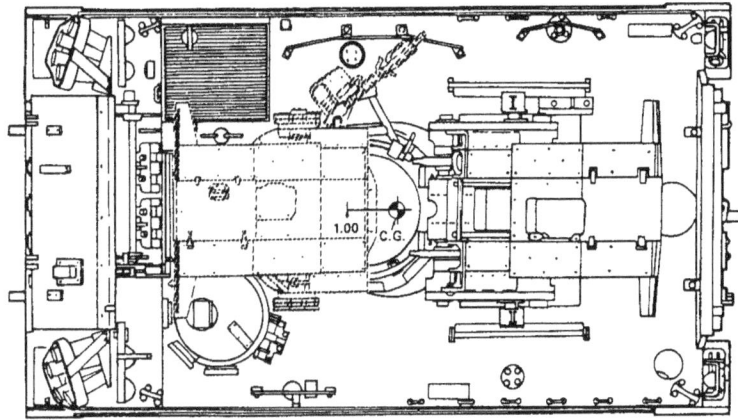

Dimensions in inches

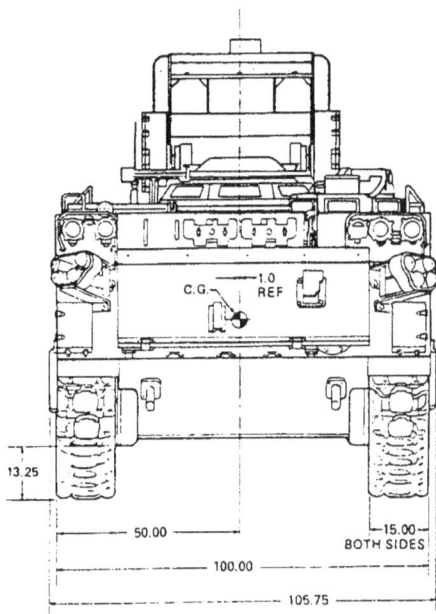

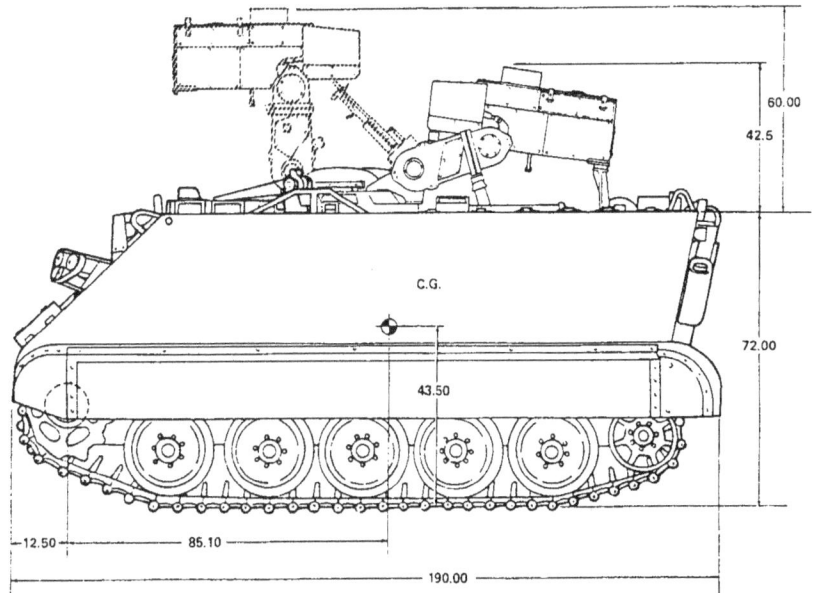

Below, the FIST-V at the left has the late M901 front plate on the hammer head. The later design can be seen below at the right.

Above and at the top right are vehicles with the two versions of the hammer head front plate. At the right is an interior view of the FIST-V. The various components of the FIST-V are identified in the drawings below.

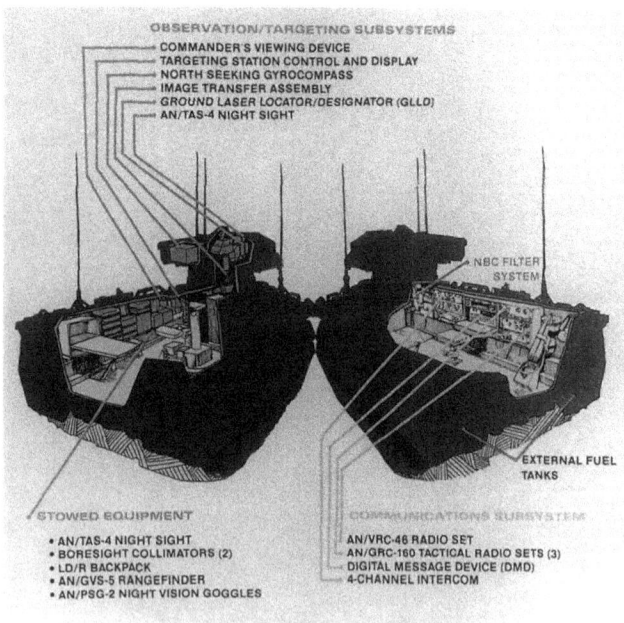

OBSERVATION/TARGETING SUBSYSTEMS
COMMANDER'S VIEWING DEVICE
TARGETING STATION CONTROL AND DISPLAY
NORTH SEEKING GYROCOMPASS
IMAGE TRANSFER ASSEMBLY
GROUND LASER LOCATOR/DESIGNATOR (GLLD)
AN/TAS-4 NIGHT SIGHT

NBC FILTER SYSTEM

EXTERNAL FUEL TANKS

STOWED EQUIPMENT
• AN/TAS-4 NIGHT SIGHT
• BORESIGHT COLLIMATORS (2)
• LD/R BACKPACK
• AN/GVS-5 RANGEFINDER
• AN/PSG-2 NIGHT VISION GOGGLES

COMMUNICATIONS SUBSYSTEM
AN/VRC-46 RADIO SET
AN/GRC-160 TACTICAL RADIO SETS (3)
DIGITAL MESSAGE DEVICE (DMD)
4-CHANNEL INTERCOM

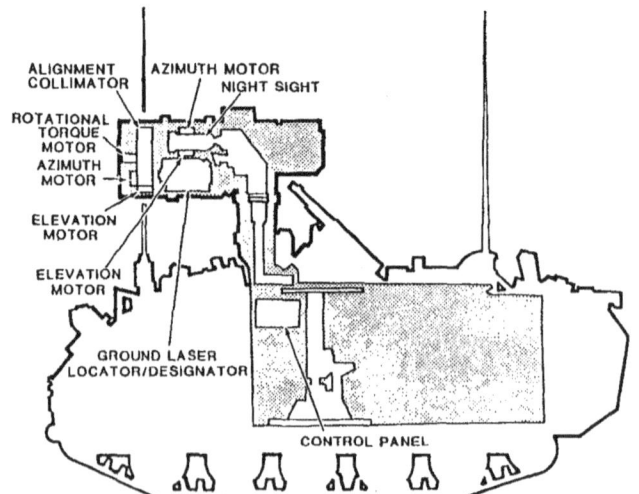

ALIGNMENT COLLIMATOR
AZIMUTH MOTOR
NIGHT SIGHT
ROTATIONAL TORQUE MOTOR
AZIMUTH MOTOR
ELEVATION MOTOR
ELEVATION MOTOR
GROUND LASER LOCATOR/DESIGNATOR
CONTROL PANEL

The AN/MPS-30 radar appears at the left on the M106 chassis. Above is the elevated target acquisition system installed on the M113A1 armored personnel carrier.

In June 1961, an XM106 mortar carrier was used to carry an AN/TPS-25 radar set. Designated as the AN/MPS-30 mobile surveillance radar, it was intended to provide armor protected, mobile, electronic ground surveillance for the field forces. The roof door on the XM106 allowed the radome and antenna to be easily raised and retracted.

The elevated target acquisition system (ETAS) with a 50 foot quick erecting mast was evaluated on the M113A1 armored personnel carrier.

The M1015 electronic warfare shelter carrier was developed under a Signal Warfare Laboratory contract. It utilized the chassis of the M548 cargo carrier modified with 16 components. The major two of these were a ground rod driver and a 60 kW, 400 Hz, generator driven off of the vehicle transfer case. Originally intended as a carrier for the TACJAM electronic warfare system, the program was expanded to include the TEAMPACK and TRAILBLAZER systems.

When the cooling and suspension improvements from the M113A2 armored personnel carrier were applied, the vehicle was designated as the M1015A1.

Approximately 250 vehicles were converted under the M1015 program. After the TACJAM and TEAM-PACK systems were declared obsolete and the TRAIL-BLAZER system shifted to a wheeled carrier, the M1015 and M1015A1 were no longer in use.

At the right, the TEAMPACK is mounted on the M1015 electronic warfare shelter carrier.

Above is the mock-up of the T115 wheeled carrier armed with the 106mm recoilless rifle to meet the requirement for the battalion antitank weapon system.

ANTITANK VEHICLES

As mentioned previously, when the lightweight carriers were originally proposed, both the T114 (tracked) and the T115 (wheeled) vehicles were to be adapted as carriers for the 106mm M40A1 recoilless rifles to meet the requirement for the infantry battalion antitank (BAT) weapon system. In June 1957, mock-ups of the T114 and T115 were demonstrated. However, they were considered to be unsatisfactory because the external mount for the recoilless rifle exposed the gunner during loading and firing. At this point, the wheeled T115 vehicle was dropped, but further development continued on the T114. Six pilots of the T114 were

ordered with the first four to be completed as command and reconnaissance vehicles and the last two to be armed with the 106mm recoilless rifle equipped with a semiautomatic loader. Both types had the same chassis and the large cupola with the .50 caliber machine gun on the command and reconnaissance vehicle was replaced by a one man turret mounting the 106mm recoilless rifle for the BAT weapon system. Pilot number 5 was shipped to Aberdeen Proving Ground for firing tests with the 106mm rifle and the semiautomatic loader. These tests began on 2 December 1960 and continued until 13 January 1961.

Below is the T114, pilot number 5, with the 106mm recoilless rifle and semiautomatic loader. These photographs were taken at Aberdeen Proving Ground during the firing tests.

Two types of ammunition were provided for the 106mm rifle. These were the high explosive plastic with tracer (HEP-T) M346 and the high explosive antitank (HEAT) M344. The first of these was a squash head projectile with the plastic explosive enclosed inside a thin metal case. It was fitted with a pre-engraved rotating band to engage the rifling in the cannon. Two indexing buttons on the shell were used to align the rotating band with the rifling grooves. The HEAT M344 round was a fin stabilized, shaped charge, projectile that did not have a rotating band. It was fitted with rear bourrelet to support the projectile in the tube.

On the two man BAT vehicle, the modified M40A1 106mm recoilless rifle was mounted in an armored pod on the right side of the one man turret along with a .50 caliber M8 spotting rifle. The semiautomatic loader with a three round magazine was installed in the pod outboard of the rifle and there was access to the weapon and the loader from inside the turret. The ammunition in the loader's three round magazine could be replenished from inside the turret and the rifle could be loaded manually. In fact, it was necessary to load the HEP round manually in order to engage the rifling grooves with the pre-engraved rotating band. A .50 caliber T175E1 machine gun was specified for installation on an external mount at the commander's station. Two M14 rifles were stowed internally. Ammunition stowage included 26 rounds 106mm, 150 rounds .50 caliber spotting rifle, and 480 rounds .50 caliber machine gun.

These are additional views of T114 pilot number 5, registration number 12M653, at Aberdeen Proving ground. The openings in the breech of the recoilless rifle for the back blast are clearly visible.

Below are the two types of ammunition available for the modified M40A1 recoilless rifle. The HEP-T M346 is at the left and the HEAT M344 is at the right.

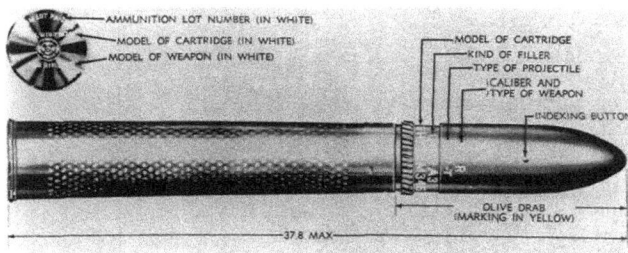

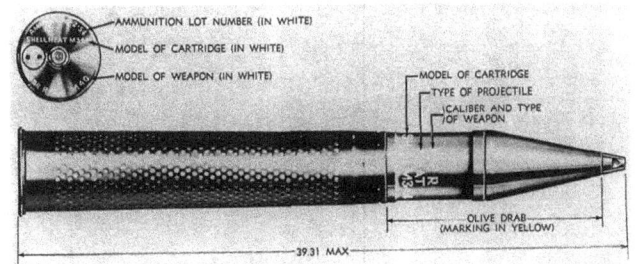

Although there were some problems with the semiautomatic loader, the tests at Aberdeen concluded that there was no degradation of accuracy when firing the turret mounted recoilless rifle. However, in March 1961, the Materiel Requirement Review Committee decided that there was no longer a requirement that justified the development of an antitank version of the T114 and that all time and assets should be allocated to the development of the command and reconnaissance vehicle. The two antitank pilots were then converted to the revised design of the command and reconnaissance vehicle for user evaluation.

After introduction of the M113 armored personnel carrier, there were several proposals for the installation of recoilless rifles as armament. In Vietnam, the 106mm recoilless rifle was mounted on the M113 as a field modification. However, all of these installations were to provide support firepower and were not directly intended for antitank use.

In November 1951, the Army established a requirement for an antitank guided missile. The development program for the Dart missile was in response to this requirement. The Dart was a wire guided missile with a seven inch diameter, 20 pound, shaped charge warhead. It flew at about 350 feet per second with a maximum range of approximately 6,000 yards. However, the effective range was probably about 2,000 yards. As mentioned before, it was proposed for installation on the early T113 armored personnel carrier. However, the Dart was canceled in 1958 because of increasing weight and complexity.

The operating sequence of the semiautomatic loader is illustrated at the right. Below is an artist's concept (left) and an actual installation of the 106mm recoilless rifle on the M113 armored personnel carrier.

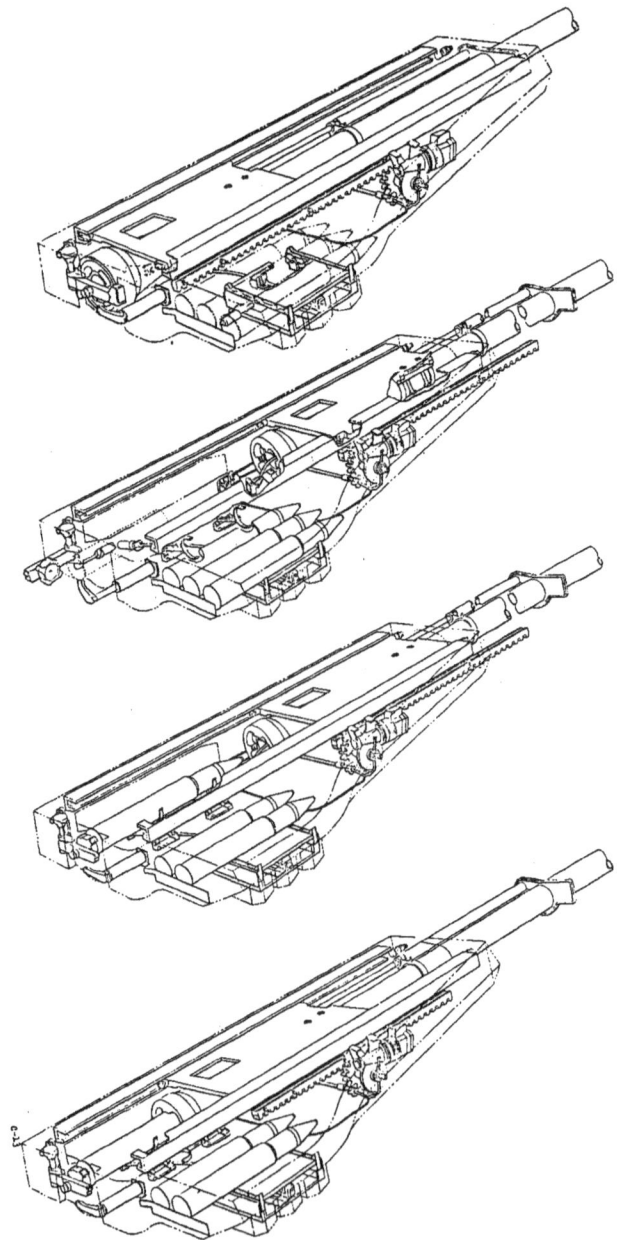

Above, the Dart antitank missile is on an early launcher at the White Sands Missile range on 28 October 1955. At the right, the SS-11 (upper) and SS-10 (lower) missiles can be seen.

The French SS-10 wire guided missile was built under license in the United States by the General Electric Company. The 6.46 inch diameter, 32.75 inch long missile had a wingspan of 29.55 inches. Weighing only 33.1 pounds, it had an maximum range of 1,500 meters at a speed of about 180 mile per hour. The SS-10 was installed experimentally on the M59 armored personnel carrier.

The French SS-11 missile also was produced in the United States by the General Electric Company. Larger than the SS-10, it was 42.5 inches long, but the wingspan was reduced to 20.45 inches. Weighing 63 pounds, the wire guided missile had an effective range of about 3,600 meters traveling at a speed of approximately 425 miles per hour. Intended for use on both helicopters and ground vehicles, it was installed experimentally on the M59 and T113E2 armored personnel carriers.

The photographs below and at the right above show the SS-11 missiles installed on the T113E2 armored personnel carrier. The .50 caliber machine gun was retained on a new mount.

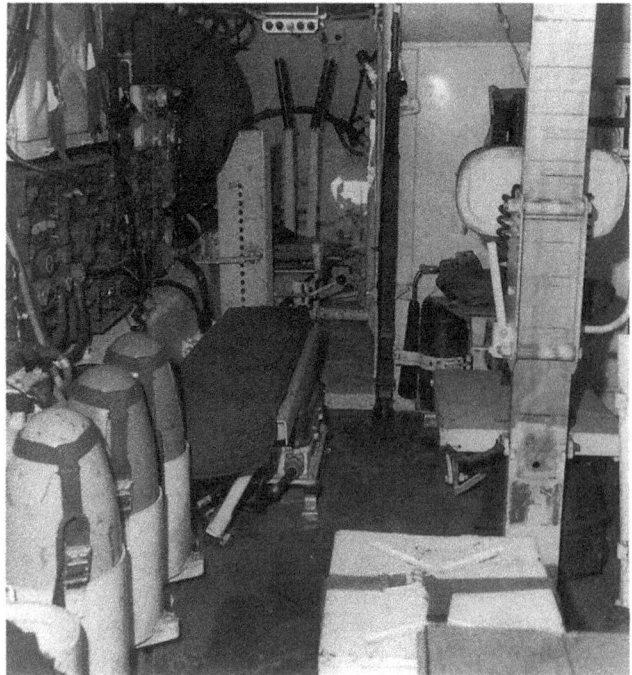

The top of the T113E2 carrier armed with the SS-11 missiles is above. Note the new hatch and machine gun mount. The stowage of the SS-11 missiles inside the T113E2 can be seen at the right.

The development of the TOW (tube launched, optically tracked, wire guided) missile provided a powerful antitank weapon for both ground mounts and helicopters. Introduced late in the Vietnam War, it was an obvious candidate for a vehicle mount that would provide mobility and armor protection for its crew. Development of the XM233E1 kit allowed the TOW system to be installed in an M113 and fired from the open cargo hatch. Standardized as the M233, this kit consisted of a pedestal mount for the TOW system that could be retracted inside the vehicle when not in use. Also, the TOW could be quickly removed from the vehicle and fired from the ground.

A problem with the M233 installation was that the gunner was exposed in the open vehicle hatch when loading or firing the missile. A protective framework with ballistic nylon/Kevlar panels was developed as an interim solution by the Army Natick Laboratories. Mounted on top of the vehicle, it gave some top and side protection, but it was open at the front and rear. It also restricted the traverse of the launcher. Designated as the TOW CAP (TOW cover, artillery protection), it was obviously a stopgap measure.

The TOW missile launcher installed on the M113 is shown above and below. At the bottom left, the missile has been launched.

178

Efforts to provide a better carrier mount for the TOW were in progress by early 1972. In March, a twin tube launcher was installed on an M113 using a pivoting, ten foot, vertical boom. Named the Elevated Antitank Missile Launcher Test Rig, it was intended to permit firing the missile with vehicle in hull defilade. The boom pivoted down to ground level for reloading, but it could only be done from outside the vehicle. By late 1975, a version of the M113A1 modified as the armored cavalry TOW vehicle (ACTV) was being evaluated at Fort Knox along with the armored cavalry cannon vehicle (ACCV) described in a later section. The ACTV carried an XM65 dual pod TOW launcher, normally used on the AH-1Q Cobra helicopter, on an elevating mount. It was attached on top of the M27 cupola from the M114A1E1 command and reconnaissance vehicle. A flexible fiber optic bundle transferred the image from the sight to the gunner in the cupola. It could operate from behind terrain features with only the launch tubes exposed and could be reloaded from within the vehicle by tilting the launcher back over the open cargo hatch. By the Spring of 1976, contracts had been awarded for the development of a new TOW vehicle. The ACTV configuration was assigned to Northrop Corporation and Chrysler designed a prototype with two launchers in a rotating turret, but they could not be extended above the vehicle. Since an elevating mount was considered to be a desirable feature, the Chrysler design was a backup in the event of failure with the more complicated arrangement.

At the top right is the Elevated Antitank Missile Launcher Test Rig. At the right center and bottom is the armored cavalry TOW vehicle and below is the Chrysler TOW vehicle.

179

Above, a new elevating launcher, similar to the ACTV, is at the left and the Emerson Electric Company TOW vehicle is at the right. The latter photograph was dated 15 December 1976.

A third candidate was proposed by the Emerson Electric Company. It also made use of a modified M27 cupola. This version had an armored launcher assembly with two missile tubes and sighting equipment attached to the top of a vertical pivoting arm installed on the M27 cupola. This armored launcher assembly was aimed toward the front and tilted back for reloading from the partially open cargo hatch. During loading, the operator was protected by the hatch cover and armor side flaps. For stowage, the launcher assembly was aimed toward the rear and tilted down to minimize the height. The firing position was at the maximum height. This permitted the missiles to be launched from hull defilade with only the launcher assembly exposed. The appearance of the launcher assembly resulted in it being referred to as the hammerhead.

The second Emerson Electric prototype vehicle can be seen below.

Above is the M901 improved TOW vehicle. Note the smoke grenade launchers on the front hull.

After evaluation, the Emerson Electric Company design entered production and was designated as the M901 improved TOW vehicle (ITV). With the armored launcher assembly installed on the chassis of the M113A1 or the M113A2, the M901 was manned by a crew of four in the mechanized infantry and by a crew of five in the armored cavalry. In both cases that included the driver. Secondary armament on the M901 was an M60 7.62mm machine gun on an external mount attached to a traversing rail around the M27 cupola. Stowage space was provided for ten TOW missiles in the hull. In addition, two missiles could be carried in the launcher. The performance of the M901 was essentially the same as that of the M113A1 or the M113A2. The height of the vehicle with the launcher erected in the firing position was 133¼ inches on the M113A1 chassis and it increased to 134¼ on the M113A2 because of the 1 inch increase in ground clearance. Combat loaded, both vehicles weighed about 13 tons.

When the launch system was upgraded to handle the later TOW 2 and TOW 2A missiles, the vehicle was designated as the M901A1. Like the M901, the M901A1 could be based upon either the M113A1 or the M113A2. Installation of the RISE power pack and new driver's controls resulted in another change. It now became the M901A3 improved TOW vehicle.

181

The armored infantry fighting vehicle can be seen above and at the right equipped with the TOW missile launcher.

The TOW vehicle launcher also was installed on other vehicles. These included the armored infantry fighting vehicle and the mechanized infantry combat vehicle, both developed by FMC. FMC in collaboration with Hughes and the Cadillac Gage Company designed a two tube turret launcher for installation on the M113 series as well as other chassis. In Norway, Kvaerner Eureka produced a twin tube armored launching turret for installation on the M113 series vehicles in the Norwegian Army. Neither of these turrets had the elevating feature and were exposed during firing.

Below, the FMC-Hughes TOW launcher is at the left and the Kvaerner Eureka launcher is at the right. Both were installed on the M113 series armored personnel carrier.

Labels (top left diagram):
SMOKE GRENADE STOWAGE BOX
AIR INTAKE GRILLE
EXHAUST GRILLE
SPARE TRACK SHOE
M19A1 PERISCOPE
TURRET
M26 PERISCOPE
M60 MACHINE GUN
M26 PERISCOPE
LOADER'S PROTECTION
SQUAD LEADER PERISCOPE
TRIM VANE
SMOKE GRENADE LAUNCHER
SMOKE GRENADE STOWAGE BOX
LEFT HEADLIGHT CLUSTER
DRIVER HATCH
FIRE EXTINGUISHER PULL HANDLE (FOR ENGINE COMPARTMENT ONLY)

Labels (top right diagram):
TARGETING HEAD (ERECTED)
TELEPHONE CONNECTOR
VENT COVER
HATCH COVER
RADIO ANTENNA
REAR LIFTNG EYE
WATER CAN
RAMP
REAR DOOR VISION BLOCK
RAMP DOOR
TOW PINTLE
TOW CABLE
RIGHT TAILLIGHT ASSEMBLY
IDLER WHEEL
TIEDOWN HOOK
ROAD WHEELS
TRACK
DRIVE SPROCKET
TRACK SHROUD

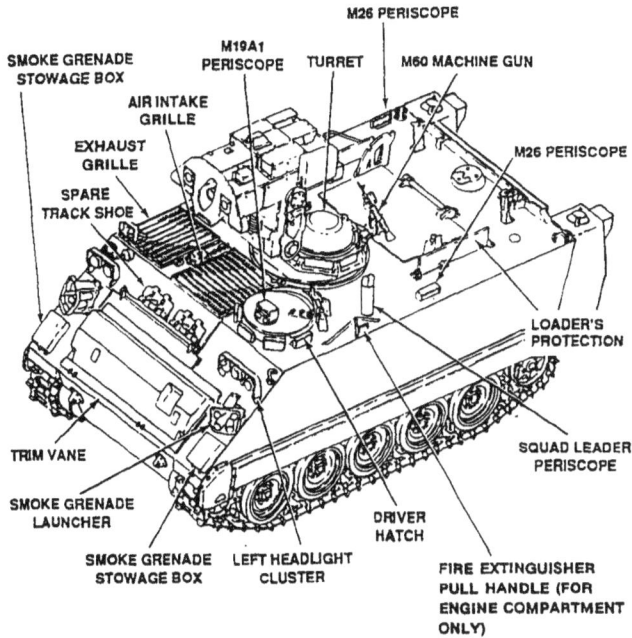

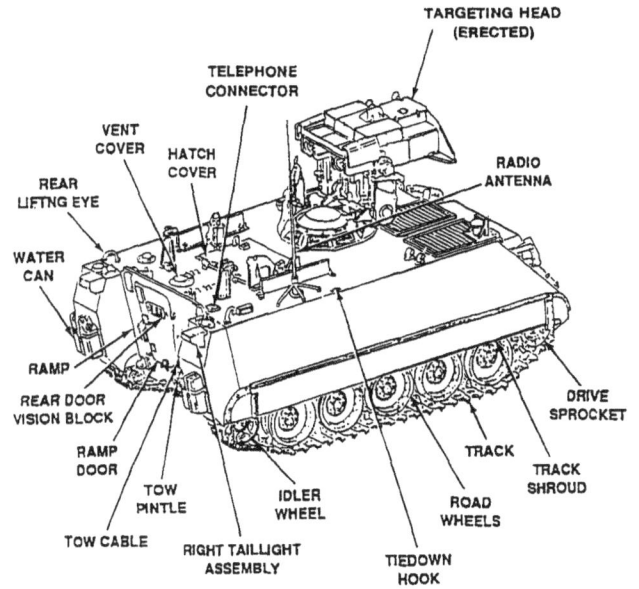

The two drawings at the top identify the components on the M901A3 improved TOW vehicle. Note the external fuel tanks on the rear. The rear view at the right shows the M901 or M901A1 improved TOW vehicle without the external fuel tanks. Note the vision block in the ramp door. This was a characteristic of the improved TOW vehicles.

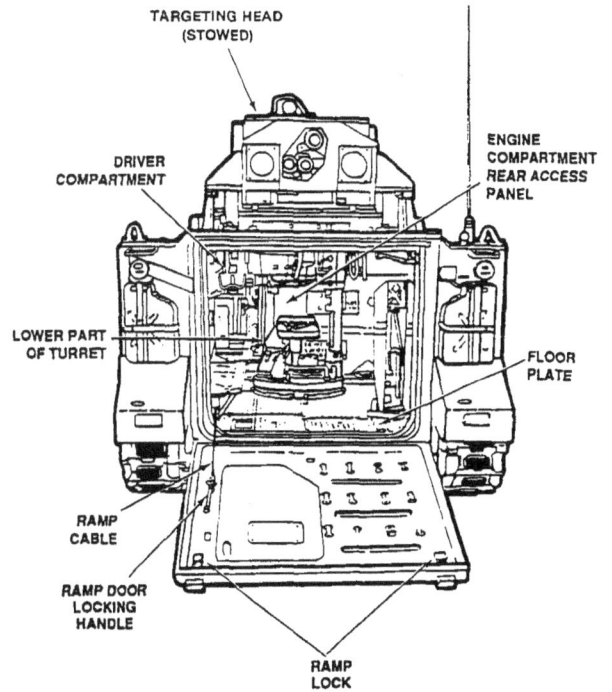

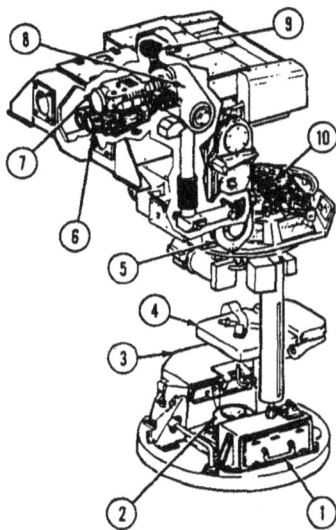

Labels (rear view diagram):
TARGETING HEAD (STOWED)
DRIVER COMPARTMENT
ENGINE COMPARTMENT REAR ACCESS PANEL
LOWER PART OF TURRET
FLOOR PLATE
RAMP CABLE
RAMP DOOR LOCKING HANDLE
RAMP LOCK

Components of the turret assembly can be seen at the left and are identified below.

1. Weapon station emergency power battery
2. Slip ring assembly
3. Missile guidance set
4. Gunner's seat
5. Erection arm
6. Daysight/tracker
7. Nightsight
8. Image transfer assembly (3 channel periscope)
9. 3x telescope
10. Gunner's hatch cover
11. Machine gun traversing rail
12. Vision block (7)
13. Machine gun pintle mount

183

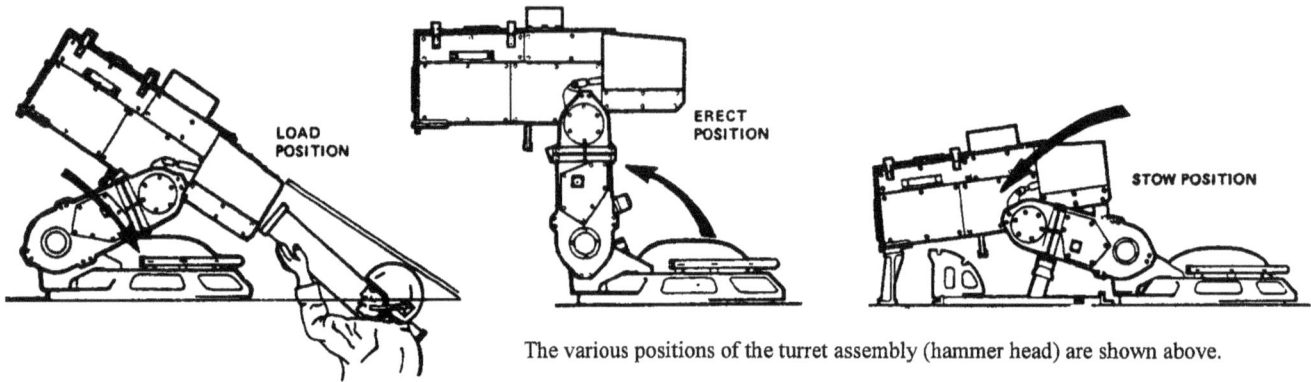

The various positions of the turret assembly (hammer head) are shown above.

Early TOW missiles, then designated as the XM26, were deployed to Vietnam during the Spring of 1972. Launched from a Bell UH-1B helicopter, they went into action for the first time on 9 May destroying three PT76 light tanks. During its long service, the TOW went through numerous modifications. The most important production missiles were the BGM-71A basic TOW with its 5 inch diameter warhead, the BGM-71C improved TOW with an improved 5 inch diameter warhead and an extensible stand-off probe, the BGM-71D TOW 2 with a new 6 inch diameter warhead and an extensible stand-off probe, and the BGM-71E TOW 2A with a small precursor warhead in the stand-off probe to defeat explosive reactive armor. Beginning with the TOW 2 in 1983, improvements in the guidance system permitted operations by day or night and through dust or smoke. A high intensity thermal beacon was added to the aft end of the TOW 2 to provide an infrared tracking source. A more powerful flight motor compensated for the increased weight of the TOW 2.

The TOW 2B differed from the earlier versions of the missile as it was designed to overfly the target and fire two downward aimed 6 inch diameter warheads. It was easily identified by the lack of a stand-off probe.

The armor protection while loading the missile launcher is shown above. At the right are various models of the TOW missile. Below is the TOW 2B top attack missile.

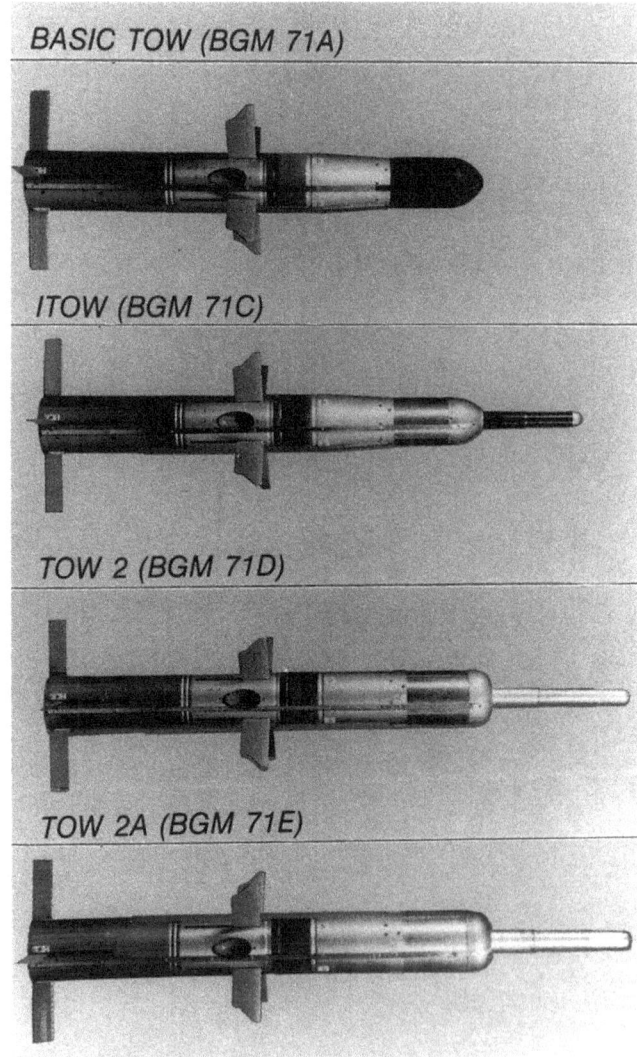

BASIC TOW (BGM 71A)

ITOW (BGM 71C)

TOW 2 (BGM 71D)

TOW 2A (BGM 71E)

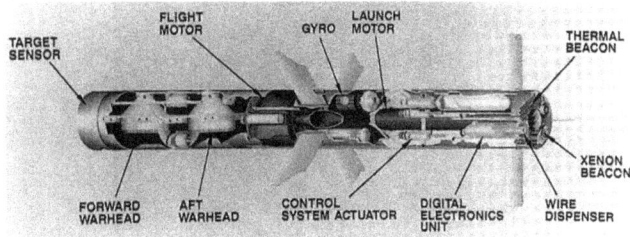
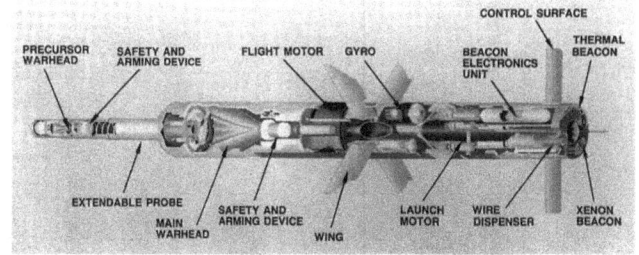

Above are sectional views of the TOW 2B at the left and the TOW 2A at the right.

The air to ground Hellfire missile was effectively employed during the war in the Persian Gulf. This AGM-114A laser guided missile with its 7 inch diameter shaped charge warhead also was adapted for ground launching from mounts on trucks, trailers, and the M113 armored personnel carrier. The Electronics and Space Corporation (ESCO) installed a turret carrying eight Hellfire missiles in two armored pods on the hull of an M113 series vehicle. The same turret also was proposed for installation on the Bradley fighting vehicle and the wheeled light armored vehicle. This two man turret carried the gunner on the left side and the commander on the right. It was armed with a machine gun for local protection. The U. S. Army ground locator designator (GLLD) and the U. S. Marine Corps modular universal laser equipment (MULE) were used to designate targets.

The early AGM-114A Hellfire missile was 64 inches long with a diameter of 7 inches. It weighed 95 pounds. A later version, the AGM-114F, was fitted with a small precursor warhead to defeat explosive reactive armor. This increased the length a little over 7 inches and the weight went up to about 107 pounds.

The air defense antitank system (ADATS) produced by Oerlikon-Buehrle also was installed on the M113 series chassis. This was a self-contained system armed with eight missiles, four on each side of the turret. The 6 inch diameter missile was 82 inches long and weighed about 113 pounds at launch. Detonated by an impact fuze for armored targets, the dual purpose warhead combined the penetration performance of the shaped charge with fragmentation effects. Additional details of this vehicle are in the following section.

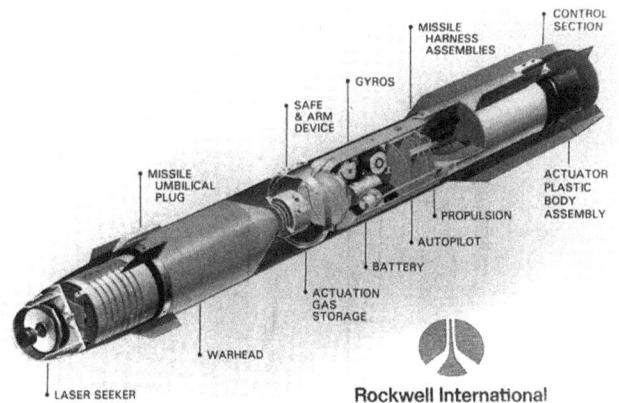

At the left, the Hellfire missile is being launched from one of the two, 4 tube, launchers on the M113 series carrier. Details of the Hellfire AGM-114A missile can be seen above. The AGM-114F is below. Note the precursor warhead on the latter.

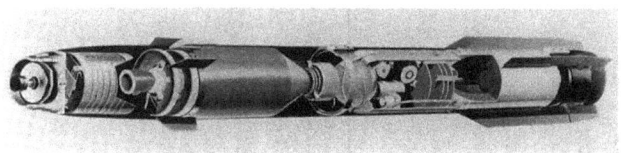

Above are two views of a pilot 37mm self-propelled antiaircraft gun T249, Vigilante B, registration number 12L701.

ANTIAIRCRAFT VEHICLES

In 1952, the Army outlined a three phase program to develop improved light antiaircraft weapons capable of dealing with high speed jet aircraft. The first phase of the program involved the upgrade of the existing twin 40mm self-propelled gun T141 (later the M42) by the addition of a range only radar. This project, nicknamed Raduster, was unsuccessful and it was finally canceled. The second phase of the program resulted in the development of a 37mm, six barrel, Gatling type gun named the Vigilante. This weapon was proposed in two versions, the towed Vigilante A and the self-propelled Vigilante B. The final phase of the search for a new antiaircraft weapon was devoted to a new self-propelled guided missile system named Mauler.

Initially, it was intended to produce four pilots of both the Vigilante A and the Vigilante B. However, the number of pilots was subsequently reduced to three in each case.

Designated as the 37mm antiaircraft, full tracked, self-propelled gun T249, the Vigilante B utilized a low silhouette armored chassis based upon the M113 armored personnel carrier. Mounted in an armored turret, the weapon itself consisted of the 37mm gun T250, the 37mm gun mount T194, the fire control system T51 including the XM17 radar target alarm group, and the XM8 constant speed control generator-

transmission. The aluminum alloy armor on the T249 was equivalent to that on the M113 armored personnel carrier. Although using the same power train components as the M113, the chassis was lengthened increasing the ground contact length to 110¾ inches and the height without the turret could be reduced to 52 inches. The suspension incorporated a lock-up system to stabilize the vehicle when firing.

The turret mounted 37mm gun T250 had a muzzle velocity of 3,000 feet per second and a firing rate of 3,000 rounds per minute. For ground targets, the latter was reduced to 120 rounds per minute. The magazine on the weapon had a maximum capacity of 192 rounds and it was reloaded manually.

By 1960, the pilots had been completed and were being evaluated. However, other automatic weapon systems were now under consideration and the Vigilante never reached production.

At the right is the self-propelled gun chassis T249.

Additional views of the Vigilante B are above and at the right. This pilot vehicle has the registration number 12L702.

Self-propelled Gun Chassis T249

Dimensions in inches

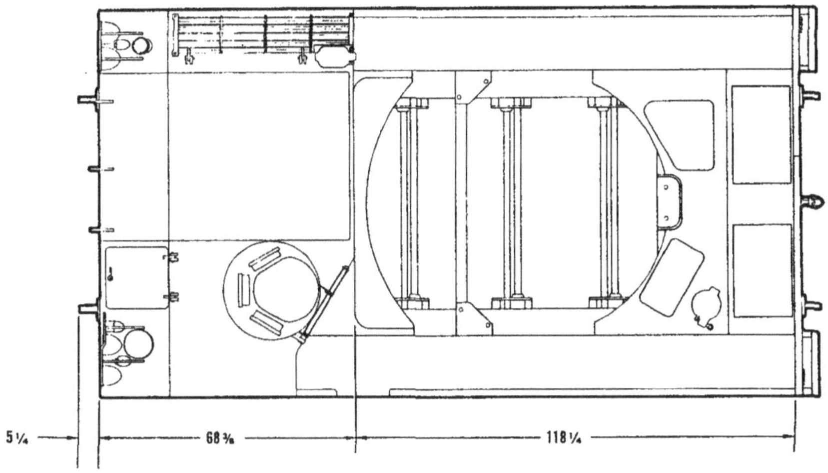
5 1/4
68 3/8
118 1/4

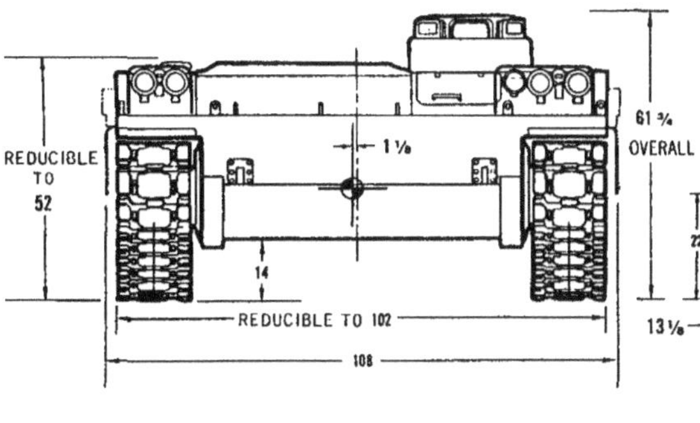
REDUCIBLE TO 52
1 1/8
14
REDUCIBLE TO 102
108

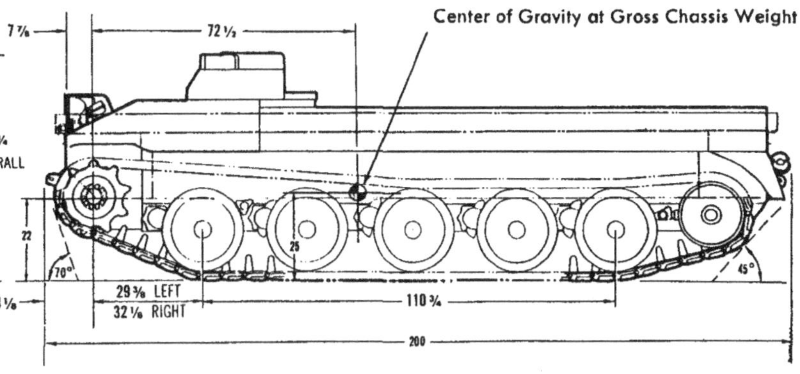
7 7/8
72 1/2
Center of Gravity at Gross Chassis Weight
61 3/4 OVERALL
22
25
70°
45°
13 1/8
29 3/8 LEFT
32 1/8 RIGHT
110 3/4
200

Above are two views of an experimental triple mount armed with Hispano Suiza 20mm guns on the M113A1. At the right is an artist's concept of a 20mm gun triple mount on a M548 cargo carrier.

A triple mount of the Hispano-Suiza HS820 20mm gun was proposed for installation on the M113 armored personnel carrier and the M548 cargo carrier. Each of the three weapons had a firing rate of 1,000 rounds per minute providing a total rate for the triple mount of 3,000 rounds per minute. Adaptation of the Air Force Vulcan 20mm Gatling type gun to a ground mount also was in progress at Rock Island Arsenal beginning in 1964. The ground version of this weapon was electrically driven with a maximum firing rate of 3,000 rounds per minute. Installed on a modified M113A1 chassis, the new weapon was designated as the 20mm self-propelled antiaircraft artillery gun XM163. It consisted of the 20mm gun XM168, the gun mount XM157, the radar set AN/VPS-2, and the automatic lead computing sight XM61. It also was referred to as the Vulcan Air Defense System (VADS).

Standardized as the M163, and after some modification of the gun mount as the M163A1, the VADS began production at the General Electric Company in 1967. The modified M113A1 armored vehicle used to carry the weapon system was designated as the M741. It utilized the same power train and suspension system as

Below are photographs of two 20mm self-propelled antiaircraft guns XM163.

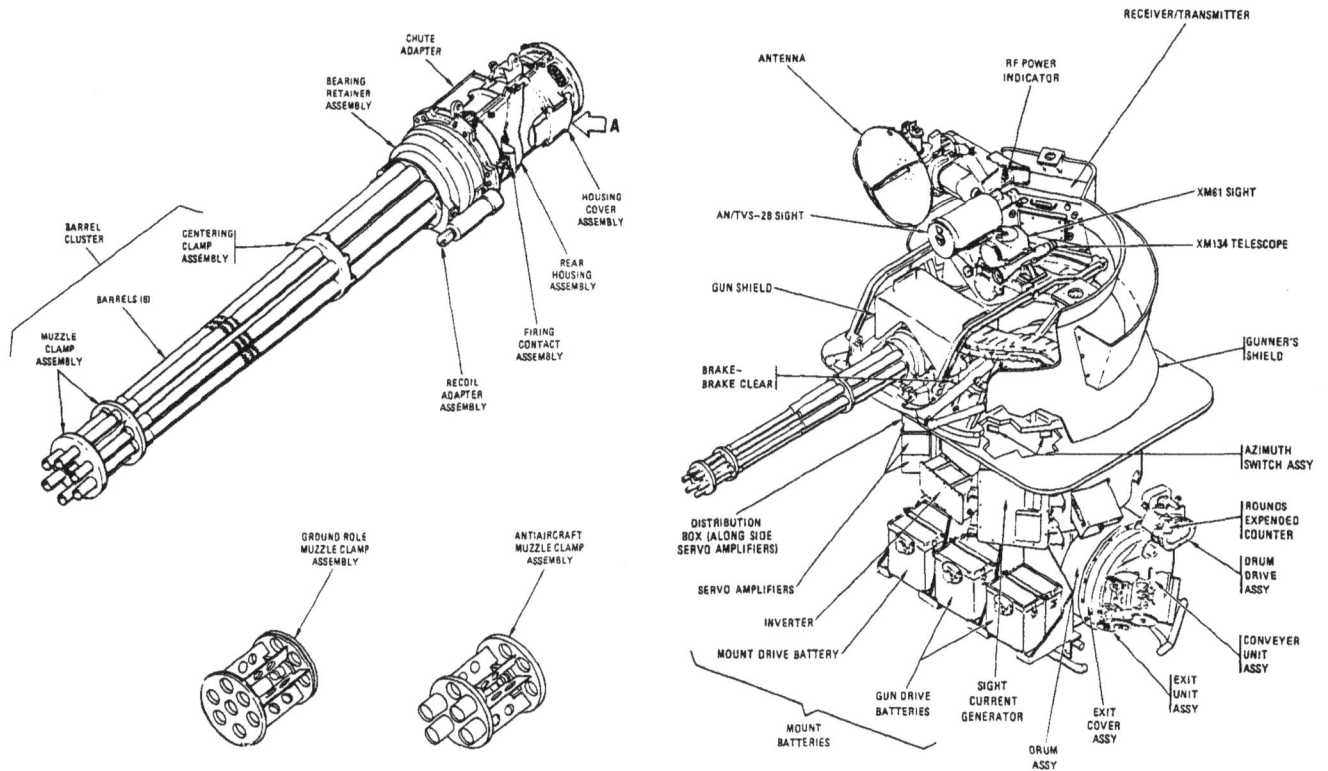

Above, details of the 20mm gun M168 can be seen at the left and the weapon is installed in the M157 mount at the right.

the M113A1 except that it was equipped with a lock-up device to stabilize the vehicle when firing. The weight of the weapon greatly reduced the freeboard of the vehicle when afloat. To retain its swimming capability, flotation cells were installed on each side of the M741 and a high displacement trim vane was fitted. When the improved engine cooling system was introduced on the M113A2 armored personnel carrier, a similar installation was made on the M741 and it was designated as the M741A1. The later suspension with the increased wheel travel was not applied to the vehicle since it was incompatible with the lock-up system required for the weapon.

The XM163 Vulcan air defense system appears above and at the left below. The standard M163 is at the right below. Note the flotation cells on the sides of the vehicles as well as the high displacement trim vane.

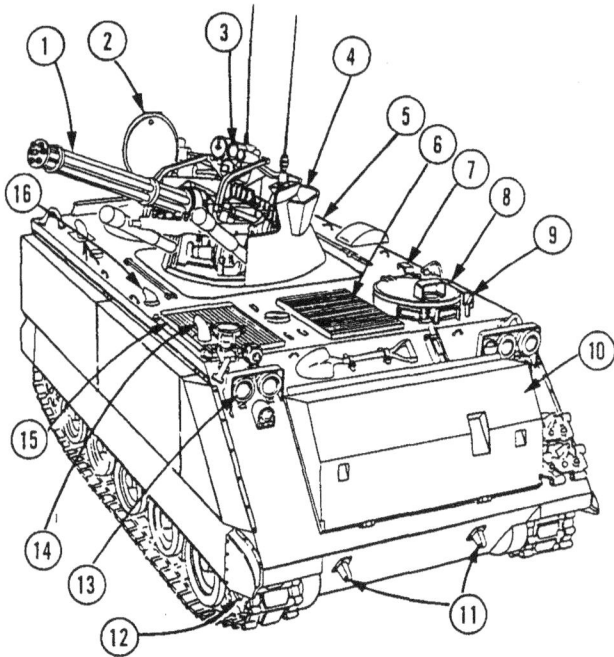

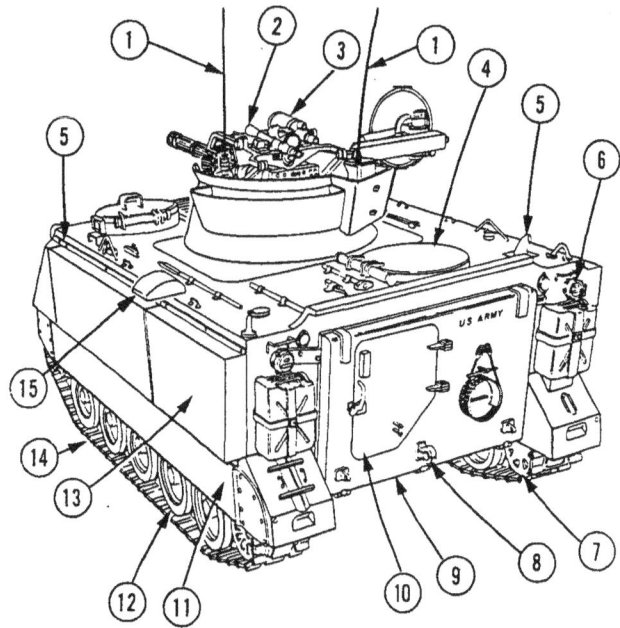

1.	20mm cannon M168	9.	Periscope M17
2.	Radar antenna (Unit 1)	10.	Trim vane
3.	Sight M61	11.	Towing eyes
4.	Mount M157A1	12.	Drive sprocket
5.	Chassis M741	13.	Lights
6.	Air intake grill	14.	Engine exhaust
7.	Fire extinguisher handle	15.	Exhaust grill
8.	Driver's hatch	16.	Heater exhaust

1.	Radio antennas	8.	Towing pintle
2.	Telescope M134	9.	Ramp
3.	Night sight AN/TVS-2B	10.	Personnel door
4.	Commander's hatch	11.	Track shroud
5.	Bilge outlet	12.	Road wheel
6.	Rear light	13.	Flotation pod
7.	Idler wheel	14.	Track
		15.	Link chute cover

Components of the M163A1 Vulcan air defense system are identified in the drawings above.

The major components of the M163A1 20mm self-propelled air defense gun consisted of the M741 or the M741A1 tracked chassis, the M168 20mm gun, the M61 sight, the M157A1 mount, and the AN/VPS-2 radar set. In addition to the high firing rate of 3,000 rounds per minute, the M168 gun could fire at a low rate of 1,000 rounds per minute. At the high rate, bursts were limited to 10, 30, 60, or 100 rounds. Full stowage of 20mm ammunition was 1,031 rounds. Combat loaded, the self-propelled Vulcan system weighed 27,542 pounds. Its performance was similar to that of the M113A1 and in water, the freeboard was 11 inches with the side flotation cells and the high displacement trim vane.

The Vulcan air defense system entered into service in 1969 parallel with the Chaparral guided missile system. Together, they armed the air defense artillery battalions in the armored divisions.

The Vulcan air defense system is firing at the left. Below is a photograph of the 20mm gun M168.

Dimensions in inches (millimeters)

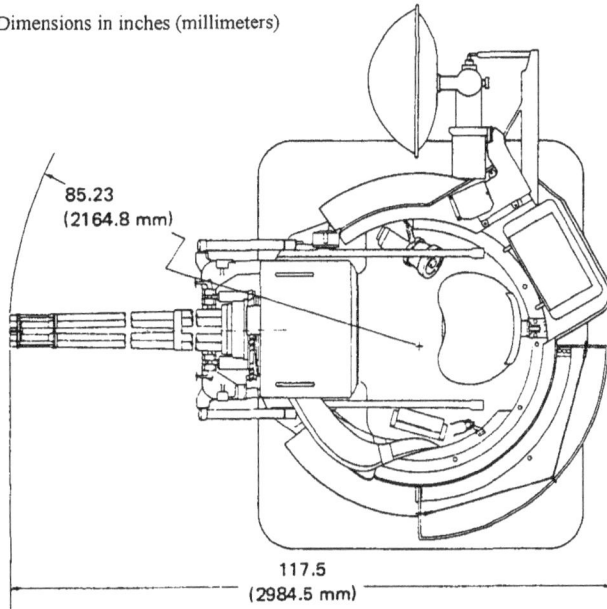

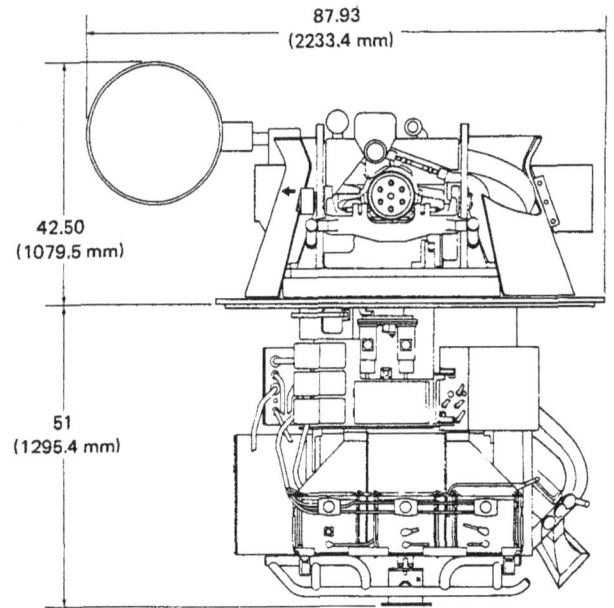

85.23
(2164.8 mm)

117.5
(2984.5 mm)

87.93
(2233.4 mm)

42.50
(1079.5 mm)

51
(1295.4 mm)

The dimensions of the M168 gun in the M157A1 mount can be seen above. Close-up views of the gun and mount are shown below.

The M163A1 Vulcan air defense system is shown in the two photographs below.

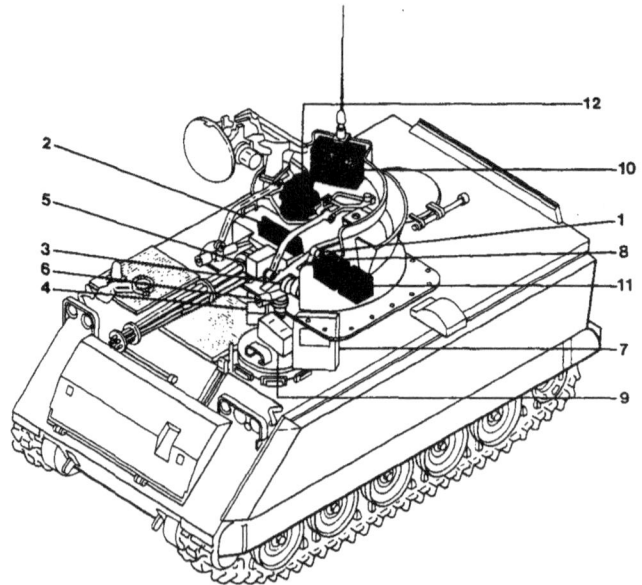

The product improved Vulcan air defense system is shown above. The components are identified by the numbers. 1. elevation synchro, 2. control panel, 3. elevation drive, 4. servo amplifiers, 5. distribution box, 6. azimuth drive, 7. electronics unit, 8. radar power supply, 9. voltage converter, 10 and 11. radar unit, 12. director gunsight.

Later modifications included the product improved Vulcan air defense system (PIVADS) which was a kit developed by the Lockheed Electronics Company. Its major components were a director type sight, a digital microprocessor, and a low backlash azimuth drive system. Installation of this kit reduced the gunner's work load and eliminated inaccuracies in the existing Vulcan system.

The Vulcan Stinger hybrid combined a four tube Stinger missile launcher with the Vulcan gun system. It eliminated the Vulcan's range only radar reducing its vulnerability to electronic countermeasures and anti-radiation missiles. It was replaced by an integrated fire control sensor package. This package included an imagery sensor from the M1 main battle tank, the M65 laser augmented airborne TOW fire control system, and the automatic video target tracker from the Maverick missile.

Other self-propelled antiaircraft gun systems included the XM166. This consisted of a modified M4A1 twin 40mm gun mount installed on the M548 cargo carrier. The modifications reduced the weight of the standard M4A1 mount by about 1,000 pounds. The standard mount was that used on the twin 40mm self-propelled gun M42. Assembled by FMC, it was amphibious and was proposed as a low cost, highly mobile, self-propelled air defense weapon.

At the right are the Vulcan Stinger hybrid (upper) and the twin 40mm self-propelled antiaircraft gun XM166 with the M4A1 mount on the cargo carrier M548 (lower).

The twin 40mm self-propelled gun XM166 is shown above and in the drawing at the right.

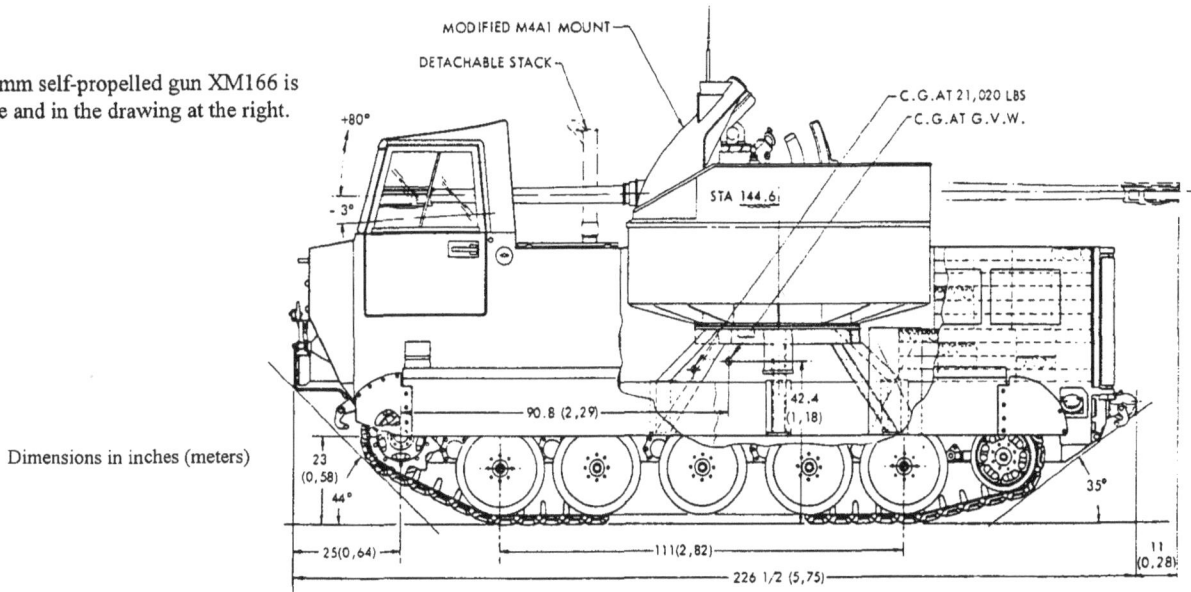

MODIFIED M4A1 MOUNT

DETACHABLE STACK

C.G. AT 21,020 LBS
C.G. AT G.V.W.

STA 144.6

+80°
-3°

90.8 (2,29)

42.4 (1,18)

Dimensions in inches (meters)

23 (0,58)
44°

25(0,64)

111(2,82)

35°

11 (0,28)

226 1/2 (5,75)

In 1979, Ares, Incorporated completed a prototype of the Eagle air defense system. Armed with two high velocity Talon 35mm guns, it utilized the modified chassis of the M548 cargo carrier. With a muzzle velocity of 1,175 meters per second (3,855 feet per second), the two guns had a combined firing rate of 1,200 rounds per minute. The self-propelled weapon was equipped with an optical sight, a laser range finder, a digital ballistic computer, a digital fire control system, and a hydraulic servo drive system.

The photographs below show the Ares Eagle air defense system armed with the twin Talon 35mm guns on the M548 cargo carrier.

Above, Hawk missiles on the launcher M754 are shown installed on the M727 carrier.

The development program that resulted in the Hawk surface to air missile system began in 1954 and it became operational in the U. S. Army in 1959. The MIM-23A Hawk was a radar guided missile propelled by a dual grain solid rocket motor. The fast burning center grain boosted the missile to about Mach 2.5. When it burned out, the slower burning outer grain sustained the missile for the remainder of the flight. The MIM-23A was 198 inches long with a diameter of 14 inches and it weighed 1,295 pounds. A later version, the improved Hawk, became operational in November 1972. The improved Hawk, the MIM-23B, was increased in length to 201.6 inches and in weight to 1,380 pounds. The maximum ranges for the MIM-23A and the MIM-23B were 22 miles and 25 miles respectively.

The complete Hawk system included a battery control center with two radar units as well as trailer mounted triple launchers for the missiles. The small tracked M501 (XM501E2) loader-transporter was used to move the missiles from the preparation site to the launcher. Later, the guided missile launcher M754 was installed on the M727 carrier vehicle using the power train and suspension from the M548 cargo carrier. This unarmored vehicle had a lock-up system on the suspension to provide stability when loading and firing the missiles. An upper steel blast deflector protected the cab top and the power plant compartment during firing. The lower steel blast deflector was hinged to and folded against the upper deflector when in the travel position. When launching, the lower blast deflector was lowered onto the hull brackets. An aluminum alloy blast cover protected the windshield during launching operations and it was stowed on top of the cab for travel. The driver's seat was at the extreme left of the cab with space for three additional passengers. The vehicle was not amphibious, but all hull openings including the full tailgate were provided with watertight seals allowing the vehicle to ford water 42 inches deep.

At the left, the M727 carriers of a Hawk missile battery are on the move.

Above, the Hawk missiles are loaded on their M727 carrier. Below, the M727 guided missile carrier is shown without the M754 launcher. At the bottom right, the carrier is shown with the blast deflector raised. The Hawk loader-transporter M501 is at the bottom left.

Hawk Missile Carrier M727

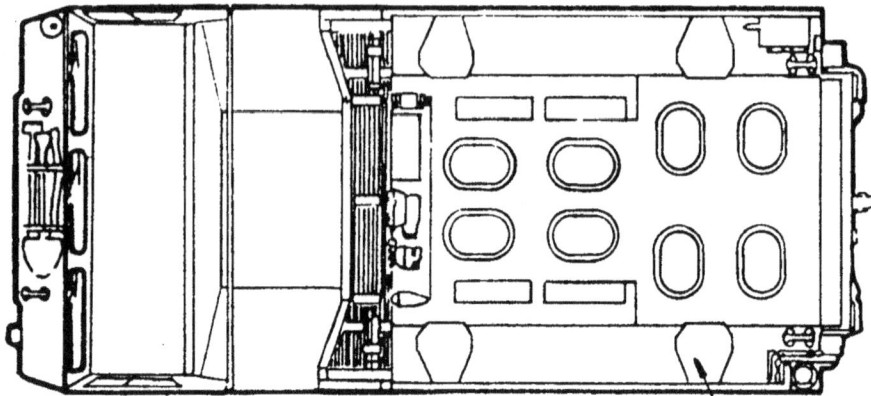

FOUR
LAUNCHER MOUNT PADS
38.44 (0,98) ABOVE
GROUND

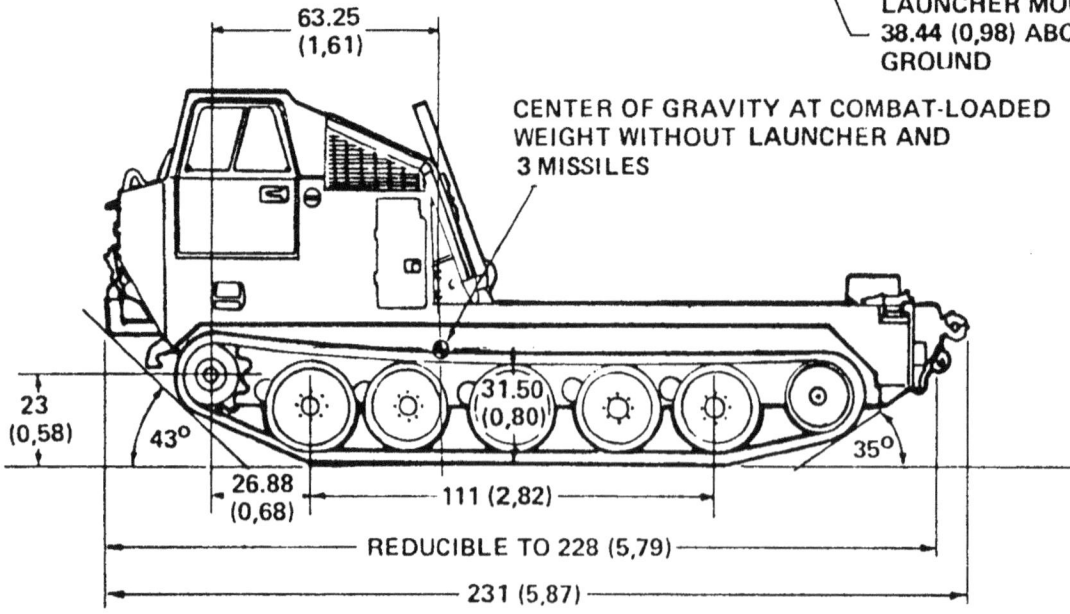

CENTER OF GRAVITY AT COMBAT-LOADED
WEIGHT WITHOUT LAUNCHER AND
3 MISSILES

63.25
(1,61)

23
(0,58)

43°

26.88
(0,68)

31.50
(0,80)

111 (2,82)

35°

REDUCIBLE TO 228 (5,79)

231 (5,87)

Dimensions in inches (meters)

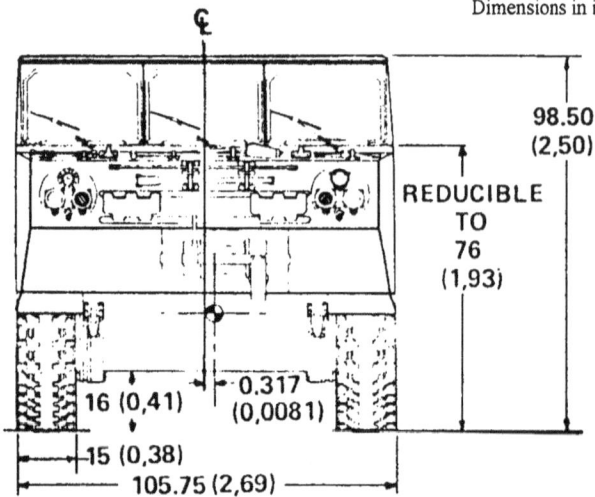

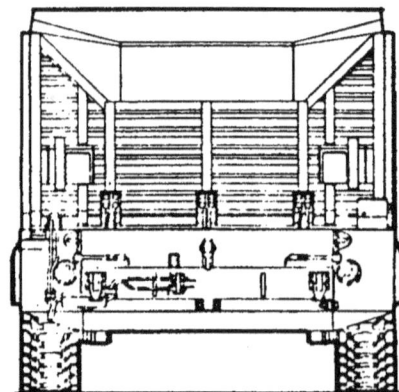

98.50
(2,50)

REDUCIBLE
TO
76
(1,93)

16 (0,41)

0.317
(0,0081)

15 (0,38)

105.75 (2,69)

Above, an early concept of the Mauler self-propelled guided missile system is at the left and at the right, a mock-up of the Mauler system is installed on the XM546 carrier.

Phase three of the Army's low altitude, lightweight, antiaircraft weapon program began in March 1960. This was the development of the Mauler self-propelled guided missile system. It was intended to provide a complete system on a single vehicle that could maneuver with the armored units and be operated by one man. The Mauler was expected to deliver a solid propellant missile with a blast fragmentation warhead out to a slant range of 5,000 meters and up to an altitude of 10,000 feet. It was to defend the field army against low altitude, high speed, aircraft and missiles. The missile system itself was under development by the Convair Division of

General Dynamics and the XM546 carrier, built by FMC, was based upon the M113 armored personnel carrier.

The configuration of the Mauler system varied during the development program. Early proposal sketches show a 12 missile launcher with two rows of six and a different radar system. By late 1963, the first engineering model was under test and it was fitted with nine missiles in three rows of three each. The radar configuration was modified, now consisting of a flat rectangular acquisition antenna on an A-frame support structure. The track-illuminator radar used two parabolic antennas. These were a 20 inch diameter transmitter antenna and a 15 inch diameter receiver antenna on the right and left sides of the launching rack respectively. An infrared scanner could be used along with the radar or independently to avoid detection by enemy counter-measures. The track evaluation computer had the capacity to handle 15 target tracks at the same time. The system was designed to operate automatically, but the firing sequence could be interrupted or modified by the one man crew.

Originally, the solid rocket Mauler missile was 5½ inches in diameter, 77 inches long, and weighed about 120 pounds. Later versions increased in length to about 81 inches.

At the left, an engineering model of the Mauler is installed on its carrier. This photograph was dated 15 November 1962.

Above are two views of the XMIM-46A Mauler system on its XM546 carrier.

The XM546 carrier was a lengthened version of the M113 armored personnel carrier with slightly thinner aluminum alloy armor. The original concept called for a stretched vehicle with six road wheels per side. However, this version was dropped at an early stage and the XM546 was fitted with the usual five road wheels per side, but they were spaced out to give a ground contact length of 111 inches. The driver and the power train were in their usual location, but the top of the cargo compartment was open and extended to the rear. The ramp in the sloped rear wall did not have a personnel door. The open cargo compartment was fitted for the installation of the Mauler weapon pod. Development of the Mauler continued into the late 1960s, but other, less complicated, systems were becoming available and it did not go into production.

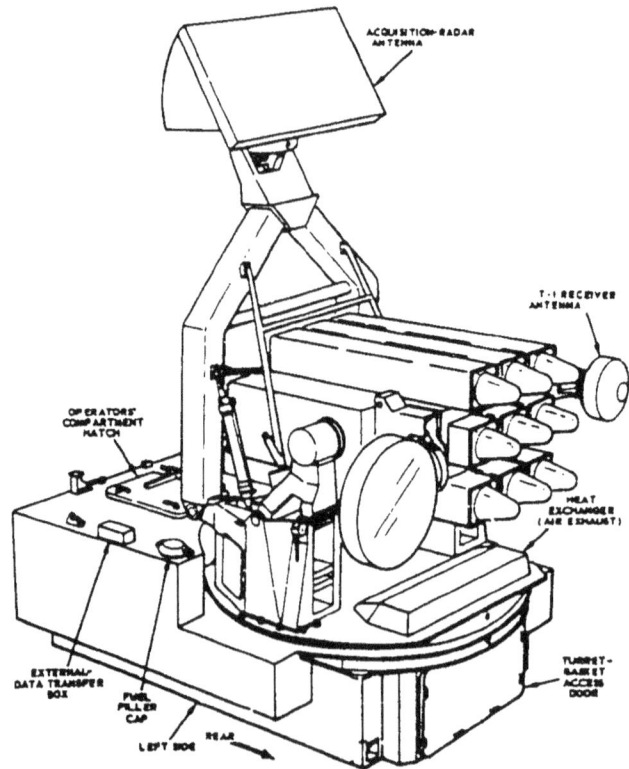

The components of the early Mauler design as on the mock-up appears in the drawing above. Compare with the final design in the photographs at the top. The dimensions of the XM546 carrier are shown in the drawing below.

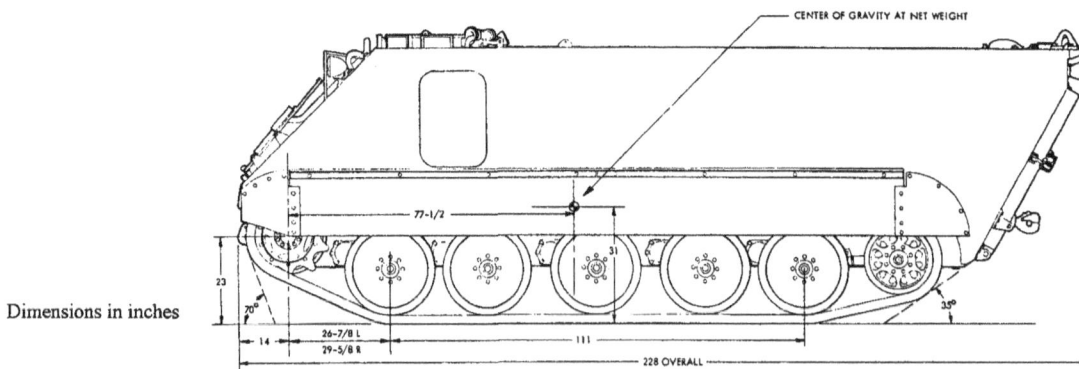

Dimensions in inches

Above, the MIM-72 Chaparral missile system is installed on the modified XM548E1 carrier which was later designated as the M750. At the bottom of the page, the gunner's station and many of the components can be seen at the left and the production model Chaparral system is at the right.

The successful development and deployment of the Sidewinder, infrared guided, air to air missile indicated that it could be employed in the ground to air role. It offered a simple solution to the low altitude air defense problem compared to the highly complex Mauler system. The Aeroneutronics Division of Ford Aerospace (then Philco-Ford) received a contract in early 1965 to adapt the Sidewinder 1C as an ground to air missile. Prototype units were under test during the Summer of 1965 and missile production began in April 1966. The missiles used a four rail launcher installed on the M730 carrier which was based upon the M548 cargo carrier. The gunner had only an optical sight to aim the system until the infrared seeker locked onto the target. When fired, the missile automatically homed onto the target without any further effort from the gunner. Named the Chaparral, the missile system installed on the M730 consisted of the M54 launch and control station with four MIM-72 missiles on launch rails. The complete unit was designated as the M48. Eight additional missiles were stowed on the carrier. The Chaparral missile itself was 113 inches long with a launch weight of 195 pounds. Its range was about 11 miles.

The unarmored M730 carrier retained the power train and suspension of the M548 cargo carrier and the cargo compartment was adapted for the installation of

A close-up view of the Chaparral launcher is at the left and the missile is launching above.

the missile system. The vehicle could ford water 40 inches deep without any preparation and with the installation of a flotation curtain, it was amphibious. A cargo compartment cover with six bows was provided to cover the missile system during travel. When prepared for action, the cargo compartment cover was removed and the bows were stowed on the front of the vehicle. A stationary blast shield in the front of the cargo compartment protected the power plant and blast covers were stowed on top of the power plant compartment. Before launching, these blast covers were unfolded to extend over the cab and the power plant compartment. The crew consisted of five men including the driver. A 20,000 pound capacity winch was installed in the front of the carrier.

When the improved engine cooling system and suspension were introduced on the M548, similar changes were made to the M730 and it was designated as the M730A1. The later introduction of the RISE power package and a nuclear, biological, chemical

(NBC) collective protection system changed the designation to the M730A2 guided missile equipment carrier. The M730A2 was the first vehicle in the M113 family to use the RISE package. It was required to transport the heavier, improved, M54A2 Chaparral aerial intercept guided missile pallet. Approximately 500 M730A1 carriers were converted to the M730A2 in a program ending in the third quarter of fiscal year 1993.

One modification to the M730A2 was the elimination of the winch in the front of the vehicle. The missile also was improved. A new M121 smokeless solid rocket motor eliminated the smoke plume of the original rocket and a more lethal M250 blast-fragmentation warhead replaced the earlier expanding rod warhead. Other improvements to the Chaparral included the introduction of a forward looking infrared (FLIR) night sight. This permitted operation at night and in all but the worst weather conditions. The advanced Chaparral multi-weapons platform was demonstrated on the XM1108 universal carrier during 1993.

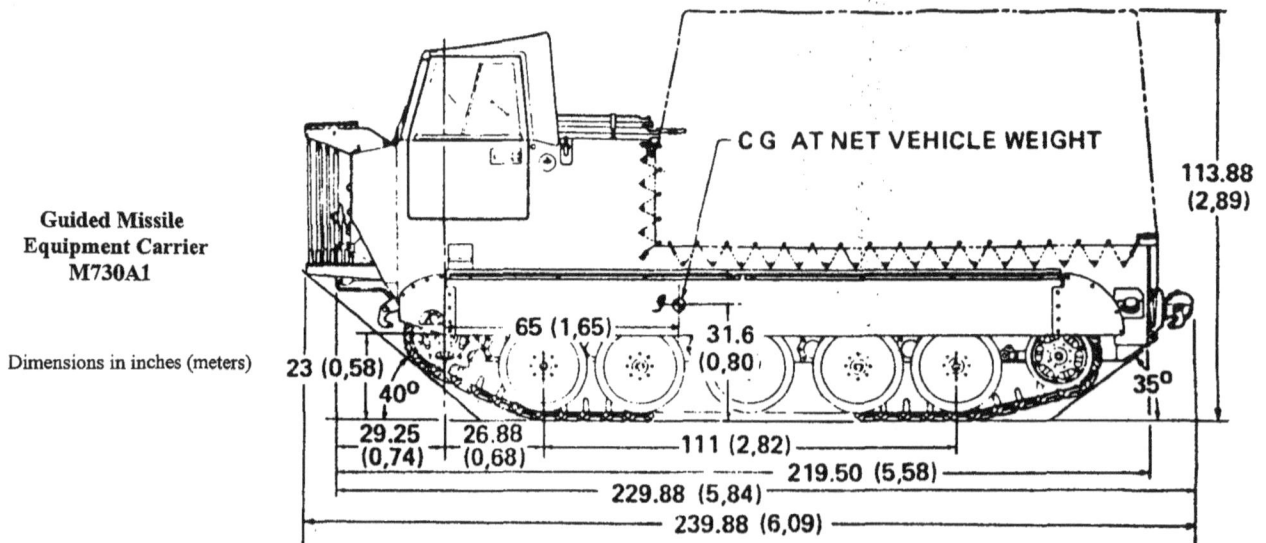

**Guided Missile
Equipment Carrier
M730A1**

Dimensions in inches (meters)

C G AT NET VEHICLE WEIGHT

113.88 (2,89)

65 (1,65)

31.6 (0,80)

23 (0,58)

40°

35°

29.25 (0,74)

26.88 (0,68)

111 (2,82)

219.50 (5,58)

229.88 (5,84)

239.88 (6,09)

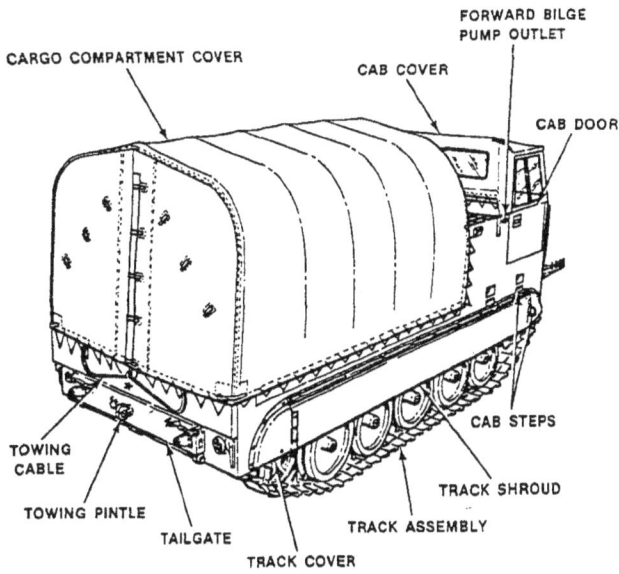

CARGO COMPARTMENT COVER
CAB COVER
FORWARD BILGE PUMP OUTLET
CAB DOOR
CAB STEPS
TRACK SHROUD
TRACK ASSEMBLY
TRACK COVER
TAILGATE
TOWING PINTLE
TOWING CABLE

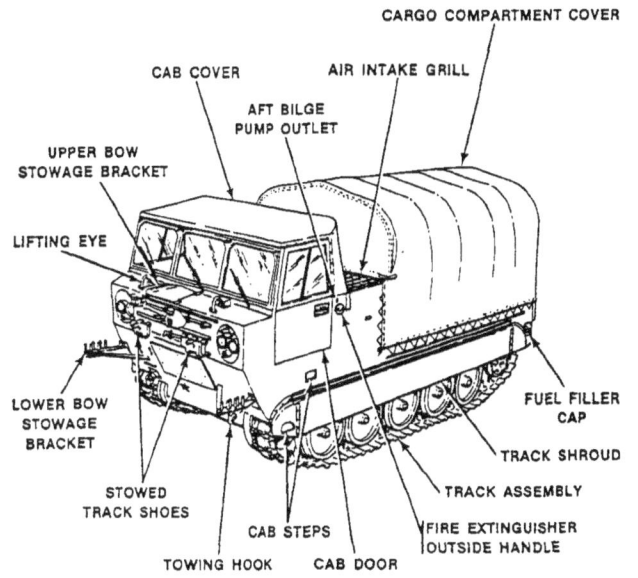

CARGO COMPARTMENT COVER
AIR INTAKE GRILL
CAB COVER
AFT BILGE PUMP OUTLET
UPPER BOW STOWAGE BRACKET
LIFTING EYE
LOWER BOW STOWAGE BRACKET
STOWED TRACK SHOES
CAB STEPS
TOWING HOOK
CAB DOOR
FIRE EXTINGUISHER OUTSIDE HANDLE
TRACK ASSEMBLY
TRACK SHROUD
FUEL FILLER CAP

The drawings on this page show the components of the guided missile equipment carrier M730A2.

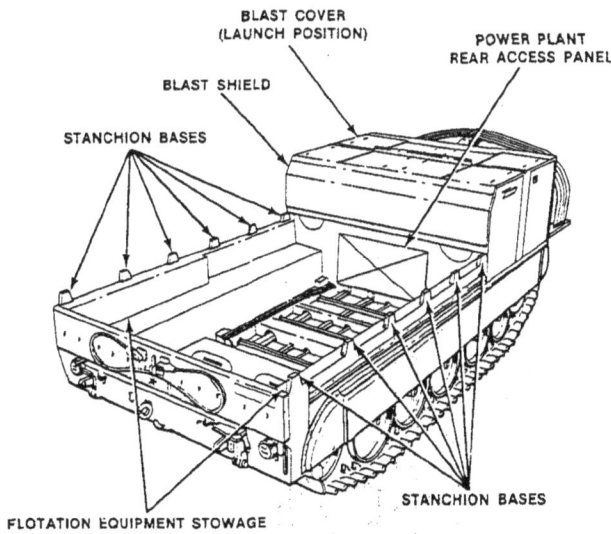

BLAST COVER (LAUNCH POSITION)
POWER PLANT REAR ACCESS PANEL
BLAST SHIELD
STANCHION BASES
FLOTATION EQUIPMENT STOWAGE
STANCHION BASES

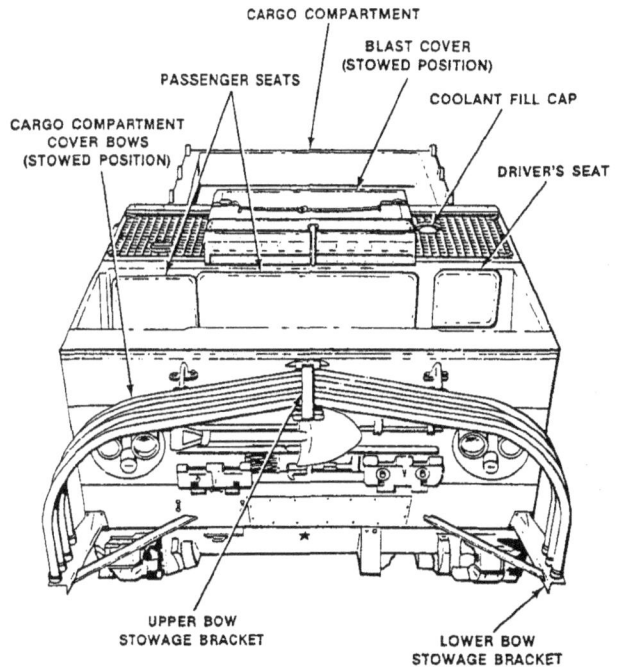

CARGO COMPARTMENT
BLAST COVER (STOWED POSITION)
PASSENGER SEATS
COOLANT FILL CAP
CARGO COMPARTMENT COVER BOWS (STOWED POSITION)
DRIVER'S SEAT
UPPER BOW STOWAGE BRACKET
LOWER BOW STOWAGE BRACKET

At the right is the product improved Chaparral system with a forward looking infrared pod on the left side of the mount.

Above is the Oerlikon-Buehrle air defense antitank system installed on the M113 series carrier. At the right, the missile has been launched.

As mentioned before, the Oerlikon-Buehrle ADATS was a combined antiaircraft antitank weapon system. Installed on the M113 series vehicle, it could use its own radar to detect low flying aircraft and pass the target to its passive optical tracking system. The Mach 3+ missile was then guided to the target by a coded carbon dioxide laser beam. A proximity fuze detonated the dual purpose warhead for airborne targets. On the M113A2 carrier, the missile system weighed over 16 tons including eight spare missiles.

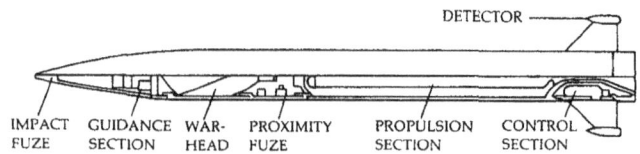

The components of the ADATS missile are shown in the sketch above. The ADATS in service with the Canadian armed forces is below at the left. The ADATS at the bottom right was produced for Saudi Arabia.

Above, the armored personnel carrier is equipped with a 104 inch wide aluminum alloy dozer blade raised into the travel position.

ENGINEER VEHICLES

Vehicles of the M113 family were adapted to perform many engineer tasks. The bulldozer conversion kits XM10 and XM11 were developed for use with the M113 and M113A1 armored personnel carriers respectively. The major differences between the two kits were in the mounting of the hydraulic pumps and the routing of the hydraulic lines. Each kit consisted of a 104 inch wide aluminum alloy blade with a bolt-on, replaceable, steel cutting edge. The hollow blade was filled with closed cell polyurethane foam and it added buoyancy to the vehicle. In the fully raised position, the blade acted as a trim vane when swimming the carrier. The blade could be raised 35 inches above or lowered 6 inches below ground level. It also could float following the ground contour for grading operations. The blade was operated by a single control in the driver's compartment and the 30 gallon capacity tank for the hydraulic system was located in the right front sponson in the personnel compartment. With the bulldozer installed and the full 13 man crew, the combat weights of the M113 and the M113A1 were 25,010 pounds and 25,720 pounds respectively.

Below at the left, the dozer blade is lowered to the ground. Note the high displacement blade that served as a trim vane when the vehicle was swimming. At the bottom right, the dozer is in operation.

The armored personnel carrier at the right is fitted with the 116 inch wide dozer blade.

Another bulldozer conversion kit for the M113 and M113A1 provided a 116 inch wide aluminum alloy blade which could be angled 30 degrees to the right side for clearing snow. As with the XM10 and the XM11, the buoyant blade was filled with polyurethane foam and fitted with a replaceable steel cutting edge. The wide blade kit used the same hydraulic system and controls as the XM10 and the XM11.

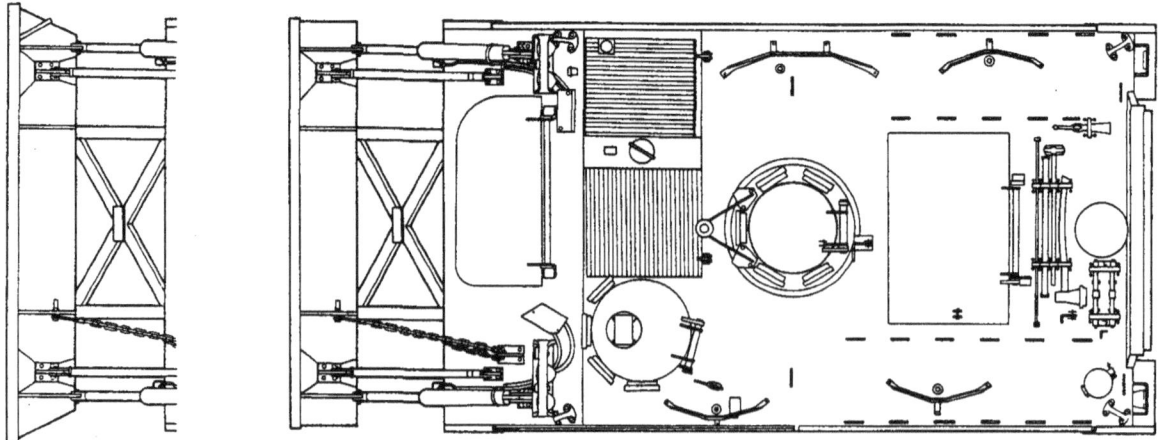

Dimensions in parentheses ( ) are in meters; all other dimensions are in inches.

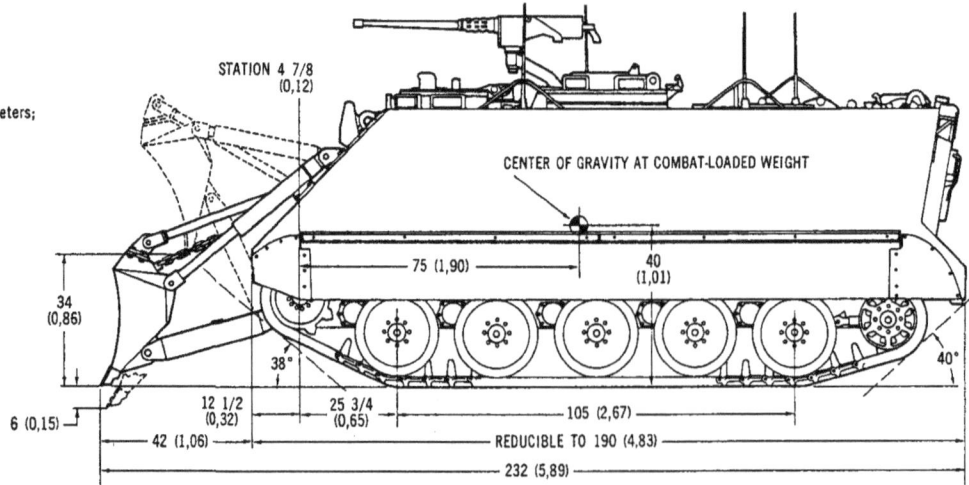

STATION 4 7/8 (0,12)

CENTER OF GRAVITY AT COMBAT-LOADED WEIGHT

75 (1,90)   40 (1,01)

34 (0,86)

38°   40°

6 (0,15)

12 1/2 (0,32)   25 3/4 (0,65)   105 (2,67)

42 (1,06)   REDUCIBLE TO 190 (4,83)

232 (5,89)

The drawing above shows the 104 inch wide blade mounted on the armored personnel carrier. The wide blade is sketched at the top left. At the right, the dozer equipped armored personnel carrier is swimming. Note the use of the dozer blade as a trim vane.

204

General Motors of Canada developed a kit to convert the M113 series of personnel carriers to an engineer vehicle. The vehicle was equipped with the wide blade bulldozer previously described and featured an improved layout for personnel, supplies, tools, and equipment. The external fuel tanks were installed leaving additional space inside the vehicle. An hydraulic powered earth auger was mounted on the left side of the roof and the hydraulic system used to power the bulldozer and the earth auger also could be used to operate hydraulic tools such as a jack hammer, a chain saw, or an impact wrench. Recognizing the fact that the crew frequently used the open ramp as a work area, a restraining system was installed capable of supporting a 1,100 pound load to prevent any damage. Box type seats used for stowage space replaced the folding seats in the original vehicle and the communication equipment was relocated to provide easy access for the driver and the commander. The converted engineer vehicle had space for a crew of eight including the driver and the commander.

The Volcano mine dispensing kit was installed on an M548A3 cargo carrier fitted with the RISE power package. The Volcano was a special purpose kit installed and removed by the operators in U.S. Army engineer units.

The photographs above show the engineer vehicle based upon the M113 series armored personnel carrier. The dozer blade is extended in the operating position and at the top right, the hydraulic powered earth auger is ready for use. Below, the ramp restraining chains are installed to permit its use as a work area.

Below, the Volcano mine dispensing kit is installed at the left on an M548A3 cargo carrier and at the right, the Giant Viper mine clearing system is being towed in its trailer by an armored personnel carrier.

The lightweight folding assault bridge can be seen above stowed for travel on its M113 carrier. The launching sequence is shown in the photographs below.

Operations in Vietnam revealed the need for a lightweight assault bridge that could be carried on the M113 armored personnel carrier. Such a bridge was developed by the U.S. Army Mobility Equipment Research and Development Center at Fort Belvoir. This aluminum alloy folding bridge weighed 2,700 pounds and it was designed to support 15 ton loads over spans up to 33 feet. Carried on top of the armored personnel carrier, it was operated hydraulically and could be placed in position in less than two minutes without exposing the personnel. With the folded bridge on board, the armored personnel carrier retained its normal performance and its amphibious capability.

The M113A3+ combat engineer squad vehicle (CESV) was developed to transport an eight man engineer squad with all of their equipment. It also could be adapted to carry the Volcano mine dispensing system, the pathfinder marking system, or for towing the mine clearing line charge (MICLC) trailer. The CESV was a stretched M113A3 with six road wheels per side. It was powered by a 6V53TA diesel engine developing 350 gross horsepower at 2,800 rpm through the Allison X200-4A transmission. The basic 5083 aluminum alloy armor was reinforced by titanium applique armor on the sides and upper front. Spaced expanded steel armor was added to the rear and the bottom was reinforced with steel laminate armor. The CESV had a gross weight of about 18 tons and a maximum road speed of 41 miles per hour. A fuel capacity of 120 gallons provided a cruising range of about 300 miles.

The combat engineer squad vehicle appears above and below. In the lower photograph, an armor shield has been installed for the .50 caliber machine gun.

The dimensions of the combat engineer squad vehicle are shown below.

Dimensions in inches

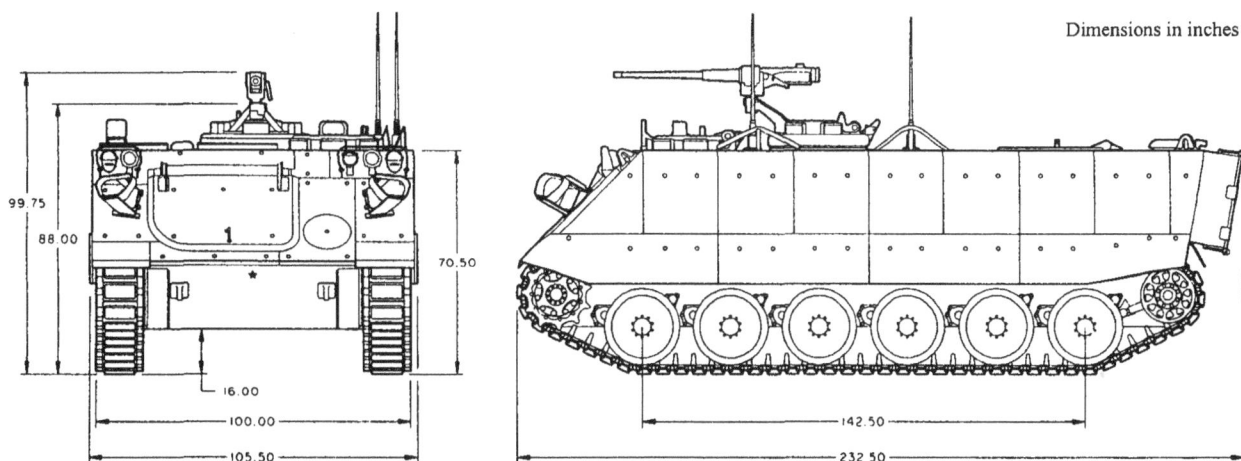

99.75
88.00
70.50
16.00
100.00
105.50
142.50
232.50

The light recovery vehicle XM696 appears above. This was the vehicle converted from the XM548 prototype.

RECOVERY AND MAINTENANCE VEHICLES

The gasoline powered XM548 prototype was rebuilt at FMC as a wrecker or recovery vehicle. Designated as the XM696 full tracked light recovery vehicle, it was intended to serve as a prototype for a recovery vehicle based upon the diesel powered M548. It was fitted with the crane from the 5 ton M543 wrecker truck and the hoist from the M578 armored recovery vehicle was used as a tow winch. It retained the XM548 winch in the front of the vehicle. The prototype weighed 24,900 pounds.

The XM806E1 recovery vehicle was a modification of the M113A1 armored personnel carrier. A 20,000 pound capacity, hydraulically driven, winch was located inside the personnel compartment for use in retrieving disabled vehicles. The vehicle was anchored during recovery operations by two spades mounted on the rear. A third spade could be installed if required in soft soil.

These views show the XM806E1 recovery vehicle. Note the expanded metal shield to protect the winch operator at the bottom left.

A 3,000 pound capacity crane was installed on top of the vehicle to handle and place heavy components. Manned by a crew of three, the combat weight of the XM806E1 was 25,200 pounds. It was armed with a .50 caliber machine gun on the commander's cupola.

STATION 4.88
(0,12)

81 (2,05)

38.50
(0,93)

23
(0,58)

70°

14 (0,36)

25.75
(0,65)

105 (2,67)

REDUCIBLE TO 195 (4,95)

210.25 (5,34)

40°

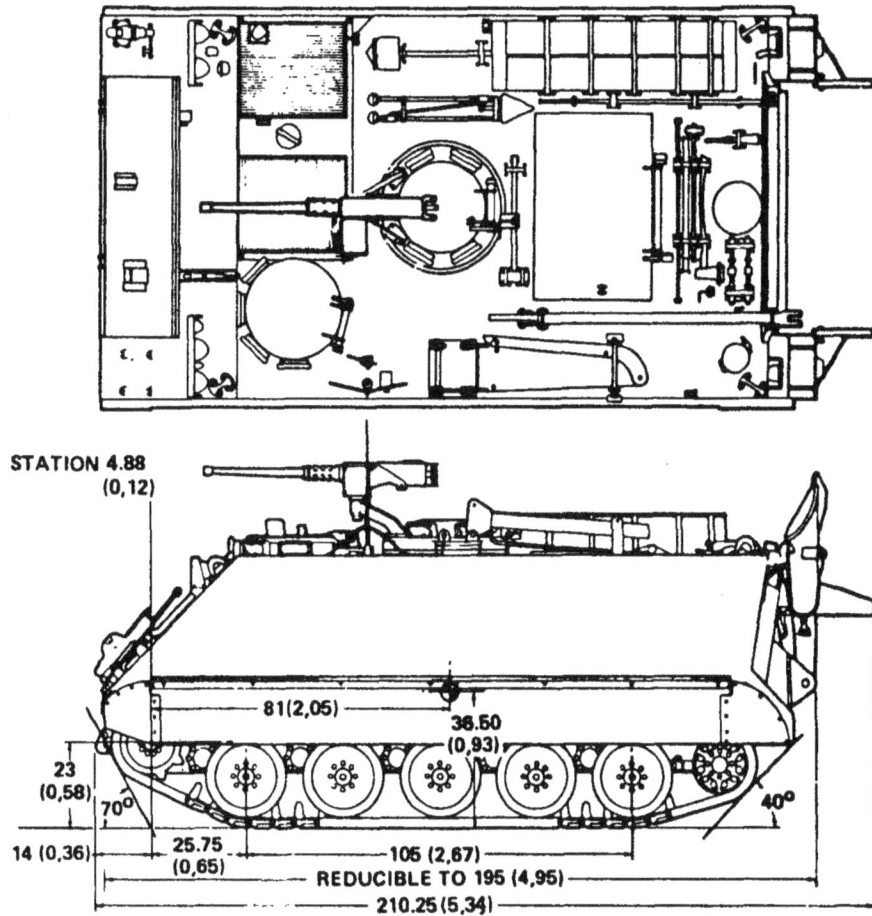

Dimensions in inches (meters)

Light Recovery Vehicle XM806E1

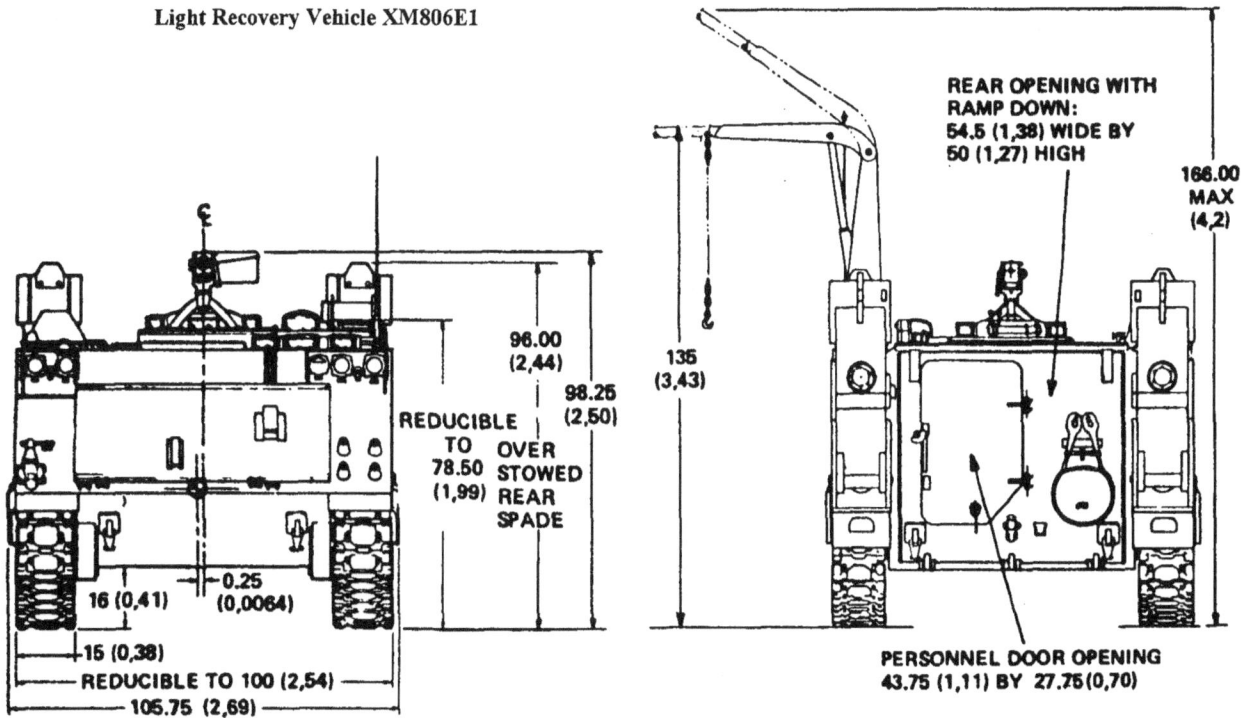

REAR OPENING WITH
RAMP DOWN:
54.5 (1,38) WIDE BY
50 (1,27) HIGH

166.00
MAX
(4,2)

96.00
(2,44)

98.25
(2,50)

REDUCIBLE
TO
78.50
(1,99)

OVER
STOWED
REAR
SPADE

135
(3,43)

16 (0,41)

0.25
(0,0064)

15 (0,38)

REDUCIBLE TO 100 (2,54)

105.75 (2,69)

PERSONNEL DOOR OPENING
43.75 (1,11) BY 27.75 (0,70)

209

The combination maintenance and recovery vehicle appears in the views above and below. Note the larger crane on this vehicle.

A combination maintenance and recovery vehicle was assembled by FMC based upon the M113A2 armored personnel carrier. Like the XM806E1, it utilized the 20,000 pound capacity winch located inside the vehicle with the rear mounted spades to anchor it in place. A larger crane was installed with a capacity of 6,800 pounds. With a three man crew, the vehicle had a combat weight of 26,950 pounds and was armed with a .50 caliber machine gun at the commander's hatch.

The fitter's vehicle was a modified M113A1 armored personnel carrier equipped with a large roof hatch and a 6,800 pound capacity crane. The large roof hatch and crane allowed the vehicle to transport and handle large components such as a complete power package. Depending upon the user requirements, radios, spare parts, or special maintenance tools could be accommodated. A two man crew could operate the fitter's vehicle and its equipment. However, space was provided for nine men including the driver. The fitter's vehicle was in production at FMC from 1964 through 1978.

The fitter's vehicle is shown at the right and below. The use of the large roof hatch can be seen below.

94 (2,39) OPENING

SMALL CARGO
HATCH CLEAR
OPENING 47.50 (1,21)
BY 30.75 (0,78)

54
(2,37)
OPENING

STOWED POSITION
OF CRANE

STATION 4.88
(0,12)

72
(1,83)

23
(0,58)

70°

40°

14 (0,36)

25.75
(0,65)

105 (2,67)
REDUCIBLE TO 190 (4,83)

191.50 (4,86)

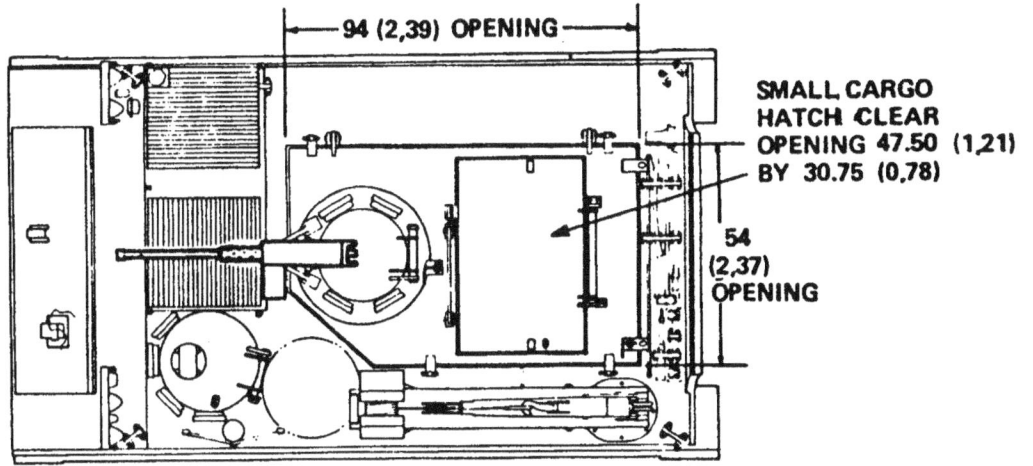

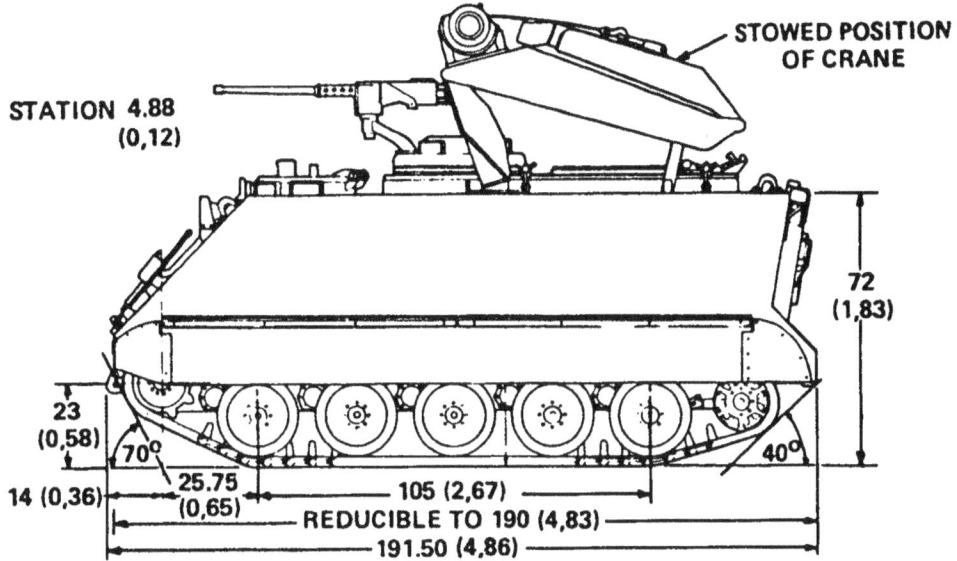

Fitter's Vehicle

Dimensions in inches (meters)

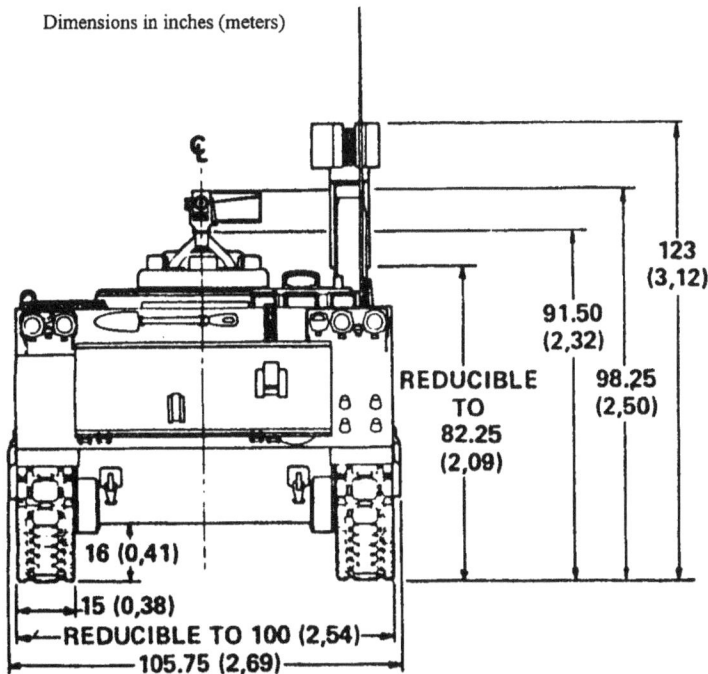

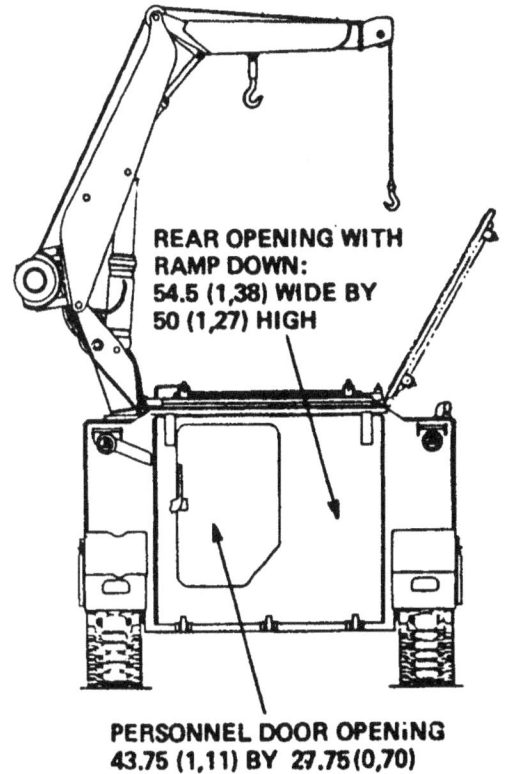

REAR OPENING WITH
RAMP DOWN:
54.5 (1,38) WIDE BY
50 (1,27) HIGH

123
(3,12)

91.50
(2,32)

REDUCIBLE
TO
82.25
(2,09)

98.25
(2,50)

16 (0,41)

15 (0,38)

REDUCIBLE TO 100 (2,54)

105.75 (2,69)

PERSONNEL DOOR OPENING
43.75 (1,11) BY 27.75 (0,70)

Above, the maintenance and recovery vehicle is lifting the turret off of an armored infantry fighting vehicle at the left and at the right, a fitter's vehicle is swimming.

In June 1985, a new armored maintenance vehicle was evaluated by the Army at Fort Hood, Texas. Two prototype vehicles were available with one based upon the M113A1E1 stretched vehicle and the other upon the fighting vehicle system (FVS) carrier. A module was installed on each with a telescoping crane capable of lifting over 5 tons. This permitted the vehicle to replace the power packs from heavy vehicles as well as handle other very heavy components. In addition to the crane, the vehicle was equipped with a work bench, an air compressor, an hydraulic pump, and a stowage area for parts and tools.

The armored maintenance vehicle based upon the stretched M113A1E1 is shown in the drawing below.

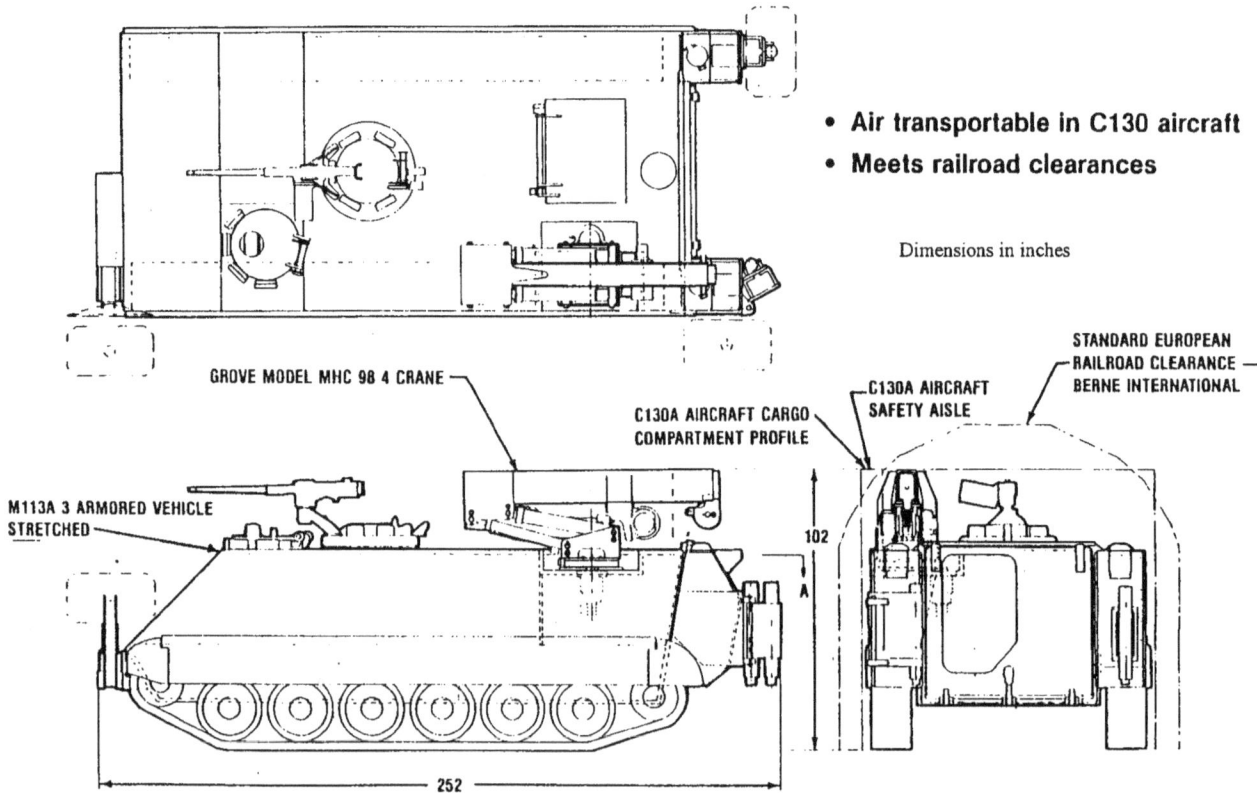

• **Air transportable in C130 aircraft**
• **Meets railroad clearances**

Dimensions in inches

GROVE MODEL MHC 98 4 CRANE

C130A AIRCRAFT CARGO COMPARTMENT PROFILE

C130A AIRCRAFT SAFETY AISLE

STANDARD EUROPEAN RAILROAD CLEARANCE — BERNE INTERNATIONAL

M113A 3 ARMORED VEHICLE STRETCHED

102

252

The self-propelled flame thrower M132 is shown above. Note the coaxial 7.62mm M73 machine gun.

CHEMICAL WARFARE VEHICLES

In June 1954, the Chemical Research and Development Laboratories (CRDL) began a study of a mechanized flame thrower based upon tanks or other armored vehicles. The development of the E31-E36 flame thrower kit was a result of this study. This nomenclature indicated that the flame thrower consisted of the E31 fuel and pressure unit and the E36 flame gun or cupola group. Three of the kits were installed in M59 tracked armored infantry vehicles for evaluation. The flame gun fuel capacity was 400 gallons in the M59 providing about 70 seconds of firing time. As a result of the test program, some modifications were made and the improved flame thrower was designated as the E31R1-E36R1. This version was designed for installation in either the M59 or the new M113 armored personnel carrier. In April 1959, a contract was awarded for the installation of three E31R1-E36R1 flame throwers in M113s and the work on the M59 installation was terminated.

The three pilot vehicles were armed with the E36R1 flame gun in a cupola mount with a .50 caliber M85 machine gun. This mount replaced the commander's cupola on the M113. The E31R1 fuel and pressure unit was installed in the personnel compartment. Two of the pilots were tested at Fort Benning and Fort Greely during 1961-61. The cupola mounted M85 .50 caliber machine gun was unsatisfactory in this application and it was replaced by a 7.62mm M73 machine gun. On 20 March 1962, the Chemical Corps Technical Committee standardized the E31R1-E36R1 as the main armament mechanized flame thrower M10-8. As before, the nomenclature indicated the M10 fuel and pressure unit and the M8 cupola group with the flame gun. On 21 March 1963, the AMCTC type classified the self-propelled flame thrower M132 as Standard A.

Since the increased cruising range of the M113A1 armored personnel carrier was a desirable feature, the M10-8 flame thrower was installed in the diesel powered vehicle. On 16 December 1963, the AMCTC type classified the self-propelled flame thrower M132A1 as Standard A and reclassified the M132 as Standard B. The M10-8 flame thrower was identical in both vehicles.

213

The cupola mounting the flame gun and the coaxial machine gun can be clearly seen in these views of the self-propelled flame thrower M132.

Combat loaded with the two man crew, the M132 weighed 23,330 pounds compared to 23,895 pounds for the M132A1. The flame gun fuel capacity was 200 gallons providing a firing time of 32 seconds. The cupola was traversed manually through 360 degrees and the flame gun had an elevation range of +55 to -15 degrees. The range for the flame gun extended from about 11 meters to 200 meters. The cupola was equipped with four vision blocks and an M28D sight for the flame gun operator.

Self-propelled Flame Thrower M132A1

Dimensions in inches (meters)

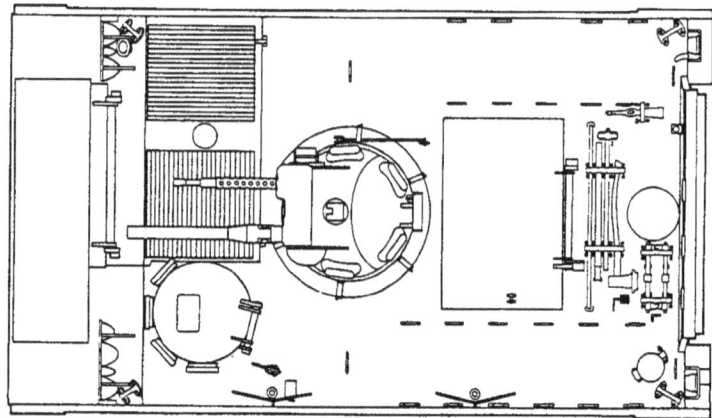

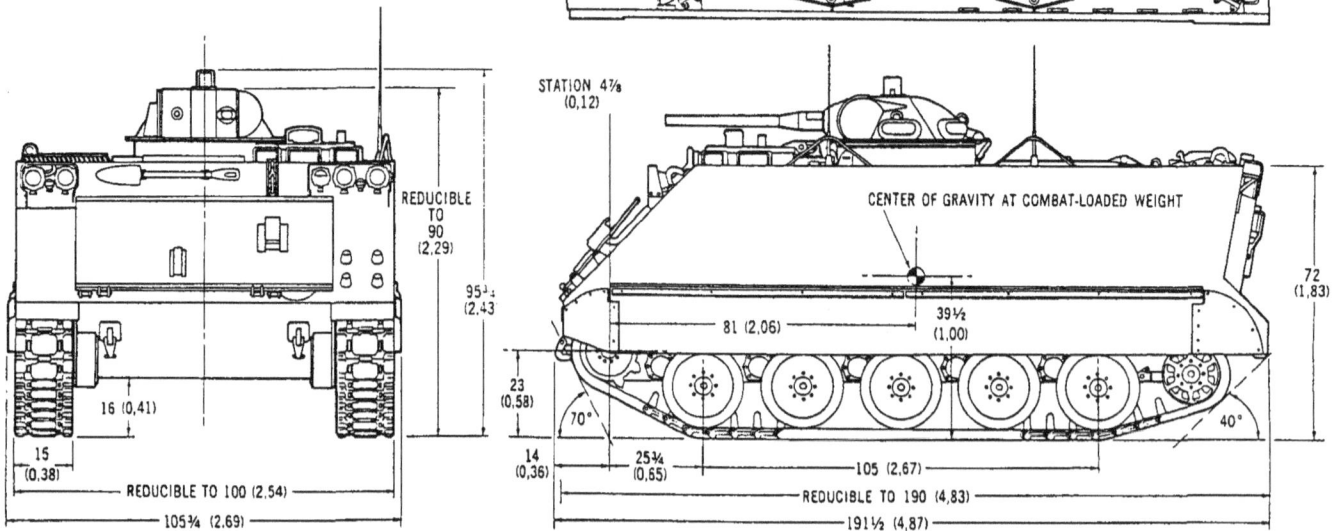

214

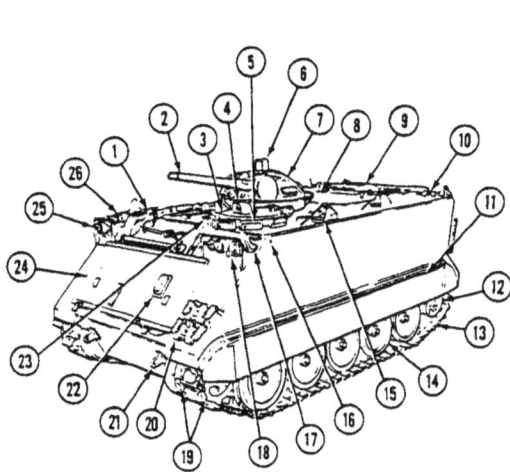

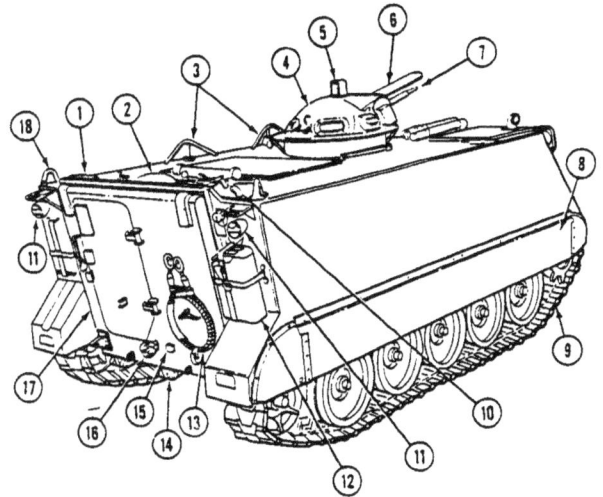

1. Exhaust grill
2. M10-8 flame gun
3. Driver's hatch
4. I.R. periscope (M19)
5. Driver's periscope (M17)
6. Cupola periscope guard
7. Flame thrower cupola
8. Cargo hatch
9. Personnel compartment air inlet ventilator
10. Fuel cap cover
11. Track shroud
12. Track idler wheel
13. Track
14. Road wheel
15. Fire extinguisher pull handle
16. Front bilge pump outlet
17. Lifting eye
18. Left headlight cluster
19. Drive sprockets
20. Spare track shoe
21. Towing eye
22. Trim vane latch
23. Trim vane control handle
24. Trim vane
25. Right headlight cluster
26. Horn

1. Fuel filler
2. Personnel compartment air inlet ventilator
3. Antenna guard
4. Flame thrower cupola
5. Cupola periscope guard
6. Flame thrower M10-8
7. 7.62 machine gun
8. Track shroud
9. Track
10. Rear bilge pump outlet
11. Tail light
12. Water can
13. Tow cable
14. Ramp hinge
15. Trailer electrical connection
16. Towing pintle
17. Rear ramp
18. Lifting eye

Components of the self-propelled flame throwers M132 and M132A1 are identified in the drawings above. Below, the self-propelled flame thrower is firing.

215

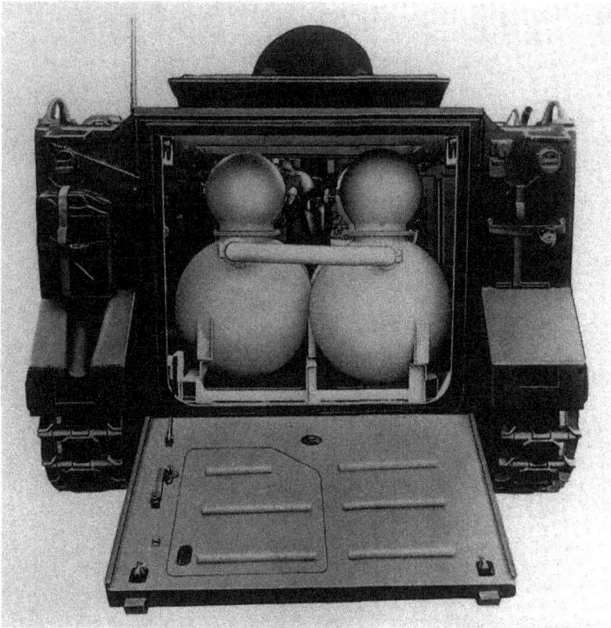

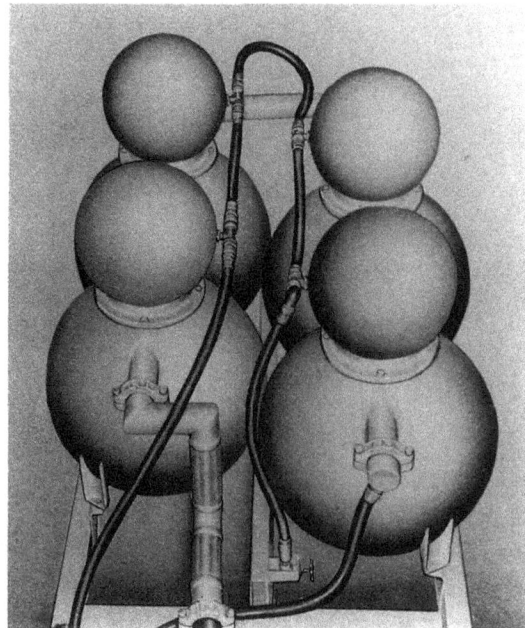

The M10 fuel and pressure unit installed in the personnel compartment consisted of four spherical 50 gallon fuel tanks, each with a spherical compressed air tank on top. The latter were pressurized to 3,000 pounds per square inch. The fuel tanks were pressurized to about 325 pounds per square inch and connected in series with the last tank connected to the rotating joint of the cupola group. The four compressed air tanks, also connected in series, supplied high pressure air to the pneumatic control unit and from it to the fuel tanks. The flame gun fuel was gasoline thickened by either M1 or M4 thickener. Production of the mechanized flame throwers by FMC totaled 201 M132s and 150 M132A1s.

Above, the open rear of the M132 appears at the left and details of the air and fuel tanks can be seen at the right. Components on the inside and on the rear of the self-propelled flame thrower M132 are identified in the drawings below.

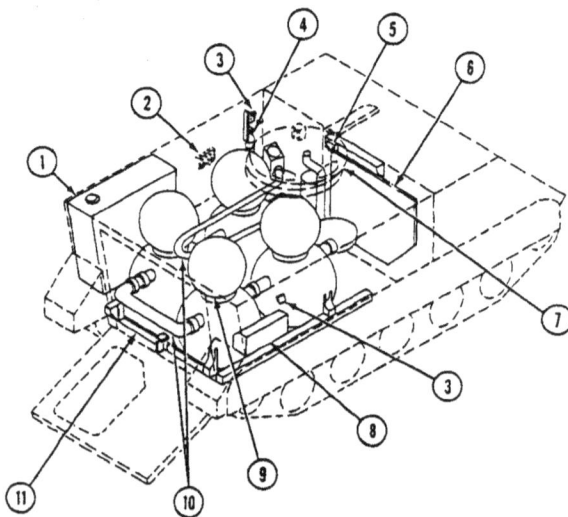

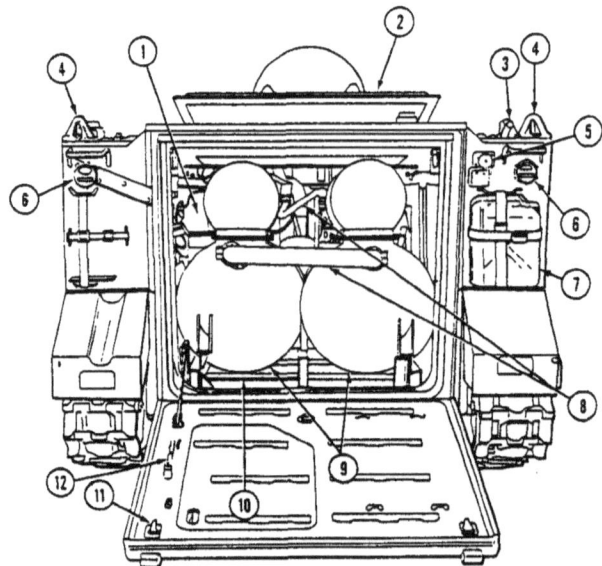

1. Carrier fuel compartment
2. Radio control box
3. Dome light
4. Fixed fire extinguisher
5. Intercom box
6. Rear power plant access panel
7. Flame thrower unit M10-8
8. Battery box
9. Flame thrower tanks
10. Flame hoses
11. Tank mounting frame

1. Driver's compartment
2. Rear cargo hatch
3. Rear bilge pump outlet
4. Lifting eye
5. Telephone connector
6. Tail light
7. Water can
8. Tank hoses
9. Flame thrower tanks
10. Tank mounting frame
11. Ramp lock
12. Ramp door handle

Above is the flame thrower service vehicle XM45E1 based upon the cargo carrier M548.

To support the mechanized flame throwers in the forward combat areas, a number of M548 cargo carriers were modified as resupply vehicles. Steel armor ¼ inch thick was added to the hull and an armored cab was installed with windshields 2¼ inches thick. Designated as the XM45E1, the vehicle was intended to mix and transfer thickened fuel and compressed air to the mechanized flame throwers. A power take-off was added to drive an air compressor and the vehicle cooling system was modified to cool the compressor and heat the fuel during mixing. A .50 caliber machine gun was mounted over the driving compartment.

Details of the XM45E1 flame thrower service vehicle can be seen below and at the right.

217

The M1059 smoke generator carrier appears above and below at the right.

The M1059 smoke generator carrier was based upon the M113A2 armored personnel carrier. It was modified to carry a single M157 smoke generator set consisting of two M54 smoke generators. The latter were mounted on the roof of the vehicle with armor protection. A 120 gallon tank inside the vehicle supplied fog oil for the smoke generators for about one hour without refueling. The two M54 smoke generators used MOGAS to fuel a pulse jet engine and produce heat to vaporize the fog oil. When the vaporized oil was released into the air, it condensed to form large clouds of white, visual obscurant, smoke. The M1059 was manned by a crew of three consisting of the commander, the driver, and the smoke generator operator. Remote controls permitted operation of the smoke generators from inside the vehicle. Fielding of the M1059 began in 1988 and was complete in 1990 for about 275 vehicles.

Components of the M1059 smoke generator carrier are identified in the drawings below.

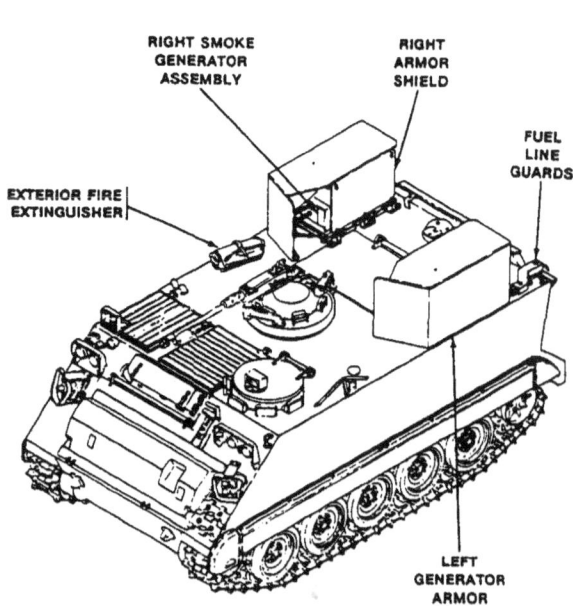

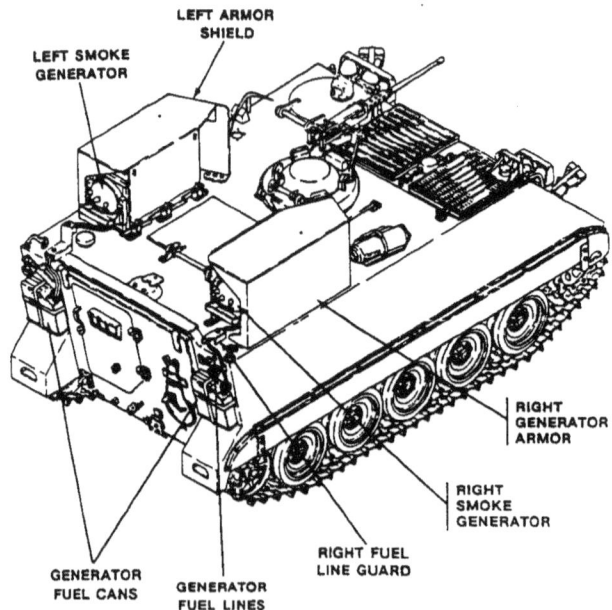

218

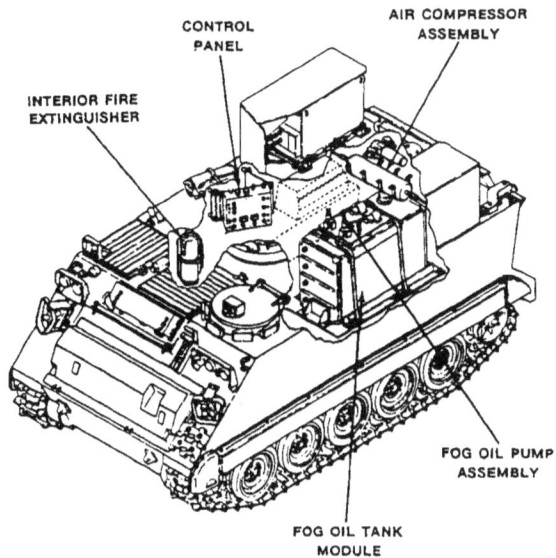

INTERIOR FIRE EXTINGUISHER

CONTROL PANEL

AIR COMPRESSOR ASSEMBLY

FOG OIL PUMP ASSEMBLY

FOG OIL TANK MODULE

Interior details of the M1059 smoke generator carrier can be seen in the cutaway drawing at the left. Below is a view of the smoke generator carrier M1059A3. Note the external fuel tanks.

The M1059A3 was an upgraded version of the M1059 smoke generator carrier. The new vehicle was based upon the M113A3 armored personnel carrier incorporating the RISE power package. Like the M113A3, it was fitted with the external fuel tanks. The M1059A3 utilized the M157A2 smoke generator set which produced multi-fuel options including diesel for smoke generation. The M1059A3 smoke generator carriers were classified as Standard A on 15 December 1994.

Below, the M1059A3 is making smoke.

Above is the large area mobile protected smoke system XM1101 after assembly at Red River Army Depot.

In early 1992, the Army initiated a project to investigate a new type of smoke generator. Referred to as the large area mobile protected smoke system (LAMPSS), it was to be capable of tailoring the smoke it produced for specific screening purposes. For example, to defeat infrared sighting equipment, thermal viewers, or image intensifiers, carbon particles were injected into the oil that was used to create the smoke. Another version was intended to provide a screen against millimeter wave seekers on some guided weapons. A demonstration vehicle was assembled in 45 days at the Red River Army Depot by converting an M901 improved TOW vehicle. The TOW launcher was replaced by an XM254 launcher for 38 2.75 inch Hydra-70 smoke rockets. These smoke rockets were intended to augment screens projected by mortars or the field artillery. In addition, a turbine smoke generator was capable of providing 90 minutes of visual or 30 minutes of infrared screening without refueling. The vehicle was operated by the same three man crew as the M1059.

The LAMPSS was now designated as the XM1101 and four prototypes were authorized for further evaluation. These vehicles were to be powered by the 350 horsepower 6V53TA engine with the Allison X200-4A transmission using the new driver's controls. These systems were road tested for over 5,000 miles at the Yuma Proving Ground, Arizona. However, the project was suspended in October 1993.

Further details of the XM1101 can be seen in the photographs below. Note the 38 tube launcher for the 2.75 inch Hydra-70 smoke rockets.

The smoke generator carrier M58 appears above with the side door open. Note the smoke grenade launchers on the front hull.

Operation Desert Storm had revealed the need to improve the mobility of the M1059 smoke generator carrier to permit it to operate with vehicles such as the Abrams tank and the Bradley fighting vehicle. In April 1993, an upgrade for both the vehicle and the smoke generating system was approved. Now designated as the M58 smoke generator carrier, the new vehicle was powered by the 275 horsepower 6V53TA RISE engine with the Allison X200-4A transmission. Named the Wolf, it was fitted with the turbine powered large area obscuration system. This was the system that provided 90 minutes of visual or 30 minutes of infrared obscuration without resupply. The millimeter wave capability was to be added as a future change. Procurement of 140 M58 systems was planned as replacements for the M1059s with the first unit to be equipped by the 4th quarter of fiscal year 1997.

Below, the M58 smoke generator carrier is moving at high speed and generating smoke. At the right is the XM87 NBC reconnaissance system.

As a possible solution to the problem of nuclear, biological, and chemical (NBC) hazards, the XM87 NBC reconnaissance system was under test during 1987. Based upon the M113A3 armored personnel carrier, the XM87 was equipped with sensors to detect and identify various NBC agents. The M113A3 was modified by the installation of two 200 ampere alternators to support the specialized equipment and the rear ramp was replaced by one that allowed the taking of surface samples. The mission of the XM87 was to be the detection, identification, and reporting of all NBC hazards.

Above is pilot number 3 of the command and reconnaissance vehicle T114, registration number 12M651, at Fort Knox during its evaluation.

COMMAND AND RECONNAISSANCE VEHICLES

The military characteristics for the new family of light armored vehicles issued by the Army Field Forces on 30 September 1954 described an 8,000 pound, four man, carrier to be used as a command and reconnaissance vehicle. It also was to provide a mount for the battalion antitank (BAT) rifle and be suitable for use as a front line litter carrier. In addition to the 8,000 pound weight limit, the little vehicle was to have a top speed of 45 miles per hour, a cruising range of at least 200 miles, and be able to tow a vehicle of its own weight on improved roads. Detroit Arsenal studied various concepts of tracked and wheeled, four man, carriers and

recommended that mock-ups be constructed of both types for evaluation. As mentioned before, the designations T114 and T115 were assigned to the tracked and wheeled design concepts respectively. The mock-up of the T114 was built by the Cadillac Division of General Motors Corporation and the T115 mockup was completed by Chrysler at Detroit Arsenal. Because of delays in negotiating the General Motors contract, both mock-ups were not available for evaluation until June 1957. At that time, the decision was made to abandon the wheeled vehicle and proceed with the development of the tracked T114. Also in June 1957, a contract was awarded to the Cadillac Division of General Motors to design the T114. It was amended in June 1958 to include the construction of six pilot vehicles. The first four were to be completed as command and reconnaissance vehicles and numbers 5 and 6 were to be carriers for the BAT weapon. The latter development is described in the section on antitank vehicles. The four command and reconnaissance vehicle pilots were completed during 1960 and one was retained by the manufacturer for tests. The other three were shipped to Aberdeen Proving Ground, the Arctic Test Board, and Fort Knox.

At the left is the mock-up of the four man, wheeled, armored carrier T115. The only armament proposed was a .50 caliber machine gun for the vehicle commander.

The photographs on this page show additional views of T114 command and reconnaissance vehicle pilot number 3. The interior of the vehicle can be seen below at the left through the open rear door.

The early production command and reconnaissance vehicle T114 appears above and below. Note the flat roof and box-like hull compared to the pilot vehicles.

In February 1961, a meeting to review the development of the T114 concluded that the estimated production costs of the T114 were excessive and recommended a redesign to reduce these costs. A mock-up of the revised design was inspected in May 1961.

Several changes were obvious. The hull roof on the original pilots was sloped at the sides and rear. This was replaced by a flat roof providing a simple box-like shape to the rear part of the hull. The vehicle was a welded assembly of 5083 aluminum alloy armor plate. On the

On the early production command and reconnaissance vehicle T114 above, a cupola with a .50 caliber machine gun has replaced the small turret on the pilots.

front hull, a wider trim vane or surfboard was installed with cutouts around the headlight groups. The rectangular door with round corners in the rear wall was replaced by a circular door. The complex two man turret armed with the M85 .50 caliber machine gun was eliminated and replaced by a simple one man turret for the commander on the left side of the hull roof behind the driver. It was armed with an externally mounted M2HB .50 caliber machine gun. The commander had a 360 degree view through eight vision blocks. A flat hatch on the right rear of the hull roof was for the observer. Periscopes were installed in the hull roof just forward and behind the observers hatch and two pedestal mounts for a .30 caliber machine gun were located on the roof adjacent to the hatch. The driver remained in his position in the left front hull. For closed hatch operation, he was provided with three hull mounted M26 periscopes and his hatch was fitted for the installation of the M19 infrared periscope.

The round rear door on the production T114 can be seen below and the arrangement of the various components is shown in the drawing at the left.

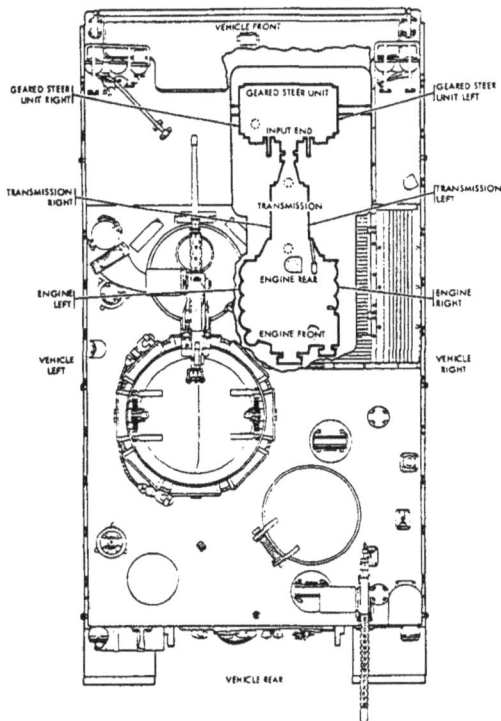

225

The command and reconnaissance vehicle M114 appears above. A 7.62mm M60 machine gun is on the rear pedestal by the observer's hatch.

Although the T114 was operated by a three man crew consisting of the driver, commander, and observer, a jump seat was provided for a fourth man. The Chevrolet V8 gasoline engine developed 160 gross horsepower at 4,200 rpm. This liquid-cooled engine was installed in the right front hull with the 305MC Hydramatic transmission and the geared steer unit. The front mounted sprockets drove the vehicle on the 16½ inch wide, band type, tracks. The flat track, torsion bar, suspension supported the vehicle on four road wheels per side. Track tension was adjusted by an idler at the rear of each track. The vehicle was amphibious without any special preparation and it was propelled in the water by the tracks.

OTCM 37970, dated 29 December 1961, classified the T114 as a limited production type and 615 of the vehicles were produced during 1962. Some additional changes were made in the production vehicle. The roof mounted periscopes for the observer were eliminated and replaced by a single rotating M13 periscope in his hatch cover. The .30 caliber pedestal mounted weapon was replaced by a 7.62mm M60 machine gun.

Below, components of the M114 are identified by the numbers in the left drawing: 1. Lights, 2. Trim vane, 3. Steer unit access, 4. Bilge pump, 5. Air inlet and exhaust grille, 6. M19 periscope, 7. Machine gun support, 8. M60 machine gun, 9. Gun mount M142, 10. M13 periscope, 11. Observers hatch, 12. Air vent, 13. Crow bar, 14. Fixed fire extinguisher, 15. Fuel cap cover, 16. M26 periscope, 17. Driver's hatch, 18. Power plant warning light, 19. Suspension and tracks, 20. Towing lug. In the right drawing the numbers indicate the following: 1. Parking brake, 2. Steering selector, 3. Radiator filler cap, 4. Batteries, 5. Engine oil dip stick, 6-7. Transmission oil filler tube, 8. Transmission shaft, 9. V-belt drives, 10. NBC filter unit, 11. Personnel heater control, 12. Observer's seat, 13.Intercom, 14. Dome light, 15. Tow pintle kit, 16. Rear door, 17. Passenger seat, 18. Radio, 19. Fire extinguisher, 20. Throttle, 21. Choke, 22. Driver's seat, 23. Steer bar, 24. Instrument panel, 25. Brakes, 26. Fuel shut-off, 27. Accelerator, 28. Auxiliary power receptacle, 29. Accessory receptacle, 30. Fuel drain, 31. Air cleaner, 32. Geared steer oil filler tube.

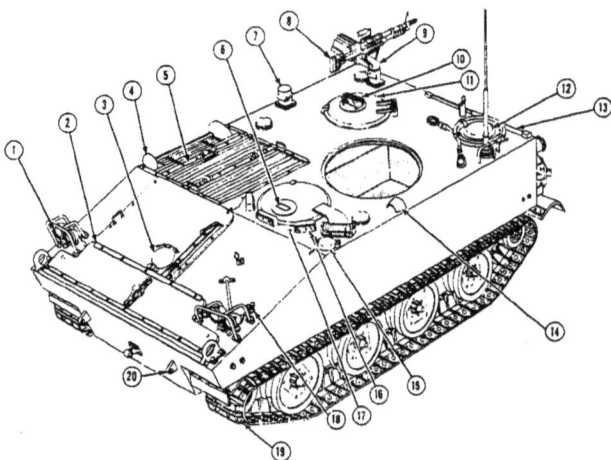

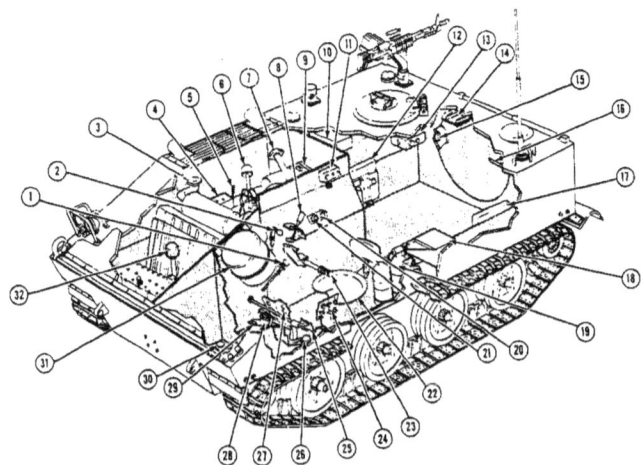

The command and reconnaissance vehicle M114 appears in these photographs during its test at Fort Knox.

Below, the command and reconnaissance vehicle M114 is swimming.

1 – M26 PERISCOPE	12 – DRIVER'S HATCH COVER LOCKING LEVER
2 – INDICATOR PANEL	13 – PERSONNEL HEATER CONTROL PANEL
3 – STEER BAR	14 – HEADLIGHT DIMMER SWITCH
4 – FUEL SIGHT TUBE	15 – TRANSMISSION SHIFT LEVER
5 – ACCESSORY OUTLET RECEPTACLE	16 – CHOKE CONTROL KNOB
6 – DRIVER'S SWITCH PANEL	17 – THROTTLE CONTROL KNOB
7 – AUXILIARY POWER RECEPTACLE	18 – STEERING SELECTOR LEVER
8 – FUEL SHUT-OFF VALVE	19 – ENGINE AIR CLEANER
9 – BRAKE PEDAL	20 – PARKING BRAKE LOCK
10 – FOOT REST	21 – ACCELERATOR PEDAL
11 – FIXED FIRE EXTINGUISHER	22 – DRIVER'S SEAT

1. HATCH COVER LOCKING ASSEMBLY	6. INTERCOM BOX
2. DOME LIGHT	7. VISION BLOCK (B)
3. RADIO EQUIPMENT	8. COMMANDER'S SEAT BACKREST
4. COMMANDER'S SEAT	9. M13 SPARE PERISCOPE STOWAGE BOX
5. HEIGHT ADJUSTMENT HANDLE (PULL UP AND WITH BODY WEIGHT RAISE OR LOWER SEAT).	

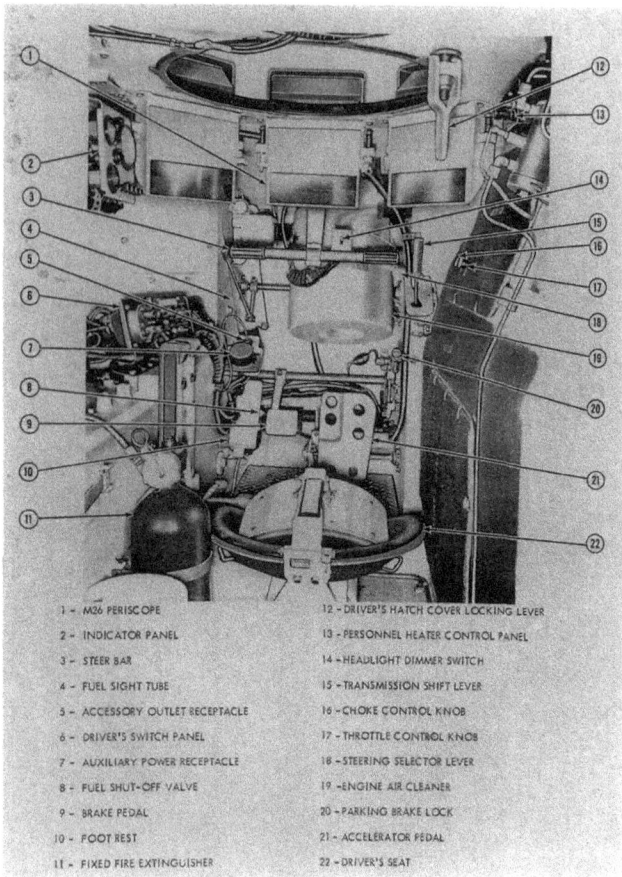

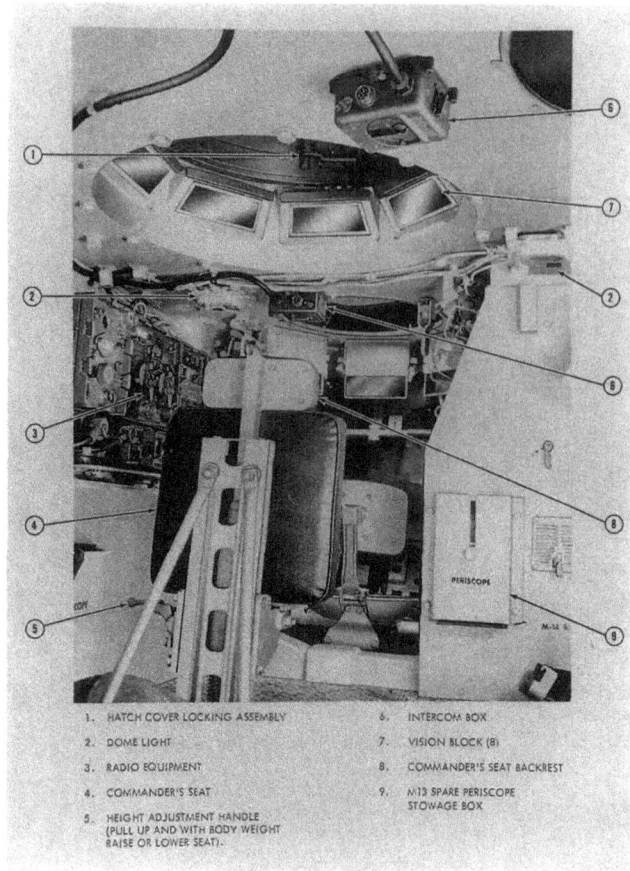

The controls and instruments for the driver (left) and the commander (right) in the M114 command and reconnaissance vehicle are identified above. The pilot T114s were used to evaluate a variety of armament. Below, the 20mm Hispano Suiza gun is installed in a ring mount (left) and in the turret mount KuKa 763 (right).

The T114 pilots were popular for testing armament. Below, T114, registration number 12M651, is armed with the 20mm Oerlikon gun in an open mount (left) and in a small turret (right).

The command and reconnaissance vehicle M114A1 appears above and the components at the commander's station are identified at the right.

Several weapons and mounts were evaluated to improve the firepower of the T114. Some of these were tested on the early pilot vehicles. This program resulted in the development of a new Model X commander's station which was introduced to replace the turret. Initially, the new station was armed with a turret type M2HB .50 caliber machine gun and it could be aimed and fired from inside the vehicle without exposing the commander. The machine gun was elevated and traversed manually and could be fired electrically or manually. When fitted with the new commander's station, the vehicle was designated as the T114E1. The remaining 600 vehicles in the original production order of 1,215 were completed as T114E1s during 1962. On 16 May 1963, AMCTCM Item 966 type classified the T114 and the T114E1 as the armored command and reconnaissance carriers M114 and M114A1 respectively. A total of 2,495 M114A1s were delivered during 1963 and 1964 bringing the total run to 3,710 vehicles.

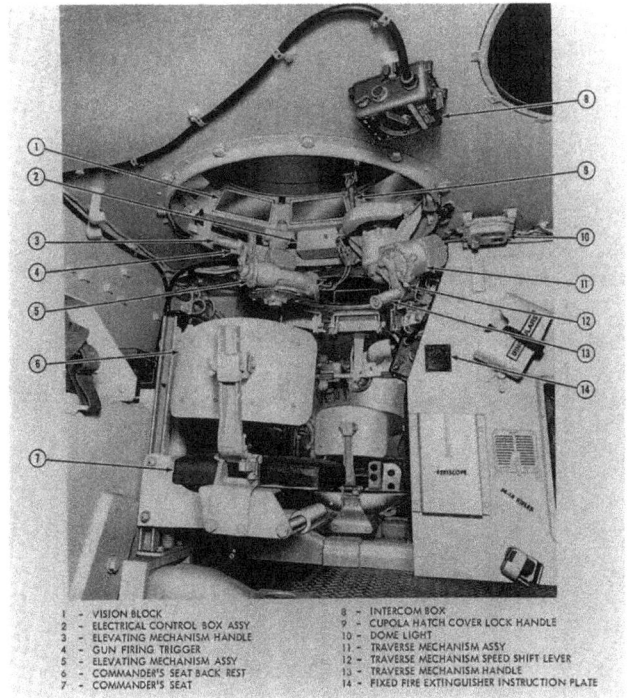

1 - VISION BLOCK
2 - ELECTRICAL CONTROL BOX ASSY
3 - ELEVATING MECHANISM HANDLE
4 - GUN FIRING TRIGGER
5 - ELEVATING MECHANISM ASSY
6 - COMMANDER'S SEAT BACK REST
7 - COMMANDER'S SEAT
8 - INTERCOM BOX
9 - CUPOLA HATCH COVER LOCK HANDLE
10 - DOME LIGHT
11 - TRAVERSE MECHANISM ASSY
12 - TRAVERSE MECHANISM SPEED SHIFT LEVER
13 - TRAVERSE MECHANISM HANDLE
14 - FIXED FIRE EXTINGUISHER INSTRUCTION PLATE

The numbers indicate the components at the commander's station below. The M114 is at the left: 1. .50 caliber flexible machine gun, 2. Commander's hatch cover, 3. Commander's seat, 4-5. Hatch cover support and ring, 6. Machine gun pintle support. The M114A1 is at the right: 1. .50 caliber turret type machine gun, 2. Traverse mechanism, 3. Commander's hatch cover, 4. Commander's seat, 5. Elevating mechanism, 6. Fire control box, 7. Machine gun travel lock.

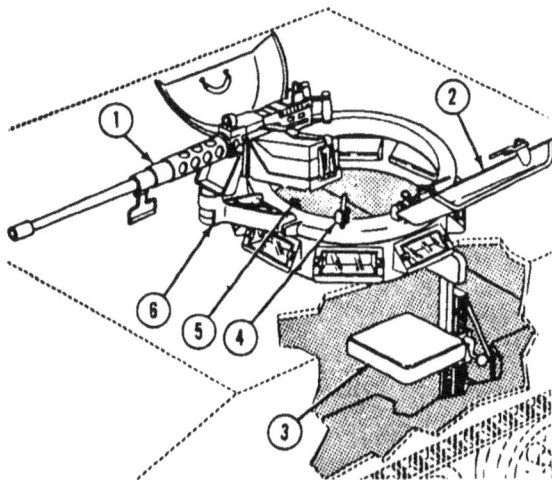

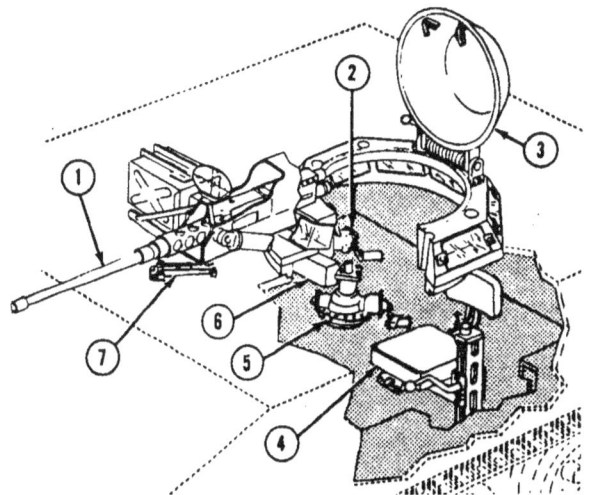

Above and at the right is the command and reconnaissance vehicle M114A1 with the turret type .50 caliber machine gun M2 HB in the XM26 commander's station. Below, the .50 caliber machine gun has been replaced by the 20mm gun M139.

Below are additional views of the command and reconnaissance vehicle armed with the 20mm gun M139. The cupola mount is now designated as the XM27.

On 2 November 1961, OTCM 37905 had initiated the development of the vehicle rapid fire weapon system (VRFWS) including a 20mm high performance gun suitable for use on armored vehicles such as the command and reconnaissance carriers. This weapon was intended for use with the new commander's station, but it was still under development when the station was authorized for installation on the T114E1. When the new 20mm gun became available, the designation of the commander's station was changed. It now became the XM26 when armed with the .50 caliber machine gun and the XM27 when fitted with the 20mm gun M139. On 22 August 1963, a meeting at Aberdeen Proving Ground decided that the manual controls on the XM26 and XM27 cupolas should be replaced by hydraulic power controls. When the new XM27 cupola with the 20mm gun was installed on the M114 and M114A1, they were designated as the armored command and reconnaissance carriers M114E2 and the M114A1E1 respectively. Later, both of these cupola mounts were standardized as the M26 and M27.

The M114 and the M114A1 had combat weights of 15,093 pounds and 15,276 pounds respectively. On the M114A1E1, the combat weight increased to 15,455 pounds. The M114 series had a maximum speed of 36 miles per hour on roads and 3.6 miles per hour in water. The cruising range on roads was about 275 miles.

The command and reconnaissance vehicle M114A1E1 is shown in these photographs.

Additional photographs of the M114A1E1 are below. Note the extension of the front hull beyond the tracks.

Above is the command and reconnaissance vehicle M114A1. At the bottom of the page, an M114 has its nose buried in a ditch. This illustrates the problem presented by the front hull overhang to cross-country mobility.

The M114 series was deployed with armored cavalry units in both the United States and Europe. They also were provided to the Army of the Republic of Vietnam (ARVN). A total of 80 M114s were shipped to Vietnam to equip four ARVN armored cavalry squadrons and to provide vehicles for the ARVN Armor School. The performance of the M114 was evaluated for a year by the Army Concept Team. Although there were no important organizational or logistical problems, a serious flaw in the design of the M114 soon became obvious. The vehicle was limited in its ability to move cross-country and had difficulty in entering and leaving waterways. Part of the problem was the fact that the front hull of the M114 extended out in front of the tracks. On a steep bank, the hull would contact the bank before the tracks and prevent the vehicle from climbing out. This greatly limited the usefulness of the vehicle in Vietnam and they were soon replaced by M113 armored personnel carriers. This performance was not lost on General Creighton Abrams and when he became Chief of Staff of the U. S. Army he ordered the vehicle retired from service in 1973.

The armored cavalry cannon vehicle appears at the right.

After the M114 reconnaissance vehicle was retired from service, the M113 was modified to provide an interim armored cavalry reconnaissance vehicle. One such modification evaluated at Fort Knox during 1975 was the armored cavalry cannon vehicle (ACCV). This was an M113A1 fitted with the M27 weapon station formerly used on the M114A1E1. On this weapon station, the 20mm gun M139 could be aimed and fired without opening the cupola and exposing the gunner.

In early 1963, FMC had presented a new armored command and reconnaissance vehicle based upon the components of the highly successful M113 armored personnel carrier. The initial design offered the choice of either the gasoline or diesel engine. However, the vehicle was developed using the diesel power package from the M113A1 armored personnel carrier. Sometimes referred to as the M113A1½, the command and recon-naissance vehicle was smaller than the M113A1 and it was manned by a crew of three. The driver and observer were in the front hull on the left and right respectively. The commander's station was in the center of the vehicle. On the early pilot, this was a mock-up of the Model X commander's cupola fitted with a dummy 20mm gun. Each crew member had a hatch above his station. The driver and observer each had four M17 periscopes in the hull roof around their hatch. The driver's hatch cover was fitted for the installation of an M19 infrared periscope. With the M26 cupola, the commander had eight vision blocks for a 360 degree view. An access door was in the right side of the hull adjacent to the commander's station.

The photographs below and the two views above at the right show the command and reconnaissance vehicle developed by FMC, the so-called M113 ½ . Note the snub nose with little front overhang.

233

The driver's station in the FMC command and reconnaissance vehicle is above and the side access door in the hull is at the right.

The command and reconnaissance vehicle was assembled by welding aluminum alloy armor plate with a thickness of 1¼ inches on the top, sides, rear, and upper front. The lower front and the floor were 1¾ inches and 1 inch thick respectively. The vehicle was amphibious without any special preparation. A trim vane was installed on the front hull and folding water shields or snorkels were fitted around the intake and exhaust grilles on the hull roof.

On this little, rear engine, front drive vehicle, the General Motors 6V53 diesel engine and the Allison TX-100 transmission were installed in the rear hull. Power was transmitted by a propeller shaft to the DS200 controlled differential in the front and then through the final drives to the sprockets for each track. The vehicle rode on the same 15 inch wide T130E1 tracks as the M113 series. The flat track, torsion bar, suspension had four road wheels per side. An adjustable idler was at the

rear of each track. The command and reconnaissance vehicle had an overall width of only 95 inches. As a result, the torsion bars were shorter than those on the M113 suspension. When completed, the armament was a .50 caliber machine gun in an M26 cupola mount for the vehicle commander. This could be replaced by the M27 cupola mount with the 20mm gun M139, the Model 74 cupola with two 7.62mm M73 machine guns, the Model 100-E cupola with a single 7.62mm machine gun, or the .50 caliber machine gun on the external mount used on the M113 armored personnel carrier. Experimental mounts also were proposed for weapons up to 25mm which could be loaded, aimed, and fired from within the vehicle. Another option was the installation of a 7.62mm M60 machine gun on an external mount in front of the observer's hatch. Other remote control 20mm gun mounts were proposed for the commander's station as well as Entac or TOW missile installations.

Below, the armor thickness on various parts of the FMC command and reconnaissance vehicle is shown at the left and the power train, suspension, and tracks are at the right.

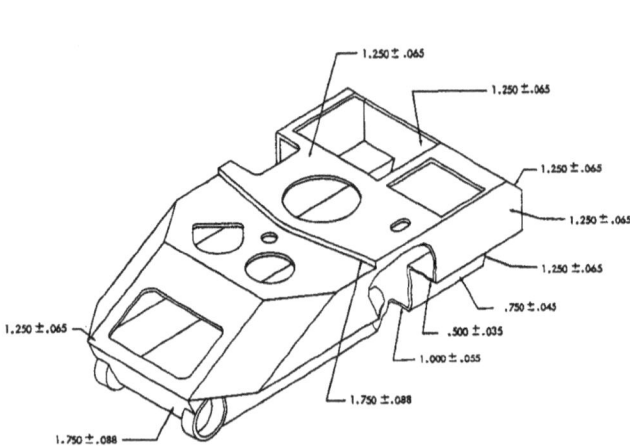

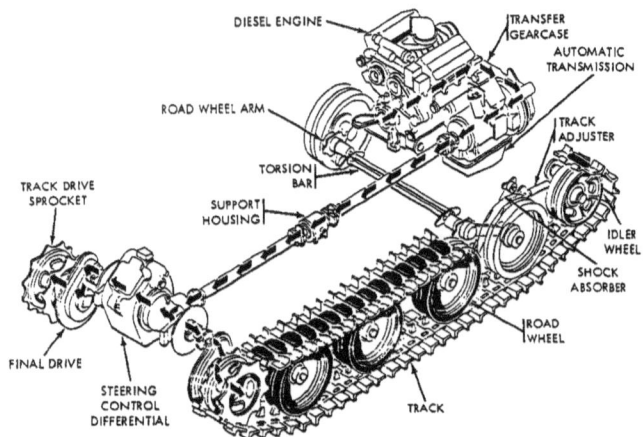

Above are two views of the FMC command and reconnaissance vehicle fitted with the XM26 cupola armed with the .50 caliber machine gun. Below at the left, the machine gun has been replaced by the 20mm gun M139.

Combat loaded, the command and reconnaissance vehicle weighed 18,650 pounds. The maximum speed was 40 miles per hour on roads and 4.1 miles per hour in water. The cruising range on roads was approximately 325 miles. The blunt front hull exposed the tracks allowing the vehicle to climb steep banks and easily exit from most waterways.

Unfortunately, by the time the FMC command and reconnaissance vehicle appeared, the M114 was already in full production and it was not procured for the U. S. Army.

FMC Command and Reconnaissance Vehicle

Dimensions in inches

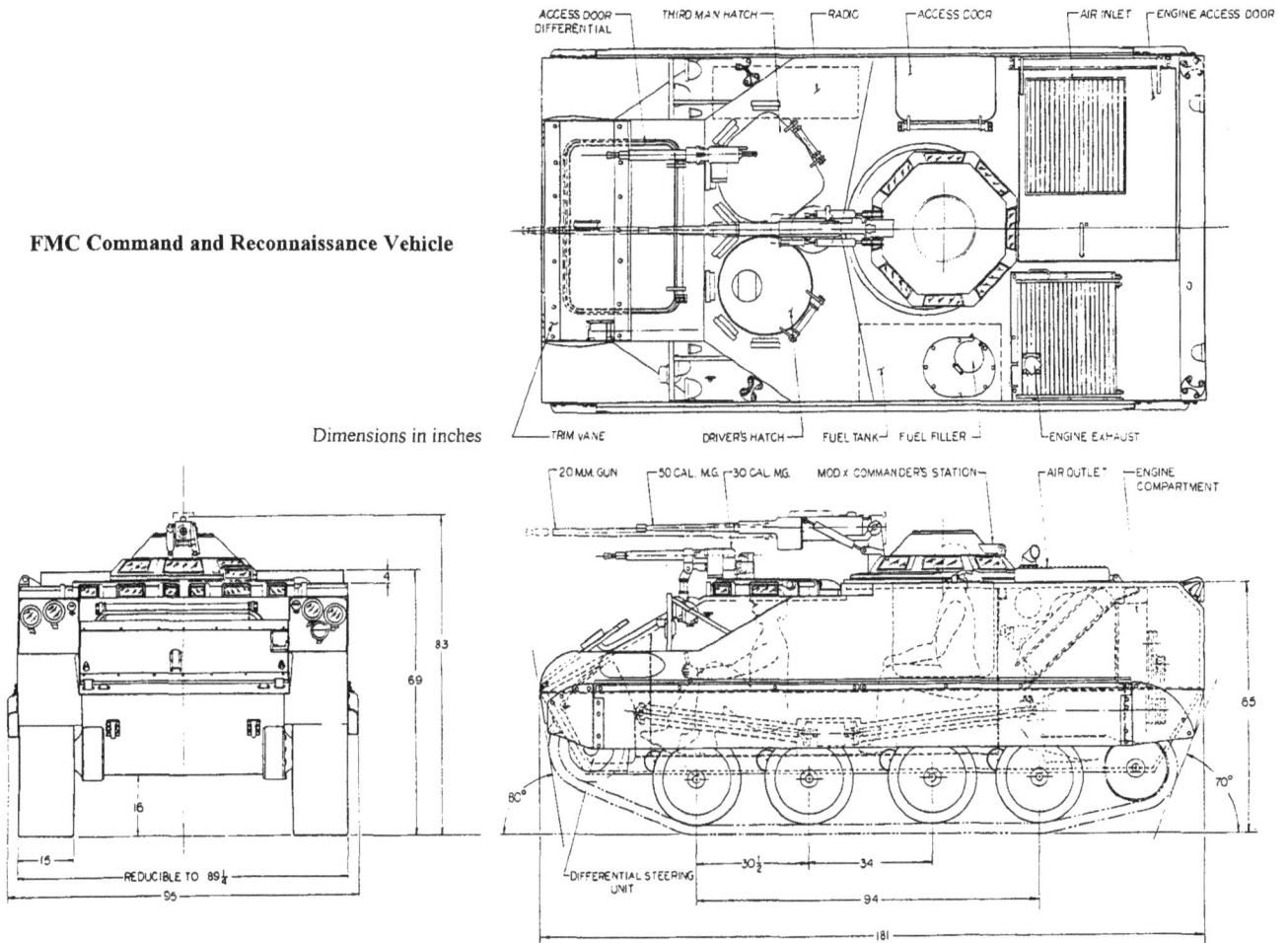

Above, the FMC command and reconnaissance vehicle is fitted with the Model 74 cupola armed with two 7.62mm machine guns. Below, the Model 100-E cupola is at the left mounting a single 7.62mm machine gun and at the right is an artist's concept of a proposed remote control .50 caliber machine gun mount that could be retracted inside the vehicle for reloading.

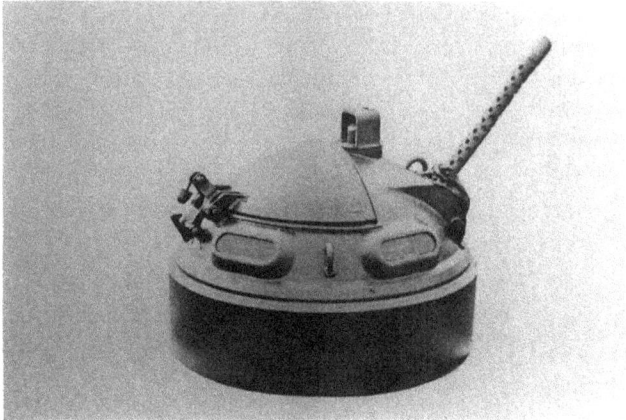
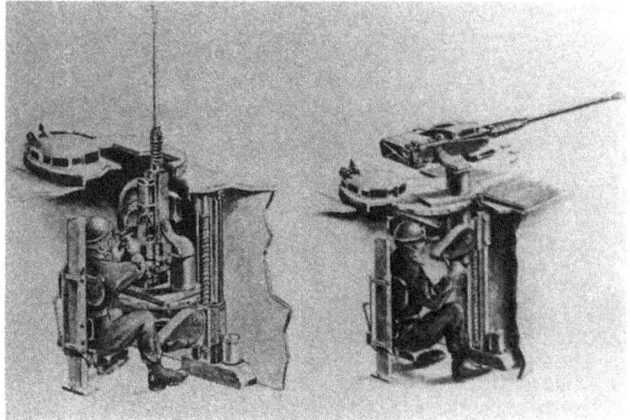

Above is an experimental 20mm gun installation on the command and reconnaissance vehicle at the left and at the right is an artist's concept of a 20mm gun cupola proposed by Cadillac Gage. Below are artist concept drawings of the vehicle armed with Entac missiles (left) and the TOW missile (right).

The FMC command and reconnaissance vehicle above has the cupola with the external mount for the .50 caliber machine gun. The snorkel water shields are erected around the grilles for swimming. Below, the vehicle is fitted with the M26 cupola mount for the .50 caliber machine gun.

The FMC command and reconnaissance vehicle Lynx appears in the photographs on this page. Note the new crew arrangement compared to the earlier vehicle. The vehicle is armed with a .50 caliber machine gun in the M26 cupola mount. In the top photograph, a .30 caliber machine gun is provided for the observer. At the bottom right, the .30 caliber weapon has been replaced by a 7.62mm M60 machine gun.

Although it never served in the U. S. Army, the FMC command and reconnaissance vehicle was purchased by the Netherlands and in a modified form by Canada. The Netherlands ordered over 260 of the little vehicles and the first was completed in September 1966. During their service, they were fitted with a variety of armament including an Oerlikon turret armed with a 25mm gun.

The last of the 174 Canadian vehicles was delivered at the end of October 1968. Named the armored, full tracked, command and reconnaissance carrier Lynx, these vehicles differed in several respects from the original FMC vehicle. The most obvious change was the relocation of the three man crew. The driver remained in his original position in the left front hull, but the observer, now called the radioman-observer, was relocated to the left side of the hull behind the driver. The driver had five M17 periscopes in the hull roof around his hatch and his hatch cover was fitted for the installation of the M19 infrared periscope. The radioman-observer had three M17 and two M17C periscopes in the hull roof around his hatch. A 7.62mm machine gun was on an external mount behind his hatch. The commander-gunner in his M26 cupola with the .50 caliber machine gun and eight vision blocks, was offset to the right center of the hull. The side access door on the earlier vehicle was eliminated and a floor escape hatch was installed in the crew compartment. Two

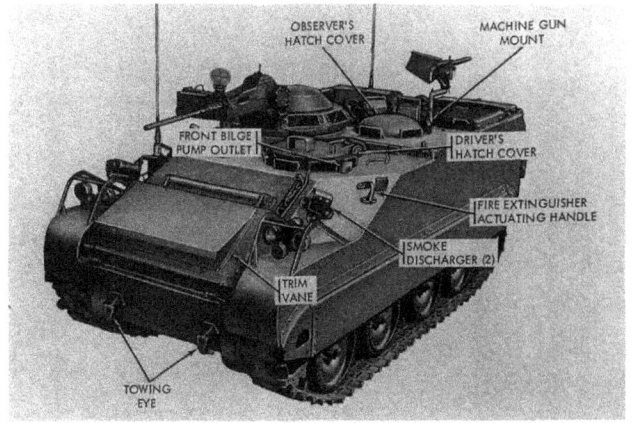

Above and at the right, components on the FMC command and reconnaissance vehicle Lynx are identified. Note that this vehicle is armed with a .30 caliber machine gun for the observer.

smoke grenade launchers were mounted, one behind each headlight group. The Lynx had a combat loaded weight of 19,340 pounds and an air-drop weight of 17,030 pounds. Its performance was essentially the same as the earlier command and reconnaissance vehicle.

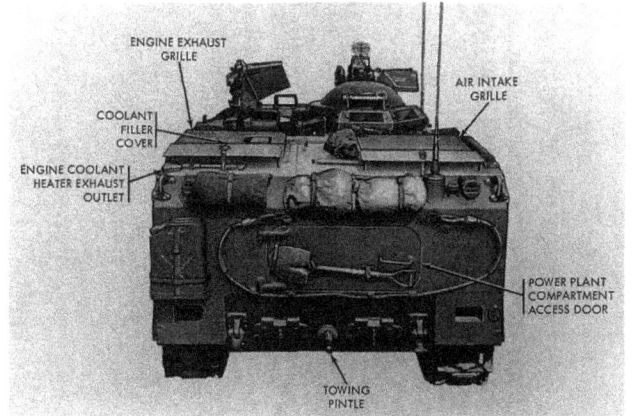

FMC Command and Reconnaissance Vehicle Lynx

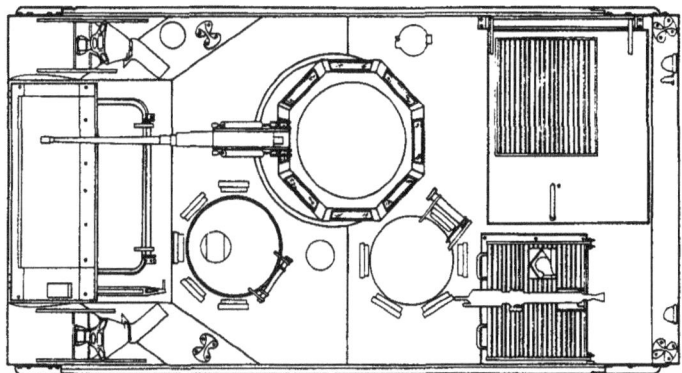

Dimensions in parentheses ( ) are in meters; all other dimensions are in inches.

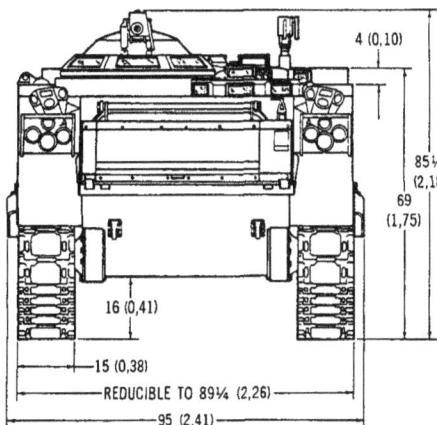

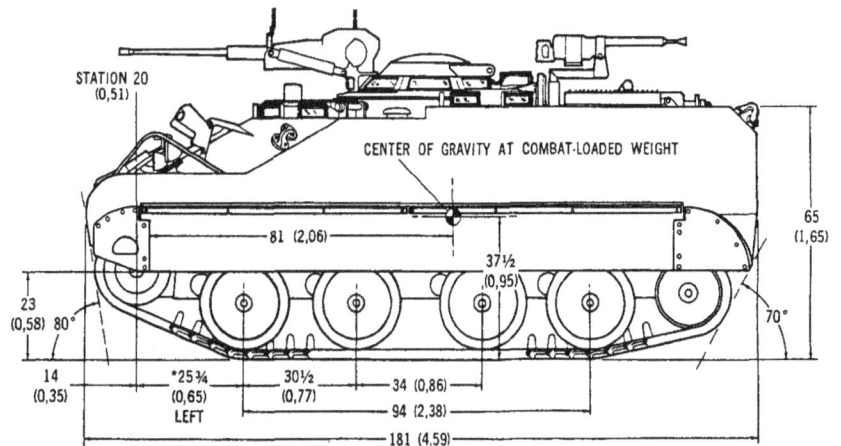

239

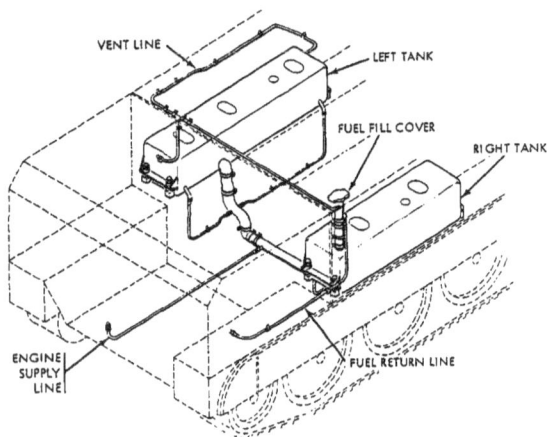

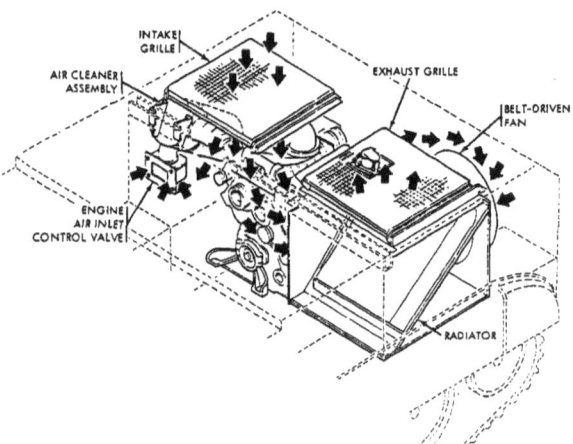

Above, the fuel system on the Lynx is sketched on the left and the air flow through the engine radiators is shown at the right.

In 1970, a prototype for a new version of the command and reconnaissance vehicle was built by FMC. Utilizing the experience obtained during the development of the PI M113A1, it was powered by the turbocharged 6V53T diesel engine developing 260 horsepower. A torsion tube-over-bar suspension was installed improving the cross-country performance. Referred to as the Recon or the PI M113A1½, it was evaluated along with other candidates for a new command and reconnaissance vehicle.

At the right is a model of a proposed FMC command and reconnaissance vehicle with a turret mounted gun. Below is the product improved M113A1½ or Recon vehicle.

The drawings above show the two versions of the armored reconnaissance scout vehicle. The wheeled XM800W is at the left and the tracked XM800T is at the right.

The disappointing performance of the M114 resulted in concept studies for a new command and reconnaissance vehicle. Beginning in 1966, these studies evolved into a requirement for a three man, lightly armored, vehicle weighing about 7 tons. It would be armed with the 20mm gun M139 and be suitable for mounting the vehicle rapid fire weapon system (VRFWS) when its development was complete. The proposed vehicle was assigned the designation XM800 armored reconnaissance scout vehicle (ARSV) and it could be either tracked or wheeled. The Army issued a request for proposals in the latter part of 1971 and received six concepts for evaluation. The six competitors were Chrysler, CONDEC, FMC, Ford, Lockheed, and Teledyne-Continental Motors. Chrysler, FMC, and Teledyne-Continental proposed a tracked vehicle while CONDEC, Ford, and Lockheed offered a wheeled design. On 23 May 1972, contracts were awarded for the development of one wheeled and one tracked design. The wheeled vehicle, now referred to as the XM800W, was that proposed by the Lockheed Missiles and Space Company. The tracked design, the XM800T, was to be developed by FMC. Each contract called for the detailed design, development, and construction of four pilot vehicles in addition to providing a hull for ballistic tests. One of the prototype vehicles would remain with the manufacturer and the other three would be submitted to the Army for evaluation. The final requirements in the request for proposal specified a vehicle weight of 17,000 pounds, a maximum speed not less than 50 miles per hour, a cruising range of 300 miles, inherent flotation with a minimum 10 inch freeboard, armament consisting of the 20mm M139 gun with the capability of mounting the 20-30mm Bushmaster when available, and a stabilized weapon system with a laser range finder.

Below are two views of the wheeled armored reconnaissance scout vehicle XM800W built by the Lockheed Missiles and Space Company.

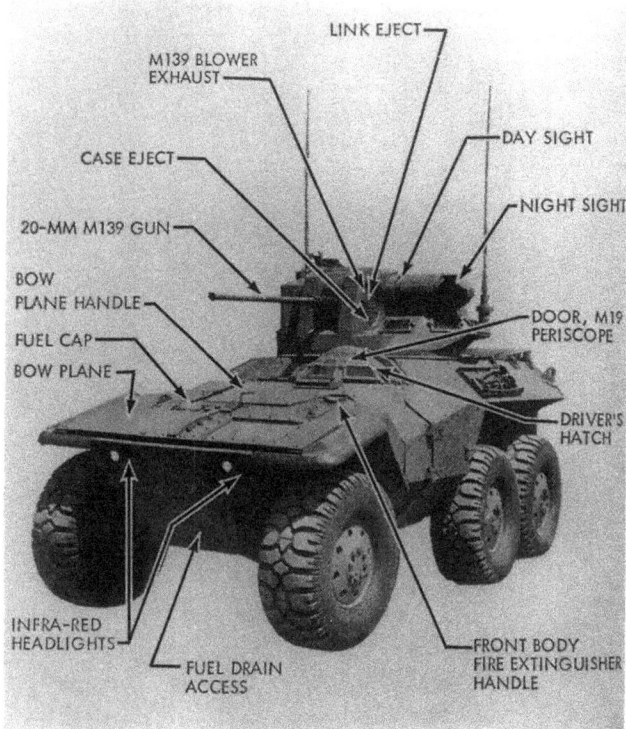

The components of the XM800W armored reconnaissance scout vehicle are identified above and the dimensions are shown at the right.

As originally proposed, the three man Lockheed XM800W was a six wheel vehicle weighing 16,972 pounds. Based upon the earlier XM808 Twister, it was a two body vehicle with roll articulation between the two bodies. The front body mounted the two wheel steerable front suspension and it contained the fuel tank, drive line, and front differential. The driver's station, turret, engine, transmission, drive lines, rear differential, and walking beam suspension with four wheels were installed in the rear body. The lightweight two man turret was armed with the 20mm gun M139. An M60D 7.62mm machine gun was carried on an external mount. The hull was assembled from aluminum alloy armor with dual hardness steel and aluminum applique. The turret was cast aluminum and dual hardness steel. The vehicle was driven by the 6V53T diesel engine with an aluminum block and it developed 300 horsepower at 2,800 rpm. This was the engine used in the M551 Sheridan. The Allison MT650 transmission had five speeds forward and two in reverse. The maximum road speed was estimated as 65 miles per hour and the 90 gallon fuel tank provided a cruising range of approximately 450 miles. The Aerojet water-jet propulsion unit was expected to drive the vehicle at a maximum speed of 5 miles per hour in water.

The prototype XM800W is swimming at the right.

Dimensions

242

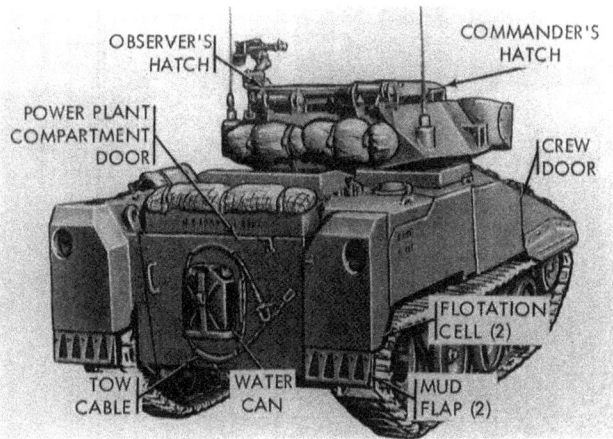

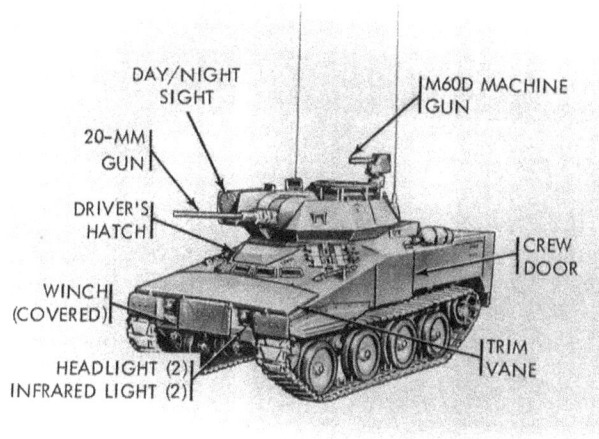

The external components of the armored reconnaissance scout vehicle XM800T built by FMC are identified in the views above. Details of the turret armed with 20mm gun M139 can be seen below at the right.

The FMC XM800T also was a three man vehicle with the driver located in the center front hull. It had an estimated weight of 18,188 pounds. The hull and turret were assembled from 5083, 7039, and Kalshield dual hardness aluminum armor. As on the Lockheed vehicle, the two man turret was armed with the M139 20mm gun and an externally mounted M60D 7.62mm machine gun. The XM800T also was powered by the Sheridan engine although it was derated to 285 horsepower to improve reliability. It drove the vehicle through the Allison X200 transmission with hydrostatic steering. The sprockets were at the rear of the 19 inch wide, double pin, tracks. The torsion bar suspension supported the vehicle on four dual road wheels per side. The estimated performance included a maximum road speed of 52 miles per hour with a cruising range of 450 miles. In water, track propulsion was expected to drive the vehicle at 4.5 miles per hour.

Details of the 7.62mm M60D machine gun and mount appear at the right. Below, the driver's station and the commander's station in the XM800T are at the left and right respectively.

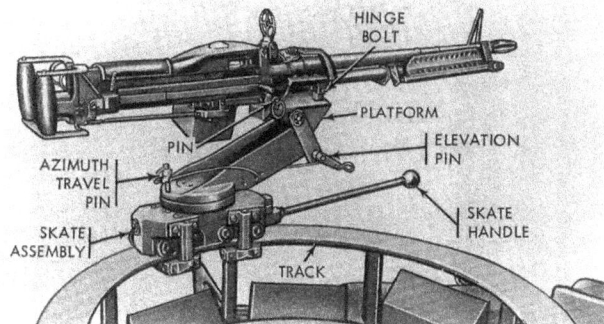

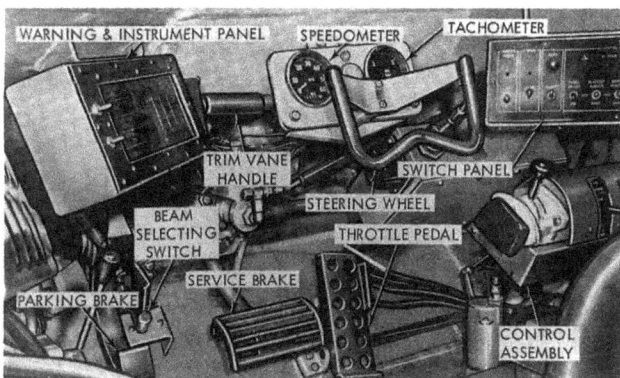

The XM800T armored reconnaissance scout vehicle is shown above during its test program. At the right it is climbing an obstacle.

From June through August 1974, the XM800T and the XM800W were compared with and without their turrets against several other vehicles during the force development test and evaluation (FDTE). The other vehicles in the comparison test included the M113A1 armored personnel carrier which was used to provide a baseline, the M113A1 AIFV, the Canadian Lynx, the PI M113A1½ with the turbocharged engine and the tube-over-bar suspension, a British Scimitar reconnaissance vehicle, a modified M551 Sheridan, an XR-311 Dune Buggy four wheel drive vehicle, and a V-150 armored car. In comparing the two XM800 vehicles with the baseline M113A1, the test report concluded that the XM800T was superior to both the M113A1 and the XM800W in overall performance as an ARSV. The XM800W performed well on roads and its quiet operation and high road speed were goals to be achieved for future scout vehicles. However, its limited cross-country capability and safety hazards associated with lateral instability and directional control made it less effective than the M113A1.

At the right, the XM800T is crossing a simulated trench.

The photographs on this page show the armored reconnaissance scout vehicle XM800T during the tests by the Army Test and Evaluation Command. Note the open driver's hatch in the view above.

As frequently happens during a development program, the final version of the XM800T was heavier and its performance somewhat different from the original estimates. The combat weight had increased to 19,600 pounds decreasing the freeboard to about 8 inches when afloat. The maximum road speed was 55 miles per hour and the maximum water speed was 4.2 miles per hour. The cruising range remained at about 450 miles.

Despite its successful performance, the ARSV did not go into production. Budget limitations required the cavalry scout vehicle program to be merged with that for the new infantry fighting vehicle. Thus the new scout eventually emerged as the M3 cavalry fighting vehicle.

The XM800T armored reconnaissance scout vehicle is shown in both of these photographs. The vehicle is swimming below.

Details of the top and rear can be seen on the XM800T armored reconnaissance scout vehicle in these two photographs. Note the external sight for the 20mm gun.

Another pilot XM800T armored reconnaissance scout vehicle is shown in these photographs. The vehicle appears to be fully stowed and the 7.62mm machine gun is mounted on top of the turret.

PART IV

INFANTRY AND CAVALRY FIGHTING VEHICLES

The armored personnel carrier M113A1 above has the kit installed for use as the armored cavalry assault vehicle (ACAV). The shields in this prototype kit are soft steel not armor.

IMPROVISED FIGHTING VEHICLES

The M113 armored personnel carrier was introduced into Vietnam when two ARVN (Army of the Republic of Vietnam) companies were organized and equipped in April 1962. Although it violated the doctrine promulgated by their American advisors, the Vietnamese insisted on fighting mounted in their vehicles, using them as light tanks. The M113's normal armament of one .50 caliber machine gun was supplemented by additional weapons mounted on top of the vehicle. This method of operation proved to be extremely effective under the conditions prevailing in Vietnam and eventually it was adopted by the American advisors. However, by early 1963, it was obvious that shields had to be provided to protect the gunner using the externally mounted weapons. At first, they were fabricated locally and consisted of a variety of configurations. Later, they were manufactured on Okinawa and shipped to Vietnam. On 6 April 1966, FMC was authorized to design and manufacture two prototype shield kits from mild steel. Design work began under Group Leader Cal Walker and Senior Project

Engineer John Giacomazzi using drawings with Vietnamese captions and penciled in translations. They also had a description of the combat requirements and five photographs of the shields improvised in Vietnam. That shield consisted of three pieces of armor, one in front and one on each side of the commander's hatch. The rear was protected by the open hatch cover. The FMC kit followed this arrangement, but the two pieces of flat angular side armor were replaced by curved sections. The design team also added a 7.62mm machine gun mount on each side of the cargo hatch. A shield was provided for each bringing the total number of armor components to five. One of the 7.62mm machine guns could be shifted with its shield to a removable pintle mount on the open cargo hatch cover allowing fire to the rear of the vehicle. This complete set of armor shields was designated as the A kit. Another version, consisting of only the three piece shield for the commander's hatch and the .50 caliber machine gun, was designated as the B kit. The latter was provided for installation on the mortar carriers.

After inspection of the two soft steel prototypes, FMC received a contract to produce 385 kits. Later, this initial contract was increased to 476 kits and all were shipped by 15 July 1966. Some of these were installed on the M113A1 armored personnel carriers being converted for use as fighting vehicles by the 11th Armored Cavalry Regiment. With this armament, they were referred to as the armored cavalry assault vehicle (ACAV) and were used by the 11th ACR when they deployed to Vietnam. The kits were then used to convert many of the M113 series vehicles to the ACAV configuration. The normal crew for the ACAV was five or six consisting of the driver, commander, two or three machine gunners, and a grenadier armed with a 40mm M79 grenade launcher. Other weapons that could be installed on the ACAV included the 40mm XM175 automatic grenade launcher and various recoilless rifles.

Above are additional views of the soft steel prototype shield kit installed on the M113A1. Below, a 7.62mm M60 machine gun is mounted on the bottom of the open roof hatch.

Below, the ACAV kit is installed on the armored personnel carrier with the crew in place.

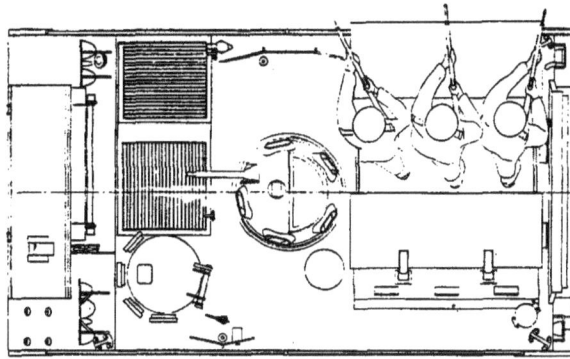

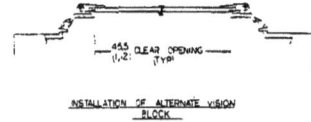

Dimensions in inches (meters)

INSTALLATION OF ALTERNATE VISION BLOCK

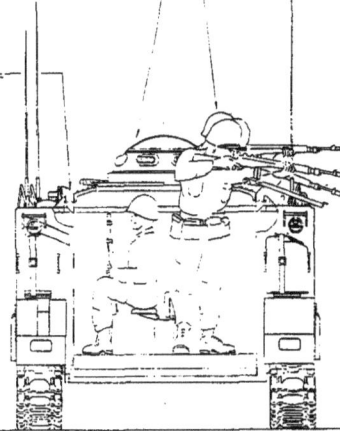

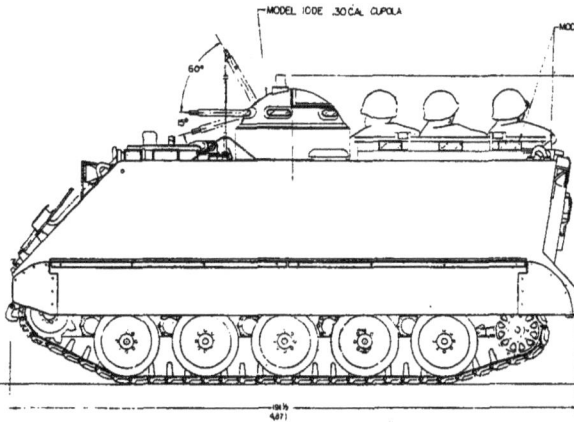

During 1964, FMC studied several concepts of the M113 modified as a fighting vehicle for counterinsurgency use. Some of these mounted the Model 100E cupola armed with a single .30 caliber M1919A4 machine gun. Vision blocks or periscopes were added to permit a view to the sides or rear. Two of the concepts are shown on this page.

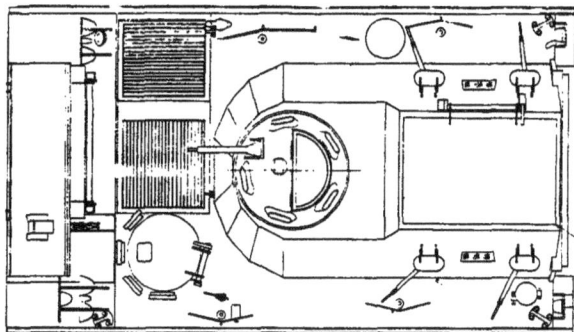

Dimensions in inches (meters)

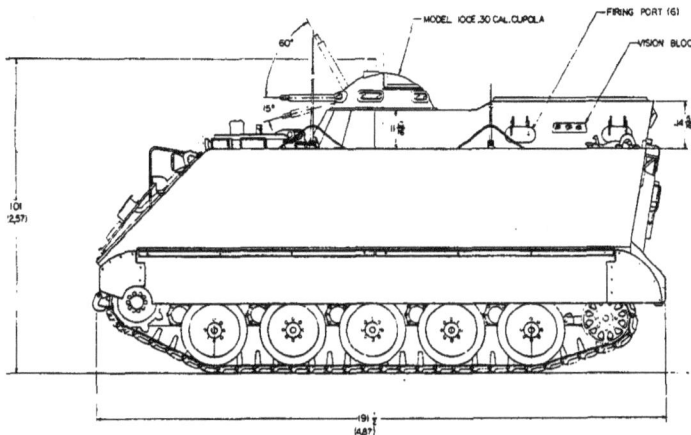

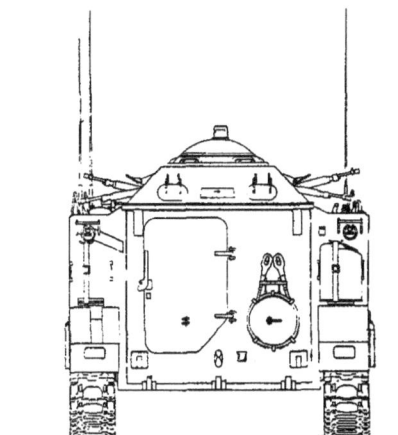

WEIGHT SUMMARY

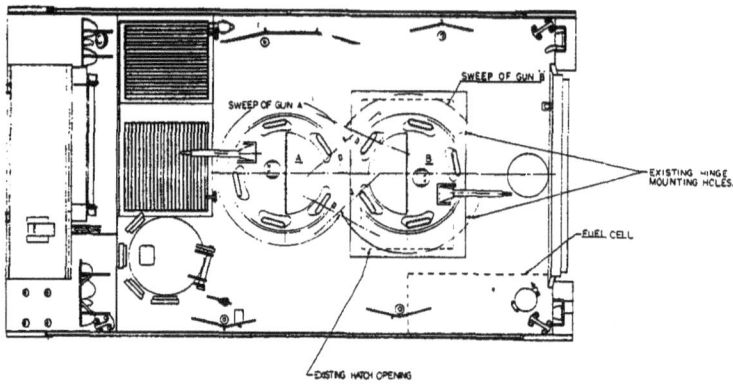

1. ITEMS ADDED TO VEHICLE	WT LBS
A. TWO 100E CUPOLAS @ 450 LBS. EA.	900
B. FRONT RISER ASSY.	180
C. REAR RISER ASSY.	300
D. MISC. HARDWARE	40
TOTAL WEIGHT ADDED	1420

2. ITEMS REMOVED FROM VEHICLE	
A. COMMANDER'S STATION	220
B. .50 CAL. MACHINE GUN	84
C. .50 CAL. GUN MOUNT	46
D. REAR TOP HATCH	225
TOTAL WEIGHT REMOVED	575

NET WEIGHT INCREASE 1420 - 575 = 845 LBS

CURRENTLY THE M113 CARRIES 2000 ROUNDS OF .50 CAL. AMMO. = 705 LBS. 705 LBS OF .30 CAL. AMMO. = 9400 ROUNDS

Dimensions in inches (meters)

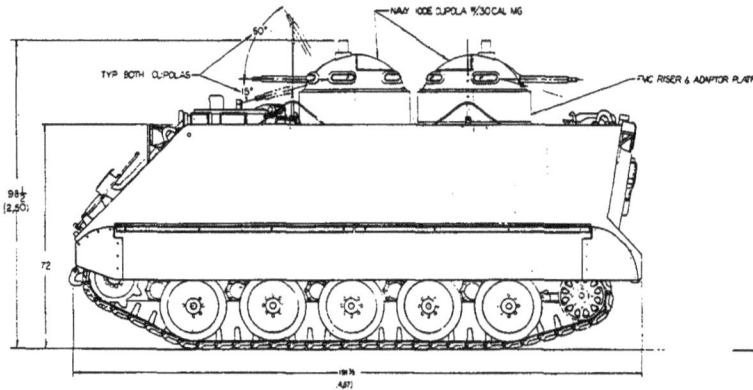

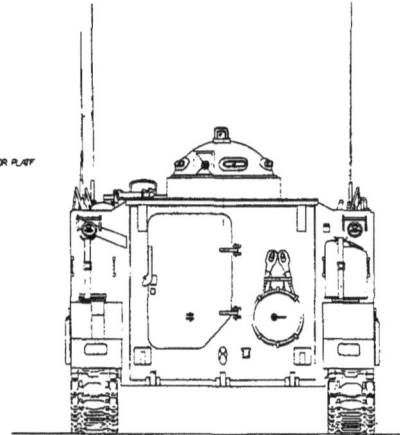

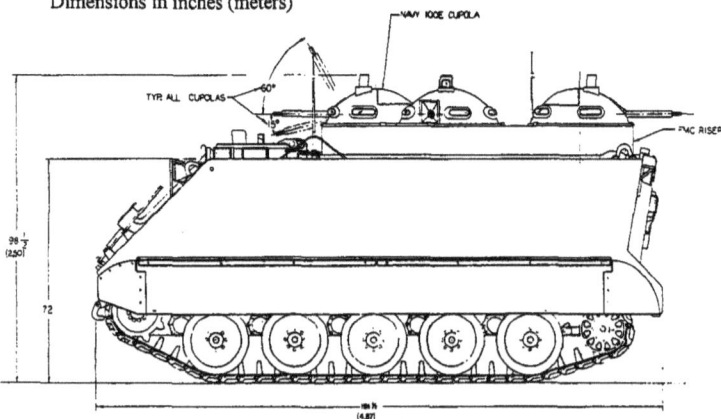

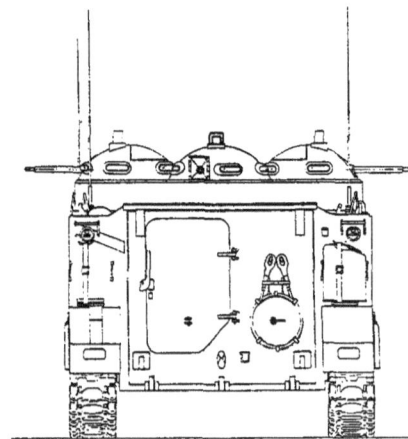

If one Model 100E cupola was good, perhaps more would be better. Here the M113 is equipped with two or four of the machine gun cupolas.

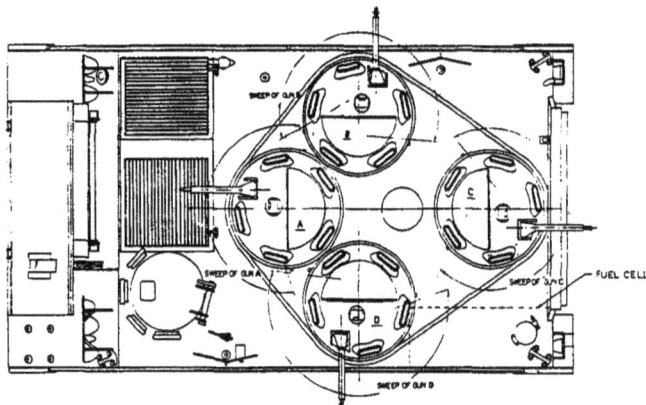

WEIGHT SUMMARY

1. ITEMS ADDED TO VEHICLE	WT LBS
A. FOUR 100E CUPOLAS @ 450 LBS. EA.	1800
B. RISER ASSY	938
C. MISC. HARDWARE	70
TOTAL WEIGHT ADDED	2808

2. ITEMS REMOVED FROM VEHICLE	
A. COMMANDERS STATION	220
B. .50 CAL. MACHINE GUN	84
C. .50 CAL. GUN MOUNT	46
D. REAR TOP HATCH	225
E. HULL CUTOUT	615
TOTAL WEIGHT REMOVED	1,190

NET WEIGHT INCREASE = 2808 - 1,190 = 1,618 LBS.

CURRENTLY, THE M113 CARRIES 2000 ROUNDS OF .50 CAL. AMMO. = 705 LBS. 705 LBS OF .30 CAL. AMMO. = 9,400 ROUNDS.

Dimensions in inches (meters)

Above the ACAV has the .50 caliber machine gun replaced by a 40mm automatic grenade launcher. At the right is a close-up view of the 7.62mm M60 machine gun with its shield.

As described earlier, a vulnerability reduction kit was developed to reduce the damage from mine explosions. This kit included a steel armor plate installed on the bottom front of the vehicle. It was attached by four bolts to the lower front armor and it extended back under the hull past the second road wheel suspension arm. Brackets secured the armor plate to the hull side armor.

Below are two views of the searchlight installation on the ACAV. Note that this vehicle is fitted with a high displacement trim vane.

Above, the early XM734 with wooden mock-up firing ports is at the left and two views of the actual XM734, registration number 12FC53, are at the right.

In December 1963, the Army Combat Developments Command published "A Study of Alternatives for a Post 1965 Infantry Combat Carrier". Later, this became the mechanized infantry combat vehicle (MICV). Since this vehicle would not be available for several years, the M113 received numerous modifications to provide an interim fighting vehicle. The first approach was to install firing ports and vision blocks in the side armor and rear ramp of the M113 to allow the troops to orient themselves and use their individual weapons. The first of the modified vehicles had four sets of firing ports and vision blocks on the right side, but only three sets on the left because of the interference of the fuel tank in the left rear sponson. Two sets of firing ports and vision blocks were in the rear ramp. The firing ports and vision blocks protruded from the side armor slightly increasing the width of the vehicle. Several mounts were installed at the commander's station. These included the standard commander's cupola from the

M113 armed with the external .50 caliber machine gun. The A kit of armor shields was mounted on at least one vehicle. The Model 74 twin machine gun cupola also was installed as well as the M27 cupola armed with the 20mm gun M139. This vehicle was the first of several designated as the XM734.

Below are two photographs of the XM734. Note that there are only three firing ports on the left side. This was because of the space occupied by the fuel tank in the left sponson.

Above and below are views of XM734, registration number 12FC55, fitted with the armor shield kit used on the armored cavalry assault vehicle. The firing ports in the rear ramp are visible at the right below.

Below is XM734, registration number 12FC31, fitted with the Model 74 cupola armed with two 7.62mm machine guns.

Additional photographs of XM734, registration number 12FC31, are shown above and below.

Below is another view of XM734, registration number 12FC53, but it is now armed with a 20mm gun in the cupola mount.

Above and at the right are photographs of the M113 converted to an infantry fighting vehicle. With the new center fuel tank there are now four firing ports and vision blocks on each side. The machine guns have not been installed in the Model 74 cupola.

In December 1965, FMC received work directives to modify six M113s as infantry fighting vehicles. The modifications included the installation of four sets of vision blocks and firing ports in each side and two sets in the rear ramp. The side mounted vision blocks and firing ports were recessed so that they did not increase the vehicle width. The original sponson mounted fuel tank was replaced by a 100 gallon fuel tank installed in the center of the vehicle. The crew seats were relocated along the center facing outward toward the gun ports and the personnel capacity was reduced from 13 to 12 men. A personnel compartment ventilation blower was installed. The original M113 type cupola was removed and replaced by a Model 74c cupola armed with two manually operated .30 caliber machine guns. Two of the six vehicles were fitted with bar armor for protection against shaped charge (HEAT) projectiles. The conversion of the six vehicles was completed in March 1966 and they were submitted for troop evaluation.

At the right is a rear view of the infantry fighting vehicle converted from the M113 armored personnel carrier. Note the firing ports and vision blocks in the rear ramp.

Interior views of the infantry fighting vehicle converted from the M113 are shown above. The new fuel tank can be seen under the center seats. The view at the top right shows the bar armor installed on some of the vehicles. At the right the vehicle is swimming.

The bar armor has been installed on the converted infantry fighting vehicle in these photographs. Above at the right, the bar armor and trim vane have been folded forward to permit access to the power train. At the right, the vehicle with the bar armor is swimming.

The original version of the XM765 infantry fighting vehicle is shown above. Note the M27 cupola mount for the 20mm gun M139 and the spaced laminate armor bolted onto the vertical sides and front.

A NEW ARMORED INFANTRY FIGHTING VEHICLE

In 1967, FMC received a contract to modify two M113A1 armored personnel carriers as prototypes to evaluate the "fight-from-the-vehicle" concept. Designated as the XM765, these vehicles retained the power train, suspension, and tracks of the standard M113A1. The upper rear side armor was sloped and four sets of firing ports and vision blocks were installed on each side. Two more sets were in the rear ramp bringing the total to ten. An M27 powered weapon station armed with the 20mm gun M139 was provided for the commander. Protection was increased by the installation of spaced laminate steel armor on the front and sides. A

100 gallon fuel tank was located in the center of the vehicle under the troop seats. After completion, each of these vehicles was tested for 4,000 miles, one at Aberdeen Proving Ground and the other at Fort Benning.

These are additional views of the XM765 infantry fighting vehicle in its original configuration. The sloped upper side armor with four vision blocks and firing ports per side are clearly visible in the top view at the right. Below are photographs of the original XM765 during test operations.

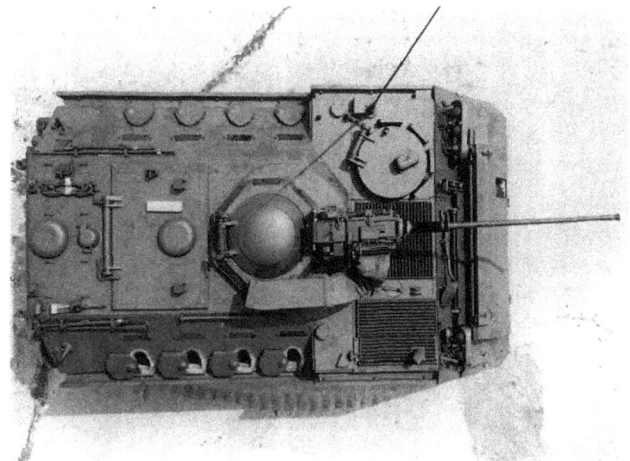

The modified XM765 infantry fighting vehicle is shown on this page. The spaced laminate armor has been removed and the 20mm gun is replaced by a .50 caliber, turret type, machine gun in the cupola mount.

Although the tests were generally successful, the automotive performance was reduced because of the additional weight. Two other problems also were noted. The first was the inadequate removal of fumes when weapons were fired from inside the crew compartment and the second was the danger presented by the internally mounted fuel tank. As a result, the XM765 at Aberdeen was returned to FMC for further modification.

To reduce the weight, the bolt-on, spaced laminate, armor was removed leaving the vehicle with the same

level of protection as the M113A1. The 20mm gun was replaced by a .50 caliber M2 machine gun in the same mount. The 100 gallon fuel tank was removed from the crew compartment and two armored external tanks were installed at the rear, one on each side of the ramp. The troop seats were replaced by ten individual seats that were rotatable and adjustable in height. Two additional ventilation fans were installed in the crew compartment bringing the total to four. The modified XM765 was then returned to the Army for further evaluation.

Details of the modified XM765 can be seen above and the interior with the crew in place appears in the view through the open ramp below.

The product improved M113A1 appears above and at the right. Note the redesigned hull with two firing ports per side and the 20mm gun on the weapon station. The commander's cupola can be seen just behind the weapon station.

Based upon the lessons learned with the XM765, a new prototype was built by FMC during 1970. Referred to as the product improved (PI) M113A1, it was powered by the turbocharged 6V53T diesel engine developing 260 gross horsepower. A new torsion tube-over-bar suspension increased the road wheel travel improving the cross-country performance. Lightweight bolt-on armor increased the protection over that of the M113A1. Polyurethane foam inside this spaced laminate steel armor increased the displacement to maintain the vehicle flotation in water. A high displacement trim vane also was installed. The driver remained in his usual position in the left front hull with four periscopes around his hatch. A low silhouette weapon station, operated from inside the vehicle, was installed in the center of the roof just behind the power plant compartment. It was either a manually operated mount with a .50 caliber machine gun or a power operated station fitted with a 20mm gun M139. The gunner in this station was provided with a unity/6 power periscopic monocular sight and seven M27 periscopes. The vehicle commander was located under a cupola in the center rear of the vehicle just behind the weapon station. His cupola was equipped with a 6 power binocular sight and

four wide angle, unity power, periscopes. The commander's forward view was blocked only to a limited extent by the weapon station. Individual seats were provided for ten men with four on each side and two in the center. Three vision blocks and two firing ports were installed on each side of the hull. Experience with the XM765 had shown that no more than two men could fire from each side without interfering with each other. A single firing port was located in the right side of the rear ramp and a periscope was installed in the hull roof above this port. A personnel access door was located in the left side of the rear ramp. The external fuel tanks were mounted on the hull at each side of the ramp.

Two additional views of the product improved M113A1 are shown below. Note the extremely high displacement trim vane.

Above, the product improved M113A1 at the left is armed with a .50 caliber M85 machine gun on a remote control mount and at the right, the interior of the vehicle can be seen through the open ramp.

The PI M113A1 was developed as a private venture by FMC and four were ordered for evaluation by the Royal Netherlands Army. Delivered in 1974, they were tested extensively to determine the modifications necessary to meet the Dutch requirements. These changes included shifting the weapon station to the right side of the vehicle and relocating the commander alongside it just behind the driver. The driver had four M27 periscopes around his hatch and one Type UA9630 universal passive periscope for night operations. The commander's cupola was slightly raised to permit forward vision over the driver's station. It was equipped with four M17 periscopes and one M20 or M20A1 periscope. Now referred to as the armored infantry fighting vehicle (AIFV), it was armed with the 25mm Oerlikon KBA cannon coaxial with a 7.62mm MAG machine gun in a one man turret. The latter also was referred to as the enclosed weapon station (EWS). The Rheinmetall RH202 20mm gun was offered as an alternate weapon. The gunner was provided with a Phillips sight, an open antiaircraft emergency gun sight, and a 150 watt infrared/white searchlight. He had four M27 periscopes in

The FMC armored infantry fighting vehicle (AIFV) appears below armed with the 25mm Oerlikon cannon and a 7.62mm MAG machine gun in the enclosed weapon station.

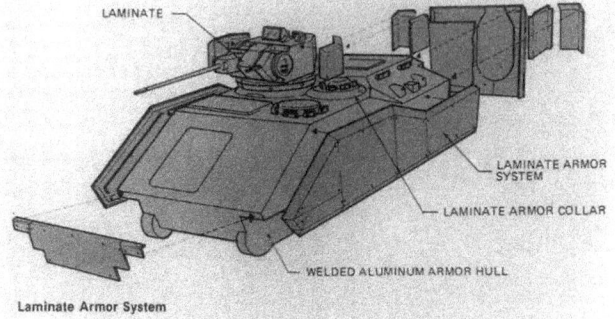

Laminate Armor System

The FMC armored infantry fighting vehicle can be seen in the photographs on this page. The armor arrangement on the AIFV is shown above.

the weapon station. With a ten man crew, the combat loaded weight of the AIFV was 30,175 pounds. Two firing ports and two M17 periscopes were installed in the sloped armor on each side. An additional firing port was located in the right side of the rear ramp. Later, it was relocated to the personnel door in the left side of the ramp. An M27 periscope was mounted in the hull roof above the rear firing port. The usual cargo hatch was installed in the hull roof behind the weapon station and the commander's cupola.

The FMC armored infantry fighting vehicle is above. The seating arrangement inside the vehicle can be seen in the drawing below.

FMC Armored Infantry Fighting Vehicle

Dimensions in millimeters (inches)

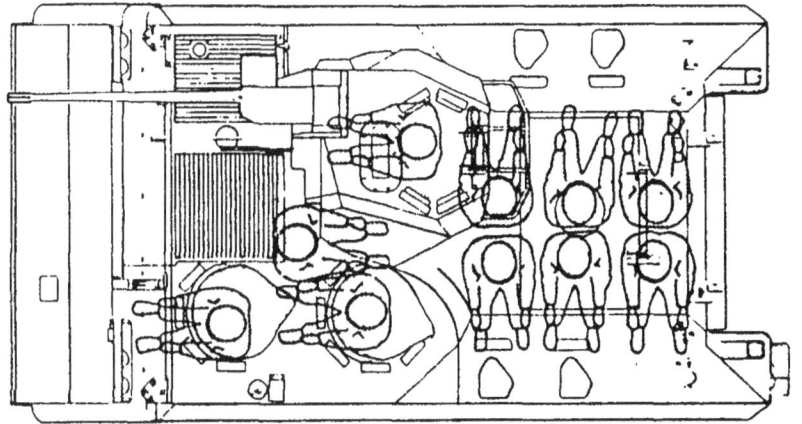

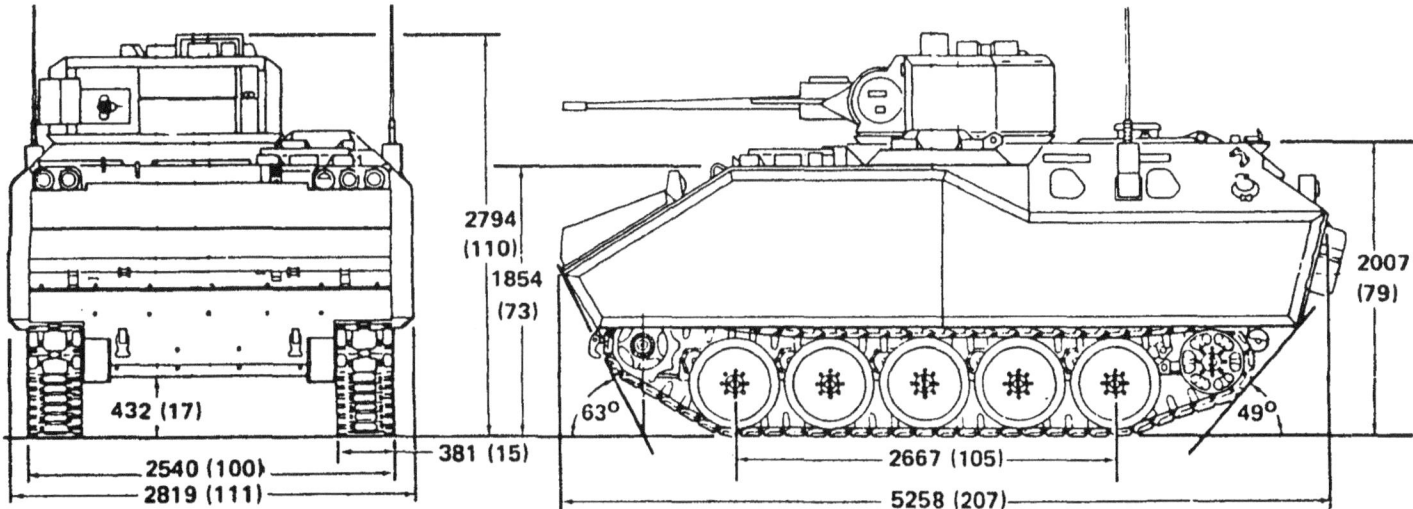

2794 (110)
1854 (73)
432 (17)
381 (15)
2540 (100)
2819 (111)

2007 (79)
63°
2667 (105)
49°
5258 (207)

The FMC armored infantry fighting vehicle is above. Note the variation in the gun mounts on these vehicles. At the bottom of the page is the family of vehicles based upon the armored infantry fighting vehicle.

In April 1975, FMC received a production contract from the Netherlands with the first deliveries in 1977. The AIFV was designated as the YPR-765 in reference to the XM765 that started this line of development. The Y indicated Dutch production. The P was for pantser (armored) and the R was for rups (tracked). FMC delivered the basic vehicle, but Dutch companies made the final assembly, produced certain components, and completed any modifications. Although the AIFV was the prime vehicle, the chassis provided the basis for a family of vehicles. Initially, there were ten variants of the basic vehicle. These were the AIFV or squad vehicle, the company command vehicle, the C1 battalion commander's vehicle, the C2 command vehicle for the battalion gunnery center, the C3 command vehicle for mortar fire control, the C4 command vehicle for antiaircraft artillery, the radar vehicle, the mortar vehicle, the cargo vehicle, and the ambulance vehicle.

the AIFV Universal Vehicle Family

AIFV
Prime Family Member

TOW Missile Vehicle

the Universal Vehicle
(the Basic Building Block)

Command Post

Ambulance

Mortar Prime Mover

Cargo Carrier

Recovery Vehicle

1 Left turn signal/navigation light	7 Right fuel tank vent	13 Fold-out hand grip
2 Left taillight and stop light	8 Enclosed weapon station	14 Ramp door gun port
3 Left fuel tank vent	9 Weapon station periscopes	15 Pintle
4 Rear periscope	10 25-mm cannon	16 Ramp door
5 Right taillight and stop light	11 Side gun port periscope	17 Tow hook
6 Right turn signal/navigation light	12 Right fuel filler armor cover	

1 Trim vane	8 Weapons sight	15 Gun port periscopes
2 Right rear view mirror	9 Ejection chute	16 Fold-out hand grip
3 25-mm cannon	10 Left rear view mirror	17 Left fuel filler armor cover
4 Right driving lights	11 Enclosed weapon station	18 Left taillight and stoplight
5 Grenade launcher	12 Driver's hatch	19 Left rear turn signal
6 Left driving lights	13 Commander's hatch	20 Left front turn signal
7 7.62-mm machine gun	14 Ventilator	21 Towing hooks

The FMC armored infantry fighting vehicle produced for the Netherlands is shown above and its components are identified. Below, the enclosed weapon station armed with the 25mm Oerlikon cannon is at the left and a drawing of the power train, suspension, and tracks is at the right.

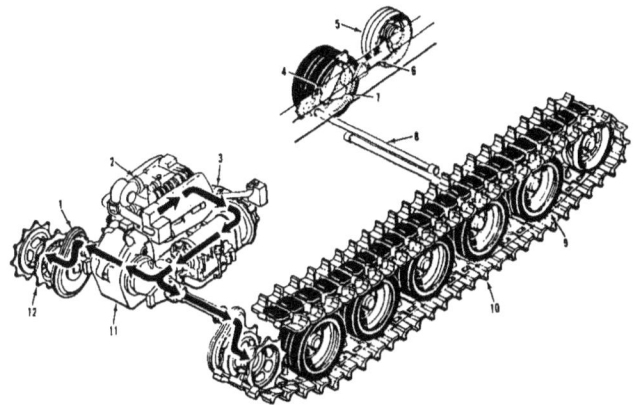

1 Final drive	8 Torsion tubes
2 Engine	9 Road wheel
3 Transfer gearcase	10 Track
4 Shock absorber	11 Steering control differential
5 Idler wheel	12 Track drive sprocket
6 Track adjuster	13 Transmission
7 Road wheel arm	

The squad vehicle and the company command vehicle both carried the EWS armed with the 25mm cannon and the coaxial 7.62mm machine gun. Both vehicles had a safety screen around the rotating weapon station. The squad vehicle carried the driver, commander, gunner, and a seven man infantry squad. The double bench type troop seats were in the center of the vehicle with stowage space underneath. The company command vehicle was fitted with a tip-up work table in the center between four individual seats. An additional seat was installed between the weapon station and the commander's seat.

The drawings below show the internal arrangement of the squad vehicle (left) and the company command vehicle (right).

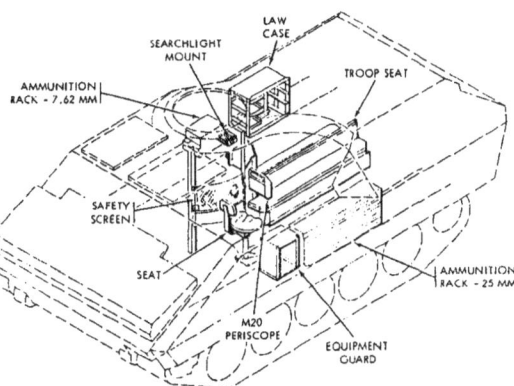

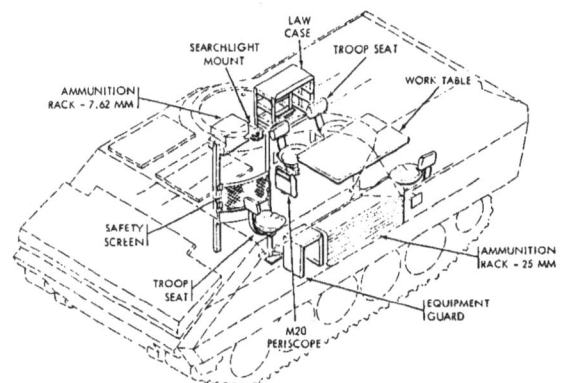

269

One of the command vehicles is above. The interior arrangement of the various command vehicles, C1 through C4, can be seen in the drawings from top to bottom at the right.

The C1 battalion commander's vehicle was armed with a .50 caliber machine gun in an M26 cupola mount. An adapter ring was required to fit the 34 inch diameter cupola mount into the 40 inch diameter opening for the EWS. A fold-up work table was in the center with two side mounted troop seats on the right and a bench seat on the left. A single floor mounted troop seat was installed between the gunner's seat and the commander's station. Either the M20 or the M27 periscope could be mounted in the front port of the commander's hatch ring.

The C2 battalion fire direction center command vehicle also was armed with the .50 caliber machine gun in the M26 cupola mount. As on the C1, the fold-up work table had two seats on the right and a bench type seat on the left. Also, either the M20 or the M27 periscope could be installed for the commander.

The C3 mortar fire control command vehicle also was adapted to carry the M26 cupola mount for the .50 caliber machine gun. A tip-up work table in the center had two seats on the left. The normal crew was the driver plus three men.

The C4 antiaircraft artillery command vehicle had the tip-up work table in the center with a bench type troop seat on the left. A floor mounted troop seat was located between the gunner's seat and the commander's station. It also carried the M26 cupola mount. The normal crew consisted of the driver, commander, gunner, and four squad members.

The radar vehicle was intended to transport the radar unit including a radar mast assembly. The vehicle was armed with the .50 caliber machine gun in the M26 cupola mount. A tip-up work table was in the center with two troop seats on the left. An additional troop seat was installed between the gunner and the commander.

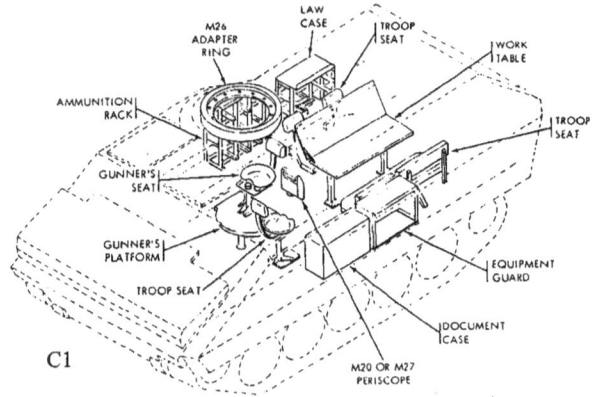

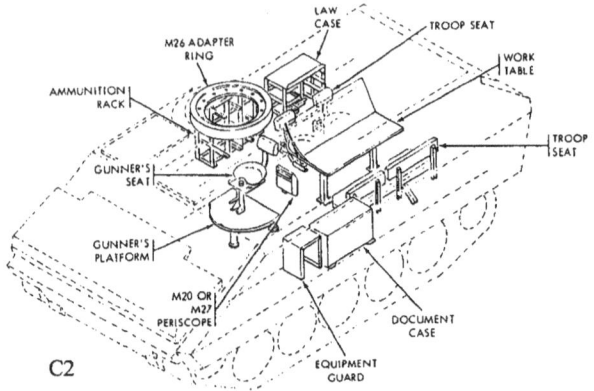

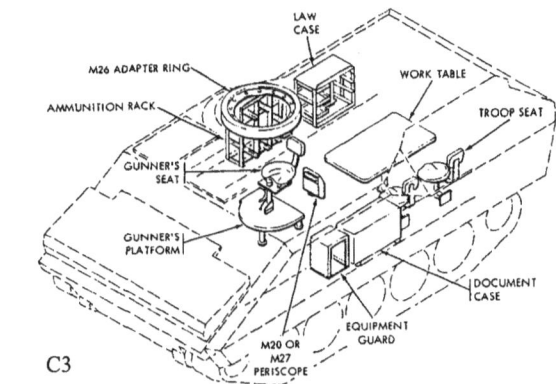

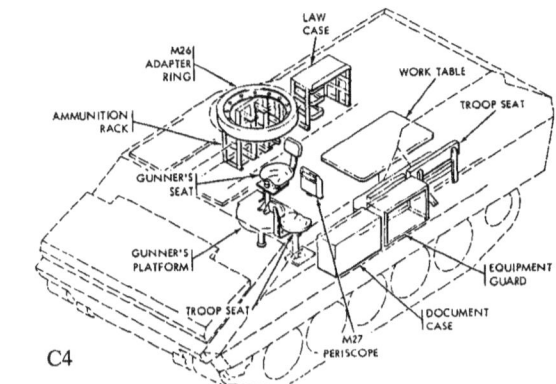

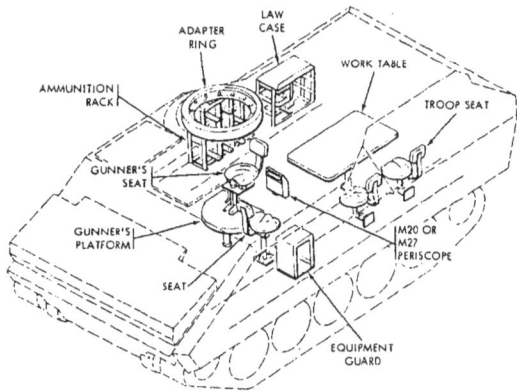
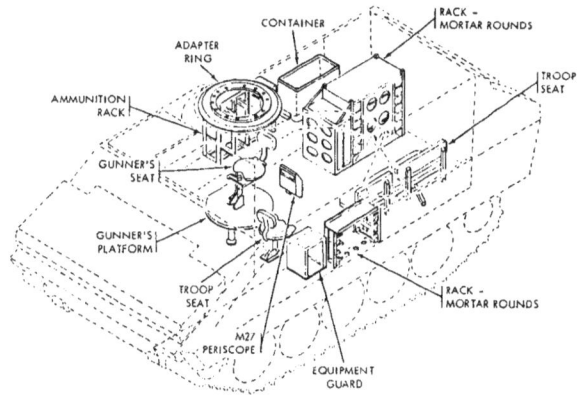

The drawings above show the internal arrangement of the radar vehicle (left) and the mortar vehicle (right). Below, the interior of the mortar vehicle is at the left and at the right, the vehicle serves as a prime mover for the 120mm mortar.

The mortar vehicle was armed with a .50 caliber machine gun in the M26 cupola mount or on the external mount from the M113A1 armored personnel carrier. If the latter was used, another adapter ring was required to install the 30 inch diameter mount into the 40 inch diameter opening. Needless to say, the M26 mount required the 34 to 40 inch diameter adapter ring as on the C1 through C4 command vehicles. The main function of the mortar vehicle was to act as prime mover and carry ammunition for the 120mm towed mortar. The main ammunition rack in the right rear of the vehicle stowed 45 rounds. An additional six rounds were carried in a smaller rack in the left sponson. A bench type troop seat was on the left side of the crew compartment and a single troop seat was located between the gunner and the commander.

At the right is the mortar vehicle. Note the cupola from the M113A1 armored personnel carrier armed with the .50 caliber machine gun on the external mount.

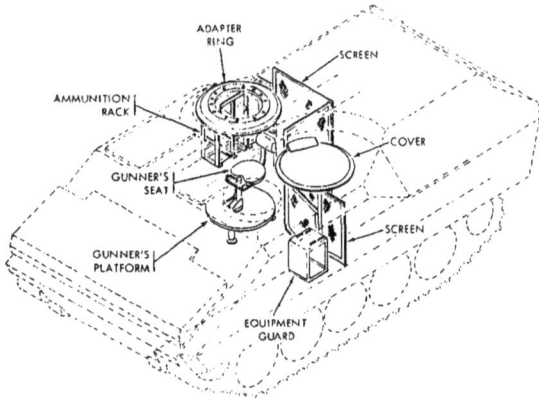

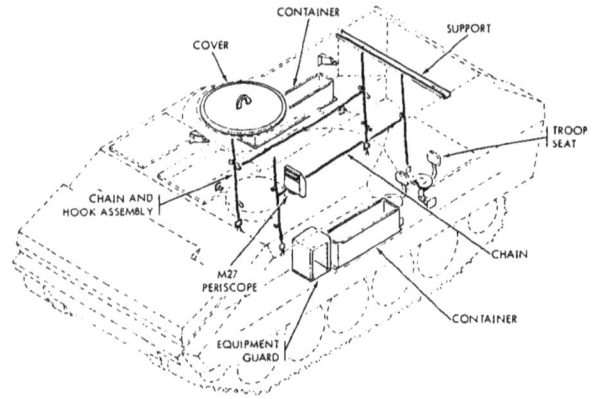

The drawings above show the cargo vehicle at the left and the ambulance at the right. Below, the cargo vehicle is being loaded at the left and a view of the ambulance is at the right.

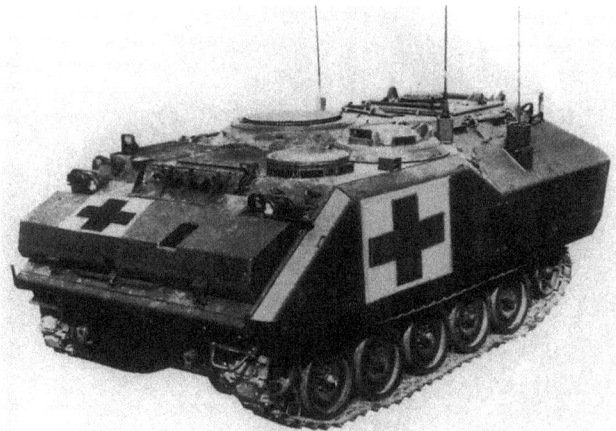

The cargo vehicle was armed with a .50 caliber machine gun on the external mount of the M113A1 type cupola. The 30 to 40 inch diameter adapter ring was required. The commander's cupola was removed and a cover installed over the opening. The two man crew consisted of the driver and the gunner. A security screen separated both men from the cargo area.

The ambulance vehicle was, of course, unarmed and the weapon station opening was closed by a cover plate. An M27 periscope was installed in the front of the commander's hatch ring. There were two supports and four chain and hook assemblies to carry four litters in the crew compartment. Two forward facing seats were located on the left side behind the commander's station.

Two additional vehicles were part of the AIFV family. These were the antitank vehicle with two TOW missile launchers and the recovery vehicle. Several experimental configurations for the TOW launcher were considered prior to the adoption of the Emerson Electric type launcher as used on the M901 improved TOW vehicle. The modification of the vehicles and the installation of the launchers were carried out by RSV Defense Engineering in Rotterdam.

At the right is the antitank vehicle equipped with the twin TOW missile launcher designed by FMC.

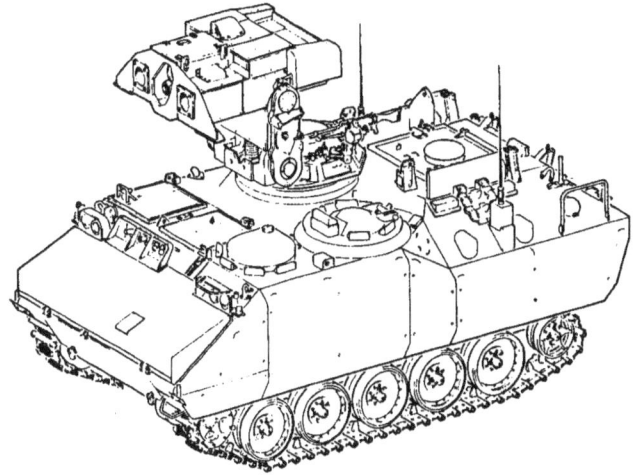

Above, the antitank vehicle mounting the Emerson Electric type launcher is shown in the photograph at the left and the drawing at the right. The recovery vehicle appears in the view below at the right.

Although the recovery vehicle was considered to be part of the family, it had the same appearance as the standard M113A1 recovery vehicle with the vertical side armor. In fact, the only difference was the AIFV power train and improved suspension. The recovery vehicle was fitted with side flotation cells and a high displacement trim vane to increase the freeboard when afloat.

Other variants of the AIFV proposed by FMC included the installation of one or two man turrets armed with the Bushmaster M242 25mm cannon. A turret armed with the Cockerill 90mm gun also was mounted upon the AIFV chassis.

Below are two photographs of the FMC armored infantry fighting vehicle mounting the two man turret similar to that on the Bradley fighting vehicle, but without the TOW missile launcher. It is armed with the same 25mm cannon M242.

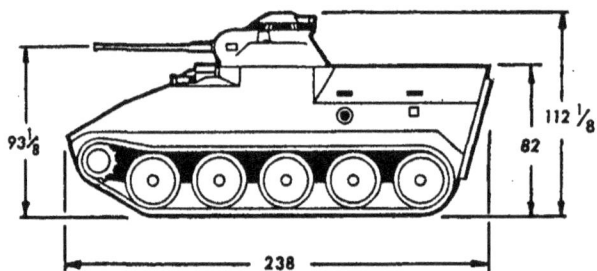

Dimensions in inches

Above, the artist's concept at the left and the drawing at the right show the proposed mechanized infantry combat vehicle 1965 (MICV-65).

THE MECHANIZED INFANTRY COMBAT VEHICLE

In August 1963, the United States and the Federal Republic of Germany signed an agreement to develop a new main battle tank for use by both nations. Referred to as the MBT70, its high performance was expected to far exceed that of the M113 armored personnel carrier. The 1963 publication "A Study of Alternatives for a Post 1965 Infantry Combat Vehicle" envisioned a new high performance mechanized infantry combat vehicle (MICV) that could operate with the new main battle tank. Dubbed the MICV-70, it was not expected to be available before the end of the decade. In the meantime, the Combat Developments Command submitted a requirement for an interim MICV to be built prior to the availability of the MICV-70. This was approved in March 1964 and in June, a contract was awarded to the Pacific Car and Foundry Company for the design and fabrication of a prototype and five pilots of the mechanized infantry combat vehicle XM701. It also was referred to as the MICV-65.

Using many components from the M107/M110 self-propelled artillery vehicles then in production, Pacific Car and Foundry was able to deliver the prototype in May 1965 followed shortly by the pilot vehicles. The prototype, registration number 12FP14, differed in some respects from the five pilot vehicles,

At the right is the prototype XM701 mechanized infantry vehicle pilot number 4, registration number 12FP14.

registration numbers 12FP15 through 12FP19. Obvious recognition differences included the steps near the front of the left side armor and the location of the headlights at the extreme edge of the vehicle front.

The first three pilots (P1 through P3) had steel hulls while P4 and P5 had aluminum alloy hulls. The turret was steel armor for all five vehicles. This hydraulically operated two man turret was armed with a 20mm M139 gun and a 7.62mm M73 coaxial machine gun. Other armament included two 7.62mm M60 machine guns and five 7.62mm M14 rifles at seven firing stations in the squad compartment. Two firing stations were located on each side, one in each rear corner, and one in the rear ramp. Vision blocks were mounted above the side and corner firing stations and an M27 periscope was installed in the hull roof above the ramp firing station.

Above is the prototype XM701 mechanized infantry combat vehicle armed with the 20mm gun M139. Note the suspension components from the M107/M110 self-propelled artillery vehicles.

Pilots P1 and P4 had ball mounts for the rifles and the remainder had firing ports. The M60 machine guns had ball mounts on all of the pilots. The 12 man crew consisted of the commander and the gunner in the turret, the driver, and nine infantrymen. The driver had one M27 and four M17 periscopes around his hatch and his hatch cover was fitted for the installation of an M24 infrared periscope. In addition to the door in the rear ramp, hatches were provided for the driver and on top of the turret and the squad compartment. The commander had a 360 degree view through eight vision blocks in the turret and the gunner was equipped with an M34C periscope. Combat loaded, the steel hull version had a weight of 54,050 pounds and the aluminum hull vehicle weighed 50,750 pounds. The XM701 was powered by

an 8V71T diesel engine with an Allison XTG-411-2A transmission. The turbocharged engine developed 425 gross horsepower at 2,300 rpm. When the vehicle was afloat, the engine compartment was completely sealed and water from the outside was used to cool the engine. The flat track torsion bar suspension had five dual road wheels per side. The rear road wheel served as an adjustable trailing idler to maintain track tension. The 15 inch wide, double pin, tracks were driven by the sprockets at the front of the vehicle. The XM701 was fitted with the E51 NBC gas particulate filter unit. This was a modification of the collective protector designed for the MBT70. The unit pressurized the entire vehicle so that the crew did not need to wear individual masks. An auxiliary power unit was installed in the engine

The mechanized infantry combat vehicle XM701, pilot number 2, appears below. Note the differences compared to the prototype vehicle at the top of the page.

Above and at the right are views of the XM701 mechanized infantry combat vehicle pilot number 4, registration number 12FP18.

compartment for starting the engine in cold weather and charging the batteries during radio watches. Because the XM701 was expected to operate for 24 hours in the buttoned-up condition, a stove and toilet were included. Although weight and space restrictions eliminated the installation of a winch, two capstans were provided for attachment to the drive sprockets. When not in use, they were stowed on the rear of the vehicle to the left of the ramp. Two 150 foot lengths of nylon line were carried in the squad compartment for use with the capstans.

At the right is another photograph of XM701 pilot number 4. Pilot number 5 is below. Note the machine gun in the ball mount on the latter vehicle.

Mechanized infantry combat vehicle XM701 pilot number 5, registration number 12FP19, appears above and at the right. Details of the turret and gun mount can be seen in the top view.

The vehicle had a maximum speed of 40 miles per hour on roads and 3.8 miles per hour in water. When afloat, the freeboard varied from about 6 inches for the steel hull vehicles to approximately 10 inches for the aluminum hull version. The cruising range on roads was about 350 miles.

Pilots P1 and P5 arrived at Aberdeen Proving Ground during the Summer of 1965 for engineering tests. P5 was then shipped to Fort Benning. P2 and P4 went to Fort Knox and P3 was shipped to the Yuma Proving Ground for desert tests. Although the XM701 was an interim vehicle that never went into production, its development and test provided useful information for the MICV-70 program.

XM701, pilot number 5, is on a test run at the right. Below, the rear ramp and the power train compartment cover are open on this XM701.

The mock-up of the mechanized infantry combat vehicle XM723 is shown in the two photographs above.

A project with the title "Mechanized Infantry Combat Vehicle Family - 1970" was initiated in October 1965. As the title indicated, this project envisioned the development of a vehicle family paralleling that based upon the M113 armored personnel carrier. It would include weapon carriers, command posts, communication vehicles, ambulances, and other vehicles in addition to the basic MICV. A design and cost effectiveness study began immediately with the objective of type classification in 1970 parallel with the MBT70. Like its ill fated companion, the MICV-70 program was to be drastically modified. The diversion of resources to support the war in Vietnam resulted in serious delays to all of the modernization programs. Unlike other wars that accelerated weapons development, this one slowed its progress.

After the end of the MBT70/XM803 program in 1972, the Army Tank Automotive Command (TACOM) requested proposals from industry for the new MICV which much earlier had been designated as the mechanized infantry combat vehicle XM723. Pacific Car and Foundry Company, Chrysler Corporation, and FMC responded with proposals. The FMC concept was accepted and they were awarded a contract in late 1972 to construct three prototypes, a ballistic vehicle, and 12 pilots.

The first prototype was delivered in 1974 and all three were completed by the Summer of 1975. In many ways, they resembled the FMC AIFV, although they were larger and heavier. The Cummins VTA903 diesel engine was installed in the right front hull. Developing 450 gross horsepower at 2,600 rpm, it was coupled to a General Electric HMPT-500 hydromechanical transmission. The front mounted final drives and sprockets drove the 21 inch wide, single pin, tracks. A torsion tube-over-bar suspension supported the vehicle on six dual road wheels per side with return rollers.

The XM723 was assembled using 5083 aluminum alloy armor except for the sloped sides. These were 7039 aluminum alloy armor. Spaced laminate steel

Below is a pilot mechanized infantry combat vehicle XM723.

Additional views of the pilot mechanized infantry combat vehicle XM723 are shown on this page.

armor was added to the vertical sides and rear. Frontal protection was enhanced by a the trim vane. At least two different high displacement trim vanes were installed on the prototypes. The crew arrangement was similar to that in the AIFV with the driver in the left front hull and the commander immediately behind him. The driver was provided with four periscopes and the commander with five periscopes in the hull roof around their hatches. The gunner was in the turret just behind the power plant compartment. He had six periscopes around his hatch

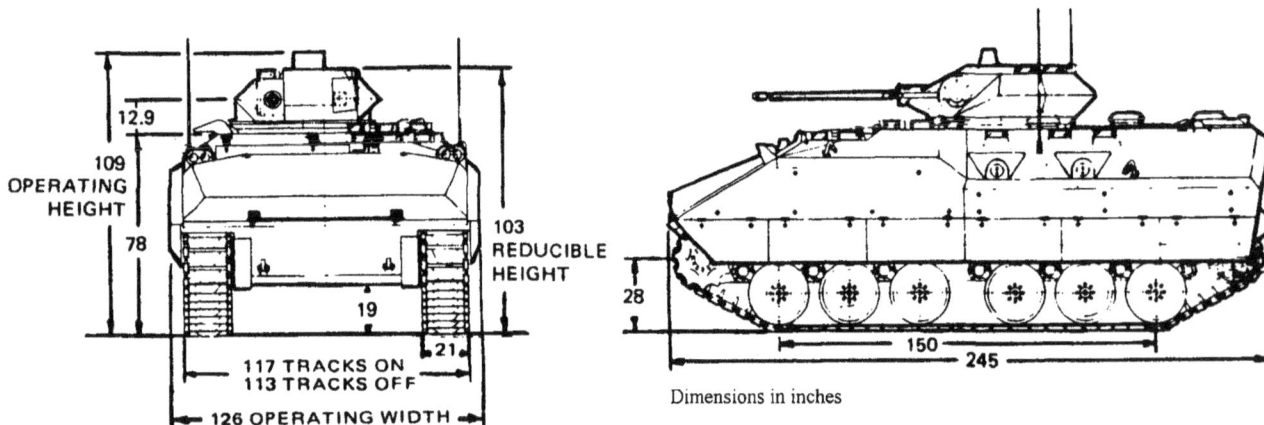

12.9

109
OPERATING
HEIGHT

78

103
REDUCIBLE
HEIGHT

19

21

117 TRACKS ON
113 TRACKS OFF

126 OPERATING WIDTH

28

150

245

Dimensions in inches

The dimensions of the mechanized infantry combat vehicle XM723 are shown in the drawing above. The design of the tube-over-bar suspension can be seen in the drawing below at the right.

and an M36E2 periscopic sight. Eight infantry soldiers were seated in the squad compartment. Six firing ports and periscopes were located two on each side and two in the rear ramp. Six .45 caliber M3A1 submachine guns were carried for use as firing port weapons. The turret armament varied during the test program. Although the 25mm Bushmaster was specified as the main armament, it was not yet available. The 20mm gun M139 and the 20mm gun XM236 were installed at different times. In a similar manner, the 7.62mm M219 coaxial machine gun was replaced by the 7.62mm XM238 machine gun. The latter was a modified version of the 7.62mm M60 machine gun.

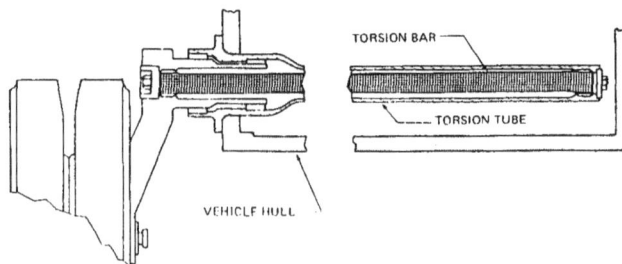

TORSION BAR

TORSION TUBE

VEHICLE HULL

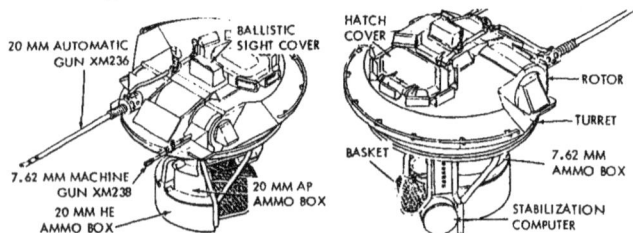

20 MM AUTOMATIC GUN XM236

BALLISTIC SIGHT COVER

HATCH COVER

ROTOR

TURRET

7.62 MM MACHINE GUN XM238

BASKET

7.62 MM AMMO BOX

20 MM AP AMMO BOX

20 MM HE AMMO BOX

STABILIZATION COMPUTER

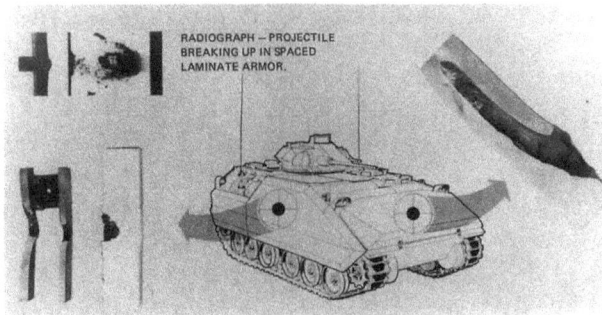

RADIOGRAPH – PROJECTILE BREAKING UP IN SPACED LAMINATE ARMOR.

Above, the drawing of the turret on the XM723 is at the left and the performance of the spaced laminate armor is illustrated at the right. The drawing below shows the seating arrangement inside the XM723.

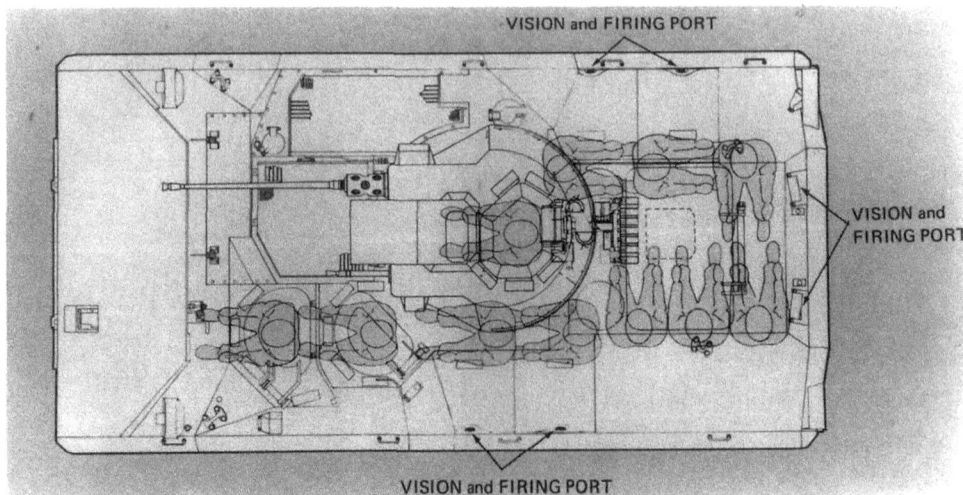

VISION and FIRING PORT

VISION and FIRING PORT

VISION and FIRING PORT

280

The XM723 appears above and at the right. The bottom photograph shows a TOW missile launcher on the turret.

Combat loaded, the XM723 weighed about 43,000 pounds. Its maximum road speed was 45 miles per hour and it could swim, propelled by track action, at about 5 miles per hour. The cruising range on roads was approximately 300 miles.

The tests of the XM723 continued into 1976. Although there were some problems with the new transmission and suspension system, they were resolved after a few months delay in the program. However, the greatest concern was the high cost of the new vehicle compared to the familiar M113 series.

The mock-up of the T-BAT-II is shown above, but the missile launcher can hardly be seen from this side.

THE APPEARANCE OF THE BRADLEY

In August 1976, a Task Force was established to evaluate the MICV program and to determine if it would meet the future requirements of the Army. The recommendations of the Task Force were accepted by the Army in October 1976. In response to these recommendations, a common vehicle would be developed to meet the requirements for a mechanized infantry vehicle and a cavalry scout vehicle. By this time, the XM800 armored reconnaissance scout vehicle had been canceled. The new vehicle would be armed with the TOW antitank guided missile system in addition to the Bushmaster 25mm cannon in a two man turret. This new turret was referred to as the TBAT-II (TOW Bushmaster armored turret, two man). The new vehicle would retain the hull firing ports, have the same armor protection as the XM723, and it would be amphibious. With the acceptance of these recommendations, the XM723 served as the basis for a new program that would eventually produce the Bradley fighting vehicle.

The T-BAT-II can be seen at the right with the TOW launcher in the firing position.

In November 1976, the development of the MICV TBAT-II was approved and FMC began the design and mock-up construction of the new vehicle. The mock-up review was at FMC in March 1977 and the basic design was approved. The MICV TBAT-II was designed to carry nine men. In the new vehicle, the driver remained in the left front alongside the power plant compartment. The commander was relocated to the right side of the new, two man, turret with the gunner on the left. In his

282

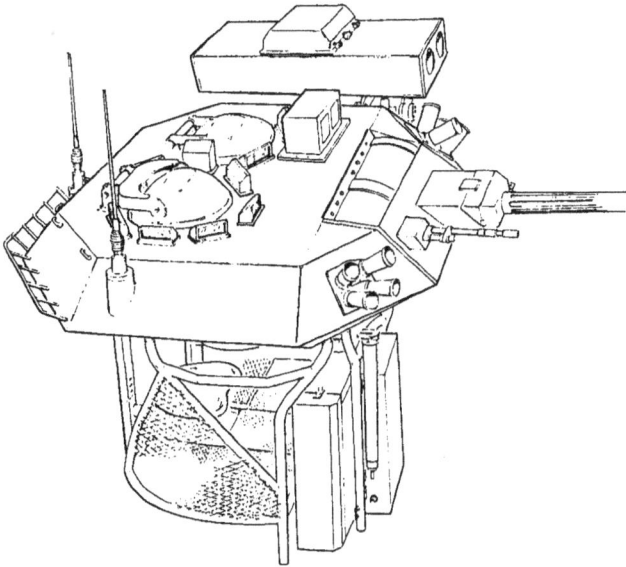

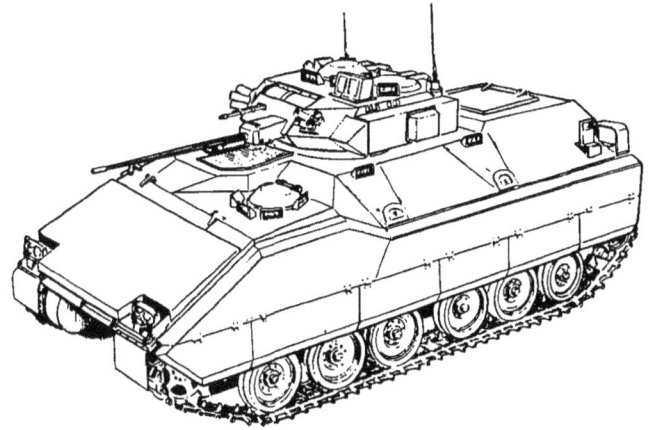

The T-BAT-II turret configuration appears in the drawing at the left and the vehicle itself is above.

new position, the commander had 360 degree vision through eight unity power periscopes around his hatch. The integrated day/night sight and two adjacent periscopes in front of his hatch provided frontal vision for the gunner. The primary armament was the 25mm automatic gun. At this time, two weapons were under consideration. These were the self powered 25mm gun XM241 and the externally powered 25mm gun XM242. The latter was the Chain Gun developed by Hughes Helicopter Company (later the McDonnell Douglas Helicopter Company). A coaxial 7.62mm MAG58 (later M240) machine gun was installed on the right side of the gun mount. An armored, twin tube, launcher on the left side of the turret carried two TOW missiles. Folded down against the side of the turret for travel, this launcher was raised to the firing position by an electric actuator. The launcher could be reloaded under partial protection by tilting it back toward the cargo hatch. The electric powered turret was stabilized in both azimuth and elevation. The radio equipment was located in the turret with two antennas mounted on the left rear and the right side. A four tube smoke grenade launcher was installed on each side of the turret front.

The new turret was shifted to the right and as far forward as possible to maximize the space in the squad compartment. The right fuel tank was reshaped and the engine cooling fan and radiator were moved forward two inches. The hull top plate and exhaust grille were modified to fit the new turret. The fuel was relocated from the left rear to a forward tank below the turret. One squad member was seated just to the rear of the driver and the remaining five were to the rear of the turret. The firing ports and periscopes were relocated, but the number remained the same with two on each side and two in the rear ramp. Five dual purpose stowage racks for TOW or Dragon missiles were in the left rear of the squad compartment with three horizontal and two vertical. Three light antitank weapon (LAW) missiles also were stowed horizontally on the left side. The cargo hatch was reshaped and lengthened by two inches to permit reloading the TOW launcher. It also was shifted toward the rear to accommodate the larger turret.

The scout version of the new vehicle was similar to the MICV except that the firing ports were deleted. The crew was reduced to five men consisting of the driver, commander, gunner, and two observers. A small scout motorcycle was stowed on the left side of the squad compartment. The TOW missile stowage was increased to ten plus the two in the launcher and the 25mm ammunition supply was increased from 900 to 1,500 rounds.

The XM 241 self powered 25mm cannon is at the left and the externally powered XM242 25mm weapon is below.

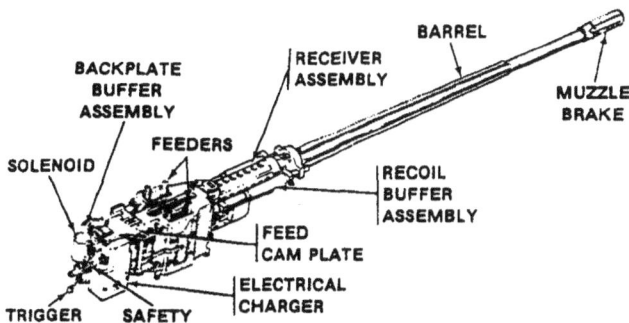

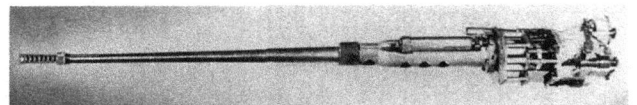

In May 1977, the program received a new name. The two versions were now referred to as the fighting vehicle systems (FVS) and the MICV and the scout vehicle became the XM2 infantry fighting vehicle (IFV) and the XM3 cavalry fighting vehicle (CFV) respectively. In June 1977, the program was expanded to include the ground support rocket system (GSRS). Later, this became the multiple launch rocket system (MLRS). The basic vehicle was referred to as the fighting vehicle system (FVS) carrier. It had the same power train and suspension as the two fighting vehicles.

The XM2 and XM3 differed slightly from the TBAT-II design. The configuration along the left side was changed. The top of the spaced laminate armor was now a straight line extending to the rear and the armor around the firing ports was modified. On the right side, the spaced laminate armor had the same stepped configuration as on the XM723, but it did not extend as high alongside the engine compartment and the turret.

Above are photographs of the mock-up for the XM2 infantry fighting vehicle and below the automotive test rig is moving at high speed.

A later mock-up of the XM2 infantry fighting vehicle is shown below. Note the change in the hull configuration compared to the XM723.

Above is an early XM2 infantry fighting vehicle and its dimensions can be seen in the drawing at the bottom of the page.

A larger driver's hatch was installed with the four periscopes mounted in the hatch cover itself. A new torsion bar suspension replaced the tube-over-bar design on the XM723. The front three road wheels were moved four inches forward increasing the ground contact length from 150 to 154 inches. Modifications also were made to the idler wheel mounting, shock absorbers, track return rollers, track guides, final drives, and sprockets.

The first two XM2 prototypes were delivered by FMC in December 1978 and six more followed in March of 1979. After further tests, the vehicles were type classified as the infantry fighting vehicle M2 and the cavalry fighting vehicle M3 in December 1979. Full production was authorized in January 1980 and the first production vehicle was delivered in May 1981. Initially, it was proposed to name the M2 after General of the Army Omar N. Bradley and to name the M3 after General Jacob L. Devers. However, because of the great similarity between the two vehicles, both were named the Bradley fighting vehicle in October 1981.

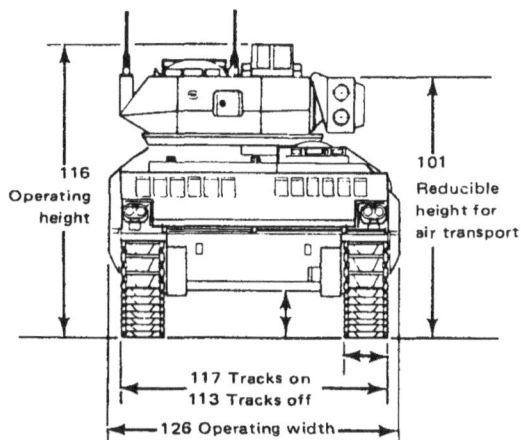

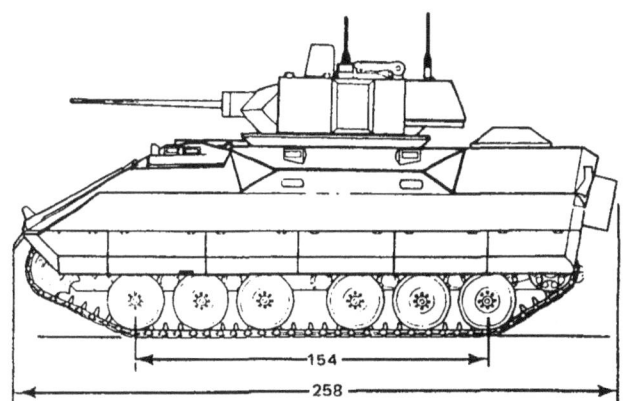

Dimensions in inches

285

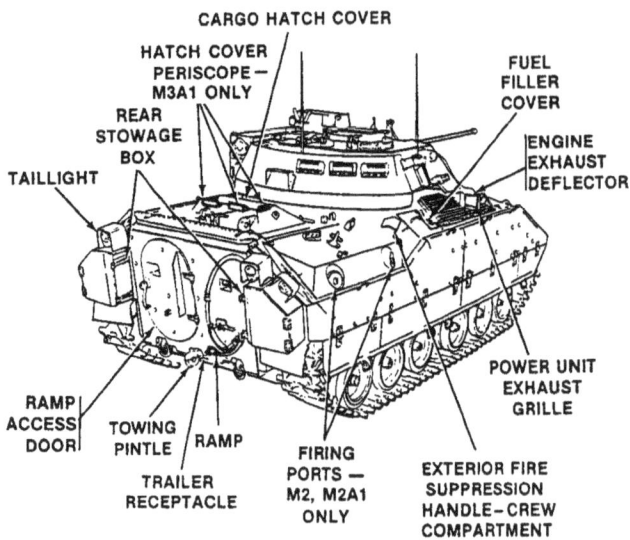

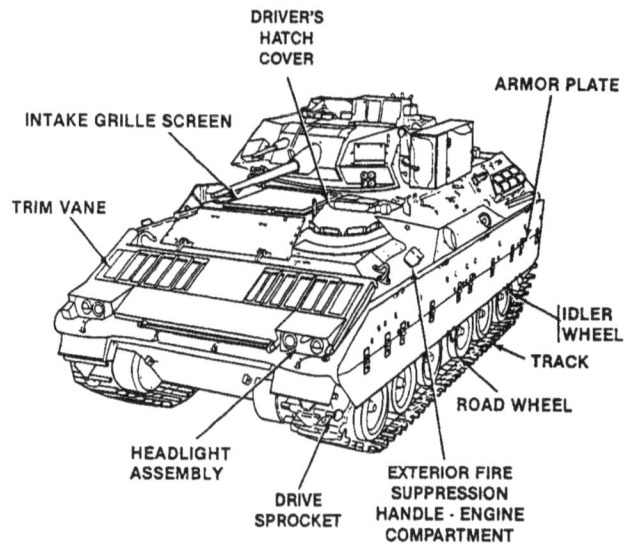

Features of the M2, M2A1, M3 and M3A1 Bradley fighting vehicles are combined in the drawings above from the technical manual. Note, for example, the periscopes in the troop compartment hatch cover that apply only to the M3A1. The gunner's station in the Bradley is shown in the photograph below.

The production M2 and M3 fighting vehicles were almost identical in outward appearance. The only obvious difference was the blanked off firing ports on the M3. The driver in both vehicles was in the left front with the four periscope hatch cover. The gunner and commander rode in the turret on the left and right respectively. The 25mm M242 Bushmaster cannon was mounted to the left of the coaxial 7.62mm M240C machine gun. The Bushmaster was an externally powered weapon which fired single shots or at rates of 100 or 200 rounds per minute. The weapon was chain driven by a 1.5 horsepower electric motor. Hence the name Chain Gun, a registered trade mark of the manufacturer. The M242 fired both high explosive (HE) and armor piercing discarding sabot (APDS) ammunition. With a muzzle velocity of 4,460 feet per second, the APDS round could defeat the armor on the Soviet BMP turret at 45 degrees obliquity and at a distance greater than the 800 meter effective range of that vehicle's 73mm gun. The dual feed mechanism of the M242 permitted the gunner to switch instantly from one type of ammunition to the other. The 25mm gun had an elevation range from +59 to -9 degrees and a high speed slew rate of 60 degrees per second to permit the rapid engagement of alternate targets such as helicopters. The all electric turret drive and stabilization system allowed the accurate engagement of targets while the vehicle was moving. A total of 300 25mm ready rounds were carried in the turret. The M2 and M3 stowed an additional 600 and 1,200 25mm rounds respectively in the hull.

The drawing at the right shows the dimensions of the 25mm XM242 cannon.

The armored, two tube, TOW missile launcher was hinged to the left side of the turret. When raised to the firing position, the launcher had a separate elevating mechanism with a range of +29 to -19 degrees. Two TOW missiles were carried in the launcher. The M2 carried five TOW or Dragon missiles in the hull stowage racks at the left rear of the squad compartment. On the M3, ten TOW missiles were stowed in the hull racks. The TOW missiles could be launched only when the vehicle was stopped. Both vehicles carried three LAW missiles in the hull.

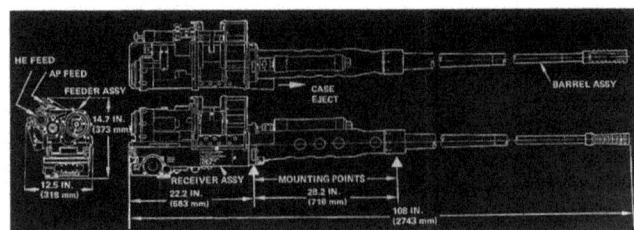

Close-up views of the Bradley turret with the TOW launcher in the firing position are above. The vehicle commander's station is below on the right side of the turret.

The gunner had a 180 degree view toward the front through two unity power periscopes in front of his hatch. A 4x and 12x day/night (thermal) sight was provided as the primary sight for the guns and the TOW missile. A 5x auxiliary sight also was installed to be shared by the gunner and the vehicle commander. The vehicle commander had 360 degree vision through seven unity power periscopes around his hatch and an optical relay permitted him to see through the gunner's day/night (thermal) sight. An external sight also was installed for open hatch firing. The vehicle radios were located in the turret bustle.

Below is a cutaway drawing of the Bradley turret.

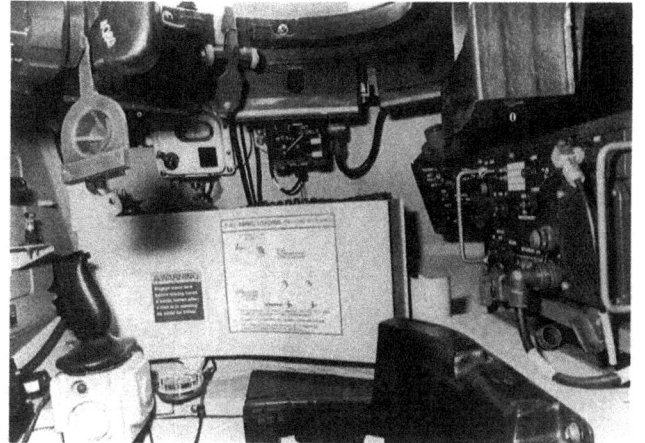

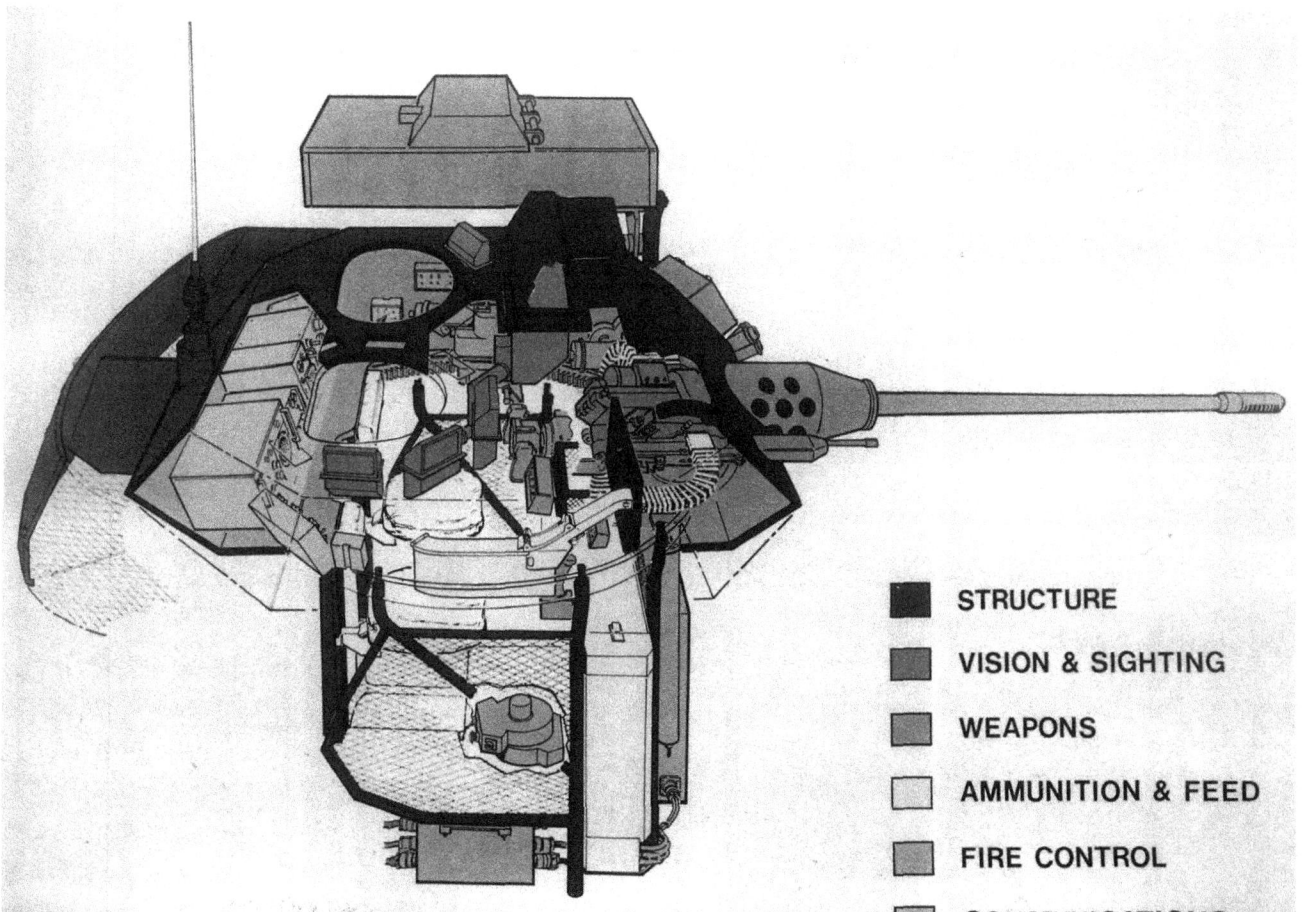

STRUCTURE

VISION & SIGHTING

WEAPONS

AMMUNITION & FEED

FIRE CONTROL

COMMUNICATIONS

Above, a view through the open rear ramp shows the seating arrangement inside the M2 (left) and the M3 (right) Bradley fighting vehicles. These seating arrangements also are shown in the drawings below.

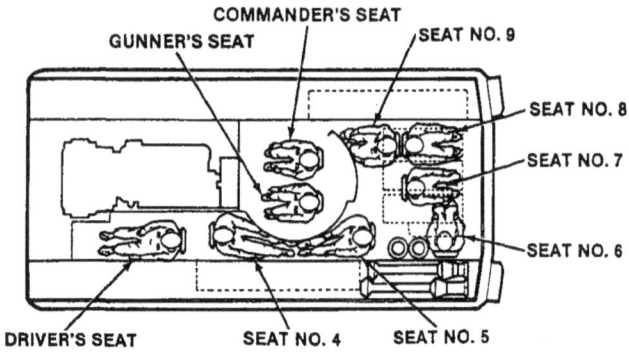

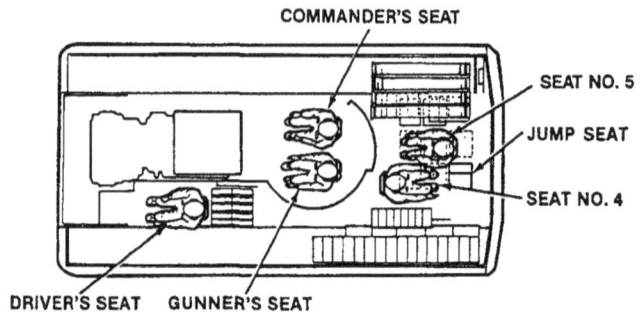

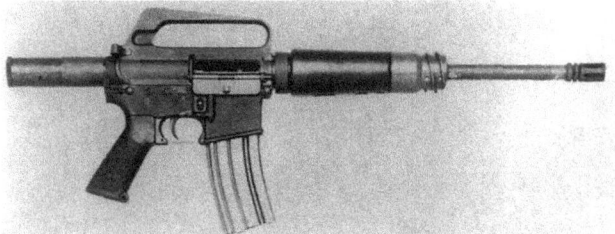

The 5.56mm M231 firing port weapon used in the M2 infantry fighting vehicle can be seen at the left. The fire from the various weapons in the M2 Bradley is illustrated below.

The type of armor on the Bradley fighting vehicle is shown in the sketch at the right.

The M2 infantry fighting vehicle provided space for six soldiers in the squad compartment. Six M231 5.56mm firing port weapons were carried for use in the six firing ports. Two of these were located in each side and two in the rear ramp. The M3 cavalry fighting vehicle had space for two observers in the rear hull. No firing port weapons were carried and the ports were blanked off. On both vehicles, three periscopes were installed in the hull roof between the cargo hatch and the rear ramp. Two additional periscopes were mounted on each side above the firing port locations. The small scout motorcycle originally proposed for the M3 was eliminated.

The hull and turret on the Bradley were assembled from 5083 and 7039 aluminum alloy armor combined with steel spaced laminate armor. The latter consisted of two ¼ inch thick, high hardness, steel plates spaced one inch apart and mounted 3½ inches outboard of the one inch thick aluminum armor. This provided protection against the Soviet 14.5mm armor piercing round and fragments from the 152mm high explosive shell. A $^3/_8$ inch thick steel armor plate was installed on the front third of the hull bottom for mine protection.

The Cummins VTA-903 diesel engine, now rated at 500 gross horsepower at 2,600 rpm, was installed in the right front with the General Electric HMPT-500 hydromechanical transmission. The final drives and sprockets were at the front and the vehicle was supported on a torsion bar suspension with six dual road wheels per side. One dual and two single track return rollers supported the upper run of the 21 inch wide, single pin, track. The vehicle was amphibious after erection of the trim vane and the water barrier. Combat loaded, the M2 and M3 weighed 50,259 pounds and 49,945 pounds respectively. The maximum speed was 41 miles per hour on roads and 4.5 miles per hour in water. The cruising range on roads was about 300 miles.

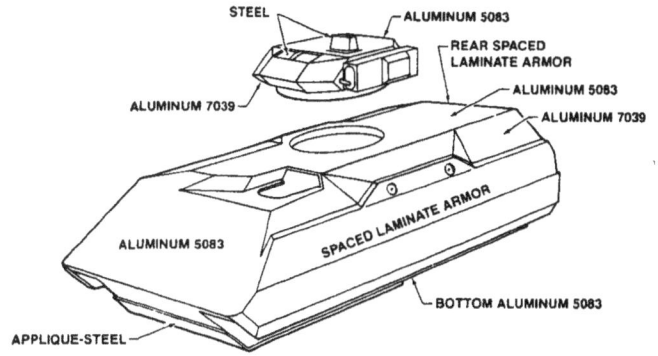

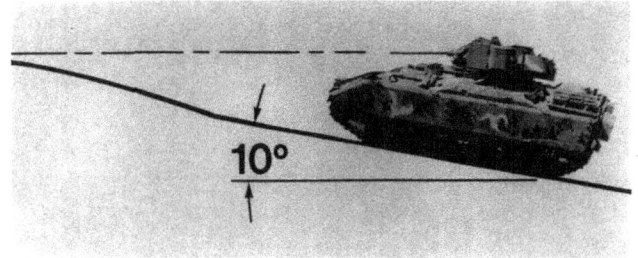

The advantage of a large depression angle in seeking cover behind terrain features is shown above. However, the Bradley had a maximum depression of 9 degrees. The drawing below shows the power train, suspension, and tracks in the Bradley.

Below, the driver's station in the Bradley appears at the left and the vehicle is swimming at the right.

289

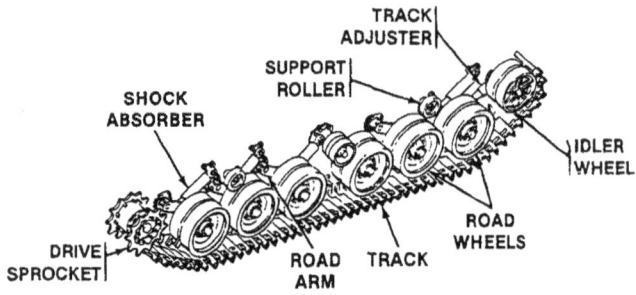

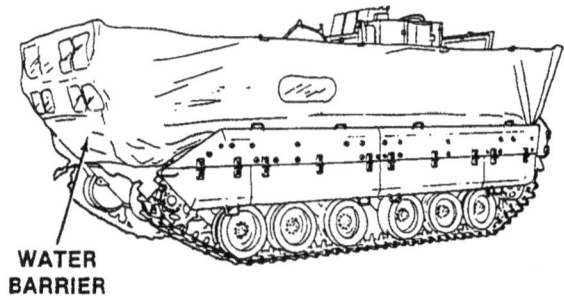

SHOCK
ABSORBER
SUPPORT
ROLLER
TRACK
ADJUSTER
IDLER
WHEEL
ROAD
WHEELS
DRIVE
SPROCKET
ROAD
ARM
TRACK

WATER
BARRIER

Above, suspension and track components on the Bradley are shown in the drawing at the left and at the right, the water barrier is erected to prepare the vehicle for swimming. The trim vane supports the water barrier in front.

Above, the water barrier is being erected on the Bradley at the left and the vehicle is swimming at the right. Below, the Bradley is fording in shallow water.

Above, the Bradley is launching a TOW missile.

With the Bradley in full production and service experience available, work continued to develop improvements to the basic vehicle. These were the Block I modifications that combined three major development programs with some minor modifications. The three major programs were the introduction of the TOW 2 missile system, the installation of a gas particulate filter unit (GPFU) for NBC protection, and various IFV/CFV design changes. The minor modifications were primarily electrical to reduce the logistics problem resulting from multiple configurations. By late 1984, prototypes incorporating many of the improvements were under evaluation at Aberdeen Proving Ground. These prototypes, converted from M2 and M3 vehicles, were designated as the M2E1 and M3E1 and they combined a mixture of old and new features.

The new TOW 2 missile system was installed on both the M2E1 and the M3E1. The TOW 2, with its full six inch diameter warhead, provided a more powerful weapon for frontal attack on the latest Soviet main battle tanks. The earlier TOW missiles also could be launched from the new system. At one time, the replacement of the twin tube launcher by two separate launchers was under consideration. These new armored launchers were of cast construction compared to the riveted assembly of the twin tube launcher and they were to be installed with one on each side of the turret. The twin tube launcher was canted slightly inward to allow the tracker to acquire the missile more rapidly. However, tests indicated that this was unnecessary and the two separate launchers were not to be canted. After further evaluation, the separate launchers were not adopted and the twin tube design was retained.

At the right is the M2E1 infantry fighting vehicle during tests at Aberdeen Proving Ground. Below are two views of the M3E1 cavalry fighting vehicle. Note that the firing ports have been covered on this vehicle.

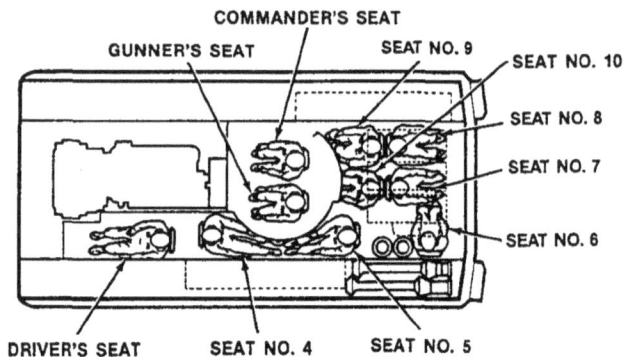

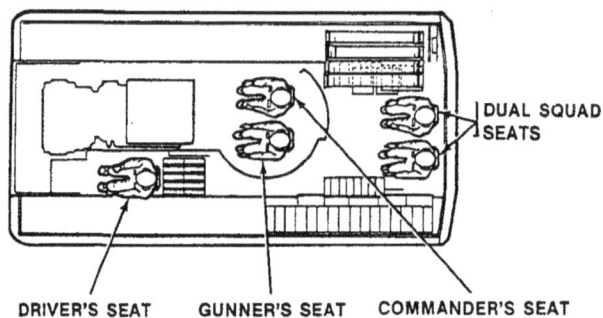

The drawings above show the seating arrangement in the M2A1 infantry fighting vehicle (left) and the M3A1 cavalry fighting vehicle (right). At the bottom of the page are two views of the M3A1 Bradley. Note the extra smoke grenade launchers on these vehicles.

The installation of the gas particulate filter unit differed on the IFV and the CFV. On the M2E1, face masks and hoses connected to the central filter unit were provided only for the driver, gunner and vehicle commander. The infantrymen in the squad compartment had individual masks and filter units built into their protective suits. This permitted them to leave the vehicle for dismounted operations. On the M3E1, all five crew members had masks connected to the central filter unit.

On the new CFV, the three periscopes in the hull roof behind the cargo hatch were eliminated and replaced by four periscopes installed in the cargo hatch cover. The two periscopes on the right side of the squad

compartment also were deleted on the CFV. The new IFV retained the original periscope arrangement, but the number of soldiers in the squad compartment was increased from six to seven. The stowage and seating arrangements in both vehicles also were modified.

The first of the Block I improvements introduced on the production line was the central gas particulate filter unit beginning in May 1986. The TOW 2 missile system appeared on new vehicles in early 1987. When both improvements were installed, the vehicles were designated as the M2A1 IFV and the M3A1 CFV. The new equipment was retrofitted on many of the earlier vehicles.

292

Above, the M3A1 cavalry fighting vehicle is shown with the hatches open and the missile launcher retracted into the travel position (left) and with the hatches closed and the launcher erected in the firing position (right). Note that this is the same M3A1 with the extra smoke grenade launchers.

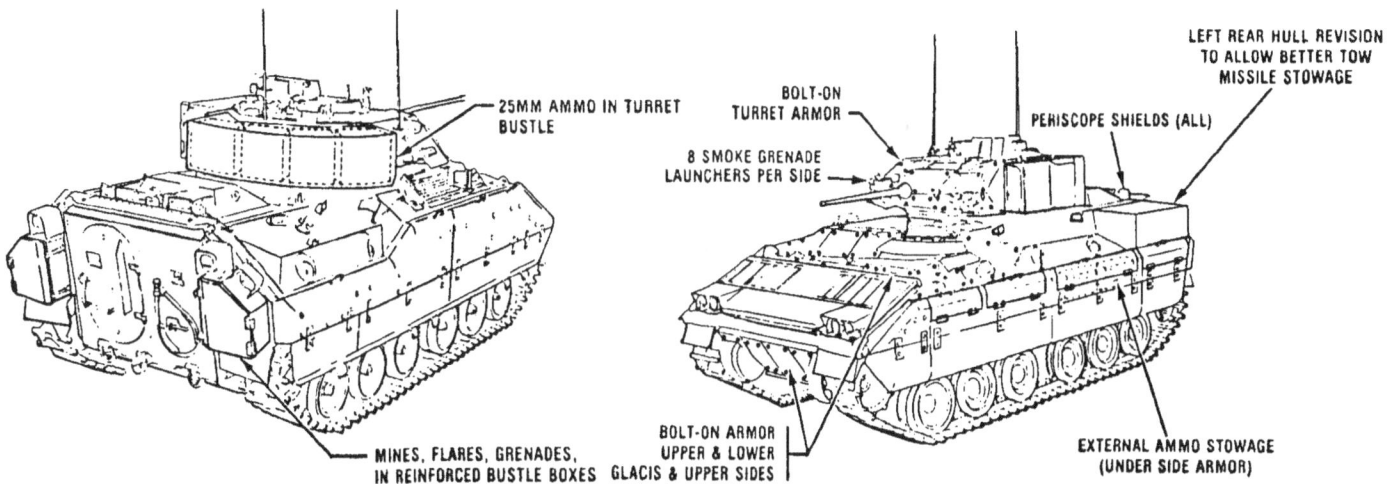

25MM AMMO IN TURRET BUSTLE

MINES, FLARES, GRENADES, IN REINFORCED BUSTLE BOXES

LEFT REAR HULL REVISION TO ALLOW BETTER TOW MISSILE STOWAGE

BOLT-ON TURRET ARMOR

PERISCOPE SHIELDS (ALL)

8 SMOKE GRENADE LAUNCHERS PER SIDE

BOLT-ON ARMOR UPPER & LOWER GLACIS & UPPER SIDES

EXTERNAL AMMO STOWAGE (UNDER SIDE ARMOR)

These drawings show the survivability improvements applied to the M2A1 and M3A1 Bradley fighting vehicles.

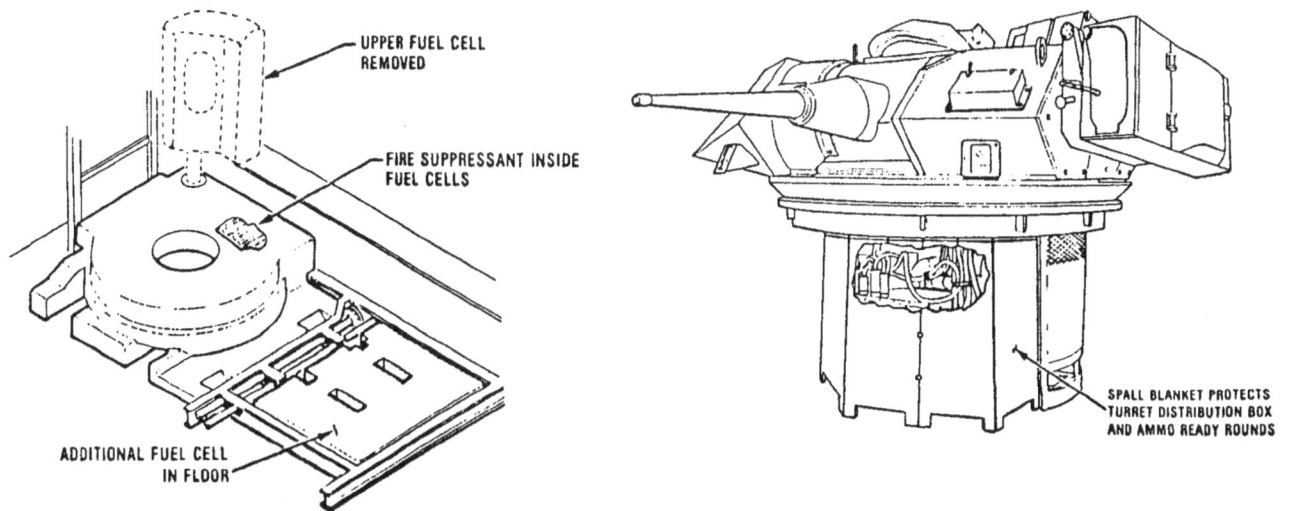

UPPER FUEL CELL REMOVED

FIRE SUPPRESSANT INSIDE FUEL CELLS

ADDITIONAL FUEL CELL IN FLOOR

SPALL BLANKET PROTECTS TURRET DISTRIBUTION BOX AND AMMO READY ROUNDS

POWER UNIT EXHAUST GRILLE

CARGO HATCH COVER

HATCH COVER PERISCOPE

REAR STOWAGE BOX

TAILLIGHT

RAMP ACCESS DOOR

FIRING PORTS

TOWING PINTLE

RAMP

ENGINE EXHAUST DEFLECTOR

EXTERIOR FIRE SUPPRESSION HANDLE – SQUAD COMPARTMENT

FUEL FILLER COVER

DRIVER'S HATCH COVER

HEADLIGHT ASSEMBLY

INTAKE GRILLE SCREEN

PERISCOPE

ARMOR PLATE

IDLER WHEEL

TRACK

ROAD WHEEL

EXTERIOR FIRE SUPPRESSION HANDLE – ENGINE COMPARTMENT

DRIVE SPROCKET

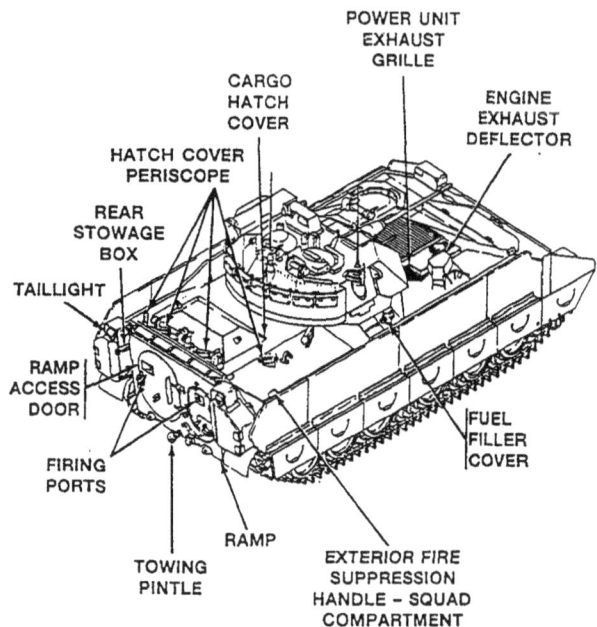

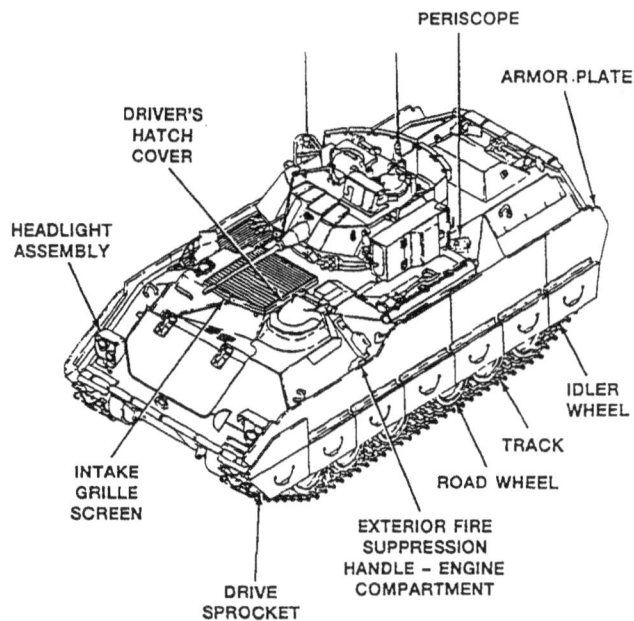

The drawings above identify the components on the M2A2 infantry fighting vehicle. The photograph at the bottom of the page shows the M3A2 cavalry fighting vehicle. Note the periscope in the troop compartment hatch cover.

During 1983/84, the Bradley was subjected to considerable criticism in Congress, much of which was directed at the armor protection. Some people felt that if it was going to operate with tanks, it should be armored like one. This was, of course, impractical within reasonable cost and weight limitations. However, a program was launched to improve the survivability of the Bradley in the battlefield environment. This resulted in the Block II or A2 modifications.

Unlike the Block I improvements, the new modifications changed the external appearance of the Bradley. On the hull, the trim vane was omitted and steel applique armor was installed on the front, vertical sides, and the bottom. Spaced laminate steel armor was added to the rear and spaced laminate steel skirts protected the lower sides. Sections of steel applique armor were attached to the front and sides of the turret and the shroud was removed from the barrel of the 7.62mm

M240C coaxial machine gun. A circular shield attached to the bustle acted as spaced armor for the rear of the turret and provided additional stowage space. Fittings on the steel applique armor were provide for the attachment of passive armor tiles or explosive reactive armor. Kevlar spall liners were installed to protect critical areas inside the new vehicle.

Above are views of the Bradley A2 turret without (left) and with (right) reactive armor tiles. The attachment points for the tiles can be seen on the Bradley below.

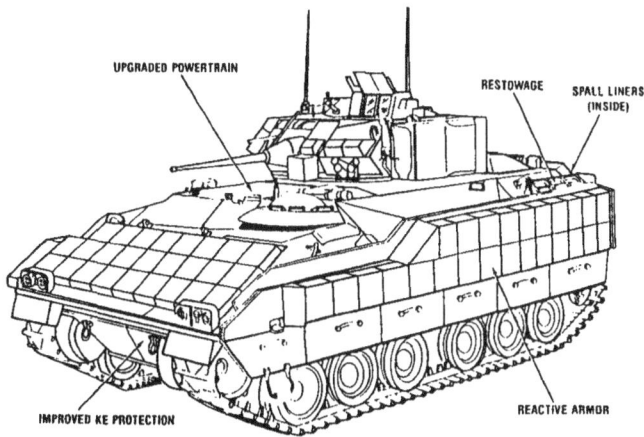

The reactive armor tiles are installed on the A2 Bradley in the drawing at the left, however, the headlights are incorrect. The A2 Bradley below also has the attachment points for the tiles. Note the lack of a trim vane.

UPGRADED POWERTRAIN
RESTOWAGE
SPALL LINERS (INSIDE)
IMPROVED KE PROTECTION
REACTIVE ARMOR

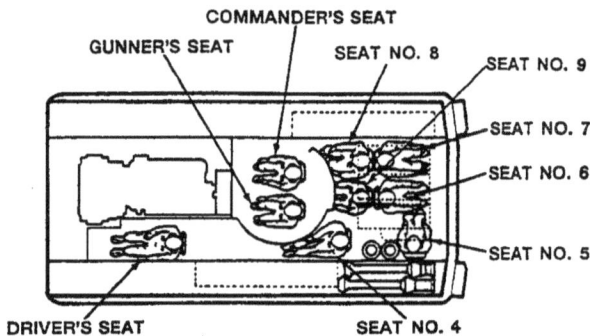

The original seating arrangement in the M2A2 (left) and the M3A2 (right) Bradley fighting vehicles can be seen in the drawings above.

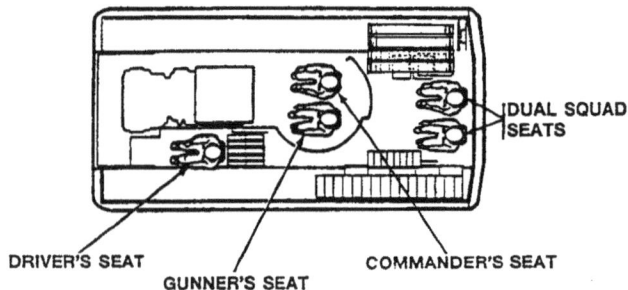

Designated as the M2A2 IFV and the M3A2 CFV, the new vehicles initially retained the same seating arrangement as in the M2A1 and the M3A1. However, the infantry squad in the M2A2 was reduced to six men with the elimination of the soldier seated with his back to the driver. The steel armor on the sides blanked off the side firing ports, but the two in the rear ramp were retained. The front periscope on the left side was deleted, but the one at the left rear was retained. Both periscopes on the right side of the squad compartment remained as on the M2A1 as well as the three roof periscopes behind the cargo hatch. A protective metal cover was installed over the driver's periscopes and wire guards were on each side in front of his hatch. These

guards were to protect the driver from wires stretched across the road when he was driving with an open hatch. Since there was no trim vane to serve as a work platform for the power plant compartment, a maintenance stand was provided. This stand was stowed on the left side of the hull alongside the turret. The stand could be mounted on the front of the vehicle to provide a working surface during maintenance in the power plant compartment. The flotation or water barrier was stowed around the periphery of the vehicle and the components of its support structure were stowed under the maintenance stand on the left side and on the top right rear of the hull. New headlight groups also were installed on the M2A2 and the M3A2.

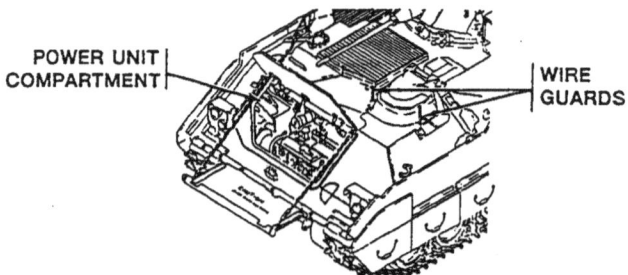

The maintenance stand on the A2 Bradley above can be compared with the trim vane as a work stand on the earlier vehicles at the left. Dimensions of the A2 Bradley are shown below.

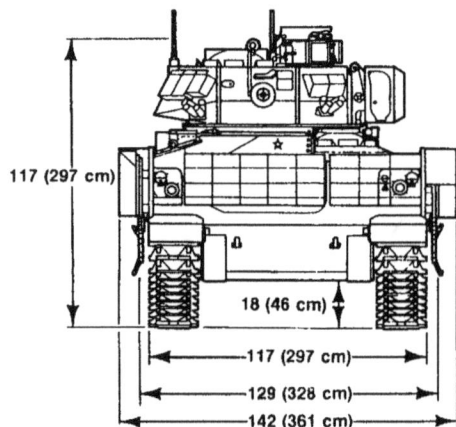

Dimensions in inches (centimeters)

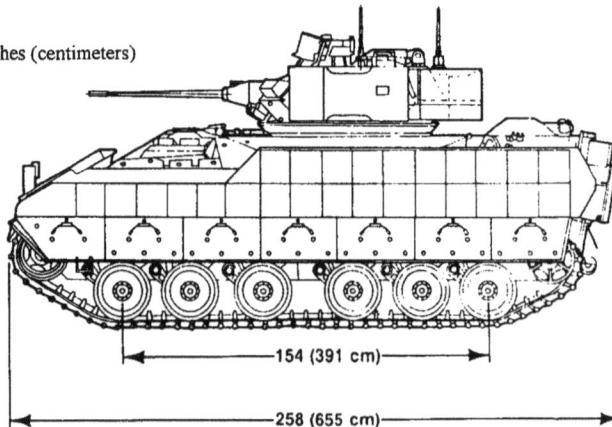

In these views, the applique armor kit has been installed on the A2 Bradley fighting vehicle. In the bottom photograph, the troop compartment hatch cover periscopes and the lack of ramp firing ports identify the M3A2 cavalry fighting vehicle.

SEAT NO. 8
SEAT NO. 9
GUNNER'S SEAT
COMMANDER'S SEAT
SEAT NO. 7
COMMANDER'S SEAT
DRIVER'S SEAT
SEAT NO. 10
SEAT NO. 4
SEAT NO. 6
SEAT NO. 5
DRIVER'S SEAT
GUNNER'S SEAT
SEAT NO. 4
SEAT NO. 5

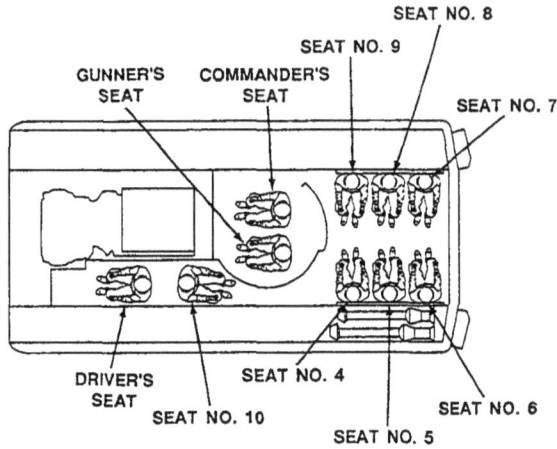

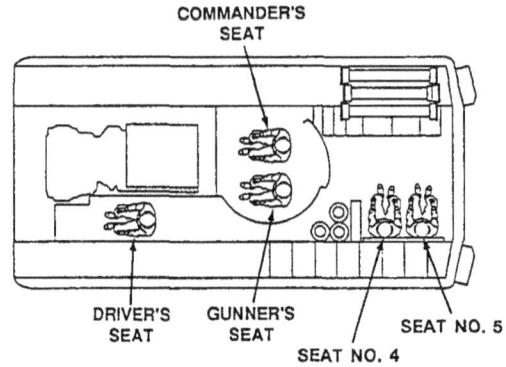

Above, the new seating arrangements in the M2A2 (left) and M3A2 (right) Bradley fighting vehicles are shown in these drawings after the vehicles were restowed.

Combat loaded, the weight of the M2A2 and the M3A2 was about 30 tons. This increased to approximately 33 tons with the installation of the armor tile kit. The weight could be reduced to 22 tons for air transportation. Obviously, this heavier weight would have reduced the performance without an increase in the available power. Such an increase was obtained by the installation of an upgraded power train. This consisted of the Cummins VTA-903T diesel engine now developing 600 gross horsepower. It was coupled to the improved HMPT-500-3 transmission. With this power train, the maximum speed was 35 miles per hour on roads and 4 miles per hour in water. The 175 gallon fuel tank provided a cruising range of about 250 miles on roads.

Live fire testing of the Bradley, completed in early 1987, evaluated the improvements introduced in the M2A1 and M3A1. The test program also included a so-called minimum casualty version of the vehicle. This vehicle stowed much of the fuel and ammunition outside of the crew compartment to protect them from secondary fires and explosions.

The results of the live fire tests led to further modification of the M2A2 and the M3A2. The seating was rearranged in the squad compartment and the vehicle was restowed to reduce the vulnerability. The infantry squad in the M2A2 was increased again from six to seven men. Three men were located on bench seats on each side of the squad compartment. The seventh man was seated facing toward the rear just behind the driver. In the M3A2, the two observers were moved to a bench seat on the left side of the squad compartment. The missile stowage was changed with the introduction of the Javelin to replace the earlier Dragon missile. In the restowed M2A2, either five TOW missiles or three TOWs and two Javelins could be carried. In a similar manner, the three LAW missiles were replaced by three AT4 missiles in the restowed M2A2 and M3A2. On later M2A2 vehicles with the MANPAD kit, six Stinger missiles were carried in the right rear of the squad compartment.

When production ended in 1994, United Defense Limited Partnership (formerly FMC Corporation) had produced 6,724 Bradleys. They consisted of 4,641 IFVs and 2,083 CFVs. The production run included 2,300 M2s and M3s, 1,371 M2A1s and M3A1s, and 3,053 M2A2s and M3A2s. Many of the earlier vehicles were converted to the -A2 configuration.

Below are views of the minimum casualty version of the Bradley with external stowage of the fuel and ammunition.

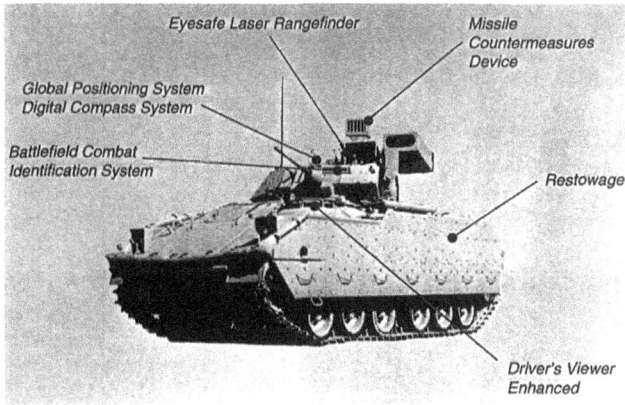

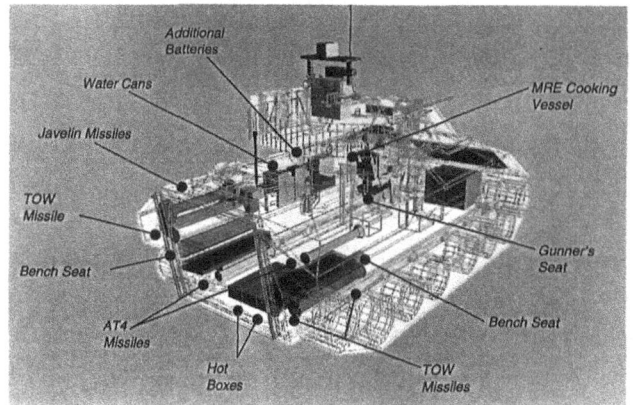

The improvements introduced on the Bradley after Operation Desert Storm can be seen in the views above of the A2ODS. Note the missile countermeasures device on top of the turret.

Following Operation Desert Storm (ODS) in the Persian Gulf, several field retrofits were made on the Bradley. Referred to as the ODS upgrade, these modifications included an eye-safe carbon dioxide laser rangefinder, a global positioning system with a compass, a combat identification system, a driver's thermal viewer, and a missile countermeasure device. With these changes, the vehicles were designated as the M2A2ODS and the M3A2ODS.

The lessons of Operation Desert Storm combined with the fielding of the fully digitized M1A2 main battle tank clearly indicated the need for an improved version of the Bradley. In Fiscal Year 1995, a development contract was awarded to United Defense Limited Partnership (UDLP) to upgrade the Bradley. A new vehicle designated as the M2A3 infantry fighting vehicle incorporated a databus, a central processing unit, a mass memory unit, and information displays for the commander, driver, and squad leader. This data system was

compatible with the intervehicular information system in the M1A2 tank and the Apache Longbow helicopter. The upgrade included an independent thermal viewer for the vehicle commander permitting the use of hunter-killer tactics. An improved integrated sight unit was installed and it featured automatic dual target tracking, automatic gun target adjustment, and automatic bore sighting. Titanium alloy roof armor provided fragmentation protection.

The M2A3 infantry fighting vehicle appears at the right and in the drawing below. The commander's independent thermal viewer is mounted on the right side of the turret and the hull is fitted with attachment points for applique armor.

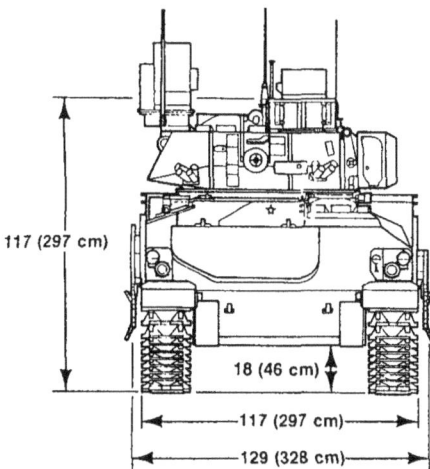

Dimensions in inches (centimeters)

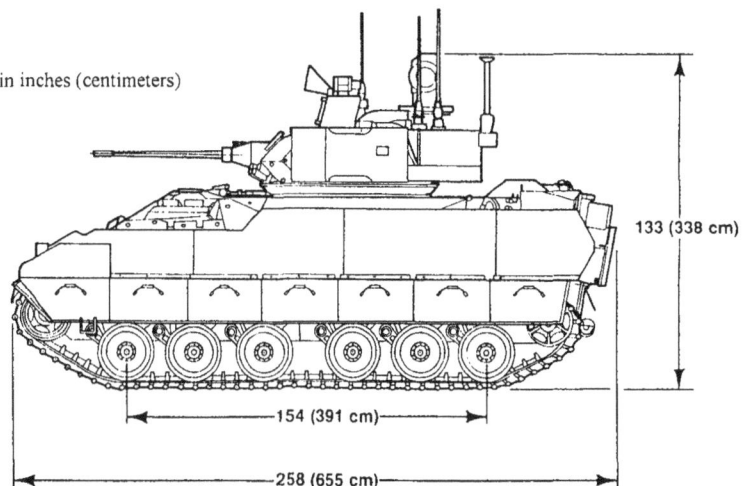

117 (297 cm)

18 (46 cm)

117 (297 cm)

129 (328 cm)

133 (338 cm)

154 (391 cm)

258 (655 cm)

Additional views of the M2A3 Bradley are above and at the right. Below at the right is the Bradley Stingray. The armored housing for the laser can be seen on the right side of the turret.

As of late 1996, it was planned to remanufacture 1,602 -A2 vehicles to the -A3 configuration with fielding to begin in 2001.

In early 1990, a strap-on prototype of a battlefield laser was tested at Fort Bliss, Texas. Referred to as the Stingray laser self-protection system, it was intended to blind or disable the optical sights and periscopes of enemy vehicles. Under development by Martin Marietta since 1982, a demonstration version was tested during 1986/87. The strap-on prototype was enclosed in an armored housing hinged to the right side of the Bradley turret. It was equipped with an internal display and a joystick control. The Stingray was under development as part of the Army Armored Systems Modernization Program.

As early as 1981, studies began to upgrade the firepower of the Bradley. Initially, this involved the installation of the 35mm Talon automatic cannon developed by Ares, Incorporated as part of the Eagle air defense system. Under the Combat Vehicle Armament System Technology (CVAST) program, the 35mm weapon was installed in a two man cleft turret on the Bradley chassis. Use of the cleft turret minimized the vehicle height and allowed the Bradley with the CVAST turret to be carried in the C130 transport aircraft.

Although the experimental CVAST turret was fired successfully during tests, it did not go into full scale development. Interest had now shifted to the use of case telescoped ammunition (CTA) for the new weapon and the project was now designated as the Combat Vehicle Armament Technology (COMVAT) program. The new telescoped ammunition had the projectile inside a control tube surrounded by the main propellant charge.

The CVAST test bed demonstrator is shown in the drawing below.

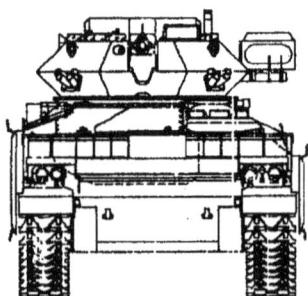

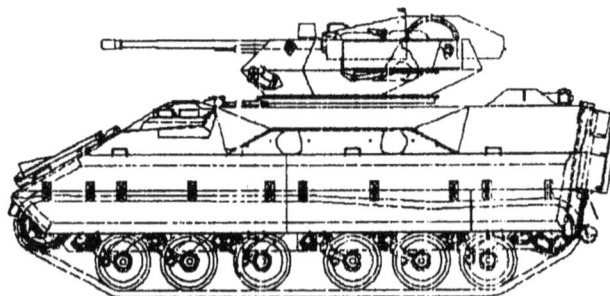

Above are two views of the Bradley fitted with the two man cleft CVAST turret. At the right, the COMVAT turret is installed with the 45mm weapon using case telescoped ammunition.

This complete assembly was inserted into a cylindrical cartridge case. When fired, the primer ignited a booster charge in the base of the control tube forcing the projectile into the bore of the gun. Once the projectile cleared the control tube, the main propellant charge ignited. Use of the cylindrical rounds greatly simplified feeding the ammunition to the gun. With the rotating breech designed by Ares, the ammunition could be fed into the weapon through the hollow trunnions. After each round was fired, the empty case was pushed out as the next round was loaded. The initial development was for a 30mm weapon, but this was soon changed to a 45mm gun. A two man turret armed with the new 45mm weapon was designed to fit the 60 inch diameter ring on the Bradley. The studies also included an unmanned turret for the same installation.

Another approach to improving the firepower of the Bradley was the installation of the 30mm Bushmaster II

cannon using GAU-8 ammunition. Although the 30mm weapon was 137¾ inches long compared to 108 inches for the 25mm M242, it could be installed in the standard Bradley turret with very little modification. The Bushmaster II weighed 325 pounds compared to 244 pounds for the 25mm weapon. With the armor piercing round, the muzzle velocity of the 30mm cannon was about 4,000 feet per second.

In addition to the Bushmaster II, a 35mm Bushmaster III and a 40mm Bushmaster IV were under development. A 50mm barrel also was under development for installation on the Bushmaster III permitting it to fire an APFSDS round referred to as the 50mm Supershot. The 35mm Bushmaster III was installed in the Bradley turret for evaluation.

Below, the standard 25mm M242 Bushmaster cannon and its ammunition are compared with the 30mm Bushmaster II. At the right, the Bradley is armed with the 30mm weapon.

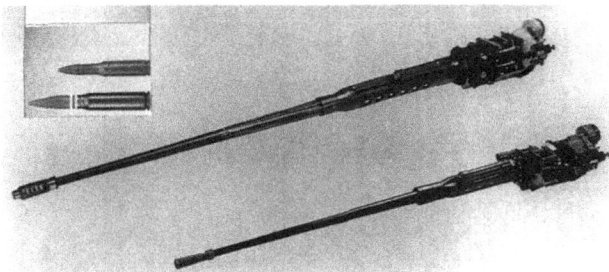

Above, the 35mm Bushmaster III is installed in the Bradley. Below is a drawing of the weapon itself.

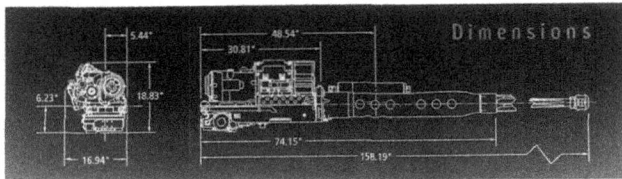

At the right, the standard 35mm Oerlikon ammunition is compared with the 50mm supershot.

35mm Oerlikon
Used by 30 Countries. Load & Pack in 6 Countries.

50mm Supershot
In Development for Future Threat.

Above is the A2ODS Bradley modified for use as a battle command vehicle. Note that it retains the full armament of the Bradley.

VEHICLES BASED UPON THE BRADLEY

The Bradley also was modified to provide a battle command vehicle (BCV) for use by task force, battalion, and brigade commanders. Battle experience during Operation Desert Storm (ODS) clearly indicated the need for such a vehicle that would provide the commanders the protected mobility required to operate with the fast moving combined arms units.

The external appearance of the BCV was identical to that of the -A2ODS Bradley except for the additional antennas required for the communications equipment. In addition to the vehicle crew, the BCV carried the commander and a small staff of up to three officers. These could be operations, fire support, intelligence, air liaison, or logistics officers depending upon the particular mission requirements.

The extra radio antennas are obvious on the Bradley battle command vehicle at the right.

The BCV retained its full armament of the 25mm M242 cannon and coaxial machine gun as well as the TOW missile launcher. However, the ammunition supply was reduced to provide space inside the vehicle.

The XM7 Bradley fire support team vehicle (FIST-V) appears above. Note the FIST equipment installation replacing the TOW launcher.

The M981 FIST-V based upon the M113 series armored personnel carrier provided essential field artillery support for the maneuver forces. With the introduction of the M1 main battle tank and the Bradley fighting vehicle, it was obvious that a new fire support team vehicle with comparable performance was required. Such a vehicle, designated as the XM7 FIST-V was built by United Defense based upon the Bradley - A2ODS. The XM7 was equipped with the AN/TVQ-2 ground vehicle laser locator designator, the AN/TAS-4 TOW night sight, and an inertial navigation system. The armored enclosure normally used for the TOW missile launchers was modified to house the FIST equipment except for the SINCGARS radios. The 25mm M242 cannon and the coaxial 7.62mm machine gun were retained for self-defense. A further development of the FIST-V vehicle based upon the -A3 Bradley was assigned the designation XM7A1.

The dimensions of the XM7 Bradley FIST-V can be seen below.

Dimensions in inches (centimeters)

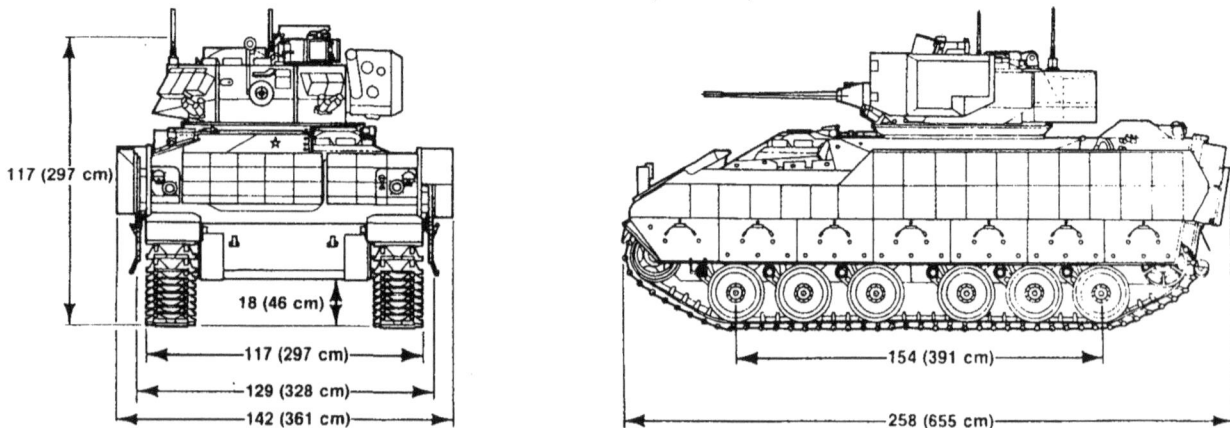

117 (297 cm)
18 (46 cm)
117 (297 cm)
129 (328 cm)
142 (361 cm)
154 (391 cm)
258 (655 cm)

The model and the artist's concept above show the line of sight antitank weapon system (LOSAT) on the Bradley chassis. In both of these views, the four tube retractable launcher is erected in the firing position.

The Bradley was considered as a carrier for the kinetic energy missile (KEM). The KEM, also referred to as the hypervelocity missile, was not fitted with an explosive warhead, but depended upon a heavy metal penetrator rod moving at about 5,000 feet per second to destroy the target. The rocket propelled, beam riding, missile was about 6.4 inches in diameter, 113 inches long, and weighed approximately 170 pounds. Installed in a four missile retractable launcher on the Bradley chassis, it was referred to as the line of sight antitank (LOSAT) weapon system. The forward looking infrared (FLIR) sensor was mounted on top of the launcher. Manned by a crew of three, the LOSAT was intended to replace the TOW vehicles in the army antitank units.

Photographs of the LOSAT trials vehicle are shown here. Below the launcher is retracted.

305

The ADATS missile is being launched in the views above and at the right.

A combined antitank, antiaircraft weapon system also was installed on the Bradley chassis. After termination of the Sergeant York program, the air defense/antitank system (ADATS) was selected to meet the requirement of the forward area air defense system (FAADS) for a line of sight-forward-heavy (LOS-F-H) weapon. The ADATS, produced by Oerlikon-Buehrle, also was mounted on the M113 series chassis and is described in that section. On the Bradley chassis, four missile launchers were installed on each side of a new turret which could be armed with an M242 25mm cannon. A radar antenna was mounted on the rear of the turret. Although the ADATS was equipped with a radar system for volume searching and could track ten targets simultaneously, its primary electro-optical fire control equipment consisted of a forward looking infrared target acquisition and tracking system, laser beam riding missile guidance, and laser ranging. As such, it was virtually immune to electronic countermeasures and antiradar missiles.

Although the initial tests of the ADATS were promising, later evaluations revealed a number of reliability problems. That, combined with a reduction in available funds, resulted in the termination of the Bradley ADATS program.

Another antiaircraft weapon based upon the Bradley chassis was the tracked Rapier. It participated in the original competition that selected the ADATS.

Below, the air defense antitank system is installed on the Bradley chassis. Note that the ADATS at the bottom right also is armed with the 25mm M242 cannon.

As an interim air defense measure, Stinger anti-aircraft missiles were deployed on the standard Bradley infantry fighting vehicle. General Electric proposed the installation of the Blazer antiaircraft turret on the Bradley chassis. This turret was armed with two, four tube, Stinger missile launchers as well as a GAU-12/U 25mm Gatling type gun. The 25mm gun was provided with 360 ready rounds and Hydra-70 2.75 inch rockets also could be carried.

During April and May 1996, a new air defense version of the Bradley armed with the Stinger missile was under limited operational testing at Eglin Air Force Base, Florida. Referred to as the Bradley Stinger fighting vehicle-enhanced (BSFV-E), it also was named Linebacker. The Army procured eight Linebackers for evaluation. Utilizing the Avenger air defense system hardware and software, the Bradley Linebacker was armed with a four missile armored launcher on the left side of the turret. It retained the 25mm Bushmaster cannon and the 7.62mm coaxial machine gun. The tests at Eglin and the following June at Roving Sands, New Mexico were highly successful and funds were available in late 1996 for the procurement of 51 additional Linebacker fire units.

At the right and below is the Bradley Linebacker armed with the four tube Stinger launcher replacing the TOW missile system.

The fighting vehicle systems carrier appears in the photographs above.

THE FIGHTING VEHICLE SYSTEM CARRIER

As mentioned earlier, the fighting vehicle systems program was expanded in June 1979 to include a third vehicle in addition to the XM2 and XM3. The new member of the family was the general support rocket system (GSRS) carrier. As its title indicated, it was intended to transport a mobile long range artillery rocket system to support fast moving combined arms forces. The new vehicle used a lengthened chassis based upon the components of the XM2 and XM3 fighting vehicles. In fact, the vehicle was soon referred to as the fighting vehicle system (FVS) carrier. It was to be the basis for a family of vehicles related to the M2 and M3 in much the same way that the M548 series was related to the M113 family, although the new carrier was armored.

The three man armored cab on the FVS carrier was located just in front of the 500 gross horsepower VTA-903 diesel engine and above the HMPT-500 transmission and final drives. The cab tilted forward to permit access to the power train. The Bradley torsion bar suspension was used with six dual road wheels per side. However, the road wheels were spaced out in groups of two increasing the ground contact length to 170.5 inches and the upper track run was supported by two dual and two single track return rollers. If required, a lock-out system could be installed on the suspension to stabilize the vehicle. The Bradley's 21 inch wide tracks were used with 89 and 88 track shoes on the left and right respectively compared to 84 and 82 on the M2 and

The lengthened chassis of the fighting vehicle systems carrier is obvious in the view below.

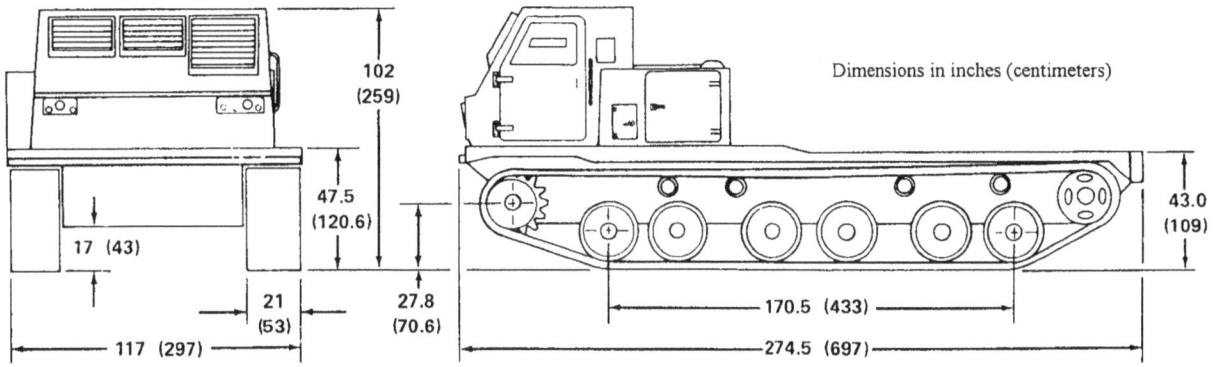

Dimensions in inches (centimeters)

102 (259)
47.5 (120.6)
17 (43)
21 (53)
27.8 (70.6)
117 (297)
43.0 (109)
170.5 (433)
274.5 (697)

The dimensions of the fighting vehicle systems carrier are shown in the drawing above. The crew location in the cab can be seen in the drawing below.

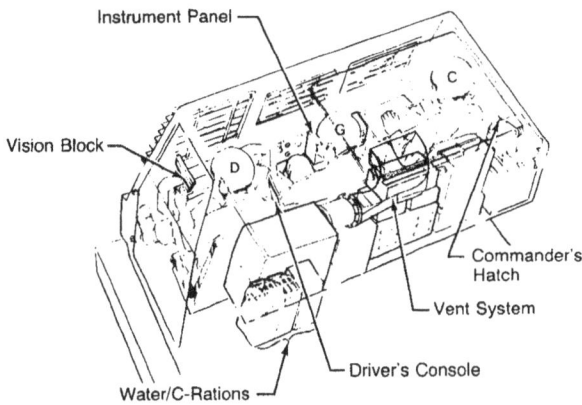

Instrument Panel
Vision Block
Commander's Hatch
Vent System
Driver's Console
Water/C-Rations

M3. The hull itself was fabricated from 5083 aluminum alloy armor. The cab was assembled using 7039 aluminum alloy armor and the windows were fitted with armor glass. Armor louvers could be lowered to protect the windows during operations. They were opened and closed by levers from inside the cab. When not in use, the louvers could be stowed to permit unobstructed vision. An overpressure ventilation system was installed in the cab for NBC protection.

Closed

Open

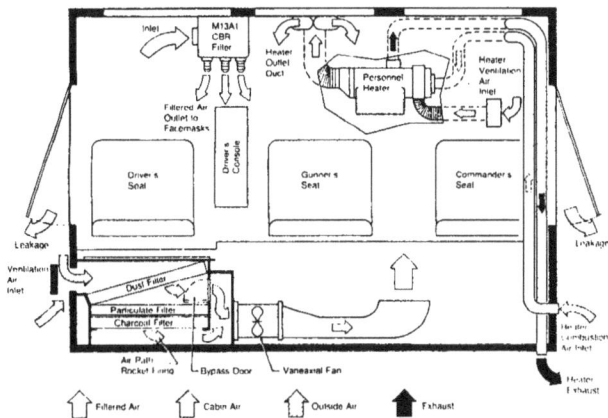
Stowed

The ventilation air flow through the cab appears in the drawing at the left and the cab louvers on the vehicle are shown in the open, closed, and stowed positions above.

309

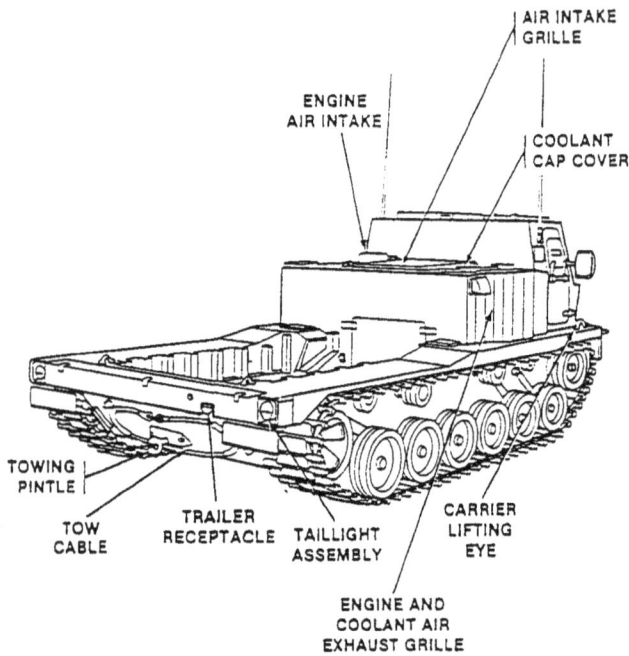

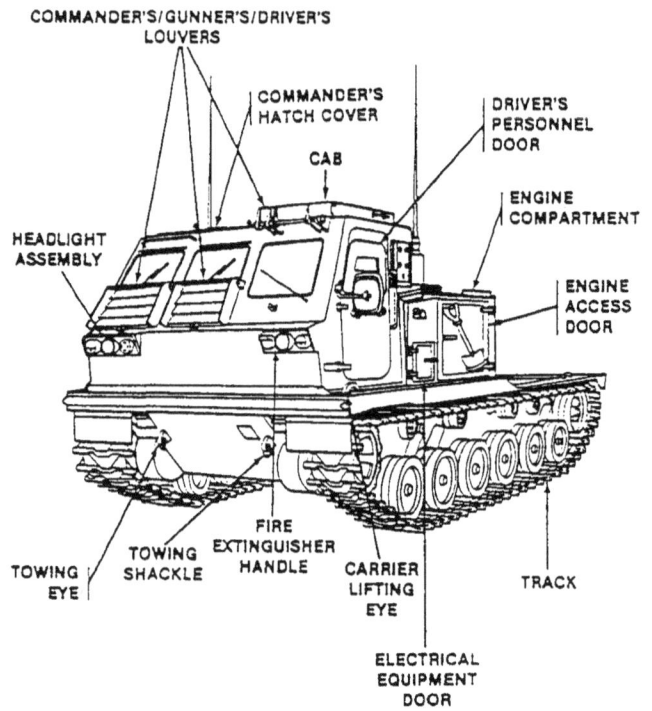

The various components on the M993 multiple launch rocket system carrier are identified in the drawings above and at the right.

The FVS carrier, now designated as the XM987, was proposed for a wide variety of applications including maintenance and recovery vehicles, forward area resupply vehicles, mine launchers, command post and communication vehicles, antiaircraft vehicles, and various missile launchers. The originally proposed general support rocket system (GSRS) was the first to be developed. As far back as September 1977, contracts had been awarded to the Boeing Aerospace Company and the Vought Corporation to develop prototypes of the GSRS. In both cases, the prototypes, as finally developed, used the FVS carrier as the basic vehicle. When modified for this application, the carrier was designated as the XM993. Originally, 210mm diameter rockets were specified for the new system. However, nego-

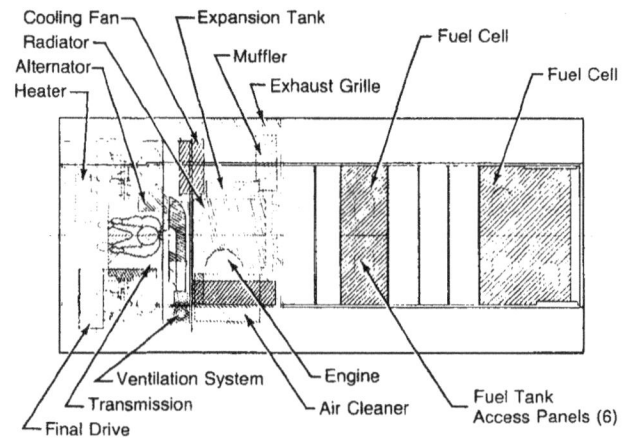

tiations with NATO resulted in changes to ensure adoption by the various member nations. The rockets were enlarged to a diameter of 227mm to permit them to carry the German AT2 antitank mine.

Below, the Boeing multiple launch rocket system (MLRS) prototype is at the left and the Vought MLRS is at the right.

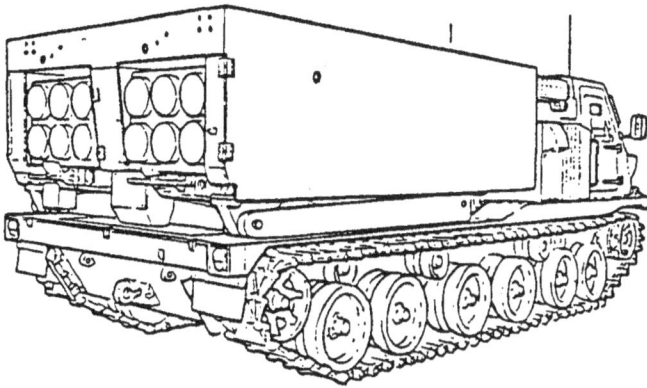

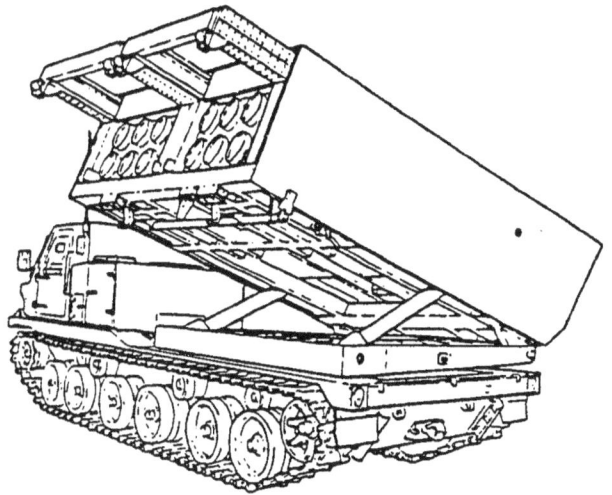

Above, the M270 ground vehicle mounted rocket launcher is installed on the M993 MLRS carrier. It is shown in the travel position (left) and the firing position (right).

The prototypes were under test from December 1979 through February 1980 and after their evaluation, the Vought candidate was selected in May 1980. The new weapon was designated as the multiple launch rocket system (MLRS) that same year. Low initial rate production began in early 1982 and the first firing batteries were equipped in the 1st Infantry Division in early 1983.

The production MLRS consisted of the M270 ground vehicle mounted rocket launcher on the multiple launch rocket system carrier M993. Suspension lock-out was installed on the 1st, 5th, and 6th road wheel arms. The M270 launcher carried two rocket pods each loaded with six 227mm M26 rockets. Each rocket was about 12.9 feet in length and weighed 676.5 pounds. Carrying 644 M77 submunitions, the M26 rocket had a range of about 32 kilometers. The solid fuel rocket was supported by a sabot and studs on helical rails inside the launch tube. This imparted a slight spin to the rocket when it was launched. Other payloads included 28 AT2 antitank mines or six SADARM (sense and destroy armor) sub-munitions. The rockets could be fired one at a time or in rapid sequence up to the full load. The combat weight of the MLRS was 52,990 pounds. It had a maximum road speed of 40 miles per hour and a cruising range of about 300 miles. The three man crew consisted of the driver, gunner, and section leader.

The multiple launch rocket system (MLRS) is at the right and below, one of the rockets is being launched.

Above, the MLRS is moving at speed and at the right the launcher is raised to the firing position. Note that one rocket pod is empty. Below, the rocket pods are being loaded into the launcher.

Below are two views of the MLRS launching the 227mm M26 rocket.

The series of photographs at the right show the launch of the M39 army tactical missile system.

The M39 army tactical missile system (ATACMS) also was designed for use with the M270 launcher on the MLRS. Development testing of the ATACMS was completed during the Spring of 1990 and the first production missiles were delivered in March. The missile measured 13 feet in length and 2 feet in diameter. Each was enclosed in a sealed pod having the same external dimensions as the six rocket pod used in the MLRS. Thus two ATACMS missiles could be loaded into the M270 launcher. The Block 1 missile with the M74 submunition warhead weighed 3,687 pounds and had a maximum range of over 100 kilometers.

As a result of studies carried out in 1984/85, the Army considered that the FVS carrier was suitable as a replacement for the M1015 electronic warfare shelter carrier based upon the M548 cargo carrier. FMC, using their own funds, designed and built a prototype based upon the M987 carrier. Tests of this vehicle began in August 1986. Designated as the XM1070 electronic fighting vehicle system (EFVS), it was fitted with a large enclosure fabricated from aluminum alloy armor. Modular armor kits also were available for increased protection. Manned by a crew of six including the driver, the XM1070 was fitted with a 6BT5.9 Cummins diesel engine in the enclosure as the primary power source for the electronic equipment. This engine drove a generator providing 60 kilowatts of AC power. The vehicle engine also could be used a back-up power source for the electronic equipment. A retractable antenna mast was installed on the left rear of the enclosure. It could be extended to a maximum height of 20 meters. The crew compartment was slightly pressurized for NBC protection.

The electronic fighting vehicle system XM1070 can be seen below.

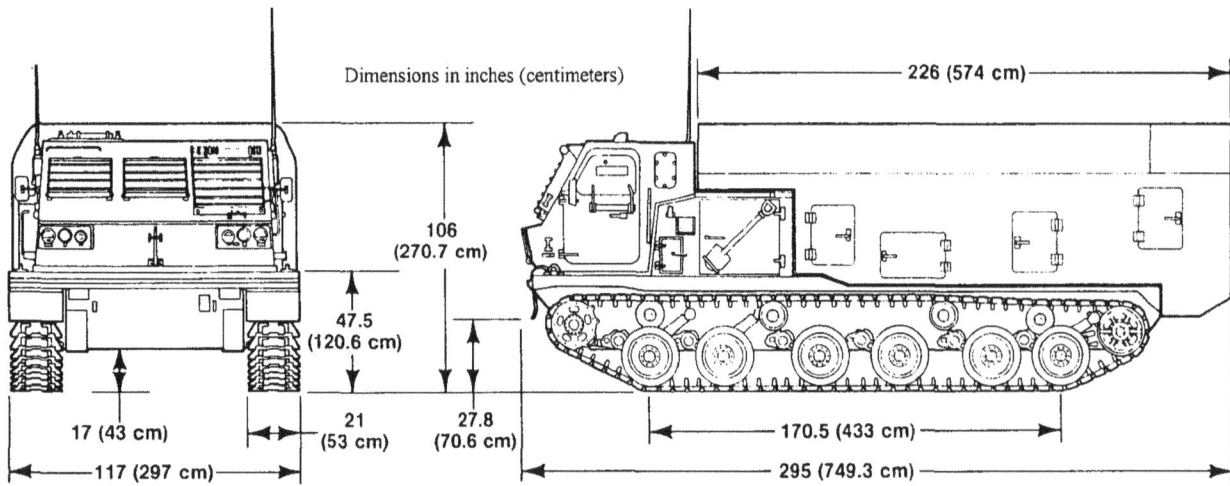

Dimensions in inches (centimeters)

226 (574 cm)

106 (270.7 cm)

47.5 (120.6 cm)

17 (43 cm)

21 (53 cm)

27.8 (70.6 cm)

117 (297 cm)

170.5 (433 cm)

295 (749.3 cm)

Dimensions of the electronic fighting vehicle system XM1070 are shown above. The vehicle itself is below and at the right is the ground based common sensor-heavy XM5.

The FVS carrier also provided a mobile platform for the Hughes VSTAR variable search and track air defense radar system. This system used a large stationary antenna with electronic scanning. The antenna rotated to provide 360 degree coverage. When not in use, the antenna was stowed in a horizontal position on top of the vehicle. For operation, it was raised to a vertical position.

Below, the Hughes VSTAR system is installed on the fighting vehicle systems carrier. The view at the bottom left shows the antenna raised for operation.

314

The XM4 command and control vehicle is shown in the photographs on this page.

The XM4 command and control vehicle (C2V) was intended to replace the M577 command post carrier in corps through battalion headquarters of heavy force units. Using the MLRS chassis, the C2V was powered by the 600 gross horsepower VTA-903T engine with the HMPT-500-3EC transmission. It provided space for nine men including the driver. A telescoped antenna mast could be extended up to a maximum of 10 meters in 30 seconds. The aluminum alloy armor enclosure was provided with a heating and cooling system as well as NBC protection. The XM4 was capable of closed operation for up to 24 hours and it was fully operational while moving. The vehicle had a maximum road speed of 40 miles per hour.

| Ambulatory Patient Configuration | Litter Patient Configuration | Future Treatment Configuration |

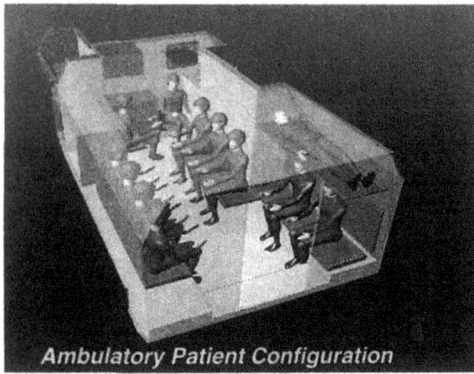

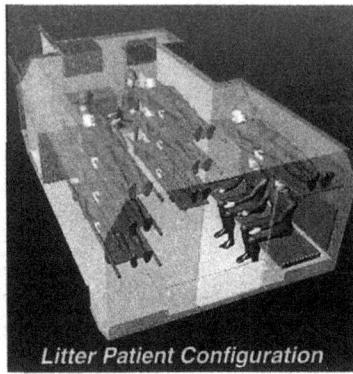

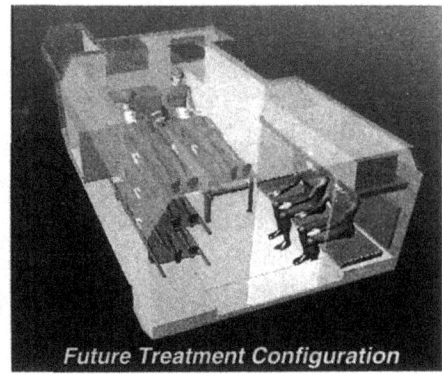

Three configurations of the proposed armored treatment and transport vehicle are illustrated above.

Another proposed application for the carrier was as an armored treatment and transport vehicle (ATTV). It was intended to serve as an armored ambulance to evacuate casualties from the forward area and provide early treatment. The ATTV used the same power train as the XM4 C2V and a similar armored enclosure. In a litter patient configuration, it could transport nine patients on litters plus a three man crew consisting of the driver and two medical personnel. Arranged for ambulatory patients, it could carry 12 plus the same crew of three.

FMC proposed three types of vehicles based upon the M987 as the forward area armored logistics system (FAALS). These three were an armored rearm vehicle, an armored refuel vehicle, and an armored maintenance vehicle. All three had a large aluminum alloy armored enclosure installed on the rear of the M987 chassis. It was fitted with a crane capable of lifting more than five tons. As an armored ammunition rearm vehicle, it could carry six pallets of tank ammunition or as an armored refuel vehicle, it could transport 2,000 gallons of fuel. It could be converted from one type to the other in only one hour. The armored maintenance vehicle carried welding equipment, an air compressor, an hydraulic pump, a work bench, and other tools and repair equipment. Seating was provided in the enclosure for three additional crewmen.

At the right is the proposed forward area armored logistics system. Below, a FAALS is rearming an M1 tank.

A side view of the proposed forward area armored logistics system is above. Below, the armored maintenance vehicle is set up for operation.

Above, a concept drawing of a composite hull for the infantry fighting vehicle appears at the left and a composite hull for the Bradley is under construction at the right.

COMPOSITE ARMORED VEHICLES

In 1983, the U. S. Marine Corps awarded contracts for the construction of two M113 type vehicles in which the aluminum armor was replaced by a non-metallic composite material. FMC in collaboration with Owens-Corning Fiberglas delivered the composite vehicle in October 1985. The hull was a sandwich structure consisting of inner and outer skins of resin bonded E-glass. The two layers were separated by closed cell polyurethane foam. Aluminum oxide ceramic tiles attached to the outside provided additional ballistic protection. The composite hull weighed about the same as the standard aluminum alloy hull and the ballistic protection was somewhat improved.

FMC also developed a composite turret for the Bradley starting in 1983 and assembled five turrets by the Summer of 1987. These turrets used a single thickness of S-glass bonded with a polyester resin. The single piece turret body retained the configuration of the standard Bradley turret. The composite Bradley turret was subjected to firing tests at Camp Roberts, California in the Fall of 1986.

In September 1986, FMC received a contract from the U. S. Army Materials Technology Laboratory to develop a composite hull for the Bradley. Completed by March 1989, the new hull had single thickness walls consisting of polyester resin bonded S-2 glass fiber. The hull was molded in two halves and the thickness varied depending upon the number of preimpregnated glass fiber sheets. This provided the maximum ballistic protection in critical areas. The two halves of the hull

The composite hull Bradley can be seen in the two photographs below. The driver's hatch is open in the right hand view of the demonstration vehicle.

The armor tiles and skirts have not yet been installed on the composite hull Bradley demonstration vehicle above.

were joined with an aluminum alloy frame and turret cage by bonding and bolting. The belly of the vehicle was formed by a single composite panel. The aluminum alloy frame provided attachment points for the road wheel suspension arms distributing the resulting stresses to the hull. The ballistic protection was enhanced by the installation of ceramic tiles in the critical areas.

Tests indicated that the composite hull provided better protection against explosive blast and greatly reduced spalling. The low thermal conductivity of the composite material simplified the problem of heating or cooling the interior of the vehicle. It was estimated that the composite hull was about 27 per cent lighter than the standard hull.

Following a composite armored vehicle (CAV) study project in 1992, United Defense received a con-

tract for the construction of an advanced technology demonstrator in December 1993. The proposed CAV-ATD was a 22 ton vehicle intended to develop manufacturing techniques that could be applied to a wide range of armored vehicles. A single production sequence was to integrate the lay up of the S-2 glass laminate hull structure and its ceramic armor with the signature management materials. The composite hull was divided into upper and lower halves. A two man, titanium alloy, crew capsule was located between the transmission in front and the engine in the center of the hull. Space for six men was provided in the rear hull. Protection from 30mm APDS ammunition was to be provided in the 60 degree frontal arc and the sides were to be proof against 14.5mm projectiles. The top and bottom were to protect against artillery submunitions and antipersonnel mines

The composite armored vehicle advanced technology demonstrator (CAV-ATD) is shown below.

319

82.50
(209.55)

REDUCIBLE
TO
94.36
(239.67)

100.36
(254.91)

16.00
(40.64)

100.00
(254.00)

107.00
(271.78)

157.32
(399.59)

246.40
(625.86)

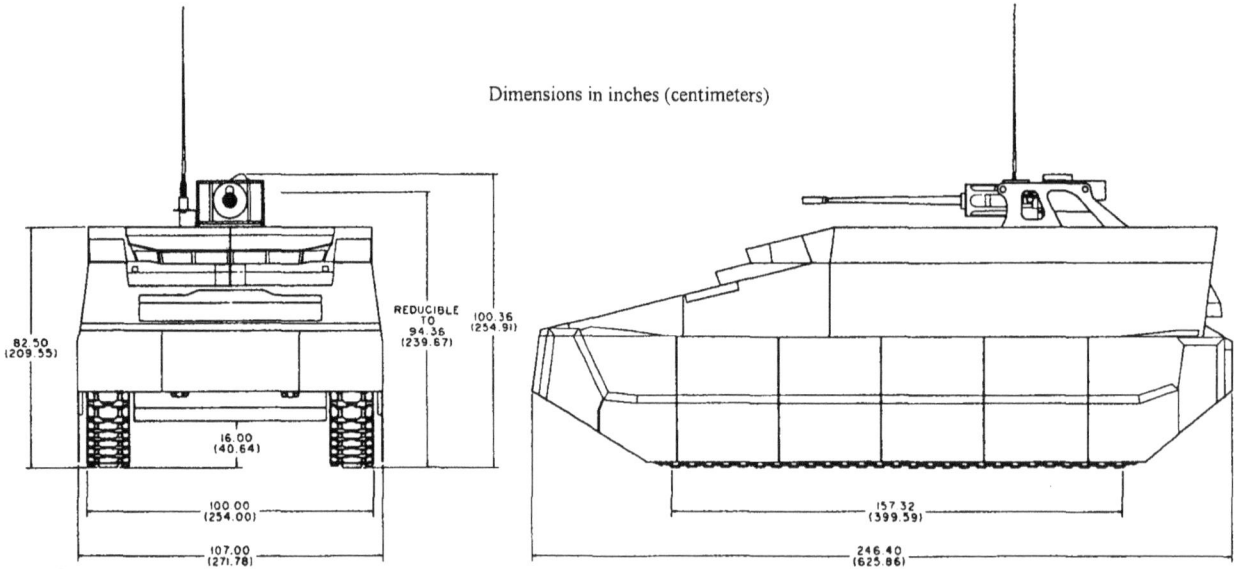

The drawing above shows the dimensions of the CAV-ATD and the layout of the vehicle appears in the drawing at the right.

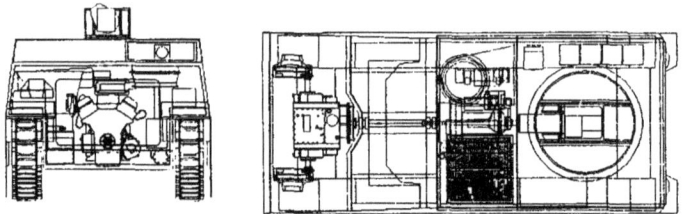

respectively. Initially, the CAV-ATD was to be powered by the 530 gross horsepower General Motors 6V92TIA diesel engine coupled to a Lockheed Martin (formerly General Electric) HMPT-500-3EC transmission. Future investigations were to include the evaluation of an electric drive system. The vehicle was provided with a flat track, in-arm, hydropneumatic suspension system with 15 inch wide T150 tracks. Two hulls were scheduled for construction with one to be used for structural and ballistic tests. The other was intended for durability testing and would initially be fitted with the 25mm

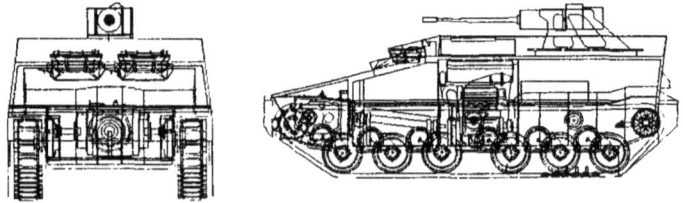

Bushmaster cannon. Later, it would be used to evaluate other weapon systems.

The roll-out of the new CAV-ATD was at the United Defense plant in San Jose, California on 19 February 1997.

The internal arrangement of the CAV-ATD can be seen in the cutaway drawing at the lower left. At the bottom right, the vehicle has been modified to reduce its detection level.

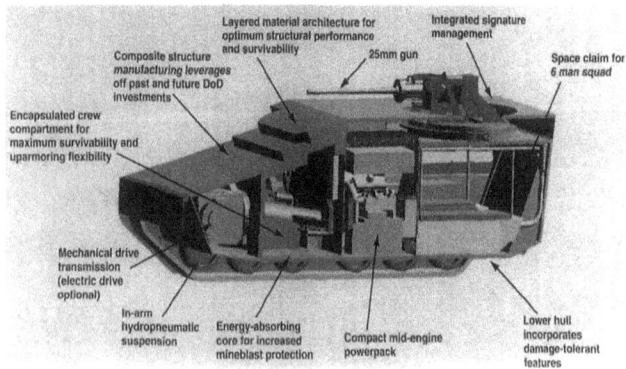

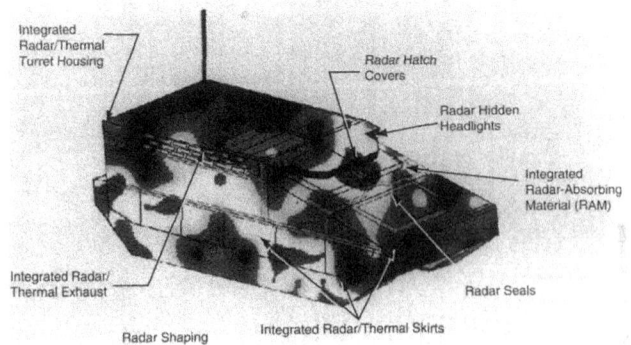

320

PART V

LANDING VEHICLES

Above, the LVT3 appears at the left and the LVTA5 is at the right.

THE LEGACY OF WORLD WAR II

At the end of World War II, large numbers of landing vehicles, tracked (LVTs) were in use by the U. S. Army, Navy, and Marine Corps. At that time, all production of new LVTs was stopped. The majority of the latest versions, the LVT3 and the LVTA5, were still in the United States. Since they represented sufficient numbers for the expected postwar requirements, most of the earlier vehicles were disposed of as surplus. However, even these new vehicles had open tops and were not only vulnerable to artillery air bursts, but also it was impossible to keep personnel and cargo dry when operating in rough water.

Two standard LVT3s were modified for arctic operations during 1947 by the U. S. Naval Engineering Experiment Station. Living quarters were installed in the

cargo compartments of both vehicles. Double tracks were fitted on one of the vehicles to lower the ground pressure and improve performance on soft terrain. Although the double tracks did reduce the ground pressure, the increased friction load resulted in a decrease in the power available for propulsion. As a result, the maneuverability of the double track vehicle was inferior to that of the standard LVT3.

At the right is a top view of the LVT3. Below are the winterized LVT3s. The single track vehicle is at the left and the double track vehicle is on the right.

The LVT3C prototype is above and the LVT3C is at the top right. At the right is the LVTA5 modified by Continental Aviation and Engineering Corporation.

In 1949, Continental Aviation and Engineering Corporation modified an LVT3 to provide overhead protection for the troops and cargo. The modifications included a higher deck, cargo covers, and escape hatches. A later change sloped the rear deck to provide clearance when leaving an LST. The power train, suspension, and tracks of the standard LVT3 were retained. In 1950, 1,200 of the LVT3s were modified to the new configuration and designated as the LVT3C. The cargo covers were fabricated from aluminum alloy and a turret armed with a machine gun was installed on the top front of the vehicle. Portable side armor plate added 860 pounds to the loaded weight of 39,190 pounds. The cargo carrying capacity was 6,100 pounds.

In 1951, the Continental Aviation and Engineering Corporation modified an LVTA5 to obtain data on underwater track return systems. They also replaced the standard 250 horsepower air-cooled engine with a Ford GAA, liquid-cooled, engine. This 500 gross horsepower engine was the same as used in the M4A3 Sherman tank. The new suspension consisted of four, two wheel,

bogies on each side with torsilastic spring assemblies. Shrouds enclosed the upper track run for the underwater track return. The test of this vehicle did not show any improvement in performance with the new suspension arrangement.

Also in 1951, FMC modified two LVTA5s in the hope of improving the performance. A large curved bow was installed on one of the vehicles extending out to the front and down to about 30 inches above the hull bottom. A fairing on top stretched from the bow back to the cab top deck. The second LVTA5 was fitted with an out-of-the-water bow cap covering the bow from the bumper back to the top deck of the cab. Detachable tanks were installed on the front simulating a rounded or a flat bow for test purposes. After evaluation an armored cover was added to the open top turret and the LVTA5s in stock were brought up to the new standard.

The LVTA5 modified by FMC is shown below. Note the new bow.

Above, the LVTPX3 appears above. At the top right, the vehicle is climbing a 60 per cent grade during the test program.

POSTWAR DEVELOPMENT

Early in 1946, FMC began a development program under contract to the Bureau of Ships for an improved LVT. Designated as the LVTPX3, it had a new submerged track suspension fitted with center guide tracks. A monocoque type hull was welded from $^3/_8$ to $^3/_4$ inch thick steel armor plate. The cargo compartment was covered and the rear consisted of a two section ramp. When closed, the upper section of this ramp sloped forward and upward from the water line. The vehicle was powered by an Allison V1710E32, liquid-cooled, engine. It drove the vehicle through a Twin Disc, direct drive, torque converter, the planetary gears of a Torqmatic transmission from an M26 tank, and the controlled differential from an M18 gun motor carriage. The latter was shortly replaced by the M26 tank controlled differential. Later, a new power train was installed consisting of the Continental AV-1790-5B air-cooled engine with the Allison CD-850-4 transmission. The maximum road speed of the LVTPX3 was 30 miles per hour. Apparently the Twin Disc torque converter and the controlled differential were more efficient in water than the CD-850-4 transmission. The maximum water speed dropped from about 8 miles per hour to 5.5 miles per hour with the later transmission. Six dual road wheels per side were suspended by internally mounted torsilastic springs. The drive sprockets were at the rear and the compensating idlers were at the front of the 25 inch wide, center guide, tracks. The single prototype LVTPX3 completed in December 1950 weighed 52,890 pounds and had a payload of 10,000 pounds. A fuel capacity of 328 gallons allowed a cruising time of ten hours on land or six hours in water. The vehicle was manned by a crew of three.

During 1946, the Bureau of Ships also awarded a contract to the Pacific Car and Foundry Company to design and develop a new LVT. Referred to as the prototype A 105mm gun carrier, it actually was armed with a turret mounted 105mm howitzer. The vehicle hull was a welded assembly of steel armor plate varying in thickness from $^3/_8$ to $^5/_8$ inches.

The single prototype A was delivered in early 1949. Like with the LVTPX3, a variety of power train components were evaluated before selecting the Continental AV-1790-5A engine and the Allison CD-850-4A transmission. Several bow designs also were tested. The torsilastic suspension supported the vehicle at nine individually sprung road wheel stations per side. Several tracks were tested with the final choice being the type VIII, center guide, track with a width of 20¾ inches. The vehicle had a combat weight of about 75,000 pounds and it was manned by a crew of five. The maximum speed was 33 miles per hour on land and 7.6 miles per hour in water.

At the left and below are photographs of the so-called 105mm gun carrier which was armed with a 105mm howitzer.

Above, the Continental Aviation and Engineering Corporation lightweight cargo carrier is shown at the left and the prototype LVTPX1 built by Baldwin-Lima-Hamilton is at the right.

In 1949, the Continental Aviation and Engineering Corporation received a contract from the Bureau of Ships to design and construct a prototype lightweight cargo carrier. The vehicle was assembled using aluminum alloy plate ranging in thickness from $3/16$ to $3/8$ inches. Delivered in 1950, the vehicle was powered by the Ford GAA engine with a three speed transmission. It rode on a torsilastic suspension with 12 road wheel stations per side. The empty weight of 26,000 pounds exceeded the Bureau Ships specification by 8,000 pounds effectively reducing the cargo capacity to zero. Only limited testing was performed, but the maximum speed was estimated to be 25 miles per hour on land and about 8 miles per hour in water. A fuel capacity of 660 gallons provided an estimated cruising range of 250 miles on land and 100 miles in water.

Under the direction of the Bureau of Ships, Baldwin-Lima-Hamilton initiated a program in 1949 to develop an improved track for heavy amphibious vehicles. Included in this program was the construction of a prototype vehicle designated as the LVTPX1. The hull of this vehicle was welded from steel armor plate using a double wall construction. This design provided spaced armor giving protection equivalent to $7/8$ to $1\frac{1}{4}$ inch thick armor at approximately two thirds the weight of a single thickness plate. Also, non-critical items could be stowed between the two shells providing additional space inside the vehicle. An hydraulically operated ramp was located at the rear and two manually operated, double folding, cargo doors were on top. The bow was cut away on each side to improve the driver's vision when operating on land. The LVTPX1 was powered by the Ford GAF, V8, liquid-cooled engine developing 500 gross horsepower. The T900, six speed, semiautomatic transmission was installed initially, but it was replaced later by a Torqmatic transmission. The torsilastic suspension had seven road wheel stations per side and it was fitted with Type VIII center guide tracks 20¾ inches in width. With a crew of four, the vehicle weight was 66,000 pounds including a payload of 7,000 pounds. The contract for the prototype was terminated in June 1952 and only limited testing was performed. The maximum

Below are additional views of the LVTPX1 prototype.

The LVTUX1 appears above and at the right. Note the hydraulically operated ramp in the rear view above.

speed was 23 miles per hour on land and 6.3 miles per hour in water. The estimated cruising range was 600 miles on land and 100 miles in water.

Another unsuccessful development program in 1951 was the LVTUX1 built by Baldwin-Lime-Hamilton for the Bureau of Ships. This unarmored cargo carrier was driven by the 500 horsepower Ford GAF engine using the T1200, eight speed, automatic transmission. The torsilastic suspension had ten road wheel stations per side and rode on $20^5/_8$ inch wide Type XIX tracks. The welded steel hull had an inverted V bottom and an internal tubular framework. Three men were accommodated in an enclosed cab at the front. The original design weight objective was 30,000 pounds, but the completed vehicle weighed 54,000 pounds empty. An hydraulically operated ramp at the rear was of two piece construction. When loading or unloading, it unfolded to provide a small angle of approach. Only a few tests were completed prior to the termination of the

contract. The maximum speed was estimated to be 30 miles per hour on land and 5 miles per hour in water. The cruising range was expected to be 300 miles on land and 100 miles in water.

Another ill-fated LVT was the 76mm gun carrier produced by the Marmon-Herrington Company in 1951. This one was not completed before the contract was terminated and it was consigned to the LVT Museum at Camp Pendleton. The hull of the vehicle was constructed of an aluminum alloy ranging in thickness from ¾ to 1½ inches. The Ford GAF engine and the Jered Model 900 transmission were proposed, but they were never installed. The torsilastic suspension had eight road wheel stations per side. The vehicle was armed with a turret mounted 76mm gun.

Below are photographs of the 76mm gun carrier produced by Marmon-Herrington.

327

FMC submitted a proposal to the Bureau of Ships in 1951 for a new LVT based upon the components of the T59/T59E1 armored infantry vehicles. Designated as the LVTPX2, five prototypes were built between 1952 and 1955. Three types of power trains were evaluated. These included two Cadillac Model 331 engines (pilot number 1) or two GMC Model 302 truck engines (pilots number 2 and 3) each with Hydramatic transmissions. As on the armored infantry vehicles, the engines were mounted with their transmissions, one in each sponson. Pilot number 4 was powered by two Chrysler V8 engines coupled to Powerflyte transmissions. Prototype number 5 was fitted with two modified Chrysler Model G 24A engines and it was considered to be the production pilot. The engines powered the vehicle through the two automatic transmissions connected to the controlled differential in front as on the T59. The front mounted sprockets drove the 21 inch wide T91E3 tracks that were fitted with side wings for water propulsion.

The torsion bar suspension had five dual road wheels per side. The cargo compartment was fitted with aluminum alloy top covers and an hydraulically operated rear ramp. Emergency escape hatches were provided in each side. The weight of the LVTPX2 was 44,000 pounds including a 7,000 payload. The maximum speed was 35 miles per hour on land and about 5 miles per hour in water. The vehicle was type classified as the LVTP6, but it was not put into production. The driver was located in the left front of the hull with a roof hatch. The vehicle commander in the right front hull was provided with the small G-1 cupola armed with a single .30 caliber machine gun.

The LVTP6 is shown above and below. The interior is visible through the open ramp in the view above.

The G1 cupola mount installed on the LVTP6 is at the left and two views of the 105mm howitzer armed LVTHX4 are shown below.

A single LVTHX4 was built based upon the LVTPX2. It had a turret armed with the 105mm howitzer T96E1 in the mount T67E1. The vehicle was manned by a crew of seven with the driver and assistant driver in the front hull. Escape hatches were located in each side of the hull. Other hatches were provided for the driver and the assistant driver, the turret crew, and on the rear deck. The LVTHX4 was powered by the same two sponson mounted Chrysler V8 engines, each developing 200 gross horsepower. The sprockets drove the same T91E3 tracks with side wings as on the

LVTP6. The loaded weight of the vehicle was 47,300 pounds for water operation and 49,300 pounds for operation on land. The maximum speed was 31 miles per hour on land and 5.7 miles per hour in water.

Another prototype based upon the LVTPX2 was the LVTAAX2. Three of these vehicles were assembled by FMC. Basically, they consisted of the LVTPX2 with the M4E1 twin 40mm gun turret installed. This was the turret normally used on the twin 40mm self-propelled gun M42. The vehicle performance was essentially the same as the LVTPX2.

The antiaircraft version of the LVTPX2, the LVTAAX2, appears in the two photographs below.

The LVTUX2 is above and at the right. It can be seen afloat at the lower right.

In 1951, the Bureau of Ships awarded a contract to the Pacific Car and Foundry Company to develop a large amphibious cargo carrier capable of transporting heavy vehicles and large quantities of supplies. Designated as the LVTUX2, it was completed in 1957. The steel hull was $1/8$ inch thick with $3/8$ inch thick steel armor on the crew compartment. The vehicle had an open top cargo well with the ramp in the front. Two 500 gross horse-power Continental AOS-895-3 engines were installed in the rear with Allison XT-1400 transmissions. The vehicle rode on four 36 inch wide tracks with each track carrying a set of 30 inch by 7.5 inch dual road wheels. The vehicle was supported by a torsilastic suspension. The weight of the LVTUX2 was 296,000 pounds including a payload of 120,000 pounds. The maximum speed was about 13 miles per hour on land and 7 miles per hour in water.

Development began at FMC in May 1960 of a lightweight, unarmored, full tracked, personnel and cargo carrier. Referred to as a landing craft, assault (LCA), its official designation was the LCAX1. It was intended to transport 30 tons of cargo from landing ships to shore through rough water, surf, reefs, sand dunes, and other obstacles. The cargo was carried in a large, unobstructed, open top well deck that was closed at the rear by a ramp. Two large propellers, one on each side of the rear ramp, folded down for water propulsion. On land, they were folded up against the hull to prevent damage. The open well deck could accommodate three M113 armored personnel carriers.

Below are photographs of the assault landing craft LCAX1. Note the three M113 armored personnel carriers in the open well deck.

A. Engine compartment hatch, B. Antenna, C. Crew compartment hatch, D. Turret, E. Cupola cover, F. Horn, G. Mooring bitt H. Driver's station, J. Ramp, K. Boarding steps, L. Starboard radiator compartment.

A. Detachable driving light, B. Turret, C. Blackout bow marker light, D. Periscope, E. Fire extinguisher pull handle, F. Hand rails, G. Port radiator compartment, H. Escape hatch, J. Boarding steps, K. Idler wheel, L. Towing eye, M. Ramp.

Components of the LVTH6 are identified in the photographs on this page.

THE LVTP5 FAMILY

After the outbreak of war in Korea during the Summer of 1950, a crash program was launched to produce a new LVT that would serve as the basis for a family of vehicles. The basic vehicle was to incorporate the best features resulting from the postwar development program. In December 1950, the Bureau of Ships selected the Ingersoll Products Division of Borg-Warner Corporation to serve as the central design agent on the crash program. Design studies began in January 1951 and the following August, the first production pilot drove off the assembly line under its own power. This was the LVTH6 armed with the turret mounted 105mm howitzer. It was the first vehicle of what was to become the LVTP5 family.

The LVTH6 was a large barge shaped vehicle with the hull assembled by welding steel armor plate varying from $\frac{1}{4}$ to $\frac{5}{8}$ inches in thickness. It had an inverted V shaped bow and bottom for more efficient water operation. An hydraulically operated ramp was installed in the bow. This ramp was composed of inner and outer steel plates separated by equally spaced webs. The crew and stowage compartment at the forward end of the hull was separated from the engine compartment at the rear by a transverse bulkhead. The driver and the crew chief (assistant driver) were located under cupolas with five periscopes at the front on the left and right respectively.

A. Howitzer travel lock, B. Lifting eye, C. Bilge outlet, D. Auxiliary Generator exhaust cover, E. Radiator intake grille cover, F. Engine air intake cover, G. Engine exhaust outlet, H. Engine compartment hatch, J. Stern light, K. Wire tow rope, L. Stern tow hitch, M. Tow hitch quick release, N. External interphone, P. Engine hatch, Q. Blackout stop light, R. Scavenger blower cover, S. Auxiliary bilge outlet, T. Engine coolant filler cover, U. Fuel filler cover, V. Crew compartment hatch.

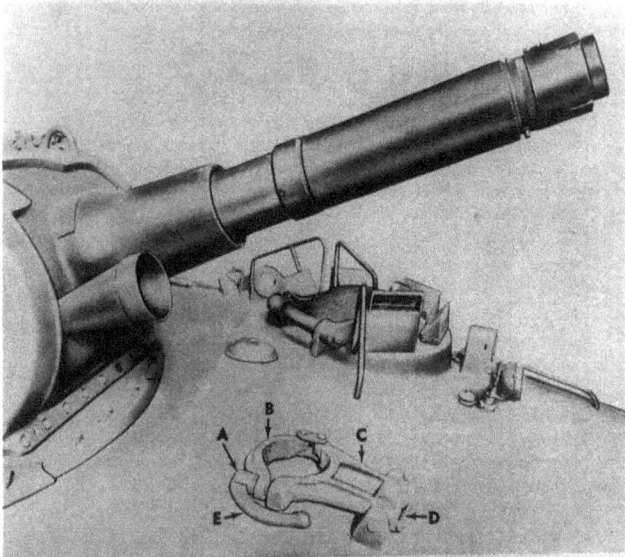

A. Lever bolt
B. Cap
C. Bracket
D. Stop pawl
E. Lever

At the top, the howitzer is shown unlocked (left) and locked (right). At the right and below are interior views of the LVTH6.

A. Trunnion
B. Rotor
C. Cradle
D. Breech recoil guard assembly
E. Breach
F. Leveling plates

A. Engine compartment access hatch
B. Crew and stowage compartment vent cover handle
C. Driver's station
D. AN/GRC-5 radio equipment
E. Turret ammunition ready rack
F. Turret basket
G. Turret slip ring
H. Turret .30 caliber machine gun foot firing pedal
J. Turret 105 mm howitzer foot firing pedal
K. Ammunition canister forward stowage rack
L. TCS radio equipment
M. Crew chief's station

The turret, armed with an M49 (T96E1) 105mm howitzer and a coaxial .30 caliber machine gun in the T172 mount, was located just forward of the vehicle center. The turret was traversed and the howitzer was elevated both manually and by electric power. The howitzer was stabilized in elevation. A .50 caliber machine gun on a pintle mount was installed on top of the turret. The seven man crew consisted of the driver, the crew chief, the vehicle commander, the gunner, the loader, and two ammunition passers. Seats for the commander, gunner, and loader were provided in the turret basket. The gunner was at the right front just forward of the commander. The gunner had a T149E2 panoramic telescope and a T150E2 direct fire telescope. The commander's cupola was fitted with five periscopes. The loader was on the left side of the turret with his own hatch. Two vision blocks and a pistol port were installed in the left side of the turret and one vision block was on the right. An escape hatch was located in each side of the hull and an access hatch was in the hull roof just behind the turret. Stowage racks were provided for 151 rounds of 105mm ammunition. A 12 round ready rack was at the loader's station in the turret and two 24 round ready racks were mounted on the transverse bulkhead, one on each side of the engine access hatch. Five additional racks stowed 91 rounds of ammunition in canisters. For water operation, the number of 105mm rounds in canisters was reduced to 40.

332

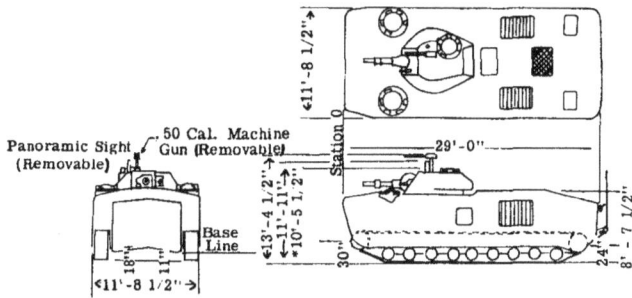

The dimensions of the LVTH6 are shown above and the liquid-cooled LV-1790-1 engine is below.

A. CARBURETOR   D. FUEL PUMPS
B. PRIMER FILTER   E. COOLANT PUMP
C. MAGNETOS   F. STARTER

A. Coolant line, B. Aspiration air duct, C. Fan drive shaft, D. Flexible connector, E. Air cleaner, F. Cross-drive oil cooler, G. Fuel fill.

The engine compartment of the LVTH6 appears above and details of the tracks and suspension are below.

The Continental LV-1790-1, V12, engine and the Allison CD-850-4B transmission powered the vehicle through the final drives at the aft end of the engine compartment. The 810 gross horsepower, liquid-cooled, engine was connected to two radiator assemblies mounted in watertight compartments on each side at the forward end of the engine compartment. When operating on land, the radiators were cooled by engine driven fans. When the vehicle was swimming, approximately two-thirds of the recessed radiator compartments were immersed and the fans were shut off by the driver. The radiators were then cooled by the water. The final drives and the sprockets at the rear of the vehicle drove the 20¾ inch wide steel tracks. Each track consisted of 134 blocks held together by inner and outer track pins. Each track block had an inverted grouser that served as a center guide and propelled the vehicle in the water. A compensating idler was at the front of each track. In addition to the transverse bulkhead hatch, access to the engine compartment was through two hatches on top of the rear hull with one on each side.

Combat loaded for land operations, the LVTH6 weighed 86,600 pounds. For water operations, the weight was reduced to 84,200 pounds by limiting the 105mm ammunition supply to 100 rounds. The maximum speed of the LVTH6 was 30 miles per hour on land and 6.8 miles per hour in water. The cruising range was about 190 miles on land or 57 miles in water.

A. IDLER WHEEL
B. ADJUSTMENT INSPECTION HOLE   D. SUSPENSION ARM
    E. ROAD WHEEL
C. TRACK ADJUSTING NUT   F. GROUSER
    G. TRACK

333

A. Engine compartment, B. Antenna, C. Crew compartment hatch, D. Cargo hatch, E. Cupola cover, F. Horn, G. Mooring bitt, H. Driver's station, J. Ramp, K. Boarding steps, L. Starboard radiator compartment.

A. Detachable driving light, B. Turret, C. Blackout bow marker light, D. Periscope, E. Fire extinguisher pull handle, F. Hand rails, G. Port radiator compartment, H. Escape hatch, J. Boarding steps, K. Idler wheel, L. Towing eye, M. Ramp.

Components on the LVTP5 are identified in the photographs above and below.

A. Cargo hatch locks, B. Lifting eye, C. Bilge outlet, D. Cargo hatch door hook, E. Aux. gen. Exhaust cover, F. Engine air intake covers, G. Radiator air intake grille, H. Engine exhaust outlet, J. Stern light, K. Wire tow rope, L. Stern tow hitch, M. Tow hitch quick release, N, Engine hatch, P. Blackout stop light, Q. Scavenger blower cover, R. Auxiliary bilge outlet, S. Engine coolant filler cover, T. Fuel filler cover, U. Crew compartment air vents, V. Crew compartment blower cover, W. Entrance hatches.

The LVTP5 was the basic vehicle of the family. The hull was similar in construction to the LVTH6 and the power train, suspension, and tracks were identical. It had the same inverted V configuration on the front ramp and the hull bottom. The driver remained in the left front above the port track channel. The vehicle commander was now located in the right front above the starboard track channel. Both had cupolas with five periscopes consisting of four short M17 periscopes on the outboard side and one longer M17C periscope on the inboard side. The latter permitted a view over the raised center of the hull roof. A cupola mount for a .30 caliber machine gun could be installed in the center between the driver and the commander. The cupola mount was manually operated and was equipped with five vision blocks and a periscopic sight. Two periscopes were installed in the front of the blister supporting the cupola mount. As on the LVTH6, an escape hatch was located in each side of the hull in the personnel and cargo compartment. Four access hatches and a large cargo hatch were in the hull roof. Two of the access hatches were over the personnel and cargo compartment just to the rear of the cargo hatch. The remaining two were in the roof over the rear of the engine compartment. The large cargo hatch had two folding covers with boarding ladders stowed on the outside. Two sets of folding troop seats were installed in the personnel and cargo compartment along the outside walls. Two additional sets of seats were mounted in the center and an adjustable stand could be installed in the front for the cupola gunner. The seats and the stand could be removed when the vehicle was used as a cargo carrier.

The crew of the LVTP5 consisted of the driver, the assistant driver, and the commander. Although space was provided for 34 troops with combat equipment, the number was reduced to 25 during operations on water.

A. FIRE EXTINGUISHER
   PULL HANDLE
B. RADIO CONTROL BOX
C. HAND THROTTLE
D. STEER AND SHIFT
   CONTROL LEVER
E. SERVICE BRAKE PEDAL
F. ACCELERATOR PEDAL
G. RAMP CONTROL LEVER

The driver's station in the LVTP5 is at the top left and the commander's station is above.

The combat loaded weight of the LVTP5 on land was 87,780 pounds. In water, the weight was limited to 81,780 pounds.

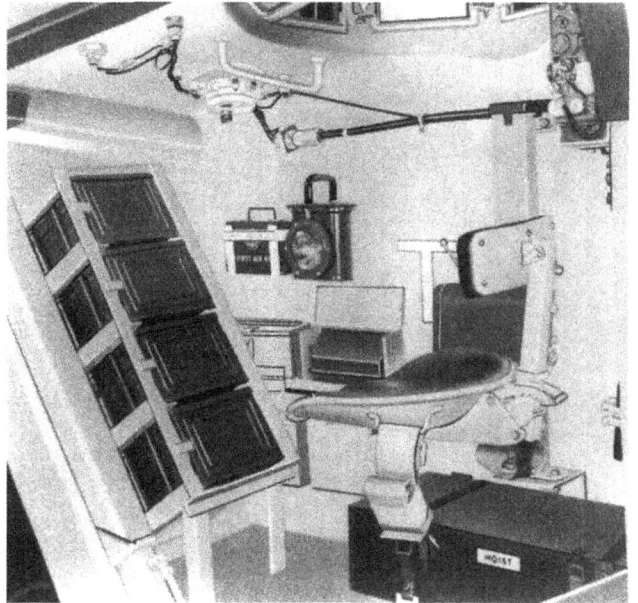

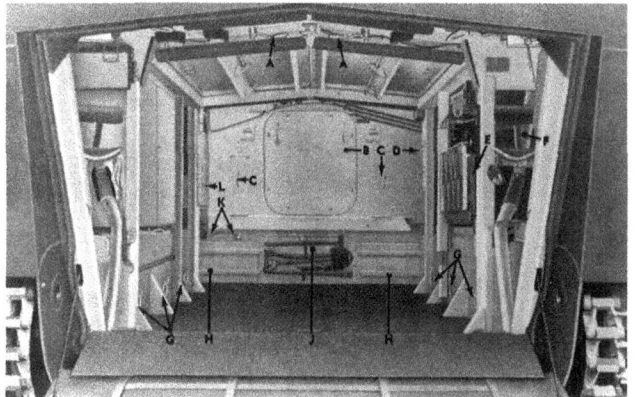

A. SIGHT GUARD
B. SIGHT
C. BODY
D. CRADLE
E. FRICTION BRAKE

A. Cargo hatch lock, B. Engine compartment access hatch, C. Cargo pad eye, D. Distribution and control box, E. Radio sets, F. Driver's station, G. Cargo lashing rings, H. Battery compartment, J. Pioneer tool rack, K. Fuel transfer valves, L. Auxiliary generator engine.

An interior view of the LVTP5 is above and exterior (upper) and interior (lower) views of the cupola are at the left. Below are the dimensions of the LVTP5.

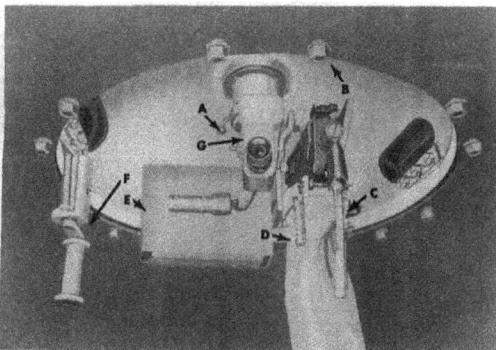

A SIGHT MOUNT LEVER
B. MOUNTING BOLT
C. ELEVATION ARM
D. CARTRIDGE EJECTION BAG
E. AMMUNITION BOX
   BRACKET
F. AZIMUTH CONTROL
G. SIGHT

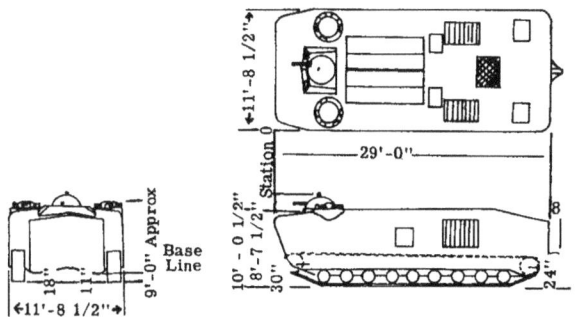

335

A. BOARDING LADDER   C. SPARE TRACK BLOCK
B. BOAT HOOK

The cargo hatch cover is shown closed (above) and open (below). Note the boarding ladders stowed on the hatch cover. At the right, the boarding ladders are ready for use.

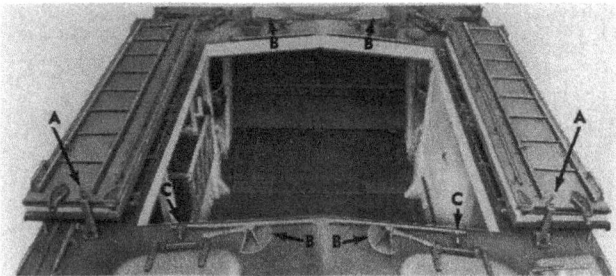

A. CARGO HATCH DOOR   B. LOCKING CAM
   HOOK   HANDLE
         C. LOCK DOG

The LVTP5 chassis provided the basis for several specialized amphibious vehicles in addition to the LVTH6. One of these was the LVTAAX1. As its designation indicated, it was an antiaircraft vehicle and it was armed with the gun mount from the twin 40mm self-propelled gun M42. The mount itself was surrounded by a rotating ready ammunition rack to facilitate the transfer of ammunition from the stowage racks in the cargo compartment. The circular ready rack had eight compartments each holding seven, four round, clips for a total of 224 rounds. The hull racks held 776 rounds bringing the total ammunition stowage to 1,000 rounds. During water operations, the total ammunition stowage was reduced to 800 rounds to improve the buoyancy. The LVTAAX1 was manned by a crew of eight including the gunners. Combat loaded, the vehicle weighed 85,760 pounds. Its performance was essentially the same as the LVTP5. A single pilot LVTAAX1 was built by Ingersoll Kalamazoo Division in 1954.

The photographs below show the LVTAAX1.

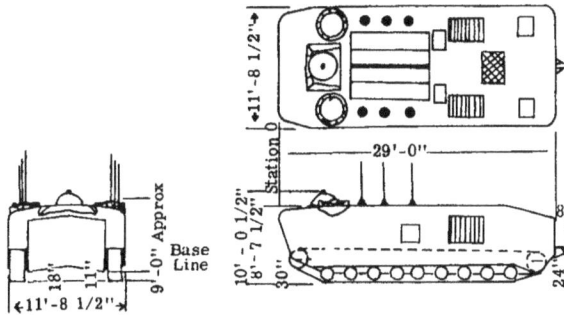

At the left is a view of the LVTCRX1 and the interior of the vehicle is above. The dimensions of the LVTP5 (CMD) can be seen in the sketch at the left.

A command radio version of the LVTP5 was designed by the Ingersoll Kalamazoo Division in 1955. Designated as the LVTCRX1, the personnel and cargo compartment was modified for the installation of nine radios in two racks. One rack was located just in front of the transverse bulkhead and the other was above the port side track channel just behind the driver. Troop seats and two stools were provided for six radio operators. Three operators sat on the outboard troop seats and the other three on the two stools and the center troop seat. The vehicle was manned by a crew of ten including the radio operators. The combat loaded weight of the LVTCRX1 was 72,460 pounds and its performance was the same as the LVTP5. Later, the command radio vehicles modified from the LVTP5 were designated as the LVTP5(CMD).

In 1954, FMC designed a modified version of the LVTP5 to serve as a maintenance and recovery vehicle. Designated as the LVTRX1, it was fitted with a 60,000 pound line pull capacity winch powered by the standard engine from the M38A1 ¼ ton truck. A boom was mounted on the front of the vehicle for use in removing or installing power packs and handling other heavy loads. A General Electric welder was installed which, in

At the right is a photograph of the LVTRX1.

addition to welding, served as a battery charger and provided extra power to the starter for cold weather operations. An air compressor with an output of 5 cubic feet per minute also was provided.

The production LVTR1 utilized the standard LVTP5 power train, suspension, and tracks. The driver remained in his usual position above the port side track channel. The crane operator (crew chief) was located in the right front over the starboard track channel. A rigger completed the three man crew. The combat weight of the LVTR1 was 82,200 pounds and its performance paralleled that of the LVTP5.

A. Mooring bitts, B. Cargo hatch, C. Sliding sheave and retrieving line, D. Kingpost, E. Side stay, F. Crane boom, G. Power traversing line, H. Vang line, J. Block, K. Ramp, L. Ramp chain, M. Crane operators cupola, N. Exterior work bench.

A. Bow fairleader, B. Kingpost erecting strut, C. Kingpost, D. Crane boom, E. Sliding sheave housing, F. Stern Fairleader, G. Hand rails, H. Port radiator compartment, J. Escape hatch, K. Boarding steps, L. Fire extinguisher pull handle, M. Towing eye, N. Ramp, P. Signalling searchlight, Q. Detachable driving light, R. Side stay.

Components on the LVTR1 are identified in the photographs above and below. The interior components are shown in the drawing below.

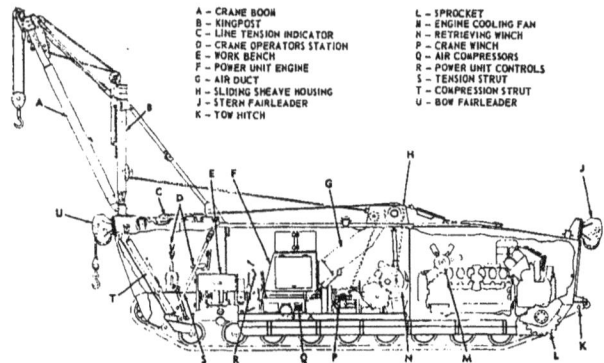

A – CRANE BOOM	L – SPROCKET
B – KINGPOST	M – ENGINE COOLING FAN
C – LINE TENSION INDICATOR	N – RETRIEVING WINCH
D – CRANE OPERATORS STATION	P – CRANE WINCH
E – WORK BENCH	Q – AIR COMPRESSORS
F – POWER UNIT ENGINE	R – POWER UNIT CONTROLS
G – AIR DUCT	S – TENSION STRUT
H – SLIDING SHEAVE HOUSING	T – COMPRESSION STRUT
J – STERN FAIRLEADER	U – BOW FAIRLEADER
K – TOW HITCH	

A. Cargo hatch, B. Driver's hatch, C. Kingpost back brace, D. Crane operator's hatch, E. Top access hatches, F. Engine air intake covers, G. Sliding sheave housing, H. Fan compartment doors, J. Power package lifting sling, K. Tow cable, L. Quick release hitch, M. Stern fairleaders, N. Engine compartment doors, P. Lifting eyes, Q. Engine hatch, R. Fuel filler covers.

At the right is the crane operator's station in the LVTR1

A. AUXILIARY LIGHT OUTLET
B. SIDE SLOPE INCLINOMETER
C. CRANE OPERATOR'S SWITCH PANEL
D. FORE AND AFT INCLINOMETER
E. INTERPHONE CONTROL BOX
F. CRANE OPERATOR'S CONTROL STAND
G. WINCH CONTROL RODS

The LVTEX1 appears in the views above. At the top right, the line charge pallet is raised into the firing position.

A single LVTEX1 was built in 1955 as an engineer version of the LVTP5. It was intended to excavate mines, fire line charges, remove obstacles, and transport combat engineers and their equipment. A large V shaped excavator blade was mounted on the front of the vehicle for clearing or detonating mines in a path 16 inches deep and about 12 feet wide. The excavator was constructed of steel and aluminum alloy and it weighed about 10,000 pounds. Flotation tanks filled with plastic foam were attached behind the blade to maintain the vehicle trim in the water. Two pallets, each containing a line charge and a rocket motor, were stowed in the personnel and cargo compartment. A lifting mechanism raised one pallet at a time through the top deck to the firing position. After firing, the empty pallet was ejected over the right side. The LVTEX1 had a combat weight of 94,470 pounds and it was manned by a crew of five. The maximum speed was 27 miles per hour on land and 5.4 miles per hour in water. The cruising range was 250 miles on land and 55 miles in water.

The later production LVTE1 was powered by the air-cooled AVI-1790-8 engine with the CD-850-5 transmission. The most obvious identification differ-ences were the lack of recessed radiator compartments on each side and the armor cover plates for the engine cooling air intake and exhaust valves on the rear deck. A .30 caliber machine gun in the cupola mount was installed on the top between the driver and the vehicle commander. The crew now consisted of the driver, the commander, and the gunner plus four or five combat engineers. The combat loaded weight of the LVTE1 was 96,200 pounds. The maximum speed was about 25 miles per hour on land and 6.2 miles per hour in water. A 560 gallon fuel tank provided a cruising range of approximately 210 miles on land. The endurance in water was 10½ hours.

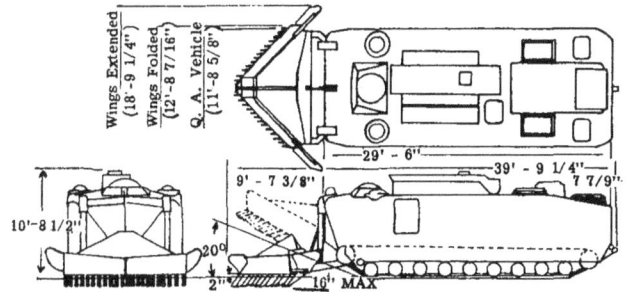

The dimensions of the LVTE1 are shown in the drawing above. Although poor in quality, the views below identify the various components of the LVTE1.

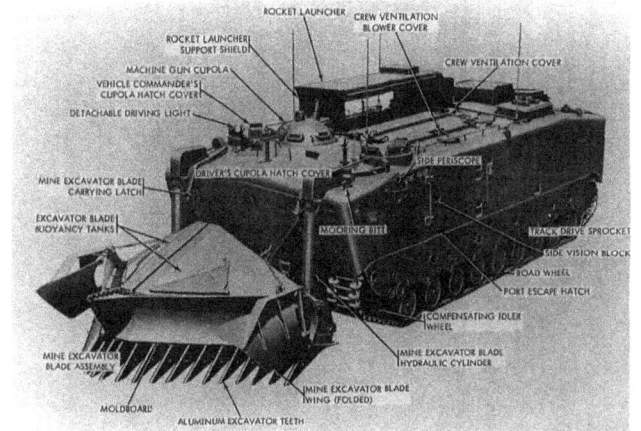

339

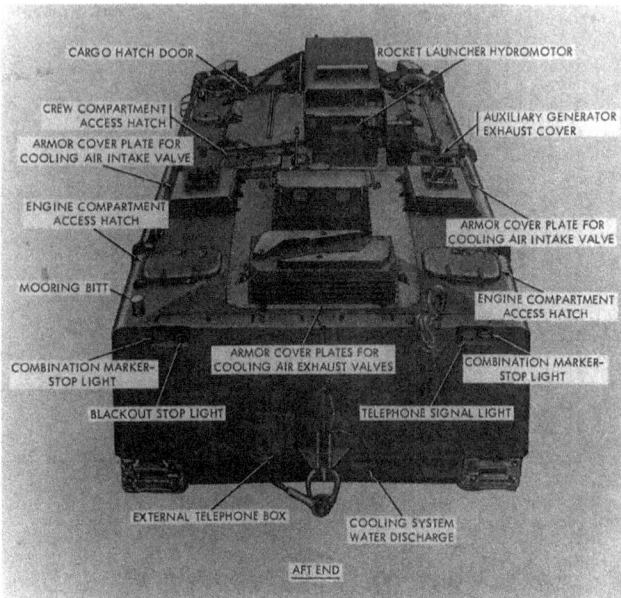

CARGO HATCH DOOR · ROCKET LAUNCHER HYDROMOTOR
CREW COMPARTMENT ACCESS HATCH · AUXILIARY GENERATOR EXHAUST COVER
ARMOR COVER PLATE FOR COOLING AIR INTAKE VALVE
ENGINE COMPARTMENT ACCESS HATCH
ARMOR COVER PLATE FOR COOLING AIR INTAKE VALVE
MOORING BITT
ENGINE COMPARTMENT ACCESS HATCH
COMBINATION MARKER-STOP LIGHT · ARMOR COVER PLATES FOR COOLING AIR EXHAUST VALVES · COMBINATION MARKER-STOP LIGHT
BLACKOUT STOP LIGHT · TELEPHONE SIGNAL LIGHT
EXTERNAL TELEPHONE BOX · COOLING SYSTEM WATER DISCHARGE
AFT END

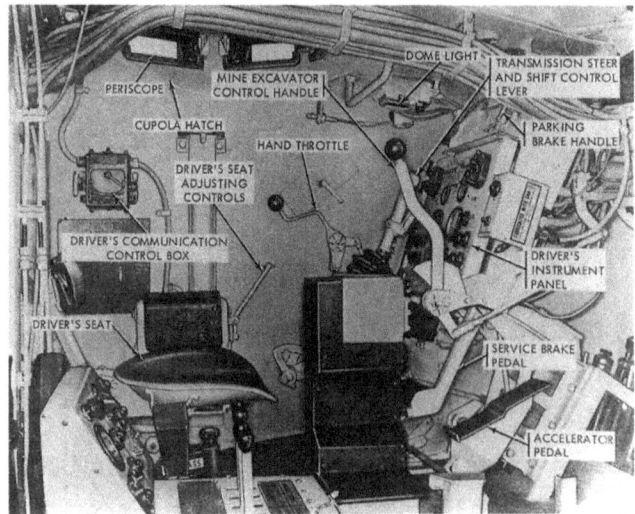

PERISCOPE · MINE EXCAVATOR CONTROL HANDLE · DOME LIGHT · TRANSMISSION STEER AND SHIFT CONTROL LEVER
CUPOLA HATCH · HAND THROTTLE · PARKING BRAKE HANDLE
DRIVER'S SEAT ADJUSTING CONTROLS
DRIVER'S COMMUNICATION CONTROL BOX · DRIVER'S INSTRUMENT PANEL
DRIVER'S SEAT · SERVICE BRAKE PEDAL
ACCELERATOR PEDAL

The components on the top rear of the LVTE1 can be seen at the top left. Above is the driver's station in the vehicle.

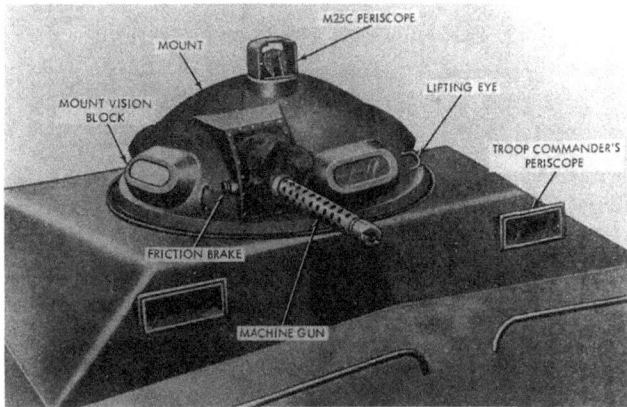

M25C PERISCOPE
MOUNT
MOUNT VISION BLOCK · LIFTING EYE
TROOP COMMANDER'S PERISCOPE
FRICTION BRAKE
MACHINE GUN

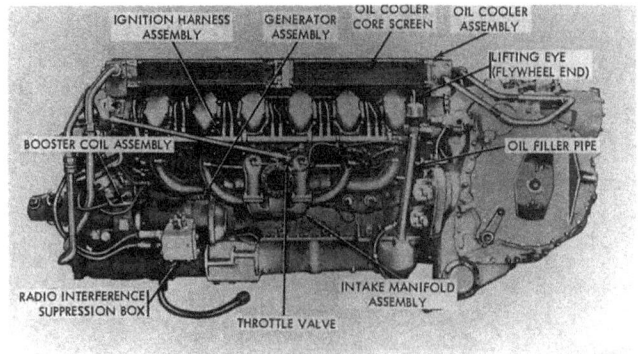

IGNITION HARNESS ASSEMBLY · GENERATOR ASSEMBLY · OIL COOLER CORE SCREEN · OIL COOLER ASSEMBLY
LIFTING EYE (FLYWHEEL END)
BOOSTER COIL ASSEMBLY · OIL FILLER PIPE
RADIO INTERFERENCE SUPPRESSION BOX · INTAKE MANIFOLD ASSEMBLY · THROTTLE VALVE

Above, the machine gun cupola on the LVTE1 is at the left and the AVI-1790-8 air-cooled engine is at the right. Below the excavator on the LVTE1 is raised into the travel position. The machine gun has not been mounted in the cupola in the latter view.

Above, one of the few U.S. Army LVTP5A1s is climbing out of the water. The LVTH6A1 is below. Note the air intake and exhaust superstructure on the rear deck of the A1 vehicles.

Later modifications to the LVTP5, the LVTH6, and the LVTR1 resulted in a change of the designations. They now became the LVTP5A1, the LVTH6A1, and the LVTR1A1. The most obvious identification feature for the modified vehicles was the superstructure on the rear deck over the engine compartment. This superstructure housed the modified air intake and exhaust system.

The LVTR1A1 appears here. The crane is in the operating position in the right view above.

In 1966, FMC modified two LVTP5A1 vehicles to use the air-cooled AVDS-1790 diesel engine. Like the LVTE1, these vehicles could be identified by the absence of the recessed radiator compartments on the sides and the armor plate covers for the cooling air intake and exhaust valves on the rear deck. They were designated as the LVTPXD1.

The LVTP5 family was successfully employed in Vietnam. However, by this time, a new development program was in progress that would produce its successor.

The diesel powered LVTPXD1 is below. Note the absence of radiator louvers on the sides and the changes in the deck superstructure.

An artist's concept (left) and a model (right) of the LVTPX12 are shown above. Note the 20mm gun provided as armament.

THE LVTP7 FAMILY

Design studies for an improved LVT started in 1964 and in February 1966, engineering development began at FMC. The first of 12 prototypes of the new vehicle, designated as the LVTPX12, was completed during the Summer of 1967. It was vastly different from the LVTP5. Assembled from 5083 aluminum armor plate, the level of protection was approximately the same as that of the M113 armored personnel carrier. The LVTPX12 was a front drive vehicle with a ramp in the rear. The interior was divided between the crew and personnel compartment and the engine compartment.

The engine compartment was located in the center at the forward end of the hull. It housed the General Motors 8V53T diesel engine with the HS-400 transmission. The liquid-cooled V8 engine developed 400 gross horsepower at 2,800 rpm. The semiautomatic transmission had four speeds forward and two in reverse with hydrostatic steering. The front mounted drive sprockets drove the 21 inch wide, single pin, steel tracks. A torsion tube-over-bar suspension supported the vehicle on six dual road wheels per side with an adjustable idler at the rear. In water, the LVTPX12 was driven by two

The LVTPX12, pilot number 3, appears below. Note the troop commander's cupola just behind the driver's station.

The LVTPX12 pilot above is moving through sandy terrain during its test program.

water-jet propulsion units at the rear. These units were connected by drive shafts through two power transfer assemblies to the power take-off on the transmission. Water-jet deflectors on the propulsion units steered the vehicle in the water.

The driver, assistant driver, vehicle commander, troop commander, and 24 troops were accommodated in the crew and personnel compartment. The driver rode in the left front on the port sponson with the troop commander just behind him. The driver and the troop commander each had a cupola with seven vision blocks around the hatch. In addition, the driver had an M24 infrared periscope and the troop commander had an M17C periscope. The latter permitted forward vision over the driver's cupola. The vehicle commander was located in the weapon station at the right front on the starboard sponson. The weapon station was armed with an M139 20mm gun and an M73E1 7.62mm coaxial machine gun. It was fitted with six vision blocks and a periscope. The assistant driver was seated on the port sponson behind the troop commander. The 24 troops were seated in three rows of eight. A seat was attached to each side wall and the third was in the center. All of these seats could be stowed away to provide a clear space for cargo. A large cargo hatch was in the hull roof above the crew and personnel compartment.

The LVTPX12 had a maximum speed of 40 miles per hour on land and 8.4 miles per hour in water. The

cruising range on land was about 300 miles and the endurance in water was 7 hours. The combat loaded weight was 48,500 pounds.

Like the LVTP5, the LVTPX12 was intended to provide the basis for a family of vehicles. Initially, four variants of the basic vehicle were proposed. These were the LVTCX2 command vehicle, the LVTRX2 recovery vehicle, the LVTEX3 engineer vehicle, and the LVTHX5 armed with a turret mounted 105mm howitzer. However, the latter was never built.

The LVTPX12 prototypes were tested at Aberdeen Proving Ground, the Yuma Proving Ground, Fort Greely, and in Panama. High surf tests were carried out at Camp Pendleton and Monterey, California. One of the specification requirements was the ability to operate in a 10 foot high plunging surf. Under high surf conditions, the vehicle frequently was fully submerged for 10 to 15 seconds.

After successful completion of the test program, the vehicle was type classified as the LVTP7 and the first production vehicle was delivered in the Fall of 1971. The first U. S. Marine Corps units were equipped with the LVTP7 in 1972.

At the right is an artist's concept drawing of the proposed LVTHX5.

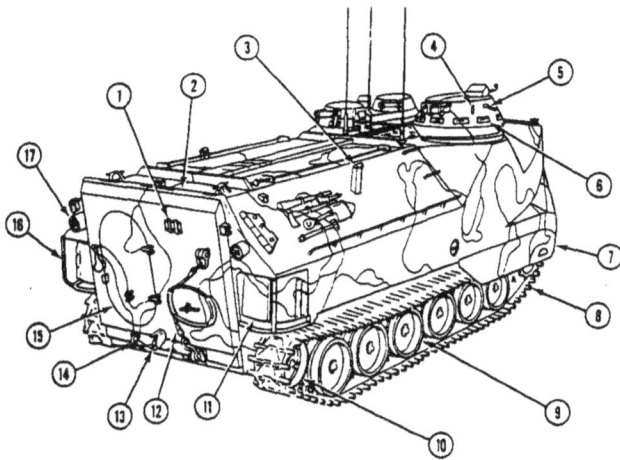

1	Stern vision block	10	Idler wheel (2)
2	Ramp	11	Water-jet deflector (2)
3	Cargo hatch support (2)	12	Tow cable
4	Armament station lifting eye (3)	13	Pintle
5	Hatch release	14	Towing eye (2)
6	Direct vision block (9)	15	Personnel door
7	Track shroud (2)	16	Deflector guard (2)
8	Track assembly (2)	17	Taillight-stop light (2)
9	Road wheel (24)		

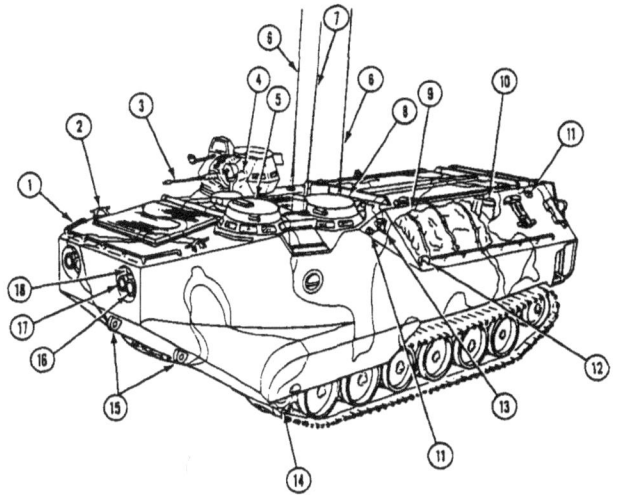

1	Handrail	10	Fuel fill cover
2	Mooring cleat (4)	11	Bilge discharge (4)
3	Caliber .50 M85 machine gun	12	Fire extinguisher external release
4	Ejection chute door	13	Heater exhaust outlet
5	Driver's hatch	14	Drive sprocket (4)
6	Antenna AS-1729/VRC (2)	15	Towing eye (2)
7	Receiving antenna	16	Service headlight (2)
8	Commander's hatch (troop commander)	17	Infrared (I. R.) headlight (2)
9	Camouflage net	18	Blackout (BO) marker light (2)

Components on the LVTP7 are identified in the drawings above and at the right.

On the production LVTP7, a new electric-hydraulic powered weapon station was installed for the vehicle commander. It was armed with a single .50 caliber M85 machine gun and fitted with nine vision blocks for a 360 degree view. An 8 power optical sight was provided for the machine gun. The combat loaded weight of the LVTP7 had increased to 50,350 pounds, but the performance remained the same. FMC produced 971 LVTP7s starting in 1971.

1	Cargo hatch cover (2)	7	Air exhaust grille
2	Sea tow quick-release	8	Ventilator-aspirator valve
3	Cargo hatch center beam	9	Air inlet grille
4	Horn	10	Ring sight
5	Engine exhaust outlet	11	Optical sight
6	Periscope	12	Armament station hatch

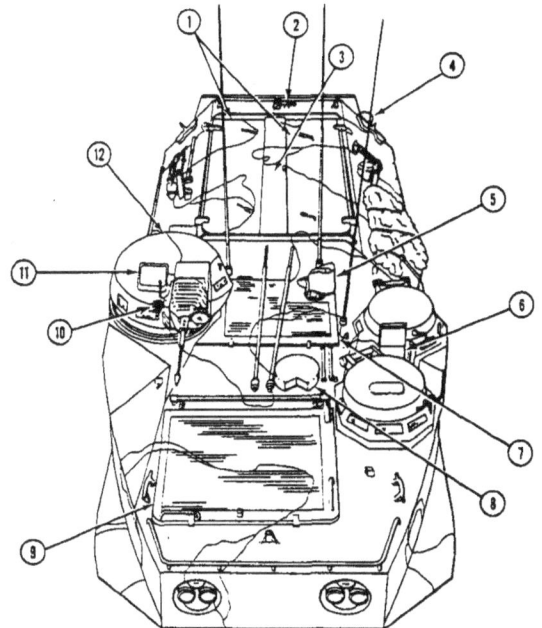

The flat track suspension on the LVTP7 is clearly visible in the photograph below.

DRIVER'S STATION

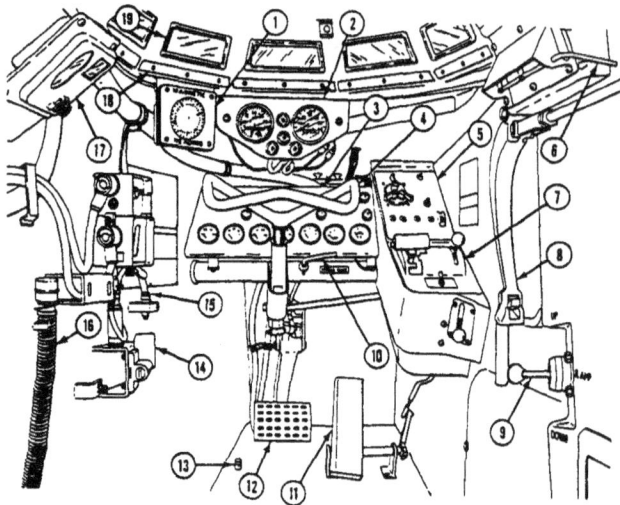

1. Magnetic navigation set (MNS) indicator, 2. Upper instrument panel, 3. Lower instrument panel, 4. Steering wheel, 5. Control console, 6. Vent control, 7. Power train control, 8. Ramp lock handle, 9. Ramp control, 10. Parking brake handle, 11. Accelerator pedal, 12. Brake pedal, 13. Beam selector switch, 14. Driver's seat latch, 15. Audio warning connector, 16. CBR hose, 17. Dome light, 18. Blackout cover hook tape, 19. Direct vision block (7)

POWER TRAIN SYSTEM

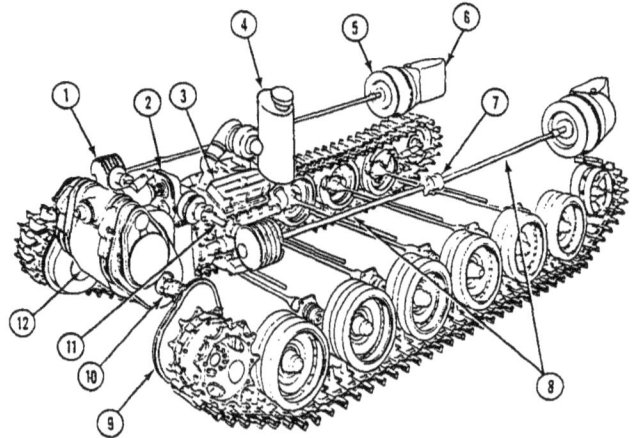

1. Right angle drive (2), 2. Power takeoff, 3. Engine, 4. Muffler, 5. Water propulsion unit (2), 6. Water-jet deflector (2), 7. Midship bearing and seal (2), 8. Longitudinal drive shaft (2), 9. Final drive (2), 10. Universal joint (2), 11. Lateral drive shaft (2), 12. Transmission

Components of the LVTP7 are identified in these drawings.

INTERIOR VIEW — PORT SIDE

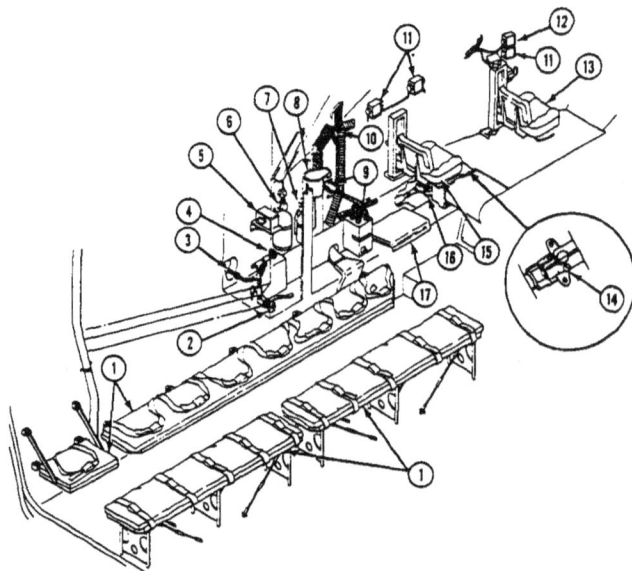

1. Troop seats, 2. Fuel tank internal drain cock, 3. Fuel tank drain plug, 4. Fuel tank external drain cock, 5. Electrical transient suppressor, 6. Fixed fire extinguisher, 7. Portable fire extinguisher, 8. Personnel heater and control box, 9. CBR unit and control panel, 10. Personnel heater exhaust closure valve, 11. Intercommunication control C-2298/VRC (3), 12. Frequency selector control C-2742/VRC, 13. Driver's seat, 14. Fuel shutoff valve, 15. Troop commander's seat, 16. Right angle drive (2), 17. Assistant driver's seat

INTERIOR VIEW — STARBOARD SIDE

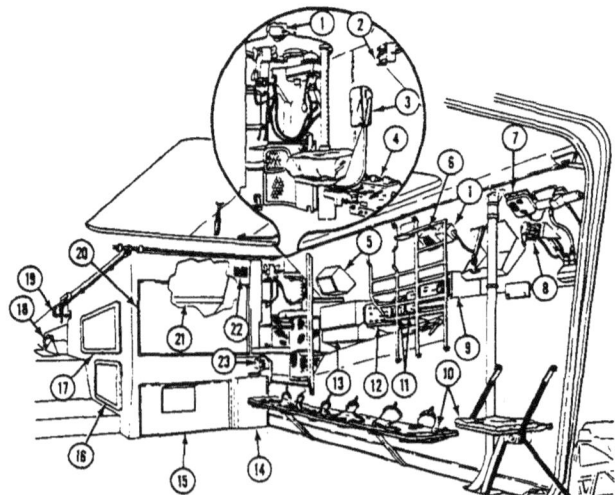

1. Intercommunication control C-2298/VRC, 2. Frequency selector control C-2742/VRC, 3. Vehicle commander's seat, 4. Audio frequency amplifier AM-1780/VRC, 5. Distribution box, 6. Loudspeaker LS-454/U, 7. Aft dome light, 8. Aft slave receptacle, 9. Receiver-transmitter RT-524/VRC, 10. Troop seat, 11. Receiver R-442/VRC (2), 12. Receiver-transmitter RT-246/VRC, 13. Battery box, 14. Air cleaner access cover, 15. Rear lower access cover, 16. Port side lower access cover, 17. Port side upper access cover, 18. Panel, 19. Cover, 20. Rear upper access cover, 21. Exhaust plenum, 22. Personnel compartment ventilation control, 23. Air cleaner indicator

346

U.S. Marine Corps LVTP7s are shown above during training operations. Note that the .50 caliber machine gun M85 has replaced the 20mm gun on the LVTPX12.

The LVTCX2 command vehicle appears above. Note the large number of antennas on this vehicle.

A single prototype LVTCX2 command vehicle was completed in 1969. The exterior of the vehicle was similar in appearance to the LVTPX12 and it was fitted with the same weapon station armed with the M139 20mm gun and the coaxial M73E1 7.62mm machine gun. Obvious differences were the extra radio antennas and the blackout shelter stowed on the top rear. The driver was in his usual position in the port sponson. The vehicle commander was under the cupola just to the rear of the driver. The assistant driver was seated on the sponson to the rear of the vehicle commander. When required, the vehicle commander manned the weapon station. In addition, the vehicle carried the unit commander, four staff personnel, and five communication systems operators. A five seat module and a folding staff desk were installed on the right side along with a sliding map board assembly. The unit commander was located at the forward end of the staff desk. A five

seat module for the communication systems operators was located on the left side along with the radios, telephone switchboard, and interchange boxes. The receiving and transmitting equipment was installed behind the sliding map board on the right side. Two jump seats were provided, one in the front and one at the left rear of the crew and personnel compartment.

The arrows in the sketch at the right indicate the various antennas on the LVTCX2.

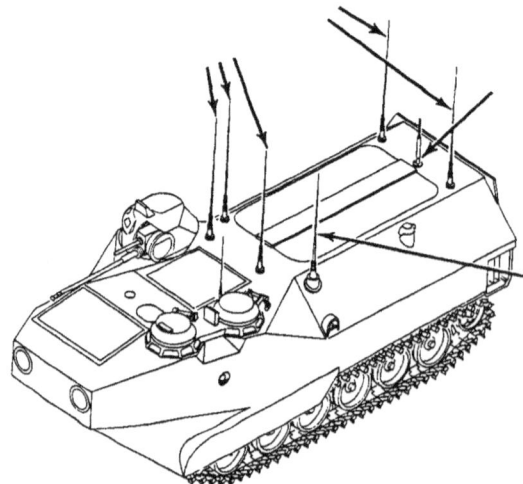

348

The photographs on this page show the LVTC7 command vehicle. As on the LVTP7, the 20mm gun has been replaced by the M85 .50 caliber machine gun.

After type classification, the LVTCX2 was designated as the LVTC7. On these production vehicles, the original weapon station was replaced by the later version armed with a single .50 caliber M85 machine gun. Later, this weapon station also was replaced by an unarmed cupola with nine vision blocks and it was occupied by the vehicle commander. The unit commander now had two alternate positions. The first was the cupola just behind the driver and the second was at the head of the staff desk on the right side of the crew and personnel compartment. In addition to the unit commander's seat, only three seats were now installed at the staff desk. The five seat module for the communication systems operators was retained on the left side of the compartment. FMC produced 85 LVTC7 command vehicles starting in 1972.

Two LVTRX2 recovery vehicles were completed based upon the LVTPX12. The recovery vehicle was manned by the usual crew consisting of the driver, assistant driver, and vehicle commander plus maintenance personnel. The weapon station on the LVTPX12 was eliminated and replaced by an unarmed cupola with nine vision blocks. A recovery winch with a maximum line pull capacity of 30,000 pounds was installed on the top rear of the vehicle. A crane with a maximum boom load capacity of 6,000 pounds was mounted on the top just behind the engine compartment. The controls and a seat for the operator were attached to one side of the crane. A portable electric welder, an air compressor, a portable AC electric generator, as well as maintenance tools and equipment were carried inside the vehicle. The LVTRX2 was type classified as the LVTR7 and 58 were built by FMC.

The prototype LVTRX2 appears above and the LVTR7 is shown below. Note the lack of armament on these vehicles.

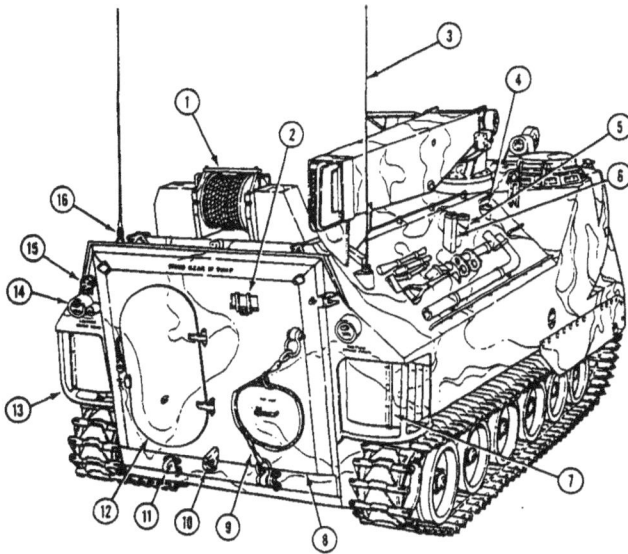

1 Winch assembly
2 Stern vision block
3 Receiving antenna
4 Engine coolant heater exhaust outlet
5 Winterization personnel heater exhaust outlet
6 Cargo hatch support (2)
7 Water jet deflector (2)
8 Ramp
9 Tow cable
10 Pintle
11 Towing eye (2)
12 Personnel door
13 Deflector guard (2)
14 Taillight—stop light (2)
15 Horn
16 Antenna AS—1729/VRC

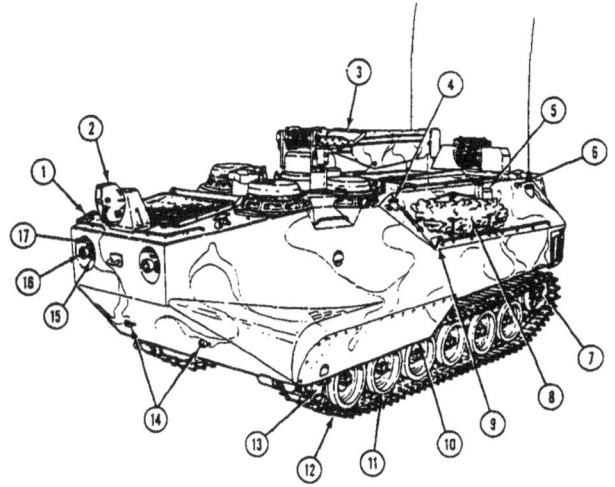

1 Handrail
2 Fairlead
3 Crane and crane operator's station
4 Heater exhaust outlet
5 Fuel fill cover
6 Bilge discharge (4)
7 Idler wheel (2)
8 Camouflage net
9 Fire extinguisher external release
10 Track shroud (2)
11 Road wheel (24)
12 Track assembly (2)
13 Drive sprocket (4)
14 Towing eye (2)
15 Infrared (I.R.) headlight (2)
16 Service headlight (2)
17 Blackout (BC) marker light (2)

The drawings on this page identify the components on the LVTR7. A photograph of the LVTR7 is shown below.

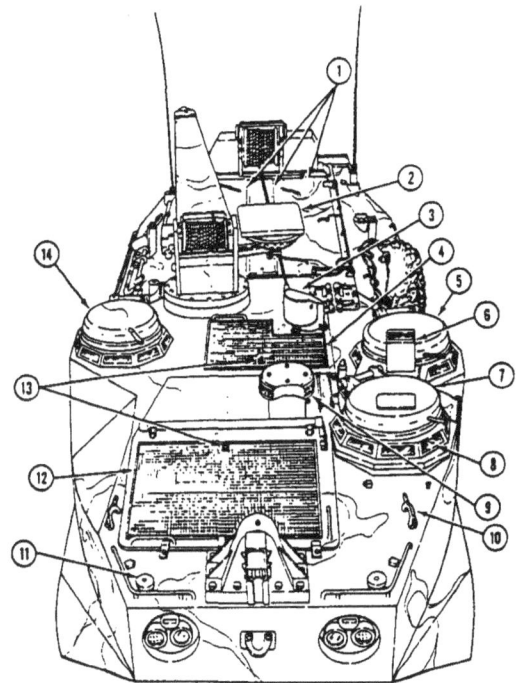

1 Cargo hatch cover (3)
2 Crane controls
3 Engine exhaust outlet
4 Air exhaust grille
5 Vehicle commander's hatch
6 Periscope
7 Driver's hatch
8 Direct vision block (9)
9 Ventilator—aspirator valve
10 Mooring cleat (4)
11 Floodlight mount (2)
12 Air inlet grille
13 Plenum cover position indicator (2)
14 Auxiliary station hatch

351

The prototype LVTEX3 can be seen above. Note that it has the same 20mm gun as the LVTPX12. At the bottom of the page is the LVTE7 with the rocket propelled line charges raised for launching.

Two prototype LVTEX3 engineer vehicles were assembled by FMC in 1970. The original pilots were fitted with the weapon station from the LVTPX12 armed with the M139 20mm gun and the coaxial M73E1 7.62mm machine gun. A dozer blade was installed on the front and the vehicle carried three rocket propelled line charges that could be launched, one at a time, from the top of the vehicle. The prototypes were modified to a production standard under the designation LVTE7. As with the other family members, the original weapon station was replaced by the later version armed with the single M85 .50 caliber machine gun. However, the LVTE7 did not enter production.

The LVTP7 proved to be one of the most successful acquisition programs of that period coming in on schedule and below the estimated cost. At that time,

The LVTP7A1 appears in the photographs above with the new electric powered weapon station still armed with the .50 caliber machine gun M85. Details of the weapon station can be seen in the drawing at the bottom of the page.

the service life was projected to be ten years. However, in the early 1980s, a service life extension program (SLEP) was initiated to extend the service life of the LVTP7 family for an additional eight years. As it turned out, it was to be even longer than that. In addition to the increase in service life, SLEP included improvements in reliability, communication capability, and safety.

After the SLEP modifications, the three vehicles in the family were designated as the LVTP7A1, LVTC7A1, and LVTR7A1. A major change under the program was the installation of the Cummins multifuel VT400 diesel engine. This liquid-cooled, turbocharged, V8 engine developed 400 gross horsepower at 2,800 rpm. It was coupled to the FMC HS400-3A1 transmission. The new vehicle still used the twin water-jets for propulsion when afloat. A new electric drive weapon station armed with the M85 .50 caliber machine gun was installed on the LVTP7A1. On the LVTC7A1 and the LVTR7A1, an unarmed cupola was installed with nine vision blocks. The suspensions on all of the upgraded vehicles were fitted with improved shock absorbers and additional shock absorbers were added at the second road wheel arm. An improved instrument panel was provided for the driver. A non-integral fuel tank that was less susceptible to damage from hull flexing replaced the integral fuel tank in the earlier vehicles. A smoke generator system was installed that injected fuel into the exhaust manifold. The new weapon station on the LVTP7A1 was fitted with launchers for eight smoke grenades. A mine clearing line charge (MICLIC) kit also was developed for installation in the troop compartment

of the LVTP7A1. It consisted of three, 350 foot, rocket propelled line charges that could be launched, one at a time, from the top of the vehicle. Another device intended for installation in the LVT7A1 was the catapult launched fuel-air explosive (CATFAE) surf zone mine clearing system. It consisted of 21 fuel-air explosive rounds and a fire control system installed in the troop compartment. The equipment could fire 234 pound propylene oxide fuel-air explosive rounds onto a minefield destroying surface laid mines. It could destroy mines in the surf zone while afloat and moving up to 6.2 miles per hour or on shore while moving at up to 15 miles per hour. A cleared lane 20 meters wide and 300 meters long could be created in 90 seconds or less.

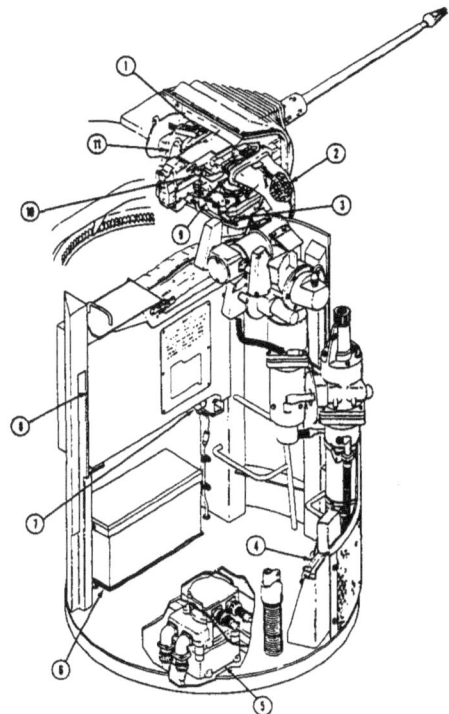

The electric drive weapon station is at the right.

1. Ejection chute door control knob, 2. Feed chute, 3. Weapon charger ratchet, 4. Electronic control unit, 5. Slip ring assembly, 6. Smoke grenade stowage box, 7. Low ammunition switch, 8. Magazine, 9. Charger, 10. Safety lever, 11. Hand charger.

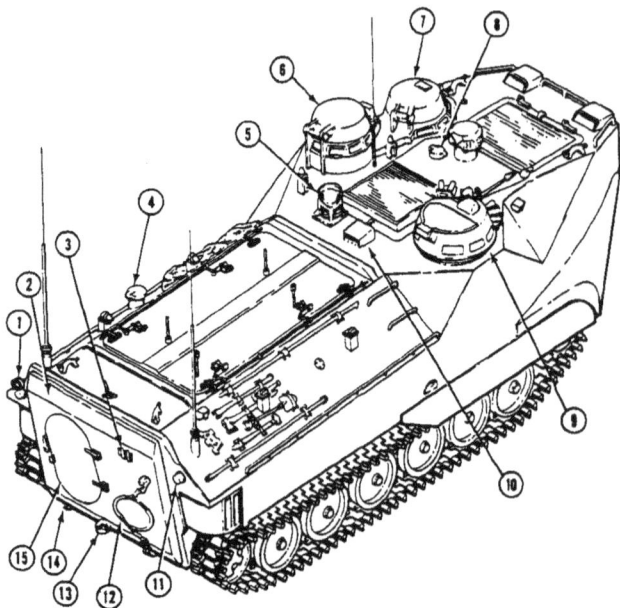

1 Horn	9 Direct vision block (8)
2 Ramp	10 Exhaust air outlet
3 Ramp vision block	11 Taillight-stoplight (2)
4 Fuel fill cover	12 Tow cable
5 Engine exhaust outlet	13 Pintle
6 Troop commander's hatch	14 Towing eye and ramp hinge (2)
7 Driver's hatch	15 Personnel door
8 Cooling system filler cap	

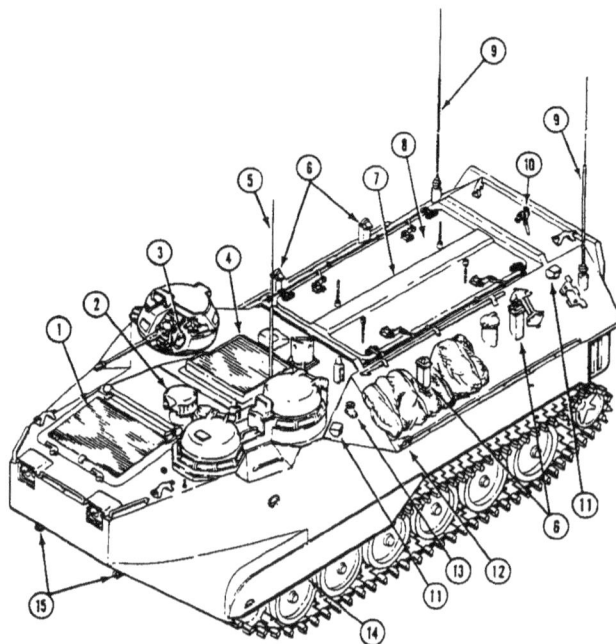

1 Air Intake grille	9 Antenna AS-1729/VRC (2)
2 Ventilator-aspirator valve	10 Sea tow quick-release
3 Ejection chute door	11 Bilge outlets (4)
4 Air exhaust grille	12 Fire extinguisher outside release handle
5 Receiving antenna	13 Personnel heater exhaust outlet
6 Cargo hatch support (4)	14 Track shroud (2)
7 Cargo hatch center beam	15 Towing eye (2)
8 Cargo hatch cover (2)	

Components on the LVTP7A1 are identified in the drawings on this page.

Other improvements included a night vision device for the driver, a new ventilation system, and improved sealing and plenum drainage to reduce and dispose of water entering the vehicles during sea operations. The combat weights of the vehicles were 52,770 pounds for the LVTP7A1, 47,517 pounds for the LVTC7A1, and 52,069 pounds for the LVTR7A1. The maximum speed for all three vehicles was 45 miles per hour on land and 8 miles per hour in water. The cruising range was about 300 miles on land and the endurance in water was 7 hours.

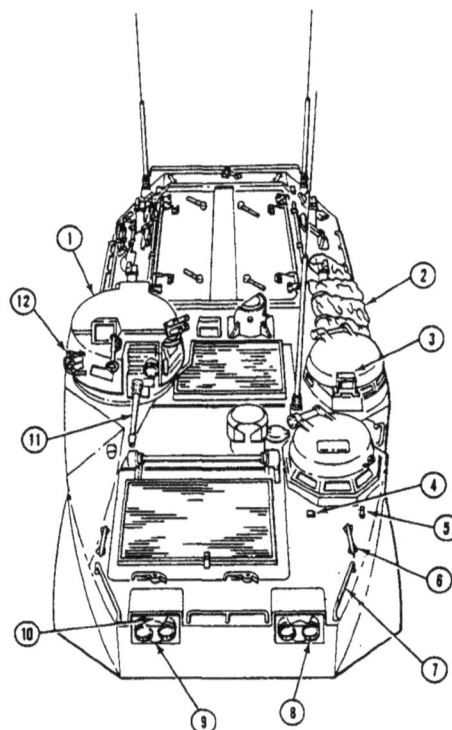

1 Weapon station hatch	7 Handrail
2 Camouflage net (stow as needed)	8 Service headlight (4)
3 Periscope	9 Blackout (BO) marker (2)
4 Searchlight power receptacle	10 Blackout (BO) light (2)
5 Searchlight pintle	11 Caliber .50 M85 machine gun
6 Mooring cleat (4)	12 Smoke grenade launcher (2)

The front of the LVTP7A1 can be seen in the photograph at the left.

354

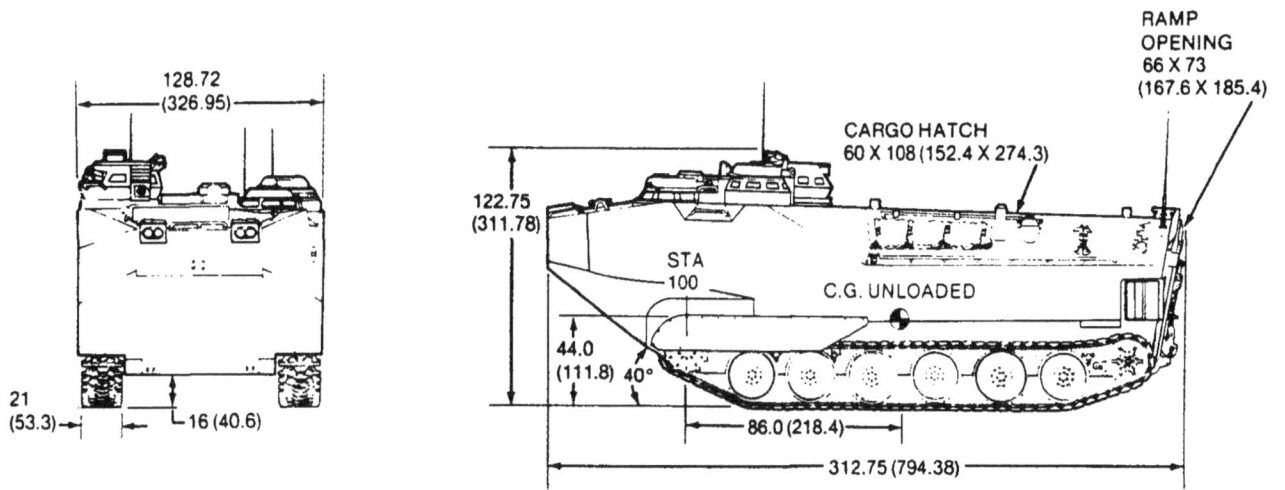

128.72
(326.95)

21
(53.3)

16 (40.6)

CARGO HATCH
60 X 108 (152.4 X 274.3)

RAMP
OPENING
66 X 73
(167.6 X 185.4)

122.75
(311.78)

STA
100

C.G. UNLOADED

44.0
(111.8) 40°

86.0 (218.4)

312.75 (794.38)

Dimensions of the LVTP7A1 (above) and the LVTC7A1 (below) are shown in these drawings.

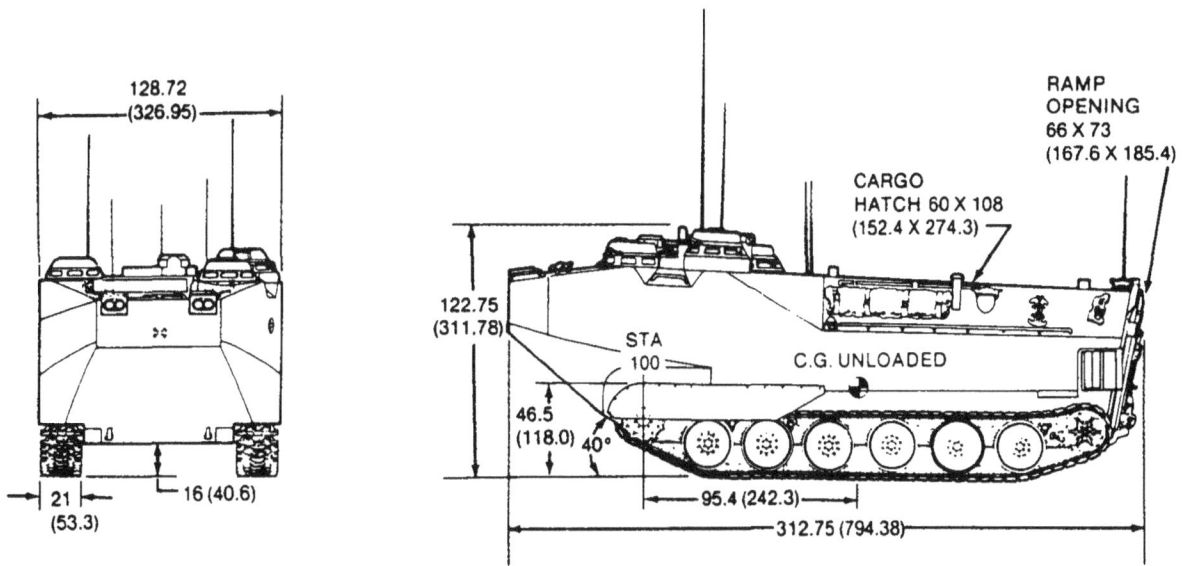

128.72
(326.95)

21
(53.3)

16 (40.6)

RAMP
OPENING
66 X 73
(167.6 X 185.4)

CARGO
HATCH 60 X 108
(152.4 X 274.3)

122.75
(311.78)

STA
100

C.G. UNLOADED

46.5
(118.0) 40°

95.4 (242.3)

312.75 (794.38)

The dimensions of the LVTR7A1 appear in the drawing below.

RAMP
OPENING
66 X 73
(167.6 X 185.4)

Dimensions in inches (centimeters)

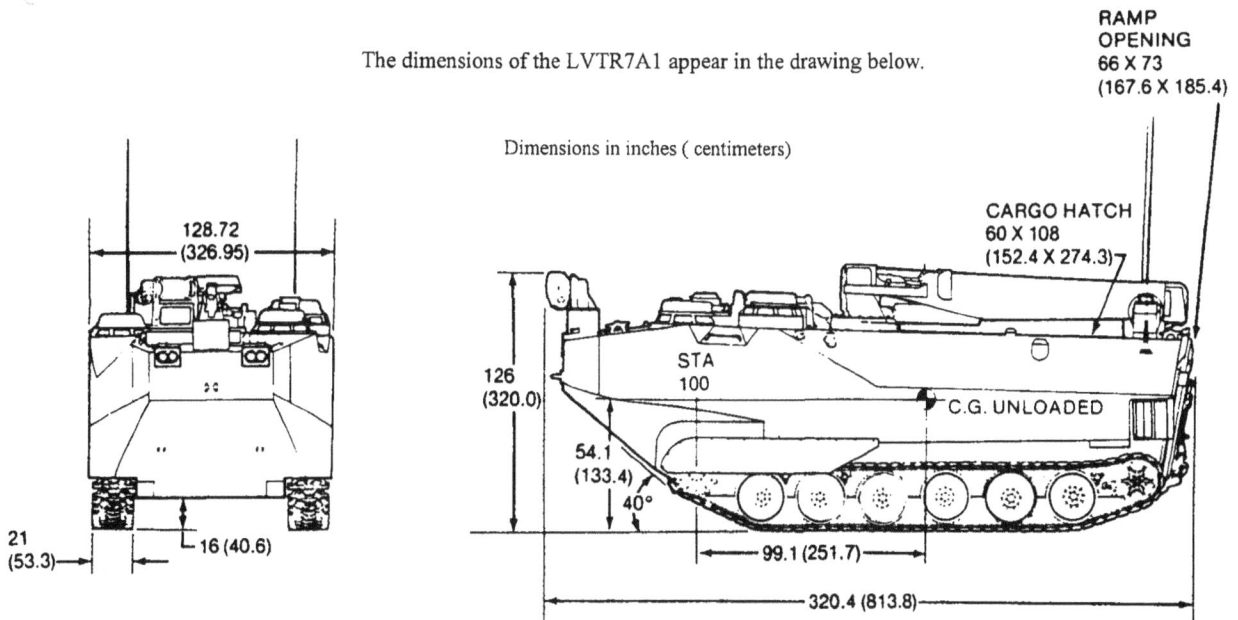

128.72
(326.95)

21
(53.3)

16 (40.6)

CARGO HATCH
60 X 108
(152.4 X 274.3)

126
(320.0)

STA
100

C.G. UNLOADED

54.1
(133.4) 40°

99.1 (251.7)

320.4 (813.8)

355

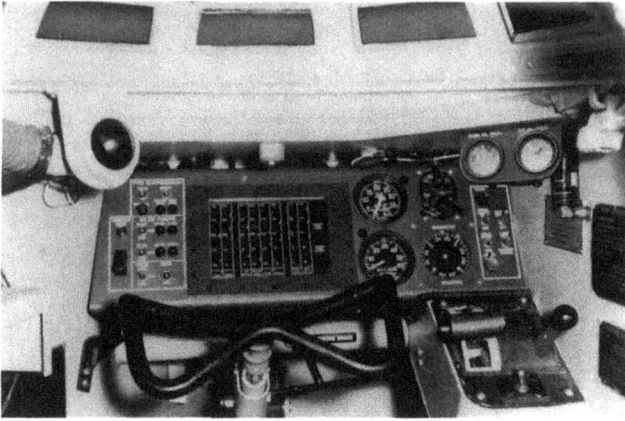

At the top left are the driver's controls of the LVTP7A1 and the mine clearance kit appears above. At the left is the seating arrangement in the LVTP7A1.

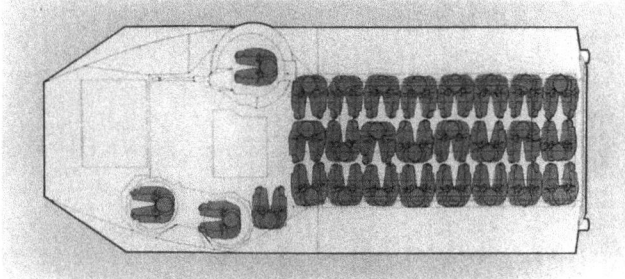

FMC built 333 new LVTP7A1s starting in 1983. They also converted 853 LVTP7s to the LVTP7A1 standard. The older command and recovery vehicles also were upgraded with 77 LVTC7s and 54 LVTR7s being converted to LVTC7A1s and LVTR7A1s.

In late 1984, the Marine Corps changed the nomenclature and the LVTP7A1 became the assault amphibian vehicle personnel Model 7A1 (AAVP7A1). The other two vehicles now became the AAVC7A1 and the AAVR7A1. A new Cadillac Gage weapon station was installed on the AAVP7A1 armed with both a .50 caliber M2 machine gun and a 40mm Mark 19 automatic grenade launcher. The combat weight had now increased to 56,552 pounds for the AAVP7A1, 48,813 pounds for the AAVC7A1, and 52,123 pounds for the AAVR7A1. The performance remained essentially the same.

Applique armor packages also were available to enhance survivability. A bow plane kit was provided to improve the performance when operating in water with the applique armor. Concept studies also considered the installation of more powerful weapons to enhance the performance as a land fighting vehicle.

Below, the LVTC7A1 is at the left and the LVTR7A1 is at the right.

130.5 (331)

16 (40)

21 (53)

128.7 (327)

155 (393)

321.3 (816)

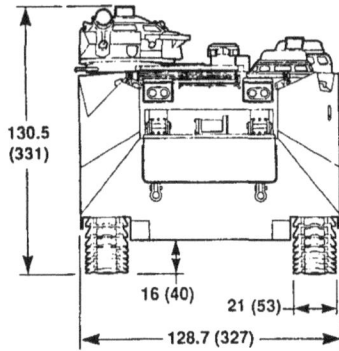

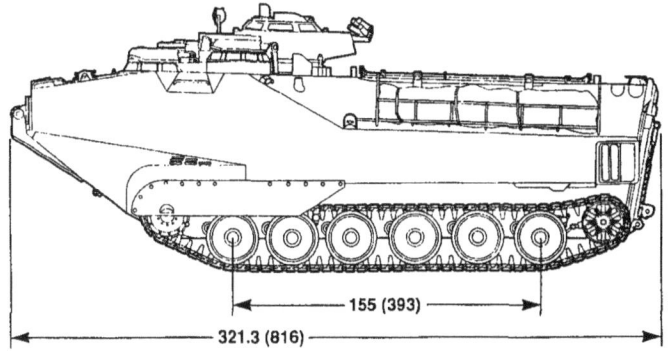

Dimensions in inches (centimeters)

Above is a dimensional sketch of the AAV7A1 and a photograph of the vehicle is at the right. Below, an applique armor kit produced by Rafael has been installed on the AAV7A1 and at the right the steel mesh applique armor has been applied. The latter photograph was taken by Greg Stewart.

Below, an artist's concept of an amphibious assault vehicle with a 25mm cannon turret is at the left and a model of an AAV with the Bradley turret is at the right.

The drawing above is an artist's concept of the LVT(X) landing on a hostile shore.

THE ADVANCED ASSAULT AMPHIBIAN VEHICLE

Under ideal conditions, an amphibious assault would move from ship to shore and then on to the final objective without a pause at the shoreline. Thus the assault vehicle would be both a landing craft and a land fighting vehicle. As early as October 1978, the U. S. Marine Corps was considering three alternative systems to meet the requirements for such an assault vehicle. The first of the three was to be a high speed amphibian with water speeds of 25 to 40 miles per hour. It was referred to as the landing vehicle, assault (LVA). However, it was not considered practical at that time to develop such a vehicle with the limited funds available. The second approach was to use the new M2 infantry fighting vehicle for this role. This was not feasible because of its limited troop capacity and its inability to operate in surf. The third alternative, referred to as the LVT(X), was then the preferred solution. It was to be a displacement hull amphibian with water speeds exceeding 10 miles per hour and was expected to provide improvements in firepower, survivability, and mobility compared to the

LVTP7 family at an affordable price. The LVT(X) was to be the basis of a family of amphibious vehicles that could operate on land as an infantry fighting vehicle and survive in an NBC environment.

Although the original LVA program was terminated, one part of it remained. This was the development of the stratified charge rotary combustion (SCRC) engine. The SCRC was a four rotor 1,500

A cutaway drawing of the stratified charge rotary combustion engine is at the right.

Above are concepts of the LVT assault personnel carrier proposed by FMC.

horsepower engine. The engine program was continued because of its possible application to other vehicles. It was expected to provide a lighter, more compact, power plant than the diesel engine for equivalent horsepower and it would be quieter and have a multi-fuel capability.

A number of concepts were studied for application to the LVT(X) requirement. FMC also proposed a vehicle based upon components of the M113A2 armored personnel carrier. It was intended to carry a squad size unit from ship to shore and on to the final objective. Although it was small and did not meet the 10 miles per hour water speed requirement of the LVT(X), it could be armed with a variety of weapons and serve both as an assault landing vehicle and as a land fighting vehicle. Dubbed the LVT assault personnel carrier, it was intended to serve as the basis of a family of vehicles. Assembled from aluminum alloy armor plate, the proposed LVT assault personnel carrier was to be manned by a crew of two and carry a 13 man squad in the rear compartment. The hydraulically operated rear ramp had a door for personnel and a cargo hatch was located over the rear compartment. The driver was in the left front and the vehicle commander was in the cupola. The latter could be anything from the M113A1 type cupola with an externally mounted .50 caliber machine gun to a one man weapon station armed with a .50 caliber machine gun or a 25mm Bushmaster cannon. The

vehicle was to be powered by the 300 horsepower 6V53T diesel engine using the Allison TX100-1A transmission with a hydrostatic steering differential. The M113A2 suspension was used on the lengthened vehicle with six road wheels per side. It rode on 17 inch wide tracks. When afloat, the vehicle was propelled by two water-jets powered by two fixed displacement hydraulic motors driven by the hydrostatic steering pump. This arrangement eliminated the need for drive shafts and gear boxes in the crew compartment. The tracks and water-jets could be operated simultaneously. Combat loaded, the estimated weight was about 33,000 pounds including a 5,000 payload. The maximum speed was expected to be 38 miles per hour on land and 7 miles per hour in water with a cruising range on roads of about 300 miles.

A larger, two man, turret also was proposed for installation on the hull of the LVT assault personnel carrier. Armed with a 90mm Cockerill cannon and a coaxial 7.62mm machine gun, the turret also had an external mount for an additional 7.62mm machine gun at the commander's hatch. Eight 90mm ready rounds were stowed in the turret with an additional 32 rounds in the hull. The crew now consisted of the driver in the left front hull with the commander and the loader in the turret. Six squad members also could be accommodated in the rear compartment.

The dimensions of the LVT assault personnel carrier proposed by FMC are shown below.

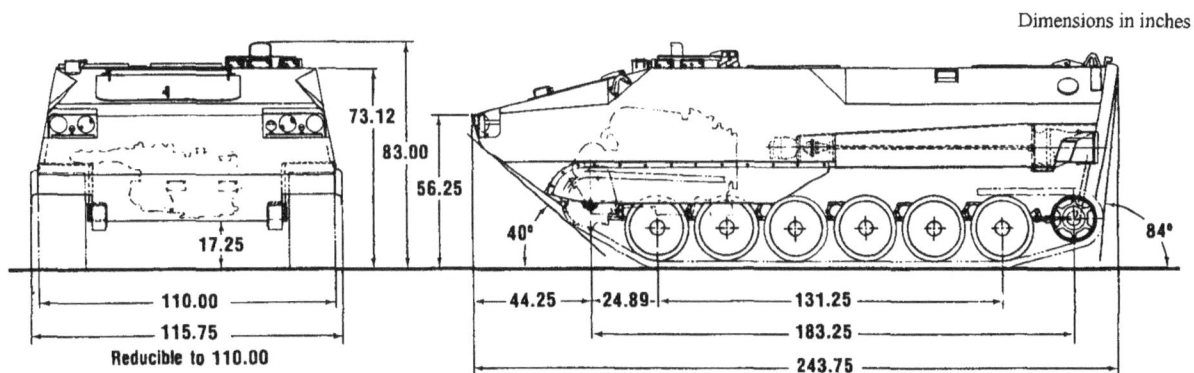

Dimensions in inches

359

The high water speed technology demonstrator is shown above and it is operating at high speed in the water at the bottom of the page.

By the late 1980s, U. S. Marine Corps doctrine envisioned over-the-horizon (OTH) landings for future amphibious assaults. The OTH concept allowed the transport vessels to remain 20 to 25 miles at sea reducing their vulnerability to shore based weapons. However, this doctrine immediately reinstated the high water speed requirement for the landing vehicles. The combination of a high water speed and the capability to operate as a fighting vehicle ashore presented a difficult design problem. A request for information was issued to industry in September 1988 followed in the Spring of 1989 by a formal request for proposals. After the review of a large number of proposals, General Dynamics Land System (GDLS) Division and the Ground Systems Division of FMC were awarded development contracts for a new advanced assault amphibian vehicle (AAAV).

The new AAAV was to be manned by a crew of three consisting of the driver, assistant driver, and the commander/gunner. It would be able to transport a reinforced Marine rifle squad of 18 men. The length and width of the vehicle could not exceed 30 and 12 feet respectively. It would have a maximum water speed of 25 to 30 miles per hour and a range of 75 miles. The maximum speed would have to be maintained in Sea State 3 conditions and the vehicle would have to survive Sea State 5 conditions including the ability to right itself after a 100 degree roll. Like the LVTP7, the AAAV would have to operate in a 10 foot plunging surf. On land, the AAAV was required to have mobility comparable to the Abrams main battle tank and its armor would have to withstand attack from 14.5mm projectiles and protection against 30mm projectiles was desired. The turret mounted main armament was to be 25mm, 30mm, or 35mm in caliber firing conventional ammunition.

Initially, the contractors followed two different approaches. GDLS working with AAI Corporation considered a planing hull to obtain the necessary water speed. Several test rigs were constructed including a high water speed technology demonstrator (HWSTD)

The photographs on this page show the advanced assault amphibian vehicle (AAAV) test rig built by United Defense (formerly FMC). At the bottom of the page, the AAAV test rig is demonstrating its high water speed.

and a propulsion system demonstrator (PSD). The HWSTD reached 29 miles per hour in some of its tests. FMC proposed a hydrofoil assisted planing hull to achieve the high speed. Early tests with a half size model had reached 37 miles per hour. This approach required the use of hydrofoils that could be extended into the water when the vehicle was underway and then retracted for land operations. The planing hull approach also was complicated, requiring the extension of a folding bow assembly and the retraction of the running gear against the hull. After further tests, the Marine Corps indicated a preference for the planing hull approach and the FMC candidate also utilized that design.

On 13 June 1996, the Marine Corps selected GDLS to develop the AAAV. Up to three prototypes will be built during the development phase of the program.

The AAAV test rig produced by General Dynamics Land Systems appears in these views.

The General Dynamics concept of an AAAV is shown above. The turret for the new AAAV appears below.

The drawing above portrays an assault landing spearheaded by the AAAV.

PART VI

ACTIVE SERVICE

During World War II, the development of the armored personnel carrier progressed from the half-track carrier at the top left to the full tracked armored utility vehicle M39 at the top right. Neither vehicle had overhead protection as originally built.

WORLD WAR II

The armored half-track vehicle was the standard personnel carrier for armored infantry during World War II. Despite its limitations, it also was used as a prime mover and a motor carriage for self-propelled weapons. A few full tracked armored personnel carriers and prime movers were improvised using the chassis of tanks and self-propelled guns. The M39 armored utility vehicle was used both as a prime mover and a personnel carrier by the tank destroyer units. The field artillery standardized the unarmored high speed tractor as the prime mover and personnel carrier for both field and antiaircraft artillery units.

The widely used high speed tractor M4 at the right provided valuable service during World War II both as a carrier and as a prime mover. Below, the M29 Weasel (left) and the amphibian M29C (right) could operate where no other vehicles could. The M29 is in France and the M29C is in the Solomon Islands.

Although it was originally designed for operations in snow, the low ground pressure of the Weasel made it extremely useful in mud or swampy areas. It was deployed to both the European and Pacific theaters of operation late in the war. On Okinawa, the Weasel was particularly effective during the heavy rains that made the roads impassable for wheeled vehicles and extremely difficult for many tracked carriers.

Above, the 7th Marine Regiment comes ashore at Wonsan, Korea with their LVT3C carriers.

KOREA

Most of the support vehicles that served in Korea were of World War II vintage or modified versions of these. Among the latter were the landing vehicles LVT3C and LVTA5. Both of these were used in Korea, particularly during the landings at Inchon and the crossing of the Han River.

At the right, an LVT3C of the 7th Marines is operating on the Imjin River in Korea. Below, an LVT3C is at the left and another is swimming in an icy Korean river at the right during December 1952.

Above, the crew of an LVT3C from the 1st Marine Division observes an air strike in Korea on 1 June 1952. At the left below is another LVT3C.

At the right and below are the U.S. Marine Corps modified LVTA5s in action in Korea.

The M75 armored infantry vehicle above has arrived in Korea. This photograph was dated 15 May 1953.

The new generation of armored personnel carriers received its baptism of fire in Korea. This was the M75, a few of which had been deployed to Korea in the Summer of 1953. Here it was used, along with the earlier open top M39, to support front line troops prior to the armistice on 27 July 1953. The importance of the M75 and the M39 was clearly illustrated in an article by Lieutenant John C. McLaughlin that appeared in the Armor Magazine issue of January-February 1954. This article described operations of the carriers in supporting the 7th Infantry Division outpost on Porkchop Hill during the last days before the armistice. The following are excerpts from that article.

"The background of how the M75 came to play its part in the battle for Porkchop during the period 6 to 11 July 1953 is worth exploring. Porkchop, a company size outpost, was located forward of the friendly Main Battle Positions. All logistical support for the elements occupying the position had to be transported along a one lane road which had only a few places

At the right, the M39 armored utility vehicle is operating along the Erie Trail in Korea.

wide enough for two vehicles to pass. The road was under direct enemy observation from positions on the Baldy hill mass and from the Hasakkol complex. By virtue of this observation, the enemy was able to place extremely accurate artillery and mortar fire, as well as direct fire, along the supply route. His direct fire weapons, located on Baldy some 1,800 meters to the southwest of Porkchop, made the use of general purpose vehicles impracticable. Further, the road, due to constant shelling, was traffic-

able only to tracked vehicles. Maintenance of the road by the engineers was normally accomplished under fire by dumping sandbags filled with rocks out of M39 personnel carriers....

Late that first night, Lt. Raymond Devereaux, platoon leader in the 17th Tank Company, received the order to move all of his M75s and M39s to an assembly area in readiness to start hauling supplies out the tortuous road to the men holding Porkchop. In a short time he had his platoon consolidated in a forward assembly area. The first calls for support were not long in coming in. 'We need ammunition of all types and grenades. We have some wounded that need to be evacuated.'

The platoon leader, through the operation of a checkpoint, controlled the flow of vehicles to and from the outpost. The M75s and M39s started to roll. Supplies went out to the defenders on 'the Chop', wounded were brought back to the checkpoint for further evacuation to forward battalion aid stations. Along the rutted road the drivers moved the vehicles in complete darkness, broken only by the flash of exploding artillery and mortar shells. The heavy rains had made the road and the steep approach to the landing on 'the Chop' a mire of mud, but the M75 was up to the test and got through....

Fresh troops were loaded aboard the M75s and transported to the outpost. The APCs provided cover for the men going out and protection for the battle weary men who were being brought off the outpost. That this saved numerous casualties is without question. Infantry units moving on foot to the beleaguered outpost would have suffered tremendous casualties from the very heavy enemy fire being laid down even as far back as the assembly area....

On the 11th of July, a decision was made that the outpost was of no further value and had served its purpose-that of stemming a Chinese attack which most certainly would have smashed into our Main Battle Positions had the

Reds been able to overrun Porkchop in their initial attack. The enemy had expended three of his regiments and, in doing so, was now unable to attempt penetration of our Main Line of Resistance....

The APCs brought out engineer demolition teams and began evacuation of elements of the 32nd Infantry Regiment, which had taken over the defense of the sector from the 17th. Friendly artillery and tanks shifted their fires closer and closer until finally they were falling on the outpost itself. As mortar and artillery rounds peppered the flaming battleground, the APCs rumbled up 'the Chop' to the remaining defenders. The steel fortresses backed right up to the caves and bunkers and the infantry climbed aboard. So skillful was the evacuation that intelligence reported the Chinese were confused to the point of believing that reinforcements were being brought to the outpost and that our artillery was mistakenly firing on our own men. The engineers set their demolitions and they too were brought back in the APCs. Porkchop had been successfully defended and successfully evacuated. Terming the bold withdrawal an 'historical example of skillful abandonment' General Trudeau said, 'with the M39 it would have been extremely difficult. Without either type it would have been impossible.'

The M75 did the job it was designed to do and performed all through the action in a manner that left little to be desired."

At the right, wounded are being loaded into an M75 armored infantry vehicle in Korea.

The problem of dismounting from the M39 armored utility vehicle is well illustrated above. This photograph was taken in Korea on 15 October 1953 after the fighting had stopped. At the bottom of the page, troops are training with the M75 armored infantry vehicle on 31 August 1953.

The greater protection provided by the M75 compared to the open top M39 was an important factor in reducing casualties in these operations. The rear doors also allowed safer entrance and exit from the vehicle compared to jumping over the sides as on the M39. The design concept of a fully enclosed armored personnel carrier with rear entrance doors was fully justified.

Above is an early M113 armored personnel carrier supplied to the Army of the Republic of Vietnam (ARVN). Note the lack of a shield on the .50 caliber machine gun.

VIETNAM

When the first U. S. Army advisors arrived in Vietnam, the Vietnamese armored units were equipped with a variety of World War II vehicles inherited from the French. These included M24 light tanks, M3 half-tracks, M3 scout cars, and M8 armored cars. To provide some modern equipment, it was recommended in late 1961 that the M113 armored personnel carrier be provided to the Vietnamese forces. As a result, two Vietnamese companies were equipped with the M113. Originally, each company consisted of three platoons with three vehicles each. Four additional armored personnel carriers were in a support platoon and two were in the headquarters section bringing the company total to 15.

Operations carried out by the two companies showed the value of the maneuverable, lightly armored, vehicles as well as the need for some modifications. The latter included an armor shield for the gunner manning the .50 caliber machine gun and additional weapons to increase the firepower. As a result of the successful performance, additional M113s as well as M114 armored reconnaissance vehicles were shipped to Vietnam. However, as described earlier, the M114 proved to be unsatisfactory and it was soon replaced by additional M113s.

At the right is an ARVN M114 armored reconnaissance vehicle. Note the shield added to the observer's machine gun. This photograph was dated 20 March 1963.

The photographs above and below show an ARVN M113 armored personnel carrier during October 1968. These photographs and those on the following four pages were taken by James Loop during his service in Vietnam.

The cupola armed with two 7.62mm machine guns has been installed on the ARVN M113 above. This is the same vehicle as on the previous page. Below, an improvised shield for the .50 caliber machine gun has been mounted on this M113.

The M113 in these photographs has been fitted with a shield for the .50 caliber machine gunner and helicopter seats have been installed on the vehicle roof. These photographs were taken by James Loop during August 1968.

These M113s also are fitted with a .50 caliber machine gun shield and helicopter seats on the roof. The track on the M113 below is being replaced. Note the M41 76mm gun tank in the background of the latter photograph.

The M113 shown here was photographed by James Loop in Vietnam during August 1968. The .50 caliber machine gun on this vehicle has been replaced by a 57mm recoilless rifle.

James Loop photographed this M132 self-propelled flame thrower in Vietnam during February 1969. The coaxial 7.62 machine gun has not been installed in the cupola. Note the gun shields mounted on each side of the roof.

Above, a U.S. Army armored cavalry assault vehicle (ACAV) is moving along a road in Vietnam during February 1969.

Prior to Vietnam, the U. S. Army regarded the armored personnel carrier as a protected means of transporting infantry to the battle area where they would fight dismounted. After the adoption of the Vietnamese tactics of fighting mounted, the vehicle itself became an important weapon. The effective use of the M113 in this new role by the U. S. Army was illustrated by the action of Troop A, 1st Squadron, 4th Cavalry at Ap Bau Bang on 11-12 November 1965. The following excerpts were taken from "Mounted Combat in Vietnam" by General Donn A. Starry.

"A task force of the 2nd Battalion, 2nd Infantry, consisting of the battalion's three rifle companies, its reconnaissance platoon, and Troop A ... was ordered to sweep and secure Highway 13 from the fire base at Lai Khe to Bau Long Pond, fifteen kilometers north. The purpose of Operation Roadrunner was to secure safe passage for a South Vietnamese infantry regiment and provide security for Battery C, 2nd Battalion, 33rd Artillery, which was moving north to support the South Vietnamese regiment.

On 10 and 11 November 1965 the road was cleared without incident; medical teams even visited the village of Bau Bang as part of a medical civic action program.

During the afternoon of 11 November, Troop A, the artillery battery, the command group, and Company A of the infantry battalion moved into a defensive position south of Bau Bang. Concertina wire was installed, individual foxholes were dug, and patrols were set up for ambushes. Dragging the hull of a destroyed armored personnel carrier around the perimeter, Troop A knocked down bushes and young rubber trees to clear fields of fire. The night

At the right, an M106 self-propelled mortar of the 4th Cavalry is operating in Vietnam.

380

The ACAV above has taken a hit from the shaped charge warhead of a rocket propelled grenade. Note the splatter pattern around the hole in the side. James Loop photographed this vehicle during July 1968.

passed with only a light enemy probe, but within minutes after the early dawn stand-to (a term applied by armored units to first-light readiness of men, vehicles, and radios) fifty to sixty mortar rounds exploded inside the perimeter. In the first few minutes, Troop A had two men wounded. Half an hour later, a violent hail of automatic weapons and small arms fire was added to the mortar fire. Under cover of this fire, the Viet Cong moved to within forty meters of the defensive positions. While the cavalry-men returned the fire, M113s of the 3rd platoon roared out and assaulted the enemy. The violence of this unexpected mounted counter-

attack disrupted the Viet Cong attack and the M113s returned to the perimeter. The troop suffered three more wounded and one killed when ammunition in a mortar carrier exploded after being hit by enemy mortar fire.

The Viet Cong made their second assault from the jungle and rubber trees south of the perimeter. Again supported by mortars and automatic weapons, they crawled through the waist high bushes of a peanut field and rushed the concertina wire. One of the M113s in that section of the perimeter was driven by Specialist 4 William D. Burnett, a mechanic. When the .50 caliber machine gun on his APC failed to

The two M106 self-propelled mortars below were photographed by James Loop in Vietnam during September 1968. Note that both vehicles have been fitted with machine gun shields.

Early example of shields fabricated for the ,50 caliber machine gun appear above. The shield at the top right is one of those manufactured on Okinawa.

function, Specialist Burnett jumped from the cover of the driver's compartment to the top of the vehicle, cleared the weapon, and opened fire on the charging Viet Cong, killing fourteen. For this and other actions during the battle, Specialist Burnett was awarded the Distinguished Service Cross. The heavy fire from Burnett's machine gun and those of the M113s near him broke the enemy assault.

The Viet Cong next attacked west from Highway 13 and again were repulsed by .50 caliber and small arms fire. Several times, M113s were moved to weak points on the perimeter so that their machine guns could fire into the enemy's ranks at point-blank range. At 0645 an air strike directed by an airborne forward air controller dropped bombs and raked the wooded area north of the task force with 20mm cannon fire....

The battle of Ap Bau Bang went on for more than six hours before the enemy withdrew to the northwest, leaving behind his wounded and dead. Troop A, commanded by Second Lieutenant John Garcia, suffered seven killed and thirty-five wounded; two M113s and three M106 mortar carriers were destroyed and three M113s were damaged. Procedures and techniques learned in training had been proven in battle. The clearing of fields of fire and the pre-dawn stand-to had insured the full application of Troop A's firepower. The 3rd Platoon's foray

into the enemy position and the positioning of M113s on the perimeter had demonstrated the unit's flexibility, and artillery and aerial fire support had provided depth to the defense. The enemy had begun the fight; the combined arms team had ended it."

The action at Ap Bau Bang illustrated the effectiveness of combined arms in jungle terrain and paved the way for the deployment of additional armored units to Vietnam.

The .50 caliber machine gun shield fabricated on Okinawa can be seen on the M113 at the right.

The armored cavalry assault vehicles on this page are fitted with the machine gun shield kit manufactured by FMC.

Above, this armored cavalry assault vehicle was photographed by James Loop during September 1969. Note the extra equipment stowed on the vehicle.

The last major tactical unit sent to Vietnam arrived in mid-1968. This was the 1st Brigade, 5th Infantry Division (Mechanized). The 1st Brigade found the use of the M113 to be particularly effective in countering the night infiltration tactics of the Viet Cong. Their operations in late 1968 are described in the following excerpt, also from "Mounted Combat in Vietnam".

"After the early encounters, combat in the 1st Brigade's area of operation was light. Since lack of vegetation made the M113 visible for hundreds of yards, particularly on a moonlit night, it was a simple matter for the enemy to bypass the vehicles. The brigade countered this tactic by saturating an area with four-man patrols. Each mechanized infantry company was required to have a minimum of twenty ambush patrols of four men each. Commanders briefed their troops at the noon meal; patrols then mounted armored personnel carriers and were taken on a ground reconnaissance of each position. At dusk, while they were still visible to enemy in the area, the M113s were again dispatched on designated reconnaissance routes. Immediately after dark, while the APCs were moving, each four-man patrol dismounted and established its ambush position. This technique made it difficult for an observing enemy to detect ambush positions because the vehicles never stopped moving during the reconnaissance.

After the patrols were in place, the M113s formed platoon night defensive positions and prepared to move to the assistance of any patrol that ambushed an enemy force. Because the ambush patrols were close together, usually separated by a rice paddy or a dike, the enemy could not bypass all of them. Establishing four-man patrols was a calculated risk since they had little staying power, but once the patrols had engaged the enemy, reinforcements moved according to plan. Upon hearing the first round fired, the vehicles nearest the ambush took the most direct route to the fight, with headlights blazing. Later, a tabulation of all fights during the use of this technique showed that the longest time lapse, from the first round fired to the arrival of the armored cavalry assault vehicles, was less than four minutes. It was the speed of the M113 that permitted the American forces to take the risk of setting up four-man patrols."

The ACAVs below are in action in Vietnam.

The armament on the ACAV above is worth noting. An M60 machine gun is mounted on the bottom of the open roof hatch cover and a .50 caliber machine gun has replaced the 7.62mm weapon on the right roof mount. The usual ACAV M60 machine gun with shield has been retained on the left roof mount as well as the .50 caliber weapon in the shielded commander's station. The ACAV in the background is armed with a 106mm recoilless rifle. Below, the five man crew of the ACAV above appear at the right. Note the 40mm M79 grenade launcher carried by the man at the right.

The firepower of the M113s was increased by numerous field installations. These included the 7.62mm Gatling type minigun, recoilless rifles ranging from 57mm to 106mm, and additional .50 caliber machine guns. A smaller version of the Claymore antipersonnel mine could be attached to the sides of the vehicle for protection against enemy infantry. The full size Claymore itself could not be used as the blast would damage the vehicle. Various techniques were developed to cross difficult terrain. As mentioned before, capstans attached to the drive sprockets were used to extricate the vehicle when it was bogged. A cooperative effort of several vehicles often allowed them to cross difficult areas and the bridge carrying M113 was particularly valuable.

The M113s below and at the right are armed with the six barrel 7.62mm Minigun. The vehicle at the right was photographed by James Loop on 9 December 1968.

The M163 Vulcan air defense system also served in Vietnam, but the lack of enemy air activity restricted it to the ground role. The M163 above was photographed by James Loop on 9 December 1968. The M163 below is climbing out of the water after swimming. Note that the trim vane is still extended.

Above, one of the XM734 infantry fighting vehicles shipped to Vietnam appears at the left and an M548 cargo carrier with a quad .50 caliber machine gun mount is at the right. Below, an 81mm self-propelled mortar can be seen at the right.

The first armored personnel carriers deployed to Vietnam were the gasoline powered M113. Later the diesel powered M113A1 was introduced and by 1 July 1968, the entire fleet was converted to diesel power reducing the fire hazard of the earlier vehicle. Other members of the M113 family in Vietnam included the M125A1 81mm mortar carrier and, as indicated in the action at Ap Bau Bang, the 4.2 inch mortar carrier M106. In August 1962, a flame thrower was installed in an M113, but it saw only limited use. In December 1964, two M132 flame thrower vehicles were assigned to the Vietnamese 1st Armored Cavalry. They were employed effectively and it was recommended that four M132s be issued to each Vietnamese armored regiment. Later, the M132A1 self-propelled flame thrower was deployed for use by the U. S. Army. The XM45E1 flame thrower resupply vehicle also was used to refuel the M132A1s.

Below, an M132 self-propelled flame thrower is being refueled at the left from an XM45E1 service vehicle. At the right, the flame thrower is firing.

387

Above, an M577 command post carrier has a roof mounted 7.62mm machine gun and shield. Note the use of the trim vane to carry supplies. Below, these M577s of the 4th Cavalry have roof mounted helicopter seats. Both photographs were taken by James Loop during June-July 1968.

As can be seen in these photographs by James Loop, the M548 cargo carrier was widely used in Vietnam. The .50 caliber machine gun ring mount was installed on both vehicles.

U.S. Marine Corps LVTP5A1 landing vehicles are shown on this page during operations in Vietnam.

The LVTP5A1 family of vehicles was deployed to Vietnam by the U. S. Marine Corps. These vehicles belonged to the 1st and 3rd Amphibian Tractor Battalions which by some strange circumstance were attached to the 3rd and 1st Marine Divisions respectively. In addition to the LVTP5A1 itself, these units were equipped with the LVTP5A1 command vehicle, the howitzer armed LVTH6A1, and the LVTE1 engineer vehicle. These vehicles were particularly vulnerable to mine explosions because of the gasoline filled fuel cells in the bottom. As a result, their use in land operations was limited.

The Marine Corps also used the M76 Otter in Vietnam. However, it could only be employed in safe areas since it was unarmored and the pneumatic tires were vulnerable to enemy fire.

Above, a U.S. Marine Corps LVTH6 is firing its howitzer in Vietnam. Below are two views of the LVTE1 supporting Marine infantry.

Below, an LVTR1 is removing a power pack from another vehicle. A U.S. Marine Corps M76 Otter is armed with a .50 caliber machine gun at the right.

Above, an A2 Bradley is concealed under a camouflage net in Saudi Arabia.

WAR IN THE PERSIAN GULF

After the invasion of Kuwait by Iraq in August 1990, Operation Desert Shield deployed United States forces to Saudi Arabia to protect that country from attack. When Iraq refused to withdraw from Kuwait, major forces were transferred to the area by the United States and the allied coalition in preparation for Operation Desert Storm. During this buildup, many earlier types of equipment in the initial deployment were replaced by the latest models direct from the United States. The ground part of Desert Storm launched in late February 1991 provided valuable information on the effectiveness of the new weapons.

At the right, an M2A2 Bradley infantry fighting vehicle is being unloaded at a port in Saudi Arabia.

Above, M3A2 cavalry fighting vehicles are being prepared for combat in Saudi Arabia. Below, the Bradleys are in the desert during Operation Desert Shield.

The photographs on this page show the A2 Bradley fighting vehicle in the desert. Note the extra stowage on the vehicles, typical of troops in combat.

Two Bradleys are shown above and below during operations in the Persian Gulf.

Above, the M113 series vehicles used by the medical units of the 3rd Infantry Division are lined up. Below, 3rd Infantry Division M577 command post carriers are operating in the desert. At the left, the vehicles with their tents are arranged to form a large command post.

Below are two views of the multiple launch rocket system. The launchers on both vehicles have their full load of twelve rockets.

Other vehicles serving in the Persian Gulf included the Chaparral missile system above and the M163 Vulcan air defense system below at the right.

Above at the left and below are photographs of the U.S. Marine Corps AAVP7A1 in the Persian Gulf.

The M59 armored infantry vehicle above is on Cold War duty near Baumholder, Germany on 20 July 1956. This vehicle belonged to Company C, 12th Armored Infantry Battalion.

WORLDWIDE SERVICE

The major point of danger during the Cold War was always considered to be in Europe. As a result, the greatest allocation of troops and equipment was to this area in support of NATO. The new series of armored personnel carriers followed this pattern with the M75, the M59, and later the M113 family. When the new generation of Bradley fighting vehicles appeared, the troops in Europe still had priority for the latest equipment. This practice of forward deployment of the latest weapon systems continued until demise of the Soviet Union and the breakup of the Warsaw Pact.

At the right, the M59s of the 3rd Armored Division are on duty near Frankfurt Germany on 21 June 1956.

The many applications of the M59 are illustrated above. Its use can be seen as an armored infantry vehicle, a medical evacuation vehicle, a supply carrier, and a prime mover. Below, the M59 shows off its swimming ability.

Above, an M113 of the 7th Infantry Division is training in Korea on 8 December 1961. Below at the left, troops are dismounting from an M113 during training.

Above at the right, M113, registration number 12L890, moves through a lightly wooded area at Fort Richardson, Alaska on 22 November 1961. Below, an M113 swims the Main River in Germany.

Above, M113s are gathered for REFORGER 86 on 2 January 1986. At the left below, an M113 crosses the Rhine River on 28 September 1983. At the right below is an M113 from the 2nd Armored Division during REFORGER 87. At the bottom right, an M113 crashes through a store front during Operation Just Cause in Panama during December 1989.

Above, this Bradley was photographed by Michael Green at Camp Roberts, California on 23 August 1988. The Bradley below has just launched a TOW missile.

The Bradleys on this page are from the 11th Armored Cavalry Regiment in Germany and they were photographed by Michael Green in May 1985.

The Bradley above was photographed by Michael Green in Germany during REFORGER 88 in September 1988. The A2 Bradley is moving at high speed in the view below.

The A2 Bradley above is fitted with dummy reactive armor. Below, a U.S. Marine Corps LVTP7A1 is operating with an M60A1 main battle tank.

The AAV7A1 with the steel mesh applique armor above is operating in the surf. It was photographed by Greg Stewart. Below, an LVTP7A1 is afloat.

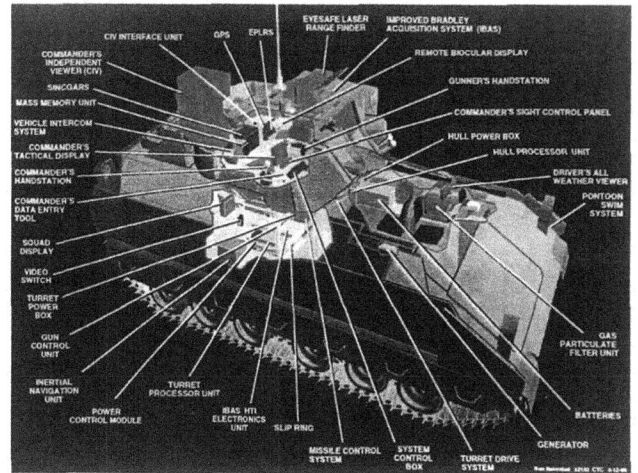

The A3 version of the Bradley fighting vehicle appears above. The commander's independent thermal viewer on the right side of the turret is an obvious point of identification. The various components installed on the A3 are identified in the cutaway drawing at the top right.

THE FUTURE

The Bradley fighting vehicle will continue to serve in the U.S. Army into the foreseeable future. The A3 version with its digital electronics architecture and improved command and control equipment will operate effectively with other 21st century weapons such as the late models of the M1A2 main battle tank and the Crusader artillery system.

The Crusader advanced field artillery system consists of the XM2001 self-propelled 155mm howitzer and the XM2002 armored resupply vehicle. Both the self-propelled howitzer and the resupply vehicle are fully automated with each requiring only a three man crew. The XM297E2, integral midwall cooled (IMC), 155mm cannon utilizes a modular charge solid propellant system. The XM2001 self-propelled howitzer is designed to fire its 60 rounds of 155mm ammunition at rates up to 10 rounds per minute with a maximum range of over 40 kilometers. The XM2002 armored resupply vehicle can transfer fuel and 60 rounds of ammunition to the self-propelled howitzer in only 12 minutes.

The Crusader advanced field artillery system is illustrated at the right. The self-propelled 155mm howitzer is followed by the armored resupply vehicle.

Four Crusader prototypes have been authorized. They consist of two self-propelled howitzers and two armored resupply vehicles. The first prototype, scheduled for delivery in June 1999, is an automotive test rig based upon the armored resupply vehicle. It is designated as the RSV(-) and is provided with only one complete crew station. Weights are added to simulate a fully loaded and equipped vehicle for the automotive tests of the hull structure and suspension system. During August 2000, delivery is scheduled for the first self-propelled howitzer (SPH1) and the complete armored resupply vehicle (RSV1). The final prototype self-propelled howitzer (SPH2) is scheduled for delivery in November 2000.

Models of the Crusader self-propelled howitzer and armored resupply vehicle can be seen above. The cutaway drawings below show the internal components of the self-propelled howitzer (left) and the armored resupply vehicle (right).

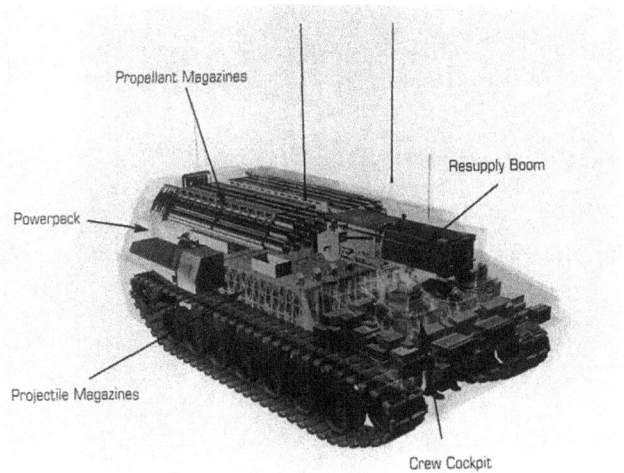

Below, the arrangement of the 155mm howitzer and the ammunition magazines appears at the left and details of the howitzer and mount can be seen at the right.

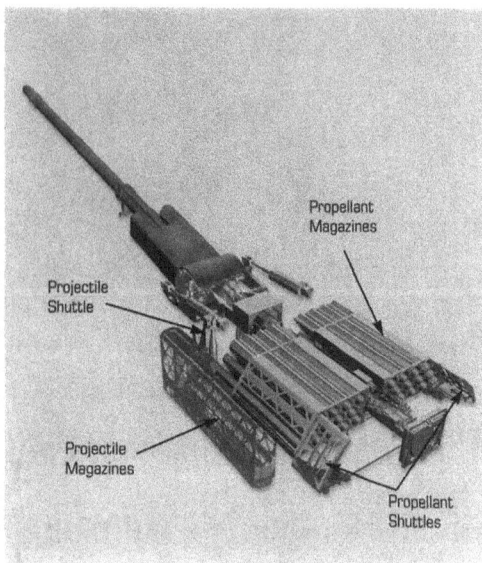

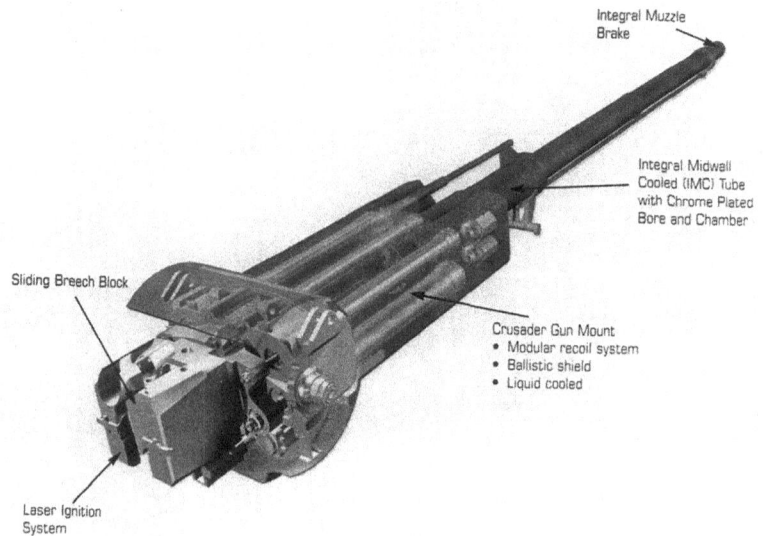

408

Above, the armored resupply vehicle is in position to transfer ammunition to the self-propelled howitzer. The arrangement of the three man crew in the Crusader is shown at the right. Below, the components of the modular artillery charge system (MACS) for the 155mm howitzer is illustrated showing the combinations required for the various range bands.

155-MM XM231	Charge-Type	155-MM XM232
Low-Zone		Top Zone
1 & 2	Crusader Zones	3, 4, 5 & 6
Green with Black Band	Color	Light Brown
152.4mm (6.00")	Length	157.5mm (6.20")
154.9mm (6.10")	Diameter	150.6mm (5.93")
Cylinder with Flat-Ends	Shape	Cylinder with Bump-Ends
Approximately 4 lbs	Weight	Approximately 6 lbs
Single-base	Propellant Type	Triple-base
4 per can	Packaging	5 per can
None	Additives	Decoppering Agent Flash Reducer Wear Reducer

MACS ZONES		DEMONSTRATED RANGE BANDS
Charge 1	XM231	3.2 - 8.0 km
Charge 2	XM231 XM231	5.6 - 12.2 km
Charge 3	XM232 XM232 XM232	7.9 - 20.2 km
Charge 4	XM232 XM232 XM232 XM232	10.4 - 24.7 km
Charge 5	XM232 XM232 XM232 XM232 XM232	15.8 - 30.5 km
Charge 6	XM232 XM232 XM232 XM232 XM232 XM232	19.0 - 40+ km

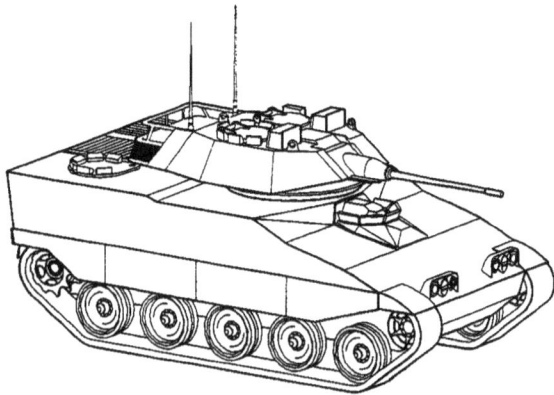

Above, the early concept of a scout vehicle based upon components from the M113 armored personnel carrier is illustrated at the left. At the right, a hypothetical future scout vehicle is depicted emphasizing its role as a carrier of sensors. Also, note the low silhouette obtainable with the use of the externally mounted main armament.

When the XM800 program was terminated, the reconnaissance role was divided between the M3 cavalry fighting vehicle and the M1114 high mobility, multipurpose, wheeled vehicle (HMMWV). Although they performed successfully, neither was designed specifically for the job. A study program was initiated to consider the design of a new future scout vehicle. Among the early concepts examined was the use of M113 components resulting in a vehicle similar in appearance to the XM800T with space for a four man crew and a rear exit from the hull.

On 7 July 1998, a memorandum of understanding was signed by the United States and the United Kingdom to jointly develop a new reconnaissance vehicle. It was referred to in the United States as the future scout and cavalry system (FSCS). In the United Kingdom, it was the tactical reconnaissance armoured combat equipment requirement (TRACER), proving that the first task for any program is to select a good acronym.

As in the old XM800 program, proposals for the scout vehicle could utilize either wheels or tracks. A single vehicle should be transportable in a C130 aircraft and the C17 should be able to carry three. Desired characteristics included a speed of 90 kilometers per hour forward and 50 kilometers per hour in reverse with a range of 600 kilometers at optimum fuel efficient speed. A fording depth of one meter was required and swimming was desirable. The vehicle should have a minimum ground clearance of 45 centimeters, the ability to cross a two meter trench and climb a 60 centimeter vertical wall. These requirements were intended to give it superior mobility compared to the main battle tanks and infantry fighting vehicles. Armament was specified for the FSCS/TRACER as an automatic medium cannon for defense in addition to secondary armament.

On 29 January 1999, contracts were awarded to two competing British-American groups for the development of prototypes for the new scout vehicle. The SIKA International group consisted of British Aerospace, Lockheed Martin Corporation, Vickers Defence Systems, and General Dynamics Land Systems Division. The LANCER group was composed of GEC-Marconi, United Defense, Raytheon-TI, and ALVIS. Based upon the results of these initial contracts, a single contractor will be selected to complete the development and bring the vehicle into production. Fielding of the new scout vehicle is anticipated during the period late 2007 to early 2008.

410

PART VII

REFERENCE DATA

This is the Bradley M2A3 infantry fighting vehicle with the twin tube TOW launcher raised into the firing position. The commander's independent thermal viewer can be seen on the opposite side of the turret from the missile launcher.

The TOW missile launcher on the A2 Bradley above is raised to the firing position. The applique armor is installed on this vehicle.

Details of the front and side armor on the A2 Bradley can be seen above. Below, the high power to weight ratio of the early Bradley is obvious in this photograph.

The early M2 Bradley infantry fighting vehicle appears in these two photographs.

An M113 armored personnel carrier, operating with its infantry, is shown above. Below, an M113 fire support vehicle is armed with a Cockerill 90mm gun in a Cadillac Gage turret.

Above the M113A3 armored personnel carrier is fitted with the external fuel tanks and a high displacement trim vane. Below, the applique armor and the shields for the .50 caliber machine gunner have been installed on this M113A3.

Above, the 20mm gun M139 has been mounted on the M114A1 command and reconnaissance vehicle changing its designation to M114A1E1. Below, the FMC command and reconnaissance vehicle "M113½" also is armed with the 20mm gun.

The U.S. Marine Corps AAVP7A1 amphibian assault vehicle appears in these two photographs by Greg Stewart. Note the pierced steel applique armor on the sides. Details of the turret and armament can be seen below.

Details of the main production vehicles described in this volume are tabulated in the following data sheets.

Vehicle dimensions were taken from the original vehicle drawings when these were available. Other sources were the official characteristics sheets, proving ground test reports, and manufacturer's data. Some dimensions such as height or ground clearance vary depending upon the vehicle load and resulting spring compression. The design reference values are quoted in these cases to permit comparison between vehicles.

Clarification of some of the terms used in the data sheets may be desirable. The tread is the distance between the track centerlines. The ground contact length is the distance between the centers of the front and rear road wheels. This value is used to calculate the ground contact area and the ground pressure of the vehicle. The vehicle combat weight is used in the latter calculation. The combat weight included the crew with a full load of fuel, stowage, and ammunition. If available, the exact weight of an experimental vehicle is shown. In some cases only approximate weights were available. For production vehicles, the average weight is often rounded off to the nearest 1000 pounds. The terms left and right are used from the perspective of someone seated in the vehicle driver's seat.

When available, the maximum values are quoted for the gross and net engine horsepower and torque. The gross horsepower and torque are the values obtained with only the accessories essential to engine operation without the effect of items such as air cleaners or generators. The net values reflect the operation of the engine as installed in the vehicle with all of its accessories. The power to weight ratios were calculated using the combat weight. During the operational life of the vehicle, the stowage items and arrangement frequently varied resulting in weight and some performance changes. In such cases, the stowage specified when the vehicle was new or during its period of greatest use is shown. Some items may have been omitted because of security restrictions.

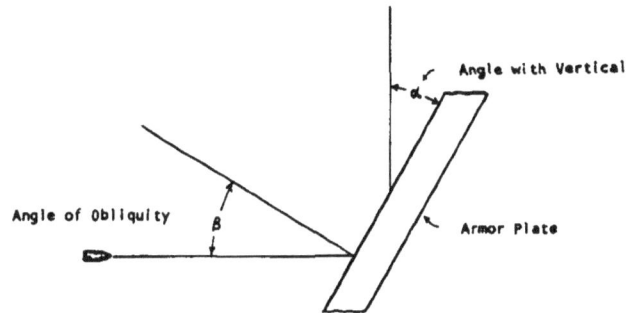

Security restrictions also limit information available on certain vehicles. This particularly applies to the use of composite special armor. On the early vehicles, the armor is specified by type, thickness, and angle with the vertical. This angle is measured between a vertical plane and the armor plate surface as indicated by the angle alpha in the sketch. In this two dimensional drawing, the angle beta is the angle of obliquity defined as the angle between a line perpendicular to the armor plate and the path of a projectile impacting the plate.

GENERAL DATA

Crew: (driver, bow gunner, commander, 24 infantry)	27	men
Length:	256.3	inches
Width:	119.6	inches
Height: Over cupola	100.0	inches
Tread:	96.0	inches
Ground Clearance:	18.3	inches
Weight, Combat Loaded:	51,000	pounds
Weight, Unstowed:	41,000	pounds
Power to Weight Ratio: Net	15.7	hp/ton
Gross	18.0	hp/ton
Ground Pressure: Zero penetration	8.1	psi

ARMOR

Type: Hull, rolled and cast homogeneous steel; Welded assembly

Hull Thickness:		Actual	Angle w/Vertical
Front,	Upper	0.375 inches (10mm)	60 degrees
	Middle	0.625 inches (16mm)	10 degrees
	Lower	0.50 inches (13mm)	0 to 68 degrees
Sides,	Upper	0.50 inches (13mm)	0 degrees
	Lower	0.375 inches (10mm)	0 degrees
Rear,	Upper	0.50 inches (13mm)	0 degrees
	Lower	0.375 inches (10mm)	53 degrees
Top,	Front	0.375 inches (10mm)	82 degrees
	Center	0.375 inches (10mm)	90 degrees
	Sides	0.375 inches (10mm)	70 degrees
Sponson Floor		0.375-0.25 inches (10-6mm)	90 degrees
Floor		0.3125 inches (8mm)	90 degrees

ARMAMENT

(1) .50 caliber MG HB M2 flexible in T107 mount on roof
(1) .30 caliber MG M1919A4 in bow mount
Provision for (2) .45 caliber SMG M3 or .30 caliber Carbine M1
Provision for (25) .30 caliber Rifle M1
(1) 2.36 inch Rocket Launcher M9

AMMUNITION

550 rounds .50 caliber
420 rounds .45 caliber or 480 rounds .30 caliber (carbine)
1000 rounds .30 caliber (machine gun)
10 rockets 2.36 inch HEAT
4 smoke pots
12 hand grenades

VISION EQUIPMENT

Vision Devices:	Direct	Indirect
Driver	Hatch	Periscope T24 (3)
Bow Gunner	Hatch	Periscope T24 (3)
Commander	Hatch	Periscope T24 (6)
Troop Compartment	Hatch and pistol ports (10)	Periscope T24 (2)

Total Periscopes: T24 (14)
Total Pistol Ports: (10)

ENGINE

Make and Model: Continental R-975-D4	
Type: 9 cylinder, 4 cycle, radial	
Cooling System: Air Ignition: Magneto	
Displacement:	973 cubic inches
Bore and Stroke:	5 x 5.5 inches
Compression Ratio:	5.7:1
Net Horsepower: (max)	400 hp at 2400 rpm
Gross Horsepower: (max)	460 hp at 2400 rpm
Net Torque: (max)	940 ft-lb at 1700 rpm
Gross Torque: (max)	1025 ft-lb at 1800 rpm
Weight:	1212 pounds, dry
Fuel: 80 octane gasoline	250 gallons
Engine Oil:	36 quarts

POWER TRAIN

Bevel Gear Ratio: 1.29:1
Transmission: Torqmatic, 3 speeds forward, 1 reverse

Gear Ratios:	1st	1.000:1	3rd	0.244:1
	2nd	0.421:1	reverse	0.756:1

Torque Converter: Detroit Transmission 900AD
  Ratio: (max) 4.7:1
Steering: Controlled differential, steering brake levers
  Bevel Gear Ratio: 3.133:1   Steering Ratio: 1.6:1
Brakes: Mechanical, external contracting
Final Drive: Spur gear   Gear Ratio: 3.82:1
Drive Sprocket: At front of vehicle with 14 teeth
  Pitch Diameter: 26.52 inches

RUNNING GEAR

Suspension: Torsion bar
  12 individually sprung dual road wheels (6/track)
  Tire Size: 25.5 x 4.5 inches
  8 dual track return rollers (4/track)
  Dual adjustable idler at rear of each track
  Idler Size: 25.7 inches, steel, no tire
  Shock absorbers on first two and last two road wheels on each side
Tracks: Center guide
  Type: Single pin, 21 inch width, cast steel chevron
  Pitch: 6.0 inches
  Shoes per Vehicle: 166 (83/track)
  Ground Contact Length: 149.1 inches

ELECTRICAL SYSTEM

Nominal Voltage: 24 volts DC
Main Generator: (1) 24 volts, 50 amperes, driven by main engine
Auxiliary Generator: (1) 24 volts, 50 amperes, driven by auxiliary engine
Battery: (2) 12 volts in series

COMMUNICATIONS

Radio: SCR 506 and SCR 508; SCR 508 and AN/VRC-3; SCR 506, SCR 510, and RC 99; SCR 508, SCR 506, AN/URC-1, and AN/VRC-3; or SCR 506, SCR 510, AN/VRC-3, and RC99

FIRE AND GAS PROTECTION

(3) 10 pound carbon dioxide, fixed
(2) 4 pound carbon dioxide, portable
(1) 1 1/2 quart decontaminating apparatus M2

PERFORMANCE

Maximum Speed: Level road	32 miles/hour
Maximum Tractive Effort: TE at stall	49,000 pounds
Per Cent of Vehicle Weight: TE/W	96 per cent
Maximum Grade:	60 per cent
Maximum Trench:	7 feet
Maximum Vertical Wall:	24 inches
Maximum Fording Depth:	40 inches
Minimum Turning Circle: (diameter)	44 feet
Cruising Range: Roads	approx. 180 miles

GENERAL DATA

Crew: (driver, commander, 10 infantry)	12	men
Length:	204.5	inches
Width: Over sandshields	112.0	inches
Height: Over cupola	108.5	inches
Tread:	97.0	inches
Ground Clearance:	18.0	inches
Weight, Combat Loaded:	41,500	pounds
Weight, Unstowed:	36,669	pounds
Power to Weight Ratio: Net	14.3	hp/ton
Gross	18.1	hp/ton
Ground Pressure: Zero penetration	8.5	psi

ARMOR

Type: Hull, rolled and cast homogeneous steel; Welded assembly

Hull Thickness:	Actual	Angle w/Vertical
Front, Upper	0.50 inches (13mm)	73 degrees
Middle	0.625 inches (16mm)	33 degrees
Lower	0.625 inches (16mm)	27 degrees
Sides	0.625 inches (16mm)	0 degrees
Rear	0.625 inches (16mm)	0 degrees
Top, Early	0.375 inches (10mm)	90 degrees
Late	0.50 inches (13mm)	90 degrees
Sponson Floor	0.50 inches (13mm)	90 degrees
Floor, Early	1.25 inches (32mm)	90 degrees
Late	1.00 inches (25mm)	90 degrees

ARMAMENT

(1) .50 caliber MG HB M2 flexible on commander's cupola
Provision for (1) .45 caliber SMG M3A1
(1) 3.5 inch Rocket Launcher M20

AMMUNITION

1800 rounds .50 caliber
180 rounds .45 caliber
10 rockets 3.5 inch M35

VISION EQUIPMENT

Vision Devices:	Direct	Indirect
Driver	Hatch	Periscope M17 (4)
		Periscope M19 (infrared) (1)
		late vehicles only
Commander	Hatch and	None
	6 vision blocks	
Troop Compartment	Roof Hatch	

Total Periscopes: M17 (4), M19 (infrared) (1)
Total Vision Blocks: (6) in cupola

ENGINE

Make and Model: Continental AO-895-4	
Type: 6 cylinder, 4 cycle, opposed	
Cooling System: Air Ignition: Magneto	
Displacement:	895.9 cubic inches
Bore and Stroke:	5.75 x 5.75 inches
Compression Ratio:	6.5:1
Net Horsepower: (max)	295 hp at 2800 rpm
Gross Horsepower: (max)	375 hp at 2800 rpm
Net Torque: (max)	640 ft-lb at 1800 rpm
Gross Torque: (max)	775 ft-lb at 2100 rpm
Weight:	1765 pounds (dry)
Fuel: 80 octane gasoline	150 gallons
Engine Oil:	44 quarts

POWER TRAIN

Transmission: Cross-drive CD-500-4, 2 ranges forward, 1 reverse
    Single stage hydraulic torque converter
    Stall multiplication: 4:1

Overall Usable Ratios:	low 15.6:1	reverse 15.6:1
	high 4.1:1	

Steering Control: Mechanical, T-bar
    Steering Rate: 6.5 rpm
Brakes: Multiple disc
Final Drive: Spur gear   Gear Ratio: 4.1:1
Drive Sprocket: At front of vehicle with 12 teeth
    Pitch Diameter: 23.422 inches

RUNNING GEAR

Suspension: Torsion bar
    10 individually sprung dual road wheels (5/track)
    Tire Size: 25.5 x 4.5 inches
    6 dual track return rollers (3/track)
    Dual compensating idler at rear of each track
    Idler Size: 22.5 x 4.5 inches, steel, no tire
    Shock absorbers on 1st 2 and last 2 road wheels per side (early vehicles)
        on 1st and last road wheels per side (late vehicles)
Tracks: Center guide, T91E3
    Type: Single pin, 21 inch width, steel w/detachable rubber pad
    Pitch: 6 inches
    Shoes per Vehicle: 140 (70/track)
    Ground Contact Length: 115.75 inches

ELECTRICAL SYSTEM

Nominal Voltage; 24 volts DC
Main Generator: (1) 24 volts, 150 amperes, driven by main engine
Auxiliary Generator: 28 volts, 300 amperes, early vehicles only, later removed
Battery: (2) 12 volts in series

COMMUNICATIONS

Radio: AN/GRC-3 thru 8, AN/VRC-13 thru 15, AN/VRC-7, or AN/VRQ-1 thru 3
Interphone: AN/UIC-1, 3 stations

FIRE PROTECTION

(2) 10 pound carbon dioxide, fixed
(1) 5 pound carbon dioxide, portable

PERFORMANCE

Maximum Speed: Level road		44 miles/hour
Maximum Tractive Effort: TE at stall		35,300 pounds
Per cent of Vehicle Weight: TE/W		85 per cent
Maximum Grade:		60 per cent
Maximum Trench:		5.5 feet
Maximum Vertical Wall:		24 inches
Maximum Fording Depth: w/o kit		48 inches
w/kit		80 inches
Minimum Turning Circle: (diameter)		pivot
Cruising Range : Roads	approx.	115 miles

GENERAL DATA

Crew: (driver, commander, 10 infantry)	12	men
Length:	212	inches
Width:	108	inches
Height: Over cupola	90	inches
Tread:	87	inches
Ground Clearance:	18	inches
Weight, Combat Loaded:	40,500	pounds
Weight, Unstowed:	36,400	pounds
Power to Weight Ratio: Net	14.6	hp/ton
Gross	18.5	hp/ton
Ground Pressure: Zero penetration	8.0	psi

ARMOR
Type: Hull, rolled and cast homogeneous steel; Welded assembly

Hull Thickness:		Actual	Angle w/Vertical
Front,	Upper	0.625 inches (16mm)	33 degrees
Driver's Cupola		0.625 inches (16mm)	46 degrees
	Lower	0.625 inches (16mm)	0 to 27 degrees
	Sides	0.625 inches (16mm)	0 degrees
	Rear	0.625 inches (16mm)	0 degrees
Top,	Front	0.5 inches (13 mm)	81 degrees
	Center	0.5 inches (13mm)	90 degrees
	Sides	0.5 inches (13mm)	80 degrees
Sponson Floor		0.375 inches (10mm)	90 degrees
Floor		1.0 inches (25mm)	90 degrees

ARMAMENT
(1) .50 caliber MG HB M2 flexible on commander's cupola
Provision for (1) .45 caliber SMG M3A1
(1) 3.5 inch Rocket Launcher M20

AMMUNITION
1800 rounds .50 caliber
180 rounds .45 caliber
10 rockets 3.5 inch M35

VISION EQUIPMENT

Vision Devices:	Direct	Indirect
Driver	Hatch	Periscope M17 (4)
Commander	Vision blocks (6) in cupola, hatch	None
Troop compartment	Hatch and pistol port (1)	None

Total Periscopes: M17 (4)
Total Vision Blocks: (6) in cupola
Total Pistol Ports: (1)

ENGINE
Make and Model: Continental AO-895-4
Type: 6 cylinder, 4 cycle, opposed
Cooling System: Air   Ignition: Magneto

Displacement:	895.9 cubic inches
Bore and Stroke:	5.75 x 5.75 inches
Compression Ratio:	6.5:1
Net Horsepower: (max)	295 hp at 2800 rpm
Gross Horsepower: (max)	375 hp at 2800 rpm
Net Torque: (max)	640 ft-lb at 1800 rpm
Gross Torque: (max)	775 ft-lb at 2100 rpm
Weight:	1765 pounds, dry
Fuel: 80 octane gasoline	150 gallons
Engine Oil:	44 quarts

POWER TRAIN
Transmission: Cross-drive XT-500-3, 3 ranges forward, 1 reverse
    Single stage hydraulic torque converter
    Stall Multiplication: 4:1

Overall Usable Ratios:	low	5.57:1	high	1.27:1
	intermediate	2.56:1	reverse	6.43:1

Steering Control: Mechanical, steering brake levers
Brakes: Multiple disc
Final Drive: Spur gear   Gear Ratio: 4:1
Drive Sprocket: At front of vehicle with 12 teeth
    Pitch Diameter: 28.8 inches

RUNNING GEAR
Suspension: Torsion bar
    10 individually sprung dual road wheels (5/track)
    Tire Size: 25.5 x 4.5 inches
    6 dual track return rollers (3/track)
    Dual compensating idler at rear of each track
    Idler Tire Size: 25.5 x 4.5 inches
    Shock absorbers on first and last road wheels on each side
Tracks: Center guide, T91E3
    Type: Single pin, 21 inch width, steel w/detachable rubber pad
    Pitch: 6 inches
    Shoes per Vehicle: 144 (72/track)
    Ground Contact Length: 120 inches

ELECTRICAL SYSTEM
Nominal Voltage: 24 volts DC
Main Generator: (1) 24 volts, 100 amperes, driven by main engine
Auxiliary Generator: None
Battery: (2) 12 volts in series

COMMUNICATIONS
Radio: AN/GRC-3 thru 8, AN/VRC-13 thru 15, AN/VRC-7, or AN/VRQ-1 thru 3
Interphone: AN/UIC-1, 2 stations

FIRE PROTECTION
(2) 10 pound carbon dioxide, fixed
(1) 5 pound carbon dioxide, portable

PERFORMANCE

Maximum Speed: Level road		38 miles/hour
Maximum Tractive Effort: TE at stall		32,200 pounds
Per Cent of Vehicle Weight: TE/W		80 per cent
Maximum Grade:		60 per cent
Maximum Trench:		6.5 feet
Maximum Vertical Wall:		18 inches
Maximum Fording Depth:		60 inches
Minimum Turning Circle: (diameter)		18 feet
Cruising Range: Roads	approx.	115 miles

GENERAL DATA

Crew: (driver, commander, 10 infantry)	12 men
Length: Overall	221.0 inches
Width: Over track shrouds	128.5 inches
Height: Vehicles F7-F1312 over MG	102.6 inches
Vehicles F1313-F2941 over MG	112.5 inches
Vehicles F2942 up, over M13 cupola	109.0 inches
Tread:	103.0 inches
Ground Clearance:	18.0 inches
Weight, Combat Loaded:	42,600 pounds
Weight, Unstowed:	39,500 pounds
Power to Weight Ratio: Net	11.9 hp/ton
	13.7 hp/ton
Ground Pressure: Zero penetration	8.4 psi

ARMOR

Type: Hull, rolled and cast homogeneous steel; Welded assembly

Hull Thickness:		Actual	Angle w/Vertical
Front,	Upper	0.625 inches (16mm)	50 degrees
	Middle	0.625 inches (16mm)	0 degrees
	Lower	0.625 inches (16mm)	55 degrees
Sides		0.625 inches (16mm)	0 degrees
Rear		0.625 inches (16mm)	7 degrees
Top		0.375 inches (10mm)	90 degrees
Floor		1.0 inches (25mm)	90 degrees

ARMAMENT

(1) .50 caliber MG HB M2 on commander's cupola, vehicles F7-F2941
(1) .50 caliber MG HB TT in M13 cupola, vehicles F2942 up
Provision for (1) .45 caliber SMG M3A1

AMMUNITION

1470 rounds .50 caliber, vehicles F7-F2941
2205 rounds .50 caliber, vehicles F2942 up
180 rounds .45 caliber

VISION EQUIPMENT

Vision Devices:	Direct	Indirect
Driver	Hatch	Periscope M17 (3)
		Periscope M19 (infrared) (1)
Commander, F7-F1312	Vision blocks (6) in cupola, hatch	None
F1313-2941	Hatch	Periscope M17 (4)
F2942 up	Vision blocks (4) in cupola, hatch	Periscope sight M28 (1)
Troop Compartment	Roof hatch (2)	None

Total Periscopes: F7-F1312, M17 (3), M19 (infrared (1)
F1313-F2941, M17 (7), M19 (infrared) (1)
F2942 up, M17 (3), M19 (infrared) (1), M28 (1)
Total Vision Blocks: F7-F1312, (6) in cupola
F1313-F2941, None
F2942 up, (4) in cupola

ENGINE

Make and Model: GMC Model 302 (two)	
Type: 6 cylinder, 4 cycle, in-line	
Cooling System: Liquid	Ignition: Delco
Displacement: (each)	301.6 cubic inches
Bore and Stroke:	4.0 x 4.0 inches
Compression Ratio:	7.2:1
Net Horsepower: (each max)	127 hp at 3350 rpm
Gross Horsepower: (each max)	146 hp at 3600 rpm
Net Torque: (each max)	254 ft-lb at 1800 rpm
Gross Torque: (each max)	265 ft-lb at 1800 rpm
Weight: (each)	630 pounds, dry
Fuel: 80 octane gasoline	135 gallons
Engine Oil: (each)	11 quarts

POWER TRAIN

Transmission: Hydramatic (two), 4 speeds forward, 1 reverse
F7-F590 Model 300MG, F591 up Model 301MG

Overall Usable Ratios:	1st	4.09:1	4th	1:1
	2nd	2.63:1	reverse	4.54:1
	3rd	1.55:1		

Right Angle Drive: (two)   Gear Ratio: 1.9:1
Steering: Controlled differential, steering brake levers
High Range: 1:1.092   Low Range: 2.359:1
Final Drive: Spur gear   Gear Ratio: 4.25:1
Drive Sprocket: At front of vehicle with 12 teeth
Pitch Diameter: 23.422 inches

RUNNING GEAR

Suspension: Torsion bar
10 individually sprung dual road wheels (5/track)
Tire Size: 25.5 x 4.5 inches
6 dual track return rollers (3/track)
Dual compensating idler at rear of each track
Idler Tire Size: 25.5 x 4.5 inches
Shock absorbers on first and last road wheels on each side
Tracks: Center guide, T91E3
Type: Single pin, 21 inch width, steel w/detachable rubber pads
Pitch: 6 inches
Shoes per Vehicle: 144 (72/track)
Ground Contact Length: 121.25 inches

ELECTRICAL SYSTEM

Nominal Voltage: 24 volts DC
Main Generator: (1) 28 volts, 100 amperes, driven by main engine
Auxiliary Generator: None
Battery: (2) 12 volts in series

COMMUNICATIONS

Radio: AN/GRC-3 thru 8 or AN/VRC-13 thru 22
Interphone: AN/UIC-1, 2 stations

FIRE PROTECTION

(2) 10 pound carbon dioxide, fixed
(1) 5 pound carbon dioxide, portable

PERFORMANCE

Maximum Speed: Level road		32 miles/hour
Water		4.3 miles/hour
Maximum Tractive Effort: TE at stall		30,000 pounds
Per Cent of Vehicle Weight: TE/W		70 per cent
Maximum Grade:		60 per cent
Maximum Trench:		5.5 feet
Maximum Vertical Wall:		18 inches
Maximum Fording Depth:		floats
Minimum Turning Circle: (diameter)		30 feet
Cruising Range: Roads	approx.	120 miles

GENERAL DATA

Crew:	6	men
Length: Overall	221.0	inches
Width: Over track shrouds	128.5	inches
Height: Over M13 cupola	109.0	inches
Tread:	103.0	inches
Ground Clearance:	18.0	inches
Weight, Combat Loaded:	47,100	pounds
Weight, Unstowed:	41,122	pounds
Power to Weight Ratio: Net	10.8	hp/ton
Gross	12.4	hp/ton
Ground Pressure: Zero penetration	9.2	psi

ARMOR

Type: Hull, rolled and cast homogeneous steel, Welded assembly

Hull Thickness:		Actual		Angle w/Vertical	
Front, Upper	0.625	inches (16mm)	50	degrees
Middle	0.625	inches (16mm)	0	degrees
Lower	0.625	inches (16mm)	55	degrees
Sides	0.625	inches (16mm)	0	degrees
Rear	0.625	inches (16mm)	7	degrees
Top	0.375	inches (10mm)	90	degrees
Floor	1.0	inches (25mm)	90	degrees

ARMAMENT

Primary: 4.2 inch mortar M30 on mount 8731738 in rear hull
　　　　M24A1 ground mount stowed on rear hull

Traverse: Manual	50 degrees (25 degrees to left or right)	
Elevation: Manual	+45 to +59 degrees	
Loading System:	Manual	

Secondary:
　(1) .50 caliber MG HB TT in M13 cupola
　(1) .30 caliber MG M1919A4
　Provision for (1) .45 caliber SMG M3A1
　Provision for (2) .30 caliber Rifles M1
　Provision for (1) .30 caliber Carbine M2
　(1) 3.5 inch Rocket Launcher M20

AMMUNITION
　88 rounds 4.2 inch
　2205 rounds .50 caliber
　3000 rounds .30 caliber (machine gun)
　180 rounds .45 caliber
　64 rounds .30 caliber (rifle)
　360 rounds .30 caliber (carbine)

FIRE CONTROL AND VISION EQUIPMENT

Primary Weapon:	Indirect	
	Sight Unit M34A2	
	Gunner's Quadrant M1A1	
	Fuze Setter M27	

Vision Devices:	Direct	Indirect
Driver	Hatch	Periscope M17 (3)
		Periscope M19 (infrared) (1)
Commander	Vision blocks (4)	Periscope sight M28 (1)
	in cupola, hatch	
Mortar Crew	Roof hatch	None

Total Periscopes: M17 (3), M19 (infrared) (1), M28 (1)
Total Vision Blocks: (4) in M19 cupola

ENGINE

Make and Model: GMC Model 302 (two)	
Type: 6 cylinder, 4 cycle, in-line	
Cooling System: Liquid	Ignition: Delco
Displacement:	301.6 cubic inches
Bore and Stroke:	4.0 x 4.0 inches
Compression Ratio:	7.2:1
Net Horsepower: (each max)	127 hp at 3350 rpm
Gross Horsepower: (each max)	146 hp at 3600 rpm
Net Torque: (each max)	254 ft-lb at 1800 rpm
Gross Torque: (each max)	265 ft-lb at 1800 rpm
Weight: (each)	630 pounds, dry
Fuel: 80 octane gasoline	135 gallons
Engine Oil: (each)	11 quarts

POWER TRAIN

Transmission: Hydramatic Model 301MG (two), 4 speeds forward, 1 reverse

Overall Usable Ratios:	1st	4.09:1	4th	1:1
	2nd	2.63:1	reverse	4.54:1
	3rd	1.55:1		

Right Angle Drive: (two)	Gear Ratio: 1.9:1	
Steering: Controlled differential, steering brake levers		
High Range: 1:1.092	Low Range: 2.359:1	
Final Drive: Spur gear	Gear Ratio: 4.25:1	
Drive Sprocket: At front of vehicle with 12 teeth		
Pitch Diameter: 23.422 inches		

RUNNING GEAR

Suspension: Torsion bar
　10 individually sprung dual road wheels (5/track)
　Tire Size: 25.5 x 4.5 inches
　6 dual track return rollers (3/track)
　Dual compensating idler at rear of each track
　Idler Tire Size: 25.5 x 4.5 inches
　Shock absorbers on first and last road wheels on each side
Tracks: Center guide, T91E3
　Type: Single pin, 21 inch width, steel w/detachable rubber pads
　Pitch: 6 inches
　Shoes per Vehicle: 144 (72/track)
　Ground Contact Length: 121.25 inches

ELECTRICAL SYSTEM

Nominal Voltage: 24 volts DC
Main Generator: (1) 28 volts, 100 amperes, driven by main engine
Auxiliary Generator: None
Battery: (2) 12 volts in series

COMMUNICATIONS

Radio: AN/GRC-3 thru 8 or AN/VRC-13 thru 22
Interphone: AN/UIC-1, 2 stations

FIRE PROTECTION

　(2) 10 pound carbon dioxide, fixed
　(1) 5 pound carbon dioxide, portable

PERFORMANCE

Maximum Speed: Level road		27 miles/hour
Water		3.5 miles/hour
Maximum Tractive Effort: TE at stall		30,000 pounds
Per Cent of Vehicle Weight: TE/W		64 per cent
Maximum Grade:		60 per cent
Maximum Trench:		5.5 feet
Maximum Vertical Wall:		18 inches
Maximum Fording Depth:		floats
Minimum Turning Circle: (diameter)		30 feet
Cruising Range: Roads	approx.	100 miles

GENERAL DATA

Crew: (driver, commander, 11 infantry)	13	men
Length: Overall	191.5	inches
Width: Over track shrouds	105.75	inches
Height: Over MG	98.25	inches
Tread:	85.0	inches
Ground Clearance:	16.1	inches
Weight, Combat Loaded:	22,900	pounds
Weight, Unstowed:	20,160	pounds
Power to Weight Ratio: Gross hp	18.8	hp/ton
Ground Pressure: Zero penetration:	7.3	psi

ARMOR

Type: Hull, rolled 5083/5086 H32 aluminum armor,
   Welded assembly

Hull Thickness:	Actual	Angle w/Vertical
Front, Upper	1.50 inches (38mm)	45 degrees
Lower	1.50 inches (38mm)	30 degrees
Sides, Upper	1.75 inches (44mm)	0 degrees
Lower	1.25 inches (32mm)	0 degrees
Rear, On Ramp	1.50 inches (38mm)	8 degrees
Off Ramp	1.50 inches (38mm)	9 degrees
Top	1.50 inches (38mm)	90 degrees
Floor	1.125 inches (29mm)	90 degrees

ARMAMENT

(1) .50 caliber MG HB M2 on commander's cupola

AMMUNITION

2000 rounds .50 caliber

VISION EQUIPMENT

Vision Devices	Direct	Indirect
Driver	Hatch	Periscope M17 (4)
		Periscope M19 (infrared) (1)
Commander	Hatch	Periscope M17 (5)
Troop Compartment	Roof Hatch	None

Total Periscopes: M17 (9), M19 (infrared) (1)

ENGINE

Make and Model: Chrysler 75M
Type: 8 cylinder, 4 cycle, vee

Cooling System: Liquid	Ignition: Battery
Displacement:	360.8 cubic inches
Bore and Stroke:	4.125 x 3.375 inches
Compression Ratio:	7.8:1
Gross Horsepower: (max)	215 hp at 4000 rpm
Gross Torque: (max)	332 ft-lb at 2800 rpm
Weight:	688 pounds, dry
Fuel: gasoline, MIL G-3056B	80 gallons
Engine Oil:	10 quarts

POWER TRAIN

Transfer Case: Gear Ratio 1.00:1
Transmission: Allison TX-200-2A
   Torque Converter: Hydraulic, single stage, multiphase w/lockup
   Torque Converter Stall Ratio: 3.0:1

Transmission Ratios:	1st	5.296:1	5th	1.39:1
	2nd	3.81:1	6th	1.00:1
	3rd	2.69:1	reverse	6.042:1
	4th	1.936:1		

Steering: DS200 controlled differential, steering brake levers
   Input Ratio: 1.28:1   Steering Ratio: 1.1 to 1.786:1
Brakes: Differential band
Final Drive: Spur gear   Gear Ratio: 3.928:1
Drive Sprockets: At front of vehicle with 10 teeth
   Pitch Diameter: 19.618 inches

RUNNING GEAR

Suspension: Flat track, torsion bar
   10 individually sprung dual road wheels (5/track)
   Tire Size: 24 x 2.1 inches
   Dual adjustable idler at rear of each track
   Idler Size: 21 x 2.1 inches
   Shock absorbers on first and last road wheels on each side
Tracks: Center guide, T130E1
   Type: Single pin, 15 inch width, steel w/detachable rubber pad
   Pitch: 6 inches
   Shoes per Vehicle: 127 (63 left, 64 right)
   Ground Contact Length: 105 inches

ELECTRICAL SYSTEM

Nominal Voltage: 24 volts DC
Main Generator: (1) 28 volts, 100 amperes, driven by main engine
Auxiliary Generator: None
Battery: (2) 12 volts in series

COMMUNICATIONS

Radio: AN/GRC-3 thru 8, AN/VRC-24, AN/GRC-19, AN/VRQ-1 thru 3,
   AN/PRC-8 thru 10, AN/GRR-5 or AN/VRC-12
Interphone: AN/UIC-1, 2 stations

FIRE PROTECTION

(1) 5 pound carbon dioxide, fixed
(1) 5 pound carbon dioxide, portable

PERFORMANCE

Maximum Speed: Level road		40 miles/hour
Water		3.5 miles/hour
Maximum Tractive Effort: TE at stall		18,000 pounds
Per Cent of Vehicle Weight: TE/W		79 per cent
Maximum Grade:		60 per cent
Maximum Trench:		5.5 feet
Maximum Vertical Wall:		24 inches
Maximum Fording Depth:		floats
Minimum Turning Circle: (diameter)		26 feet
Cruising Range: Roads	approx.	200 miles

GENERAL DATA
Crew: (driver, commander, 11 infantry)   13 men
Length: Overall   191.5 inches
Width: Over track shrouds   105.75 inches
Height: Over MG   98.25 inches
Tread: 85.0 inches
Ground Clearance: 16.1 inches
Weight, Combat Loaded: 24,080 pounds
Weight, Unstowed: 20,870 pounds
Weight, Air Drop: 18,860 pounds
Power to Weight Ratio: 17.6 hp/ton
Ground Pressure: Zero penetration   7.6 psi

ARMOR
Type: Hull, rolled 5083/5086 H32 aluminum armor,
    Welded assembly

Hull Thickness:	Actual	Angle w/Vertical
Front, Upper	1.50 inches (38mm)	45 degrees
Lower	1.50 inches (38mm)	30 degrees
Sides, Upper	1.75 inches (44mm)	0 degrees
Lower	1.25 inches (32mm)	0 degrees
Rear, On Ramp	1.50 inches (38mm)	8 degrees
Off Ramp	1.50 inches (38mm)	9 degrees
Top	1.50 inches (38mm)	90 degrees
Floor	1.125 inches (29mm)	90 degrees

ARMAMENT
   (1) 50 caliber MG HB M2 on commander's cupola
AMMUNITION
   2000 rounds .50 caliber
VISION EQUIPMENT

Vision Devices	Direct	Indirect
Driver	Hatch	Periscope M17 (4)
		Periscope M19 (infrared) (1)
Commander	Hatch	Periscope M17 (5)
Troop Compartment	Roof Hatch	None

Total Periscopes: M17 (9), M19 (infrared) (1)

ENGINE
Make and Model: General Motors 6V53
Type: 6 cylinder, 2 cycle, vee
Cooling System: Liquid   Ignition: Compression
Displacement: 318 cubic inches
Bore and Stroke: 3.875 x 4.5 inches
Compression Ratio: 17.0:1
Gross Horsepower: (max)   212 hp at 2800 rpm
Gross Torque: (max)   492 ft-lbs at 1300 rpm
Weight: 1310 pounds, dry
Fuel: diesel oil MIL-VV-F-800   95 gallons
Engine Oil: 22 quarts

POWER TRAIN
Transfer Case: Overall ratio 1.286:1 overdrive
Transmission: Allison TX-100
   Torque Converter: Hydraulic, single stage, multiphase w/lockup
   Torque Converter Stall Ratio: 3.5:1

Transmission Ratios:	1st 3.81:1	3rd 1.00:1
	2nd 1.936:1	reverse 4.35:1

Steering: DS200 controlled differential, steering brake levers
   Input ratio: 1.28:1   Steering Ratio: 1.1 to 1.786:1
Brakes: Differential band
Final Drive: Spur gear   Gear Ratio: 3.928:1
Drive Sprockets: At front of vehicle with 10 teeth
   Pitch Diameter: 19.618 inches

RUNNING GEAR
Suspension: Flat track, torsion bar
   10 individually sprung dual road wheels (5/track)
   Tire Size: 24 x 2.1 inches
   Dual adjustable idler at rear of each track
   Idler Size: 21 x 2.1 inches
   Shock absorbers on first and last road wheels on each side
Tracks: Center guide T130E1
   Type: Single pin, 15 inch width, steel w/detachable rubber pad
   Pitch: 6 inches
   Shoes per Vehicle: 127 (63 left, 64 right)
   Ground Contact Length: 105 inches

ELECTRICAL SYSTEM
Nominal Voltage: 24 volts DC
Main Generator: (1) 28 volts, 100 amperes, driven by main engine
Auxiliary Generator: None
Battery: (2) 12 volts in series

COMMUNICATIONS
Radio: AN/GRC-3 thru 8, AN/VRC-24, AN/GRC-19, AN/VRQ-1 thru 3,
   AN/PRC-8 thru 10, AN/GRR-5 or AN/VRC-12
Interphone: AN/UIC-1, 2 stations

FIRE PROTECTION
   (1) 5 pound carbon dioxide, fixed
   (1) 5 pound carbon dioxide, portable

PERFORMANCE
Maximum Speed: Level road   40 miles/hour
     Water   3.6 mile/hour
Maximum Tractive Effort: TE at stall   19,800 pounds
   Per Cent of Vehicle Weight: TE/W   82 per cent
Maximum Grade: 60 per cent
Maximum Trench: 5.5 feet
Maximum Vertical Wall: 24 inches
Maximum Fording Depth: floats
Minimum Turning Circle: (diameter)   26 feet
Cruising Range: Roads   approx. 300 miles

GENERAL DATA

Crew: (driver, commander, 11 infantry)	13	men
Length: Overall w/o external fuel tanks	191.5	inches
Overall w/external fuel tanks	208.5	inches
Width: Over track shrouds	105.75	inches
Height: Over MG	99.25	inches
Tread:	85.0	inches
Ground Clearance:	17.1	inches
Weight, Combat Loaded: w/o external tanks	25,007	pounds
w/external tanks	25,880	pounds
Weight, Unstowed: w/o external tanks	21,887	pounds
w/external tanks	22,760	pounds
Power to Weight Ratio: w/o external tanks	17.0	hp/ton
w/external tanks	16.4	hp/ton
Ground Pressure: Zero penetration w/o ext tanks	7.9	psi
w/ext tanks	8.2	psi

ARMOR

Type: Hull, rolled 5083/5086 H32 aluminum armor,
    Welded assembly

Hull Thickness:	Actual	Angle w/Vertical
Front, Upper	1.50 inches (38mm)	45 degrees
Lower	1.50 inches (38mm)	30 degrees
Sides, Upper	1.75 inches (44mm)	0 degrees
Lower	1.25 inches (32mm)	0 degrees
Rear, On Ramp	1.50 inches (38mm)	8 degrees
Off Ramp	1.50 inches (38mm)	9 degrees
Top	1.50 inches (38mm)	90 degrees
Floor	1.125 inches (29mm)	90 degrees

ARMAMENT

(1) .50 caliber MG HB M2 on commander's cupola

AMMUNITION

2000 rounds .50 caliber

VISION EQUIPMENT

Vision Devices	Direct	Indirect
Driver	Hatch	Periscope M17 (4)
		Periscope M19 (infrared) (1)
Commander	Hatch	Periscope M17 (5)
Troop Compartment	Roof Hatch	None

Total Periscopes: M17 (9), M19 (infrared) (1)

ENGINE

Make and Model: General Motors 6V53
Type: 6 cylinder, 2 cycle, vee

Cooling System: Liquid	Ignition: Compression	
Displacement:	318 cubic inches	
Bore and Stroke:	3.875 x 4.5 inches	
Compression Ratio:	17.0:1	
Gross Horsepower: (max)	212 hp at 2800 rpm	
Gross Torque: (max)	492 ft-lbs at 1300 rpm	
Weight:	1310 pounds, dry	
Fuel: diesel oil MIL-VV-F-800	95 gallons	
Engine Oil:	22 quarts	

POWER TRAIN

Transfer Case: Overall ratio 1.286:1 overdrive
Transmission: Allison TX-100
    Torque Converter: Hydraulic, single stage, multiphase w/lockup
    Torque Converter Stall Ratio: 3.5:1

Transmission Ratios:	1st	3.81:1	3rd	1.00:1
	2nd	1.936:1	reverse	4.35:1

Steering: DS200 controlled differential, steering brake levers
    Input Ratio: 1.28:1   Steering Ratio: 1.1 to 1.786:1
Brakes: Differential band
Final Drive: Spur gear   Gear Ratio: 3.928:1
Drive Sprockets: At front of vehicle with 10 teeth
    Pitch Diameter: 19.618 inches

RUNNING GEAR

Suspension: Flat track, torsion bar
    10 individually sprung dual road wheels (5/track)
    Tire Size: 24 x 2.1 inches
    Dual adjustable idler at rear of each track
    Idler Size: 21 x 2.1 inches
    Shock absorbers on first 2 and last road wheels on each side
Tracks: Center guide, T130E1 and T150
    Type: T130E1, Single pin, 15 inch width, steel w/detachable rubber pad
        T150, Double pin, 14.81 inch width, steel w/chevron rubber pad
    Pitch: 6 inches
    Shoes per Vehicle: 127 (63 left, 64 right)
    Ground Contact Length: 105 inches

ELECTRICAL SYSTEM

Nominal Voltage: 24 volts DC
Main Generator: (1) 28 volts, 100 amperes, driven by main engine
Auxiliary Generator: None
Battery: (2) 12 volts in series

COMMUNICATIONS

Radio: AN/GRC-3 thru 8, AN/VRC-24, AN/GRC-19, AN/VRQ-1 thru 3,
    AN/PRC-8 thru10, AN/GRR-5 or AN/VRC-12
Interphone: AN/UIC-1, 2 stations

FIRE AND NBC PROTECTION

    (1) 5 pound carbon dioxide, fixed
    (1) 5 pound carbon dioxide, portable
    Gas particulate filter unit

PERFORMANCE

Maximum Speed: Level road		40 miles/hour
Water		3.6 miles/hour
Maximum Tractive Effort: TE at stall		19,800 pounds
Per Cent of Vehicle Weight: TE/W		79 per cent*
Maximum Grade:		60 per cent
Maximum Trench:		5.5 feet
Maximum Vertical wall:		24 inches
Maximum Fording Depth:		floats
Minimum Turning Circle: (diameter)		26 feet
Cruising Range: Roads	approx.	300 miles

* without external fuel tanks

GENERAL DATA

Crew: (driver, commander, 11 infantry)	13	men
Length: Overall	208.5	inches
Width: Over track shrouds w/o applique armor	105.75	inches
w/applique armor	120.0	inches
Height: Over MG w/o shield	99.25	inches
Over MG w/shield	100.43	inches
Tread:	85.0	inches
Ground Clearance:	17.1	inches
Weight, Combat Loaded: w/o applique armor	27,200	pounds
w/applique armor	31,000	pounds
Weight, Unstowed: w/o applique armor	23,880	pounds
Weight, Air Drop:	22,128	pounds
Power to Weight Ratio: w/o applique armor	20.2	hp/ton
w/applique armor	17.7	hp/ton
Ground Pressure: w/o applique armor	8.6	psi
w/applique armor	9.8	psi

ARMOR

Type: Basic hull structure, rolled 5083/5086 H32 aluminum armor,
Welded assembly

Hull Thickness:	Actual	Angle w/Vertical
Front, Upper	1.50 inches (38mm)	45 degrees
Lower	1.50 inches (38mm)	30 degrees
Sides, Upper	1.75 inches (44mm)	0 degrees
Lower	1.25 inches (32mm)	0 degrees
Rear, On Ramp	1.50 inches (38mm)	8 degrees
Off Ramp	1.50 inches (38mm)	9 degrees
Top	1.50 inches (38mm)	90 degrees
Floor	1.125 inches (29mm)	90 degrees

Steel armor incorporated in floor for blast protection.
Attachment points on hull for applique armor to provide
Protection against 14.5 mm ammunition.
Spall protection liners inside crew compartment.

ARMAMENT

(1) .50 caliber MG HB M2 on commander's cupola
Gun shield kit available for installation on cupola

AMMUNITION

2000 rounds .50 caliber

VISION EQUIPMENT

Vision Devices	Direct	Indirect
Driver	Hatch	Periscope M17 (4)
		Periscope M19 (infrared) (1)
Commander	Hatch	Periscope M17 (5)
Troop Compartment	Roof Hatch	None

Total Periscopes: M17 (9), M19 (infrared) (1)

ENGINE

Make and Model: General Motors 6V53T
Type: 6 cylinder, 2 cycle, vee, turbosupercharged

Cooling System: Liquid	Ignition: Compression
Displacement:	318 cubic inches
Bore and Stroke:	3.875 x 4.5 inches
Compression Ratio:	18.0:1
Gross Horsepower: (max)	275 hp at 2800 rpm
Gross Torque: (max)	627 ft-lbs at 1600 rpm
Weight:	1495 pounds. dry
Fuel: diesel oil MIL-VV-F-800	95 gallons
Engine Oil:	22 quarts

POWER TRAIN

Transmission: Allison X-200-4 or X-200-4A
Torque Converter: Hydraulic, single stage, multiphase w/lockup
Torque Converter Stall Ratio: 2.72:1 to 3.32:1

Transmission Ratios:	1st	4.16:1	4th	1.04:1
	2nd	2.34:1	reverse 1	6.62:1
	3rd	1.46:1	reverse 2	2.16:1

Steering: Hydrostatic, steering yoke

Steering Ratios:	1st	2.15:1	4th	1.20:1
	2nd	1.52:1	reverse 1	3.76:1
	3rd	1.29:1	reverse 2	1.47:1

Brakes: Multiple plate, oil cooled
Final Drive: Spur gear   Gear Ratio: 3.928:1
Drive Sprockets: At front of vehicle with 10 teeth
Pitch Diameter: 19.618 inches

RUNNING GEAR

Suspension: Flat track, torsion bar
10 individually sprung dual road wheels (5/track)
Tire Size: 24 x 2.1 inches
Dual adjustable idler at rear of each track
Idler Size: 21 x 2.1 inches
Shock absorbers on first 2 and last road wheels on each side
Tracks: Center guide T130E1 and T150
Type: T130E1, Single pin, 15 inch width, steel w/detachable rubber pad
T150, Double pin, 14.81 inch width, steel w/chevron rubber pad
Pitch: 6 inches
Shoes per Vehicle: 127 (63 left, 64 right)
Ground Contact Length: 105 inches

ELECTRICAL SYSTEM

Nominal Voltage: 24 volts DC
Main Generator: (alternator) 28 volts, 200 amperes driven by main engine
Auxiliary Generator: None
Battery: (4) 12 volts, 2 sets of 2 in series connected in parallel

COMMUNICATIONS

Radio: AN/VRC-12
Interphone: AN/UIC-1, 2 stations

FIRE AND NBC PROTECTION

(1) 5 pound carbon dioxide, fixed
(1) 5 pound carbon dioxide, portable
Gas particulate filter unit

PERFORMANCE

Maximum Speed: Level road		40 miles/hour
Water		3.6 miles/hour
Maximum Grade:		60 per cent
Maximum Trench:		5.5 feet
Maximum Vertical Wall		24 inches
Maximum Fording Depth:	Floats, but restricted to	40 inches
Minimum Turning Circle: (diameter		pivot
Cruising Range: Roads	approx.	300 miles

ARMORED SELF-PROPELLED 81mm MORTAR CARRIER M125A2
(Based upon the APC M113A2)

GENERAL DATA
Crew: 6 men
Length: Overall 191.5 inches
Width: Overall 105.75 inches
Height: Over MG 99.25 inches
Tread: 85.0 inches
Ground Clearance: 17.1 inches
Weight, Combat Loaded: 25,410 pounds
Weight, Unstowed: 23,970 pounds
Weight, Air Transport: 20,780 pounds
Power to Weight Ratio: 16.7 hp/ton
Ground Pressure: Zero penetration 8.1 psi
ARMOR
Type: Hull, rolled 5083/5086 H32 aluminum armor,
 Welded assembly

Hull Thickness:	Actual	Angle w/Vertical
Front, Upper	1.50 inches (38mm)	45 degrees
Lower	1.50 inches (38mm)	30 degrees
Sides, Upper	1.75 inches (44mm)	0 degrees
Lower	1.25 inches (32mm)	0 degrees
Rear, On Ramp	1.50 inches (38mm)	8 degrees
Off Ramp	1.50 inches (38mm)	9 degrees
Top	1.50 inches (38mm)	90 degrees
Floor	1.125 inches (29mm)	90 degrees

ARMAMENT
Primary: 81mm mortar M252
Secondary:
 (1) .50 caliber MG HB M2 on cupola
AMMUNITION
 114 rounds 81mm
 600 rounds .50 caliber
VISION EQUIPMENT

Vision Devices:	Direct	Indirect
Driver	Hatch:	Periscope M17 (4)
		Periscope M19 (infrared) (1)
Commander	Hatch	Periscope M17 (5)
Crew Compartment	Roof Hatch	None

Total Periscopes: M17 (9), M19 (infrared) (1)

ENGINE
Make and Model: General Motors 6V53
Type: 6 cylinder, 2 cycle, vee
Cooling System: Liquid   Ignition: Compression
Displacement: 318 cubic inches
Bore and Stroke: 3.875 x 4.5 inches
Compression Ratio: 17.0:1
Gross Horsepower: (max) 212 hp at 2800 rpm
Gross Torque: (max) 492 ft-lbs at 1300 rpm
Weight: 1310 pounds, dry
Fuel: diesel oil MIL-VV-F-800 95 gallons
Engine Oil: 22 quarts
POWER TRAIN
Transfer Case: Overall ratio 1.286:1 overdrive
Transmission: Allison TX-100
 Torque Converter: Hydraulic, single stage, multiphase w/lockup
 Torque Converter Stall Ratio: 3.5:1

Transmission Ratios:	1st	3.81:1	3rd	1.00:1
	2nd	1.936:1	reverse	4.35:1

Steering: DS200 controlled differential, steering brake levers
 Input Ratio: 1.28:1   Steering Ratio: 1.1 to 1.786:1
Brakes: Differential band
Final Drive: Spur gear   Gear Ratio: 3.928:1
Drive Sprockets: At front of vehicle with 10 teeth
 Pitch Diameter: 19.618 inches
RUNNING GEAR
Suspension: Flat track, torsion bar
 10 individually sprung dual road wheels (5/track)
 Tire Size: 24 x 2.1 inches
 Dual adjustable idler at rear of each track
 Idler Size: 21 x 2.1 inches
 Shock absorbers on first 2 and last road wheels on each side
Tracks: Center guide, T130E1 and T150
 Type: T130E1, Single pin, 15 inch width, steel w/detachable rubber pad
 T150, Double pin, 14.81 inch width, steel w/chevron rubber pad
 Pitch: 6 inches
 Shoes per vehicle; 127 (63 left, 64 right)
 Ground Contact Length: 105 inches
ELECTRICAL SYSTEM
Nominal Voltage: 24 volts DC
Main generator: (1) 28 volts, 100 amperes, driven by main engine
Auxiliary Generator: None
Battery: (2) 12 volts in series
COMMUNICATIONS
Radio: AN/GRC-3 thru 8, AN/VRC-24, AN/GRC-19, AN/VRQ-1 thru 3
 AN/PRC-8 thru 10, AN/GRR-5 or AN/VRC-12
Interphone: AN/UIC-1, 4 stations
FIRE AND NBC PROTECTION
 (1) 5 pound carbon dioxide, fixed
 (1) 5 pound carbon dioxide, portable
 Gas particulate filter unit
PERFORMANCE

Maximum Speed: Level road	40 mile/hour
Water	3.6 miles/hour
Maximum Tractive Effort: TE at stall	19,800 pounds
Per Cent of Vehicle weight: TE/W	78 per cent
Maximum Grade:	60 per cent
Maximum Trench:	5.5 feet
Maximum Vertical Wall:	24 inches
Maximum Fording Depth:	floats
Minimum Turning Circle: (diameter)	26 feet
Cruising Range: Roads approx.	300 miles

ARMORED SELF-PROPELLED 107mm MORTAR CARRIER M106A2
(Based upon the APC M113A2)

GENERAL DATA

Crew:	6	men
Length: Overall	194	inches
Width: Overall	112.75	inches
Height: Over MG	99.25	inches
Tread:	85.0	inches
Ground Clearance:	17.1	inches
Weight, Combat Loaded:	26,910	pounds
Weight, Unstowed:	24,604	pounds
Weight, Air Transport:	20,620	pounds
Power to Weight Ratio:	15.8	hp/ton
Ground Pressure: Zero penetration	8.5	psi

ARMOR

Type: Hull, rolled 5083/5086 H32 aluminum armor,
      Welded assembly

Hull Thickness:	Actual	Angle w/Vertical
Front, Upper	1.50 inches (38mm)	45 degrees
Lower	1.50 inches (38mm)	30 degrees
Sides, Upper	1.75 inches (44mm)	0 degrees
Lower	1.25 inches (32mm)	0 degrees
Rear, On Ramp	1.50 inches (38mm)	8 degrees
Off Ramp	1.50 inches (38mm)	9 degrees
Top	1.50 inches (38mm)	90 degrees
Floor	1.125 inches (29mm)	90 degrees

ARMAMENT

Primary: 107mm (4.2 inch) mortar M30
Secondary:
  (1) .50 caliber MG HB M2 on cupola

AMMUNITION

  88 rounds 107mm (4.2 inch)
  600 rounds .50 caliber

VISION EQUIPMENT

Vision Devices	Direct	Indirect
Driver	Hatch	Periscope M17 (4)
		Periscope M19 (infrared) (1)
Commander	Hatch	Periscope M17 (5)
Crew Compartment	Roof Hatch	None

Total Periscopes: M17 (9), M19 (infrared) (1)

ENGINE

Make and Model: General Motors 6V53
Type: 6 cylinder, 2 cycle, vee

Cooling System: Liquid	Ignition: Compression
Displacement:	318 cubic inches
Bore and Stroke:	3.875 x 4.5 inches
Compression Ratio:	17.0:1
Gross Horsepower: (max)	212 hp at 2800 rpm
Gross Torque: (max)	492 ft-lbs at 1300 rpm
Weight:	1310 pounds, dry
Fuel: diesel oil MIL-VV-F-800	95 gallons
Engine oil:	22 quarts

POWER TRAIN

Transfer Case: Overall ratio 1.286:1 overdrive
Transmission: Allison TX-100
    Torque Converter: Hydraulic, single stage, multiphase w/lockup
    Torque Converter Stall Ratio: 3.5:1

Transmission Ratios:	1st	3.81:1	3rd	1.00:1
	2nd	1.936:1	reverse	4.35:1

Steering: DS200 controlled differential, steering brake levers
    Input Ratio: 1.28:1   Steering Ratio: 1.1 to 1.786:1
Brakes: Differential band
Final Drive: Spur gear   Gear Ratio: 3.928:1
Drive Sprockets: At front of vehicle with 10 teeth
    Pitch Diameter: 19.618 inches

RUNNING GEAR

Suspension: Flat track, torsion bar
    10 individually sprung dual road wheels (5/track)
    Tire Size: 24 x 2.1 inches
    Dual adjustable idler at rear of each track
    Idler Size: 21 x 2.1 inches
    Shock absorbers on first 2 and last road wheels on each side
Tracks: Center guide, T130E1 and T150
    Type: T130E1, Single pin, 15 inch width, steel w/detachable rubber pad
        T150, Double pin, 14.81 inch width, steel w/chevron rubber pad
    Pitch: 6 inches
    Shoes per Vehicle: 127 (63 left, 64 right)
    Ground Contact Length: 105 inches

ELECTRICAL SYSTEM

Nominal Voltage: 24 volts DC
Main Generator: (1) 28 volts, 100 amperes, driven by main engine
Auxiliary Generator: None
Battery: (2) 12 volts in series

COMMUNICATIONS

Radio: AN/GRC-3 thru 8, AN/VRC-24, AN/GRC-19, AN/VRQ-1 thru 3,
    AN/PRC-8 thru 10, AN/GRR-5 or AN/VRC-12
Interphone: AN/UIC-1, 4 stations

FIRE AND NBC PROTECTION

    (1) 5 pound carbon dioxide, fixed
    (1) 5 pound carbon dioxide, portable
    Gas particulate filter unit

PERFORMANCE

Maximum Speed: Level road	40 miles/hour
Water	3.6 miles/hour
Maximum Tractive Effort: TE at stall	19,800 pounds
Per Cent of Vehicle Weight: TE/W	74 per cent
Maximum Grade:	60 per cent
Maximum Trench:	5.5 feet
Maximum Vertical Wall:	24 inches
Maximum Fording Depth:	floats
Minimum Turning Circle: (diameter)	26 feet
Cruising Range: Roads	approx. 300 miles

GENERAL DATA
Crew: 4 men
Length: Overall   209.4 inches
Width: Over track shrouds   105.75 inches
Height: Over MG   99.25 inches
Tread: 85.0 inches
Ground Clearance: 17.1 inches
Weight, Combat Loaded: 28,900 pounds
Weight, Unstowed: 23,360 pounds
Power to Weight Ratio: 19.0 hp/ton
Ground Pressure: Zero penetration   9.2 psi
ARMOR
Type: Hull, rolled 5083/5086 H32 aluminum armor,
   Welded assembly

Hull Thickness:	Actual	Angle w/Vertical
Front, Upper	1.50 inches (38mm)	45 degrees
Lower	1.50 inches (38mm)	30 degrees
Sides, Upper	1.75 inches (44mm)	0 degrees
Lower	1.25 inches (32mm)	0 degrees
Rear, On Ramp	1.50 inches (38mm)	8 degrees
Off Ramp	1.50 inches (38mm)	9 degrees
Top	1.50 inches (38mm)	90 degrees
Floor	1.125 inches (29mm)	90 degrees

ARMAMENT
Primary: 120mm mortar M121
Secondary:
   (1) .50 caliber MG HB M2 on cupola
AMMUNITION
   69 rounds 120mm
   600 rounds .50 caliber
VISION EQUIPMENT

Vision Devices	Direct	Indirect
Driver	Hatch	Periscope M17 (4)
		Periscope M19A1 (infrared) (1)
Commander	Hatch	Periscope M17 (5)
Crew Compartment	Roof Hatch	None

Total Periscopes: M17 (9), M19A1 (infrared) (1)

ENGINE
Make and Model: General Motors 6V53T
Type: 6 cylinder, 2 cycle, vee, turbosupercharged
Cooling System: Liquid   Ignition: Compression
Displacement: 318 cubic inches
Bore and Stroke: 3.875 x 4.5 inches
Compression Ratio: 18.0:1
Gross Horsepower: (max)   275 hp at 2800 rpm
Gross Torque: (max)   627 ft-lbs at 1600 rpm
Weight: 1495 pounds, dry
Fuel: diesel oil MIL VV-F-800   95 gallons
Engine Oil: 22 quarts
POWER TRAIN
Transmission: Allison X-200-4 or X-200-4A
   Torque Converter: Hydraulic, single stage, multiphase w/lockup
   Torque Converter Stall ratio: 2.72:1 to 3.32:1

Transmission Ratios:	1st	4.16:1	4th	1.04:1
	2nd	2.34:1	reverse 1	6.62:1
	3rd	1.46:1	reverse 2	2.16:1

Steering: Hydrostatic, steering yoke

Steering Ratios:	1st	2.15:1	4th	1.20:1
	2nd	1.52:1	reverse 1	3.76:1
	3rd	1.29:1	reverse 2	1.47:1

Brakes: Multiple plate, oil cooled
Final Drive: Spur gear   Gear Ratio: 3.928:1
Drive Sprockets: At front of vehicle with 10 teeth
   Pitch Diameter: 19.618 inches
RUNNING GEAR
Suspension: Flat track, torsion bar
   10 individually sprung dual road wheels (5/track)
   Tire Size: 24 x 2.1 inches
   Dual adjustable idler at rear of each track
   Idler Size: 21 x 2.1 inches
   Shock absorbers on first 2 and last road wheels on each side
Tracks: Center guide, T130E1 and T150
   Type: T130E1, Single pin, 15 inch width, steel w/detachable rubber pad
      T150, Double pin, 14.81 inch width, steel w/chevron rubber pad
   Pitch: 6 inches
   Shoes per Vehicle: 127 (63 left, 64 right)
   Ground Contact Length: 105 inches
ELECTRICAL SYSTEM
Nominal Voltage: 24 volts DC
Main Generator: (alternator) 28 volts, 200 amperes, driven by main engine
Auxiliary Generator: None
Battery: (4) 12 volts, 2 sets of 2 in series connected in parallel
COMMUNICATIONS
Radio: AN/VRC-12
Interphone: AN/UIC-1, 2 stations
FIRE AND NBC PROTECTION
   (1) 5 pound carbon dioxide, fixed
   (1) 5 pound carbon dioxide, portable
   Gas particulate filter unit
PERFORMANCE
Maximum Speed: Level road   40 miles/hour
              Water   3.6 miles/hour
Maximum Grade: 60 per cent
Maximum Trench: 5.5 feet
Maximum Vertical Wall: 24 inches
Maximum Fording Depth: Floats, but restricted to 40 inches
Minimum Turning Circle: (diameter)   pivot
Cruising Range: Roads   approx.   300 miles

FIRE SUPPORT TEAM VEHICLE (FIST-V) M981
(Based upon the APC M113A2)

GENERAL DATA

Crew:	4	men
Length: Overall	208.5	inches
Width: Over track shrouds	105.75	inches
Height: Head erected	134.25	inches
Head stowed	115.75	inches
Air transport	103.5	inches
Tread:	85.0	inches
Ground Clearance:	17.1	inches
Weight, Combat Loaded:	27,900	pounds
Weight, Unstowed:	26,200	pounds
Power to Weight ratio:	15.2	hp/ton
Ground Pressure: Zero penetration	8.9	psi

ARMOR

Type: Hull, rolled 5083/5086 H32 aluminum armor
   Welded assembly

Hull Thickness:	Actual		Angle w/Vertical
Front, Upper	1.50 inches (38mm)		45 degrees
Lower	1.50 inches (38mm)		30 degrees
Sides, Upper	1.75 inches (44mm)		0 degrees
Lower	1.25 inches (32mm)		0 degrees
Rear, On Ramp	1.50 inches (38mm)		8 degrees
Off Ramp	1.50 inches (38mm)		9 degrees
Top	1.50 inches (38mm)		90 degrees
Floor	1.125 inches (29mm)		90 degrees

ARMAMENT

(1) 7.62mm M60 MG on operator's cupola
(3) 5.56mm M16A1 rifles stowed
(2) 4 tube smoke grenade launchers

AMMUNITION

800 rounds 7.62mm
840 rounds 5.56mm
16 smoke grenades

FIRE DIRECTION AND VISION EQUIPMENT

Ground/vehicular laser locator designator (G/VLLD)
AN/TAS-4 night sight
Targeting station controls and displays

Vision Devices	Direct	Indirect
Driver	Hatch	Periscope M17 (4)
		Periscope M19 (infrared) (1)
Commander	None	Panoramic telescope
Operator	Hatch and	Tank periscope
	7 vision blocks	
Crew Compartment	vision block in	Periscope M26 (2)
	rear ramp	

Total Periscopes: M17 (4), M19 (infrared) (1), M26 (2)
Tank Periscope (1), Panoramic Telescope (1)

ENGINE

Make and Model: General Motors 6V53
Type: 6 cylinder, 2 cycle, vee

Cooling System: Liquid	Ignition: Compression
Displacement:	318 cubic inches
Bore and Stroke:	3.875 x 4.5 inches
Compression Ratio:	17.0:1
Gross Horsepower: (max)	212 hp at 2800 rpm
Gross Torque: (max)	492 ft-lbs at 1300 rpm
Weight:	1310 pounds, dry
Fuel: diesel oil MILVV-F-800	95 gallons
Engine Oil:	22 quarts

POWER TRAIN

Transfer Case: Overall ratio 1.286:1 overdrive
Transmission: Allison TX-100
   Torque Converter: Hydraulic, single stage, multiphase w/lockup
   Torque Converter Stall Ratio: 3.5:1

Transmission Ratios:	1st	3.81:1	3rd	1.00:1
	2nd	1.936:1	reverse	4.35:1

Steering: DS200 controlled differential, steering brake levers
   Input Ratio: 1.28:1   Steering Ratio: 1.1 to 1.786:1
Brakes: Differential band
Final Drive: Spur gear   Gear Ratio: 3.928:1
Drive Sprockets: At front of vehicle with 10 teeth
   Pitch Diameter: 19.618 inches

RUNNING GEAR

Suspension: Flat track, torsion bar
   10 individually sprung dual road wheels (5/track)
   Tire Size: 24 x 2.1 inches
   Dual adjustable idler at rear of each track
   Idler Size: 21 x 2.1 inches
   Shock absorbers on first 2 and last road wheels on each side
Tracks: Center guide, T130E1 and T150
   Type: T130E1, Single pin, 15 inch width, steel w/detachable rubber pad
      T150, Double pin, 14.81 inch width, steel w/chevron rubber pad
   Pitch: 6 inches
   Shoes per Vehicle: 127 (63 left, 64 right)
   Ground Contact Length: 105 inches

ELECTRICAL SYSTEM

Nominal Voltage: 24 volts DC
Main Generator: (1) 28 volts, 100 amperes, driven by main engine
Auxiliary Generator: None
Battery: (4) 12 volts, 2 sets of 2 in series connected in parallel

COMMUNICATIONS

Radio: AN/VRC-46, AN/GRC-160 (3), AN/PSG-5, AN/PSG-2A,
AN/GRA-39, TSEC/KY-57
Interphone: 4 stations

FIRE AND NBC PROTECTION

(1) 5 pound carbon dioxide, fixed
(1) 5 pound carbon dioxide, portable
Gas particulate filter unit

PERFORMANCE

Maximum Speed: Level road		40 miles/hour
Water		3.6 miles/hour
Maximum Tractive Effort: TE at stall		19,800 pounds
Per Cent of Vehicle Weight: TE/W		71 per cent
Maximum Grade:		60 per cent
Maximum Trench:		5.5 feet
Maximum Vertical Wall:		24 inches
Maximum Fording Depth:		floats
Minimum Turning Circle: (diameter)		26 feet
Cruising Range: Roads	approx.	300 miles

GENERAL DATA

Crew:	5	men
Length: Overall	191.5	inches
Width: Over track shrouds	105.75	inches
Height: Over antenna guards	106.5	inches
Tread:	85.0	
Ground Clearance:	17.1	inches
Weight, Combat Loaded:	25,450	pounds
Weight, Unstowed:	24,247	pounds
Weight, Air Transport:	23,120	pounds
Power to Weight Ratio:	16.7	hp/ton
Ground Pressure: Zero penetration	8.1	psi

ARMOR

Type: Hull, rolled 5083/5086 H32 aluminum armor
    Welded assembly

Hull Thickness:	Actual	Angle w/Vertical
Front, Upper	1.75 inches (44mm)	0 degrees
Middle	1.50 inches (38mm)	45 degrees
Lower	1.50 inches (38mm)	30 degrees
Sides, Upper	1.75 inches (44mm)	0 degrees
Lower	1.25 inches (32mm)	0 degrees
Rear, On Ramp	1.50 inches (38mm)	8 degrees
Off Ramp	1.50 inches (38mm)	9 degrees
Top	1.50 inches (38mm)	90 degrees
Floor	1.125 inches (29mm)	90 degrees

ARMAMENT
    (6) 7.62mm M14 rifles stowed

AMMUNITION
    720 rounds 7.62mm, rifle

VISION EQUIPMENT

Vision Devices	Direct	Indirect
Driver	Hatch	Periscope M17 (4)
		Periscope M19 (infrared) (1)
Crew Compartment	Roof Hatch	None

Total Periscopes: M17 (4), M19 (infrared) (1)

ENGINE

Make and Model: General Motors 6V53
Type: 6 cylinder, 2 cycle, vee

Cooling System: Liquid		Ignition: Compression
Displacement:		318 cubic inches
Bore and Stroke:		3.875 x 4.5 inches
Compression Ratio:		17.0:1
Gross Horsepower: (max)		212 hp at 2800 rpm
Gross Torque: (max)		492 ft-lbs at 1300 rpm
Weight:		1310 pounds, dry
Fuel: diesel oil MIL-VV-F-800		120 gallons
Engine Oil:		22 quarts

POWER TRAIN

Transfer Case: Overall ratio 1.286:1 overdrive
Transmission: Allison TX-100
    Torque Converter: Hydraulic, single stage, multiphase w/lockup
    Torque Converter Stall Ratio: 3.5:1

Transmission Ratios:	1st	3.81:1	3rd	1.00:1
	2nd	1.936:1	reverse	4.35:1

Steering: DS200 controlled differential, steering brake levers
    Input Ratio: 1.28:1   Steering Ratio: 1.1 to 1.786:1
Brakes: Differential band
Final Drive: Spur gear   Gear Ratio: 3.928:1
Drive Sprockets: At front of vehicle with 10 teeth
    Pitch Diameter: 19.618 inches

RUNNING GEAR

Suspension: Flat track, torsion bar
    10 individually sprung dual road wheels (5/track)
    Tire Size: 24 x 2.1 inches
    Dual adjustable idler at rear of each track
    Idler Size: 21 x 2.1 inches
    Shock absorbers on first 2 and last road wheels on each side
Tracks: Center guide T130E1 and T150
    Type: T130E1, Single pin, 15 inch width, steel w/detachable rubber pad
        T150, Double pin, 14.81 inch width, steel w/chevron rubber pad
    Pitch: 6 inches
    Shoes per Vehicle: 127 (63 left, 64 right)
    Ground Contact Length: 105 inches

ELECTRICAL SYSTEM

Nominal Voltage: 24 volts DC
Main Generator: (1) 28 volts, 100 amperes, driven by main engine
Auxiliary Generator: (1) 28 volts, 150 amperes, driven by auxiliary engine
Battery: (2) 12 volts in series

FIRE PROTECTION
    (1) 5 pound carbon dioxide, fixed
    (1) 5 pound carbon dioxide, portable

PERFORMANC

Maximum Speed: Level road		40 miles/hour
Water		3.6 miles/hour
Maximum Tractive Effort: TE at stall		19,800 pounds
Per Cent of Vehicle Weight: TE/W		78 per cent
Maximum Grade:		60 per cent
Maximum Trench:		5.5 feet
Maximum Vertical Wall:		24 inches
Maximum Fording Depth:		floats
Minimum Turning Circle: (diameter)		26 feet
Cruising Range: Roads	approx.	370 miles

STANDARD INTEGRATED COMMAND POST SYSTEM CARRIER M1068A3

GENERAL DATA

Crew:	4	men
Length: Overall	202	inches
Width: Over track shrouds	105.75	inches
Height: Over antenna guards	106.5	inches
Tread:	85.0	inches
Ground Clearance:	17.1	inches
Weight, Combat Loaded:	27,130	pounds
Weight: Unstowed:	25,508	pounds
Power to Weight Ratio:	20.3	hp/ton
Ground Pressure: Zero penetration	8.6	psi

ARMOR

Type: Hull, rolled 5083/5086 H32 aluminum armor,
    Welded assembly

Hull Thickness:		Actual		Angle w/Vertical	
Front, Upper		1.75 inches (44mm)		0 degrees
	Middle	1.50 inches (38mm)		45 degrees
	Lower	1.50 inches (38mm)		30 degrees
Sides, Upper		1.75 inches (44mm)		0 degrees
	Lower	1.25 inches (32mm)		0 degrees
Rear, On Ramp		1.50 inches (38mm)		8 degrees
	Off Ramp	1.50 inches (38mm)		9 degrees
Top		1.50 inches (38mm)		90 degrees
Floor		1.125 inches (29mm)		90 degrees

ARMAMENT
    (4) 5.56mm M16A1 rifles stowed

AMMUNITION
    600 rounds 5.56mm

VISION EQUIPMENT

Vision Devices	Direct	Indirect
Driver	Hatch	Periscope M17 (4)
		Periscope M19A1 (infrared) (1)
Crew Compartment	Hatch	None

Total Periscopes: M17 (4), M19A1 (infrared) (1)

ENGINE

Make and Model: General Motors 6V53T
Type: 6 cylinder, 2 cycle, vee, turbosupercharged

Cooling System: Liquid	Ignition: Compression	
Displacement:		318 cubic inches
Bore and Stroke:		3.875 x 4.5 inches
Compression Ratio:		18.0:1
Gross Horsepower: (max)		275 hp at 2800 rpm
Gross Torque: (max)		627 ft-lbs at 1600 rpm
Weight:		1495 pounds, dry
Fuel: diesel oil MIL-VV-F-800		120 gallons
Engine Oil:		22 quarts

POWER TRAIN

Transmission: Allison X-200-4 or X-200-4A
    Torque Converter: Hydraulic, single stage, multiphase w/lockup
    Torque Converter Stall ratio: 2.72:1 to 3.32:1

Transmission ratios:	1st	4.16:1	4th	1.04:1
	2nd	2.34:1	reverse 1	6.62:1
	3rd	1.46:1	reverse 2	2.16:1

Steering: Hydrostatic, steering yoke

Steering Ratios:	1st	2.15:1	4th	1.20:1
	2nd	1.52:1	reverse 1	3.76:1
	3rd	1.29:1	reverse 2	1.47:1

Brakes: Multiple plate, oil cooled
Final Drive: Spur gear   Gear Ratio: 3.928:1
Drive Sprockets: At front of vehicle with 10 teeth
    Pitch Diameter: 19.618 inches

RUNNING GEAR

Suspension: Flat track, torsion bar
    10 individually sprung dual road wheels (5/track)
    Tire Size: 24 x 2.1 inches
    Dual adjustable idler at rear of each track
    Idler Size: 21 x 2.1 inches
    Shock absorbers on first 2 and last road wheels on each side
Tracks: Center guide, T130E1 and T150
    Type: T130E1, Single pin, 15 inch width, steel w/detachable rubber pad
        T150, Double pin, 14.81 inch width, steel w/chevron rubber pad
    Pitch: 6 inches
    Shoes per Vehicle: 127 (63 left, 64 right)
    Ground Contact Length: 105 inches

ELECTRICAL SYSTEM

Nominal Voltage: 24 volts DC
Main Generator: (alternator) 28 volts, 200 amperes, driven by main engine
Auxiliary Generator: (1) 28 volts, 150 amperes, driven by auxiliary engine
Battery: (4) 12 volts, 2 sets of 2 in series connected in parallel

FIRE AND NBC PROTECTION
    (1) 5 pound carbon dioxide, fixed
    (1) 5 pound carbon dioxide, portable
    Gas particulate filter unit

PERFORMANCE

Maximum Speed: Level road		40 miles/hour
Water		3.6 miles/hour
Maximum Grade:		60 per cent
Maximum Trench:		5.5 feet
Maximum Vertical Wall:		24 inches
Maximum Fording Depth:	Floats, but restricted to 40 inches
Minimum Turning Circle: (diameter)		pivot
Cruising Range: Roads		approx. 370 miles

ANTITANK COMBAT VEHICLE, IMPROVED TOW VEHICLE M901A1
(Based upon the APC M113A2)

GENERAL DATA

Crew: Mechanized infantry		4	men
Armored cavalry		5	men
Length: Overall		191.5	inches
Width: Over track shrouds		105.75	inches
Height: Head erected		134.25	inches
Head stowed		115.75	inches
Air transport		103.5	inches
Tread:		85.0	inches
Ground Clearance:		17.1	inches
Weight, Combat Loaded:	approx.	26,000	pounds
Power to Weight Ratio:		16.3	hp/ton
Ground Pressure: Zero penetration		8.3	psi

ARMOR

Type: Hull, rolled 5083/5086 H32 aluminum armor
Welded assembly

Hull Thickness:	Actual	Angle w/Vertical
Front, Upper	1.50 inches (38mm)	45 degrees
Lower	1.50 inches (38mm)	30 degrees
Sides, Upper	1.75 inches (44mm)	0 degrees
Lower	1.25 inches (32mm)	0 degrees
Rear, On Ramp	1.50 inches (38mm)	8 degrees
Off Ramp	1.50 inches (38mm)	9 degrees
Top	1.50 inches (38mm)	90 degrees
Floor	1.125 inches (29mm)	90 degrees

ARMAMENT

Primary: TOW missile launcher
Secondary:
    (1) 7.62 mm M60 MG on gunner's cupola
    (2) 5.56 mm M16A1 rifles stowed
    (2) 4 tube smoke grenade launchers

AMMUNITION
    10 TOW2 missiles
    800 rounds 7.62mm
    560 rounds 5.56mm
    16 smoke grenades

FIRE DIRECTION AND VISION EQUIPMENT

Primary Weapon: TOW2 guidance system

Vision Devices	Direct	Indirect
Driver	Hatch	Periscope M17 (4)
		Periscope M19 (infrared) (1)
Commander	None	Panoramic telescope
Gunner	Hatch and	TOW sight assembly
	7 vision blocks	
Crew compartment	vision block in	Periscope M26 (2)
	rear ramp	

Total Periscopes: M17 (4), M19 (infrared) (1), M26 (2)
TOW sight assembly, panoramic telescope
Total Vision Blocks: (8)

ENGINE

Make and Model: General Motors 6V53
Type: 6 cylinder, 2 cycle, vee

Cooling System: Liquid	Ignition: Compression
Displacement:	318 cubic inches
Bore and stroke:	3.875 x 4.5 inches
Compression Ratio:	17.0:1
Gross Horsepower: (max)	212 hp at 2800 rpm
Gross Torque: (max)	492 ft-lbs at 1300 rpm
Weight:	1310 pounds, dry
Fuel: diesel oil MIL-VV-F-800	95 gallons
Engine Oil:	22 quarts

POWER TRAIN

Transfer Case: Overall ratio 1.286:1 overdrive
Transmission: Allison TX-100
    Torque Converter: Hydraulic, single stage, multiphase w/lockup
    Torque Converter Stall Ratio: 3.5:1

Transmission Ratios:	1st	3.81:1	3rd	1.00:1
	2nd	1.936:1	reverse	4.35:1

Steering: DS200 controlled differential, steering brake levers
    Input Ratio: 1.28:1   Steering Ratio: 1.1 to 1.786:1
Brakes: Differential band
Final Drive: Spur gear   Gear Ratio: 3.928:1
Drive Sprockets: At front of vehicle with 10 teeth
    Pitch Diameter: 19.618 inches

RUNNING GEAR

Suspension: Flat track, torsion bar
    10 individually sprung dual road wheels (5/track)
    Tire Size: 24 x 2.1 inches
    Dual adjustable idler at rear of each track
    Idler Size: 21 x 2.1 inches
    Shock absorbers on first 2 and last road wheels on each side
Tracks: Center guide, T130E1 and T150
    Type: T130E1, Single pin, 15 inch width, steel w/detachable rubber pad
          T150, Double pin, 14.81 inch width, steel w/chevron rubber pad
    Pitch: 6 inches
    Shoes per Vehicle: 127 (63 left, 64 right)
    Ground Contact Length: 105 inches

ELECTRICAL SYSTEM

Nominal Voltage: 24 volts DC
Main Generator: (1) 28 volts, 100 amperes, driven by main engine
Auxiliary Generator: None
Battery: (4) 12 volts, 2 sets of 2 in series connected in parallel

COMMUNICATIONS

Radio: AN/VRC-64, AN/GRC-160
Interphone: 5 stations

FIRE AND NBC PROTECTION

    (1) 5 pound carbon dioxide, fixed
    (1) 5 pound carbon dioxide, portable
    Gas particulate filter unit

PERFORMANCE

Maximum Speed: Level road		40 miles/hour
Water		3.6 miles/hour
Maximum Tractive Effort: TE at stall		19,800 pounds
Per cent of Vehicle Weight: TE/W		76 per cent
Maximum Grade:		60 percent
Maximum Trench:		5.5 feet
Maximum Vertical Wall:		24 inches
Maximum Fording Depth:		floats
Minimum Turning Circle: (diameter)		26 feet
Cruising Range: Roads	approx.	300 miles

GENERAL DATA

Crew:	4	men
Length:	191.5	inches
Width: Over flotation pods	112.4	inches
Height: Overall	115	inches
Tread:	85	inches
Ground Clearance:	16.1	inches
Weight, Combat Loaded:	27,542	pounds
Power to Weight Ratio:	15.4	hp/ton
Ground Pressure: Zero penetration	8.7	psi

ARMOR

Type: Hull, rolled 5083/5086 H32 aluminum armor,
    Welded assembly

Hull Thickness:	Actual	Angle w/Vertical
Front, Upper	1.50 inches (38mm)	45 degrees
Lower	1.50 inches (38mm)	30 degrees
Sides, Upper	1.75 inches (44mm)	0 degrees
Lower	1.25 inches (32mm)	0 degrees
Rear, On Ramp	1.50 inches (38mm)	8 degrees
Off Ramp	1.50 inches (38mm)	9 degrees
Top	1.50 inches (38mm)	90 degrees
Floor	1.125 inches (29mm)	90 degrees

ARMAMENT

Primary: 20mm cannon M168 in mount M157A1

Traverse: Electric	360 degrees
Traverse Rate: (max)	4.8 seconds/360 degrees
Elevation: Electric	+80 to -5 degrees
Elevation Rate: (max)	60 degrees/second
Firing Rate: High	3000 rounds/minute
Low	1000 rounds/minute
Burst Limits: High rate only	10, 30, 60, or 100 rounds

Secondary:
    (4) 5.56mm M16A1 rifles stowed

AMMUNITION
    2100 rounds 20mm
    800 rounds 5.56mm

FIRE CONTROL AND VISION EQUIPMENT

Primary Weapon: Sight M61, night sight AN/TVS-2B, Telescope M134,
    AN/VPS-2 radar set

Vision Devices	Direct	Indirect
Driver	Hatch	Periscope M17 (4)
		Periscope M19 (infrared) (1)
Commander	Hatch	None
Gunner	open turret	None
Crew Compartment	None	None

Total Periscopes: M17 (4), M19 (infrared)(1)

ENGINE

Make and Model: General Motors 6V53	
Type: 6 cylinder, 2 cycle, vee	
Cooling System: Liquid	Ignition: Compression
Displacement:	318 cubic inches
Bore and Stroke:	3.875 x 4.5 inches
Compression Ratio:	17.0:1
Gross Horsepower: (max)	212 hp at 2800 rpm
Gross Torque: (max)	492 ft-lbs at 1300 rpm
Weight:	1310 pounds, dry
Fuel: diesel oil MIL-VV-F-800	95 gallons
Engine Oil:	22 quarts

POWER TRAIN

Transfer Case: Overall Ratio 1.286:1 overdrive
Transmission: Allison TX-100
    Torque Converter: Hydraulic, single stage, multiphase w/lockup
    Torque Converter Stall Ratio: 3.5:1

Transmission Ratios:	1st	3.81:1	3rd	1.00:1
	2nd	1.936:1	reverse	4.35:1

Steering: DS200 controlled differential, steering brake levers

Input Ratio: 1.28:1	Steering Ratio: 1.1 to 1.786:1

Brakes: Differential band

Final Drive: Spur gear	Gear Ratio: 3.928:1

Drive Sprockets: At front of vehicle with 10 teeth
    Pitch Diameter: 19.618 inches

RUNNING GEAR

Suspension: Flat track, torsion bar w/hydraulic lockout
    10 individually sprung dual road wheels (5/track)
    Tire Size: 24 x 2.1 inches
    Dual adjustable idler at rear of each track
    Idler Size: 21 x 2.1 inches
    Shock absorbers on first and last road wheels on each side
Tracks: Center guide, T130E1
    Type: Single pin, 15 inch width, steel w/detachable rubber pad
    Pitch: 6 inches
    Shoes per Vehicle: 127 (63 left, 64 right)
    Ground Contact Length: 105 inches

ELECTRICAL SYSTEM

M741 vehicle:
    Nominal Voltage: 24 volts DC
    Main Generator: (1) 28 volts, 100 amperes, driven by main engine
    Battery: (2) 12 volts in series
M157A1 gun mount:
    Battery: (3) 24 volts in parallel

FIRE PROTECTION
    (1) 5 pound carbon dioxide, fixed
    (1) 5 pound carbon dioxide, portable

PERFORMANCE

Maximum Speed: Level road		40 miles/hour
Water		3.6 miles/hour
Maximum Tractive Effort: TE at stall		19,800 pounds
Per Cent of Vehicle Weight: TE/W		72 per cent
Maximum Grade:		60 per cent
Maximum Trench:		5.5 feet
Maximum Vertical Wall:		24 inches
Maximum Fording Depth:		floats
Minimum Turning Circle: (diameter)		26 feet
Cruising Range: Roads	approx.	300 miles

GENERAL DATA

Crew: (in cab)	5	men
Length: Overall	239.88	inches
Width: Over track shrouds	105.75	inches
Height: Over canopy bows	113.88	inches
Tread:	85.0	inches
Ground Clearance:	17.1	inches
Weight, Combat Loaded:	29,300	pounds
Weight, Without Missile Pallet:	14,691	pounds
Power to Weight Ratio:	18.8	hp/ton
Ground Pressure: Zero penetration	8.8	psi

ARMOR
None

ARMAMENT
The M54A2 Chaparral guided missile pallet armed with 4 MIM-72 missiles

AMMUNITION
    12 MIM-72 Chaparral missiles (including 4 on launcher)

VISION EQUIPMENT
Windows in cab or open top vehicle

ENGINE
Make and Model: General Motors: 6V53T
Type: 6 cylinder, 2 cycle, vee, turbosupercharged

Cooling System: Liquid	Ignition: Compression
Displacement:	318 cubic inches
Bore and Stroke:	3.875 x 4.5 inches
Compression Ratio:	18.0:1
Gross Horsepower: (max)	275 hp at 2800 rpm
Gross Torque: (max)	627 ft-lbs at 1600 rpm
Weight:	1495 pounds, dry
Fuel: diesel oil MIL-VV-F-800	111 gallons
Engine Oil:	22 quarts

POWER TRAIN
Transmission: Allison X-200-4
    Torque Converter: Hydraulic, single stage, multiphase w/lockup
    Torque Converter Stall Ratio: 2.72:1 to 3.32:1

Transmission Ratios:	1st	4.16:1	4th	1.04:1
	2nd	2.34:1	reverse 1	6.62:1
	3rd	1.46:1	reverse 2	2.16:1

Steering: Hydrostatic, steering yoke

Steering Ratios:	1st	2.15:1	4th	1.20:1
	2nd	1.52:1	reverse 1	3.76:1
	3rd	1.29:1	reverse 2	1.47:1

Brakes: Multiple plate, oil cooled
Final Drive: Spur gear   Gear Ratio: 3.928:1
Drive Sprockets: At front of vehicle with 10 teeth
    Pitch Diameter: 19.618 inches

RUNNING GEAR
Suspension: Flat track, torsion bar
    10 individually sprung dual road wheels (5/track)
    Tire Size: 24 x 2.1 inches
    Dual adjustable idler at rear of each track
    Idler Size: 21 x 2.1 inches
    Shock absorbers on first 2 and last road wheels on each side
Tracks: Center guide, T130E1
    Type: Single pin, 15 inch width, steel w/detachable rubber pad
    Pitch: 6 inches
    Shoes per Vehicle: 130 (65/track)
    Ground Contact Length: 111 inches

ELECTRICAL SYSTEM
Nominal Voltage: 24 volts DC
Main Generator: (alternator) 28 volts, 200 amperes, driven by main engine
Auxiliary Generator: None
Battery: (4) 12 volts, 2 sets of 2 in series connected in parallel

FIRE AND NBC PROTECTION
    (1) 5 pound carbon dioxide, fixed
    (1) 5 pound carbon dioxide, portable
    Gas particulate filter unit

PERFORMANCE

Maximum Speed: Level road	37 miles/hour
Maximum Grade:	60 per cent
Maximum Trench:	5.5 feet
Maximum Vertical Wall:	24 inches
Maximum Fording Depth:	40 inches
Minimum Turning Circle: (diameter)	28 feet
Cruising Range: Roads	255 miles

SELF-PROPELLED FLAME THROWER CARRIER M132A1
(Based upon the APC M113A1)

GENERAL DATA

Crew: (driver, flame thrower operator)	2	men
Length: Overall	191.5	inches
Width: Over track shrouds	105.75	inches
Height: Over cupola periscope guard	95.75	inches
Tread:	85.0	inches
Ground Clearance:	16.1	inches
Weight, Combat Loaded:	23,895	pounds
Weight, Air Transport:	20,885	pounds
Power to Weight ratio:	17.7	hp/ton
Ground Pressure: Zero penetration	7.6	psi

ARMOR

Type: Hull, rolled 5083/5086 H32 aluminum armor,
Welded assembly

Hull Thickness:	Actual	Angle w/Vertical
Front, Upper	1.50 inches (38mm)	45 degrees
Lower	1.50 inches (38mm)	30 degrees
Sides, Upper	1.75 inches (44mm)	0 degrees
Lower	1.25 inches (32mm)	0 degrees
Rear, On Ramp	1.50 inches (38mm)	8 degrees
Off Ramp	1.50 inches (38mm)	9 degrees
Top	1.50 inches (38mm)	90 degrees
Floor	1.125 inches (29mm)	90 degrees

ARMAMENT

Primary: M10-8 flame thrower in cupola mount

Traverse:	360 degrees
Elevation:	+55 to -15 degrees
Duration of fire:	32 seconds

Secondary:
(1) 7.62mm M73 MG coaxial w/flame gun in cupola

AMMUNITION
200 gallons flame fuel
200 rounds 7.62mm

FIRE CONTROL AND VISION EQUIPMENT

Primary Weapon Sight: Periscope in cupola

Vision Devices	Direct	Indirect
Driver	Hatch	Periscope M17 (4)
Flame Operator	Hatch and	Periscope sight M28D
	(4) vision blocks	

Total Periscopes: M17 (4), Periscope sight (1)
Total Vision Blocks: (4) in cupola

ENGINE

Make and Model: General Motors 6V53
Type: 6 cylinder, 2 cycle, vee

Cooling System: Liquid	Ignition: Compression
Displacement:	318 cubic inches
Bore and Stroke:	3.875 x 4.5 inches
Compression Ratio:	17.0:1
Gross Horsepower: (max)	212 hp at 2800 rpm
Gross Torque: (max)	492 ft-lbs at 1300 rpm
Weight:	1310 pounds, dry
Fuel: diesel oil MIL-VV-F-800	95 gallons
Engine Oil:	22 quarts

POWER TRAIN

Transfer Case: Overall Ratio 1.286:1 overdrive
Transmission: Allison TX-100
Torque Converter: Hydraulic, single stage, Multiphase w/lockup
Torque Converter Stall Ratio: 3.5:1

Transmission Ratios:	1st	3.81:1	3rd	1.00:1
	2nd	1.936:1	reverse	4.35:1

Steering: DS200 controlled differential, steering brake levers

Input Ratio: 1.28:1	Steering Ratio: 1.1 to 1.786:1

Brakes: Differential band
Final Drive: Spur gear   Gear Ratio: 3.928:1
Drive Sprockets: At front of vehicle with 10 teeth
Pitch Diameter: 9.618 inches

RUNNING GEAR

Suspension: Flat track, torsion bar
10 individually sprung dual road wheels (5/track)
Tire Size: 24 x 2.1 inches
Dual adjustable idler at rear of each track
Idler Size: 21 x 2.1 inches
Shock absorbers on first and last road wheels on each side
Tracks: Center guide, T130E1
Type: Single pin, 15 inch width, steel w/detachable rubber pad
Pitch: 6 inches
Shoes per Vehicle: 127 (63 left, 64 right)
Ground Contact Length: 105 inches

ELECTRICAL SYSTEM

Nominal Voltage: 24 volts DC
Main Generator: (1) 28 volts, 100 amperes, driven by main engine
Battery: (2) 12 volts in series

COMMUNICATIONS

Radio: AN/GRC-3 thru 8, AN/VRC-24, AN/GRC-19, AN/VRQ-1 thru 3
AN/PRC-8 thru 10, AN/GRR-5 or AN/VRC-12
Interphone: AN/UIC-1, 2 stations

FIRE PROTECTION

(1) 5 pound carbon dioxide, fixed
(1) 5 pound carbon dioxide, portable

PERFORMANCE

Maximum Speed: Level road		40 miles/hour
Water		3.6 miles/hour
Maximum Tractive Effort: TE at stall		19,800 pounds
Per Cent of Vehicle Weight: TE/W		83 per cent
Maximum Grade:		60 per cent
Maximum Trench:		5.5 feet
Maximum Vertical Wall:		24 inches
Maximum Fording Depth:		floats
Minimum Turning Circle: (diameter)		26 feet
Cruising Range: Roads	approx.	300 miles

SMOKE GENERATOR CARRIER M1059
(Based upon the APC M113A2)

GENERAL DATA

Crew: (driver, commander, operator)	3	men
Length: Overall	191.5	inches
Width: Over track shrouds	105.75	inches
Height: Over smoke generators	100.75	inches
Tread:	85.0	inches
Ground Clearance:	17.1	inches
Weight, Combat Loaded:	24,400	pounds
Power to Weight Ratio:	17.4	hp/ton
Ground Pressure: Zero penetration	7.7	psi

ARMOR

Type: Hull, rolled 5083/5086 H32 aluminum armor,
    Welded assembly

Hull Thickness:	Actual	Angle w/Vertical
Front, Upper	1.50 inches (38mm)	45 degrees
Lower	1.50 inches (38mm)	30 degrees
Sides, Upper	1.75 inches (44mm)	0 degrees
Lower	1.25 inches (32mm)	0 degrees
Rear, On Ramp	1.50 inches (38mm)	8 degrees
Off Ramp	1.50 inches (38mm)	9 degrees
Top	1.50 inches (38mm)	90 degrees
Floor	1.125 inches (29mm)	90 degrees

ARMAMENT

    (1) .50 caliber MG HB M2 on commander's cupola

AMMUNITION

    2000 rounds .50 caliber

SMOKE EQUIPMENT

Smoke generator system M157
    (2) Smoke generators M54
    (1) air compressor assembly
    (1) control panel assembly
    (1) fog oil pump tank assembly, 120 gallon
    (2) 5 gallon fuel cans w/special plugs

VISION EQUIPMENT

Vision Devices:	Direct	Indirect
Driver	Hatch	Periscope M17 (4)
		Periscope M19 (infrared) (1)
Commander	Hatch	Periscope M17 (5)
Smoke Operator	Hatch and	None
	vision block (1)	

Total Periscopes: M17 (9), M19 (infrared) (1)
Total Vision Blocks: (1) in rear ramp

ENGINE

Make and Model: General Motors 6V53
Type: 6 cylinder, 2 cycle, vee

Cooling System: Liquid	Ignition: Compression
Displacement:	318 cubic inches
Bore and Stroke:	3.875 x 4.5 inches
Compression Ratio:	17.0:1
Gross Horsepower: (max)	212 hp at 2800 rpm
Gross Torque: (max)	492 ft-lbs at 1300 rpm
Weight:	1310 pounds, dry
Fuel: diesel oil MIL-VV-F-800	95 gallons
Engine Oil:	22 quarts

POWER TRAIN

Transfer Case: Overall ratio 1.286:1 overdrive
Transmission: Allison TX-100
    Torque Converter: Hydraulic, single stage, multiphase w/lockup
    Torque Converter Stall Ratio: 3.5:1

Transmission Ratios:	1st	3.81:1	3rd	1.00:1
	2nd	1.936:1	reverse	4.35:1

Steering: DS200 controlled differential, steering brake levers
    Input Ratio: 1.28:1   Steering Ratio: 1.1 to 1.786:1
Brakes: Differential band
Final Drive: Spur gear   Gear Ratio: 3.928:1
Drive Sprockets: At front of vehicle with 10 teeth
    Pitch Diameter: 19.618 inches

RUNNING GEAR

Suspension: Flat track, torsion bar
    10 individually sprung dual road wheels (5/track)
    Tire Size: 24 x 2.1 inches
    Dual adjustable idler at rear of each track
    Idler Size: 21 x 2.1 inches
    Shock absorbers on first 2 and last road wheels on each side
Tracks: Center guide, T130E1 and T150
    Type: T130E1, Single pin, 15 inch width, steel w/detachable rubber pad
    T150, Double pin, 14.81 inch width, steel w/chevron rubber pad
    Pitch: 6 inches
    Shoes per Vehicle: 127 (63 left, 64 right)
    Ground Contact Length: 105 inches

ELECTRICAL SYSTEM

Nominal Voltage: 24 volts DC
Main Generator: (1) 28 volts, 100 amperes, driven by main engine
Auxiliary Generator: None
Battery: (2) 12 volts in series

COMMUNICATIONS

Radio: AN/VRC-46, AN/GRC-160
Interphone: 2 stations

FIRE AND NBC PROTECTION

    (3) 5 pound carbon dioxide, fixed
    (1) 5 pound carbon dioxide, portable
    Gas particulate filter unit

PERFORMANCE

Maximum Speed: Level road		40 miles/hour
Water		3.6 miles/hour
Maximum Tractive Effort: TE at stall		19,800 pounds
Per Cent of Vehicle Weight: TE/W		81 per cent
Maximum Grade:		60 per cent
Maximum Trench:		5.5 feet
Maximum Vertical Wall:		24 inches
Maximum Fording Depth:		floats
Minimum Turning Circle: (diameter)		26 feet
Cruising Range: Roads	approx.	300 miles

GENERAL DATA

Crew: (driver, commander, observer)	3	men
Jump seat provided for one passenger		
Length: Overall	175.75	inches
Width: Over track shrouds	91.75	inches
Height: Over M2 MG, M114	94.1	inches
M114A1	84.9	inches
Tread:	72.75	inches
Ground Clearance:	14.25	inches
Weight, Combat Loaded: M114	15,093	pounds
M114A1	15,276	pounds
Weight, Unstowed: M114	12,354	pounds
M114A1	12,537	pounds
Power to Weight Ratio: Net M114	15.2	hp/ton
M114A1	15.0	hp/ton
Gross M114	21.2	hp/ton
M114A1	20.9	hp/ton
Ground Pressure: Zero penetration, M114	5.0	psi
M114A1	5.1	psi

ARMOR

Type: Hull, rolled aluminum armor, Welded assembly

Hull Thickness:	Actual	Angle w/Vertical
Front, Upper	1.25 inches (32mm)	70 degrees
Lower	1.75 inches (44mm)	28 degrees
Sides, Upper	1.25 inches (32mm)	0 degrees
Lower	0.75 inches (19mm)	0 degrees
Rear	1.25 inches (32mm)	0 degrees
Top	1.5 inches (38mm)	90 degrees
Floor	1.00 inches (25mm)	90 degrees

ARMAMENT

(1) .50 caliber MG HB M2 on commander's cupola, M114
(1) .50 caliber MG HB M2 TT on commander's station, M114A1
(1) 7.62mm MG M60 at observer's station
(1) 7.62mm rifle M14, stowed
(1) 40mm grenade launcher M79, stowed

AMMUNITION

1000 rounds .50 caliber   8 hand grenades
3000 rounds 7.62mm for M60 MG
150 rounds 7.62mm for M14 rifle
48 rounds 40mm for M79 grenade launcher
3 66mm M72 light antitank weapon (LAW)

VISION EQUIPMENT

Vision Devices	Direct	Indirect
Driver	Hatch	Periscope M26 (3)
		Periscope M19 (infrared)(1)
Commander	Hatch and	None
	8 vision blocks	
Observer	Hatch	Periscope M13 (1)

Total Periscopes: M13 (1), M19 (infrared) (1), M26 (3)
Total Vision Blocks: (8)

ENGINE

Make and Model: Chevrolet 283-V8 Military
Type: 8 cylinder, 4 cycle, vee

Cooling System: Liquid	Ignition: Delco	
Displacement:	283 cubic inches	
Bore and Stroke:	3.875 x 3 inches	
Compression Ratio:	8:1	
Net Horsepower: (max)	115 hp at 3600 rpm	
Gross Horsepower: (max)	160 hp at 4200 rpm	
Net Torque: (max)	210 ft-lb at 2400 rpm	
Gross Torque: (max)	240 ft-lb at 2800 rpm	
Fuel: 80 octane gasoline	110 gallons	
Engine Oil:	7 quarts	

POWER TRAIN

Transmission: Hydramatic Model 305MC
    Automatic, 4 speeds forward, 1 reverse

Gear Ratios:	1st	4.09:1	4th	1.00:1
	2nd	2.63:1	reverse	4.54:1
	3rd	1.55:1		

Steering: Allison GS 100-3, geared, clutch-brake, steer bar
Brakes: Multiple plate, oil cooled
Final Drive: Integral w/GS 100-3   Gear Ratio: 4.17:1
Drive Sprockets: At front of vehicle with 11 teeth
    Pitch Diameter: 14.00 inches

RUNNING GEAR

Suspension: Flat track, torsion bar
    8 individually sprung dual road wheels (4/track)
    Tire Size: 22 x 2 inches
    Dual adjustable idler at rear of each track
    Idler Size: 13.25 x 1.75 inches
    Shock absorbers on first and last road wheels on each side
Tracks: Double center guide
    Type: Band type, 16.5 inch width
    Pitch: Cross bar, 4 inches, 8 cross bars/32 inch section
    Sections per Vehicle: 20 (10/track)
    Ground Contact Length: 91 inches

ELECTRICAL SYSTEM

Nominal Voltage: 24 volts DC
Main Generator: (alternator) 24 volts, 100 amperes, driven by main engine
Auxiliary Generator: None
Battery: Early (2) 12 volts in series
    Late (4) 12 volts, 2 sets of 2 in series connected in parallel

COMMUNICATIONS

Radio: AN/PRC 8, 9, 10 & 25; AN/VRC 12, 24, 43, 45, 46 & 47;
    AN/VRC 8, 9, 13, 15, 16, 17, 18 & 35; AN/GRC 3, 4, 7, 8, & 15;
    AN/VRQ 1 thru 3; AN/GRR 5
Interphone: AN/UIC 1 or AN/VIC 1, 3 stations

FIRE PROTECTION

    (1) 5 pound carbon dioxide, fixed
    (1) 5 pound carbon dioxide, portable

PERFORMANCE

Maximum Speed: Level road	36 miles/hour	
Water	3.4 miles/hour	
Maximum Tractive Effort: TE at stall	9500 pounds	
Per Cent of Vehicle Weight: TE/W M114	63 per cent	
M114A1	62 per cent	
Maximum Grade:	60 per cent	
Maximum Trench:	5 feet	
Maximum Vertical Wall:	20 inches	
Maximum Fording Depth:	floats	
Minimum Turning Circle: (diameter)	16 feet	
Cruising Range: Roads	approx.	275 miles

GENERAL DATA

Crew: (driver, commander, observer)		3	men
Length: Overall		181	inches
Width: Over track shrouds		95	inches
Height: Over .50 caliber MG on M26 cupola		85.5	inches
Tread:		74.25	inches
Ground Clearance:		16.1	inches
Weight, Combat Loaded: Dutch		18,650	pounds
Lynx		19,340	pounds
Weight, Air Drop: Dutch		16,300	pounds
Lynx		17,030	pounds
Power to Weight Ratio: Dutch		22.7	hp/ton
Lynx		21.9	hp/ton
Ground Pressure: Dutch		6.6	psi
Lynx		6.9	psi

ARMOR

Type: Hull, rolled 5086 aluminum armor
     Welded assembly

Hull Thickness:	Actual		Angle w/Vertical	
Front, Upper	1.25	inches (32mm)	60	degrees
Lower	1.75	inches (44mm)	29	degrees
Sides, Upper	1.25	inches (32mm)	0	degrees
Lower	0.75	inches (19mm)	0	degrees
Rear, Upper	1.25	inches (32mm)	0	degrees
Lower	1.25	inches (32mm)	30	degrees
Top	1.25	inches (32mm)	90	degrees
Floor	1.00	inches (25mm)	90	degrees

ARMAMENT

(1) .50 caliber MG HB M2 TT on cupola M26
(1) 7.62mm MG at observers station

AMMUNITION

1155 rounds .50 caliber   8 hand grenades
2000 rounds 7.62mm
3 66mm M72 light antitank weapon (LAW)

VISION EQUIPMENT

Vision Devices	Direct	Indirect
Driver, Dutch	Hatch	Periscope M17 (4)
		Periscope M19 (infrared) (1)
Driver, Lynx	Hatch	Periscope M17 (5)
		Periscope M19 (infrared) (1)
Commander, Dutch	Hatch and 8 vision blocks	None
Commander, Lynx	Hatch and 8 vision blocks	None
Observer, Dutch	Hatch	Periscope M17 (4)
Observer, Lynx	Hatch	Periscope M17 (3)
		Periscope M17C (2)

Total Periscopes: Dutch, M17 (8), M19 (infrared) (1)
             Lynx, M17 (8), M17C (2), M19 (infrared) (1)
Total Vision Blocks: 8 in cupola M26

ENGINE

Make and Model: General Motors 6V53	
Type: 6cylinder, 2 cycle, vee	
Cooling System: Liquid	Ignition: Compression
Displacement:	318 cubic inches
Bore and Stroke:	3.875 x 4.5 inches
Compression Ratio:	17.0:1
Gross Horsepower: (max)	212 hp at 2800 rpm
Gross Torque: (max)	492 ft-lbs at 1300 rpm
Weight:	1310 pounds, dry
Fuel: diesel oil, Dutch	87 gallons
Lynx	80 gallons
Engine Oil:	22 quarts

POWER TRAIN

Transfer Case: Overall Ratio 1.286:1 overdrive
Transmission: Allison TX-100
     Torque Converter: Hydraulic, single stage, multiphase w/lockup
     Torque Converter Stall Ratio: 3.5:1

Transmission Ratios:	1st	3.81:1	3rd	1.00:1
	2nd	1.936:1	reverse	4.35:1

Steering: DS200 controlled differential, steering brake levers
     Input Ratio: 1.28:1   Steering Ratio: 1.1 to 1.786:1
Brakes: Differential band
Final Drive: Spur gear   Gear Ratio: 3.928:1
Drive Sprockets: At front of vehicle with 10 teeth
     Pitch Diameter: 19.618 inches

RUNNING GEAR

Suspension: Flat track, torsion bar
     8 individually sprung dual road wheels (4/track)
     Tire Size: 24 x 2.1 inches
     Dual adjustable idler at rear of each track
     Idler Size: 21 x 2.1 inches
     Shock absorbers on first and last road wheels on each side
Tracks: Center guide, T130E1
     Type: Single pin, 15 inch width, steel w/detachable rubber pad
     Pitch: 6 inches
     Shoes per Vehicle: 119 (59 left, 60 right)
     Ground Contact Length: 94 inches

ELECTRICAL SYSTEM

Nominal Voltage: 24 volts DC
Main Generator: (1) 28 volts, 100 amperes, driven by main engine
Auxiliary Generator: None
Battery: (2) 12 volts in series

COMMUNICATIONS

Radio: AN/VRC-46, AN/VRC-49, AN/GRC-125
Interphone: AN/UIC-1, 3 stations

FIRE PROTECTION

(1) 5 pound carbon dioxide, fixed
(1) 5 pound carbon dioxide, portable

PERFORMANCE

Maximum Speed: Level road		40 miles/hour
Water		4 miles/hour
Maximum Grade:		60 per cent
Maximum Trench:		5 feet
Maximum Vertical Wall:		24 inches
Maximum Fording Depth:		floats
Minimum Turning Circle: (diameter)		24 feet
Cruising Range: Roads	approx.	325 miles

GENERAL DATA

Crew:	5 to 9	men
Length: Overall:	191.5	inches
Width: Over track shrouds	105.75	inches
Height: Over MG shields	99.43	inches
Tread:	85.0	inches
Ground Clearance:	16.1	inches
Weight, Combat Loaded:	approx. 24,000	pounds
Weight, Unstowed:	approx. 21,000	pounds
Power to Weight ratio:	17.7	hp/ton
Ground Pressure: Zero penetration	7.6	psi

ARMOR

Type: Hull, rolled 5083/5086 H32 aluminum armor,
   Welded assembly

Hull Thickness:	Actual	Angle w/Vertical
Front, Upper	1.50 inches (38mm)	45 degrees
Lower	1.50 inches (38mm)	30 degrees
Sides, Upper	1.75 inches (44mm)	0 degrees
Lower	1.25 inches (32mm)	0 degrees
Rear, On Ramp	1.50 inches (38mm)	8 degrees
Off Ramp	1.50 inches (38mm)	9 degrees
Top	1.50 inches (38mm)	90 degrees
Floor	1.125 inches (29mm)	90 degrees

Steel armor shields provided for .50 caliber and 7.62mm MG.
Ballistic plate added to bottom front of hull.

ARMAMENT

(1) .50 caliber MG HB M2 on commander's cupola
(2) 7.62mm M60 MG on hull roof
(1) 40mm M79 grenade launcher in troop compartment

AMMUNITION

2000 rounds .50 caliber
4000 rounds 7.62mm for M60 MG
48 rounds 40mm for M79 grenade launcher

VISION EQUIPMENT

Vision Devices	Direct	Indirect
Driver	Hatch	Periscope M17 (4)
		Periscope M19 (infrared) (1)
Commander	Hatch	Periscope M17 (5)
Troop Compartment	Roof Hatch	None

Total Periscopes: M17 (9), M19 (infrared) (1)

ENGINE

Make and Model: General Motors 6V53
Type: 6 cylinder, 2 cycle, vee

Cooling System: Liquid	Ignition: Compression
Displacement:	318 cubic inches
Bore and Stroke:	3.875 x 4.5 inches
Compression Ratio:	17.0:1
Gross Horsepower: (max)	212 hp at 2800 rpm
Gross Torque: (max)	492 ft-lbs at 1300 rpm
Weight:	1310 pounds, dry
Fuel: diesel oil MIN-VV-F-800	95 gallons
Engine Oil:	22 quarts

POWER TRAIN

Transfer Case: Overall ratio 1.286:1 overdrive
Transmission: Allison TX-100
   Torque Converter: Hydraulic, single stage, multiphase w/lockup
   Torque Converter Stall Ratio: 3.5:1

Transmission Ratios:	1st	3.81:1	3rd	1.00:1
	2nd	1.936:1	reverse	4.35:1

Steering: DS200 controlled differential, steering brake levers
   Input Ratio: 1.28:1   Steering Ratio: 1.1 to 1.786:1
Brakes: Differential band
Final Drive: Spur gear   Gear Ratio: 3.928:1
Drive Sprockets: At front of vehicle with 10 teeth
   Pitch Diameter: 19.618 inches

RUNNING GEAR

Suspension: Flat track, torsion bar
   10 individually sprung dual road wheels (5/track)
   Tire Size: 24 x 2.1 inches
   Dual adjustable idler at rear of each track
   Idler Size: 21 x 2.1 inches
   Shock absorbers on first and last road wheels on each side
Tracks: Center guide T130E1
   Type: Single pin, 15 inch width, steel w/detachable rubber pad
   Pitch: 6 inches
   Shoes per Vehicle: 127 (63 left, 64 right)
   Ground Contact Length: 105 inches

ELECTRICAL SYSTEM

Nominal Voltage: 24 volts DC
Main Generator: (1) 28 volts, 100 amperes, driven by main engine
Auxiliary Generator: None
Battery: (2) 12 volts in series

COMMUNICATIONS

Radio: AN/GRC -3 thru 8, AN/VRC-24, AN/GRC-19, AN/VRQ-1 thru 3
   AN/PRC-8 thru 10, AN/GRR-5 or AN/VRC-12
Interphone: AN/UIC-1, 2 stations

FIRE PROTECTION

(1) 5 pound carbon dioxide, fixed
(1) 5 pound carbon dioxide, portable

PERFORMANCE

Maximum Speed: Level road		40 miles/hour
Water		3.6 miles/hour
Maximum Tractive Effort: TE at stall		19,800 pounds
Per cent of Vehicle Weight: TE/W		82 per cent
Maximum Grade:		60 per cent
Maximum Trench:		5.5 feet
Maximum Vertical Wall:		24 inches
Maximum Fording Depth:		floats
Minimum Turning Circle: (diameter)		26 feet
Cruising Range: Roads	appro	300 miles

FMC ARMORED INFANTRY FIGHTING VEHICLE
(Dutch configuration, Squad Vehicle)

GENERAL DATA

Crew:	10 men
Length: Overall	207 inches
Width: Overall	111 inches
Height: Over weapon station	110 inches
Tread:	85 inches
Ground Clearance:	17 inches
Weight, Combat Loaded:	30,175 pounds
Weight, Unstowed:	25,075 pounds
Power to Weight Ratio:	17.2 hp/ton
Ground Pressure: Zero penetration	9.6 psi

ARMOR
Type: Hull, rolled 5083/5086 H32 aluminum armor,
     Welded assembly

Hull Thickness:	Actual		Angle w/Vertical
Front, Upper	1.50	inches (38mm)	45 degrees
Lower	1.50	inches (38mm)	30 degrees
Sides, Upper	1.75	inches (44mm)	0 degrees
Lower	1.25	inches (32mm)	0 degrees
Rear, On Ramp	1.50	inches (38mm)	8 degrees
Off Ramp	1.50	inches (38mm)	9 degrees
Top	1.50	inches (38mm)	90 degrees
Floor	1.125	inches (29mm)	90 degrees

Steel spaced laminate armor on hull front, sides, and rear and on
the sides of the weapon station turret.

ARMAMENT
    (1) 25mm Oerlikon KBA cannon in turret weapon station
    (1) 7.62mm MAG MG in turret weapon station

AMMUNITION
    315 rounds 25mm
    1840 rounds 7.62mm

VISION EQUIPMENT

Vision Devices	Direct	Indirect
Driver	Hatch	Periscope M27 (4)
		Passive Periscope UA9630 (1)
Commander	Hatch	Periscope M17 (4)
		Periscope M20A1 (1)
Gunner	Hatch	Periscope M27 (4)
Troop Compartment	Roof Hatch	Periscope M17 (4)
	Firing Ports (5)	Periscope M27 (1)

Total Periscopes: M17 (8), M20A1 (1), M27 (9), Passive UA9630 (1)

ENGINE

Make and Model: General Motors 6V53T	
Type: 6 cylinder, 2 cycle, vee, turbosupercharged	
Cooling System: Liquid	Ignition: Compression
Displacement:	318 cubic inches
Bore and Stroke:	3.875 x 4.5 inches
Compression Ratio:	1 7.0:1
Gross Horsepower: (max)	260 hp at 2800 rpm
Gross Torque: (max)	570 ft-lbs at 1300 rpm
Weight:	1495 pounds, dry
Fuel: diesel oil MIL-VV-F-800	110 gallons
Engine Oil:	22 quarts

POWER TRAIN
Transfer Case: Overall ratio 1.286:1 overdrive
Transmission: Allison TX-100A
    Torque Converter: Hydraulic, single stage, multiphase w/lockup
    Torque Converter Stall Ratio: 3.5:1

Transmission Ratios:	1st	3.81:1	3rd	1.00:1
	2nd	1.936:1	reverse	4.35:1

Steering: DS200 controlled differential, steering brake levers
    Input Ratio: 1.28:1   Steering Ratio: 1.1 to 1.786:1
Brakes: Differential band
Final Drive: Spur gear   Gear Ratio: 4.31:1
Drive Sprockets: At front of vehicle with 10 teeth
    Pitch Diameter: 19.618 inches

RUNNING GEAR
Suspension: Flat track, torsion tube over bar
    10 individually sprung dual road wheels (5/track)
    Tire Size: 24 x 2.1 inches
    Dual adjustable idler at rear of each track
    Idler Size: 21 x 2.1 inches
    Shock absorbers on first and last road wheels on each side
Tracks: Center guide T130E1
    Type: Single pin, 15 inch width, steel w/detachable rubber pad
    Pitch: 6 inches
    Shoes per Vehicle: 127 (63 left, 64 right)
    Ground Contact Length: 105 inches

ELECTRICAL SYSTEM
Nominal Voltage: 24 volts DC
Main Generator: (1) (alternator) 28 volts, 180 amperes driven by
                main engine
Auxiliary Generator: None
Battery: (2) 12 volts in series

COMMUNICATIONS
Radio: AN/VRC-47, AN/VRC-64
Interphone: AN/VIC-1, 5 stations plus external phone

FIRE PROTECTION
    (1) 5 pound carbon dioxide, fixed
    (1) 5 pound carbon dioxide, portable

PERFORMANCE

Maximum Speed: Level road		40 miles/hour
Water		3.9 miles/hour
Maximum Grade:		60 per cent
Maximum Trench:		5.5 feet
Maximum Vertical Wall:		25 inches
Maximum Fording Depth:		floats
Minimum Turning Circle: (diameter)		28 feet
Cruising Range: Roads	approx.	300 miles

GENERAL DATA

Crew:	12	men
Length: Overall	245	inches
Width: Overall	124	inches
Height: Overall	113	inches
Tread:	106	inches
Ground Clearance:	18	inches
Weight, Combat Loaded: Steel	54,050	pounds
Aluminum	50,750	pounds
Weight, Unstowed: Steel	49,670	pounds
Aluminum	46,370	pounds
Power to Weight Ratio: Net hp, Steel	12.8	hp/ton
Aluminum	13.6	hp/ton
Power to Weight ratio: Gross hp, Steel	15.7	hp/ton
Aluminum	16.8	hp/ton
Ground Pressure: Zero penetration, Steel	11.1	psi
Aluminum	10.4	psi

ARMOR

Type: (all five pilots) Turret rolled homogeneous steel armor, Welded assembly

Turret Thickness	Actual	Angle w/Vertical
Front	0.75 inches (19mm)	60 degrees
Sides	1.625 inches (41mm)	0 to 23 degrees
Rear	1.437 inches (36mm)	30 degrees
Top	0.687 inches (17mm)	90 degrees

Type: (pilots 1,2,3) Hull rolled homogeneous steel armor Welded assembly

Hull Thickness	Actual	Angle w/Vertical
Front, Upper	0.687 inches (17mm)	60 degrees
Middle	0.625 inches (16mm)	50 degrees
Lower	0.625 inches (16mm)	70 degrees
Sides, Upper	0.687 inches (17mm)	50 degrees
Above Sponson	1.50 inches (38mm)	0 degrees
Below Sponson	1.25 inches (32mm)	0 degrees
Sponson Bottom	0.375 inches (10mm)	90 degrees
Rear	1.25 inches (32mm)	15 degrees
Top	0.625 inches (16mm)	90 degrees
Floor, Front	0.625 inches (16mm)	90 degrees
Rear	0.375 inches (10mm)	90 degrees

Type: (pilots 4,5) Hull rolled aluminum armor, Welded assembly

Hull Thickness	Actual	Angle w/Vertical
Front, Upper	1.565 inches (40mm)	60 degrees
Middle	1.75 inches (44mm)	50 degrees
Lower	1.75 inches (44mm)	70 degrees
Sides, Upper	1.687 inches (43mm)	50 degrees
Above Sponson	3.5 inches (89mm)	0 degrees
Below Sponson	2.875 inches (73mm)	0 degrees
Sponson Bottom	1.062 inches (27mm)	90 degrees
Rear	2.875 inches (73mm)	15 degrees
Top	1.656 inches (42mm)	90 degrees
Floor, Front	1.75 inches (44mm)	90 degrees
Rear	1.062 inches (27mm)	90 degrees

ARMAMENT

(1) 20mm Hispano Suiza 820 cannon in turret weapon station
(1) 7.62mm M73 MG coaxial w/20mm cannon
(2) 7.62mm M60 MG in hull ball mounts
(10) 7.62mm M14 rifled, stowed, five firing ports
(2) 40mm M79 grenade launchers, stowed
(8) fixed smoke grenade launchers

AMMUNITION

1250 rounds 20mm	16 smoke grenades
1500 rounds 7.62mm for M73 MG	16 hand grenades
3000 rounds 7.62mm for M60 MG	
1200 rounds 7.62mm for M14 rifles	
96 rounds 40mm for M79 launcher	

VISION EQUIPMENT

Vision Devices	Direct	Indirect
Driver	Hatch	Periscope M17 (4)
		Periscope M27 (1)
		Periscope M24 (infrared) (1)
Commander	Hatch and 8 vision blocks	None
Gunner	Hatch	Periscope M34C
Troop Compartment	Roof hatch and 6 vision blocks	Periscope M27 (1)

Total Periscopes: M17 (4), M24 (infrared) (1), M27 (2), M34C (1)
Total Vision Blocks: (14)

ENGINE

Make and Model: General Motors 8V71T
Type: 8 cylinder, 2 cycle, vee, turbosupercharged

Cooling System: Liquid	Ignition: Compression
Displacement:	567.5 cubic inches
Bore and Stroke:	4.25 x 5 inches
Compression Ratio:	17:1
Net Horsepower: (max)	345 hp at 2300 rpm
Gross Horsepower: (max)	425 hp at 2300 rpm
Net Torque: (max)	895 ft-lbs at 1700 rpm
Gross Torque: (max)	980 ft-lbs at 1700 rpm
Weight:	2422 pounds, dry
Fuel: diesel oil MIL-VV-F-800	275 gallons
Engine Oil:	37 quarts

POWER TRAIN

Transmission: Allison XTG-411-2A
Torque Converter: Hydraulic, single stage, multiphase w/lockup
Torque Converter Stall Ratio: 3.5:1

Transmission Ratios:	1st	4.69:1	4th	0.794:1
	2nd	3.18:	reverse 1	5.6:1
	3rd	1.59:1	reverse 2	3.79:1

Steering: 1st, 2nd and 1st reverse, clutch-brake steering
3rd, 4th and 2nd reverse, geared steering
Brakes: Multiple plate, oil cooled
Final Drive: Planetary gear   Gear Ratio: 5.35:1
Drive Sprockets: At front of vehicle with 11 teeth
Pitch Diameter: 21 inches

RUNNING GEAR

Suspension: Flat track, torsion bar
10 individually sprung dual road wheels (5/track)
Tire Size: 32 x 4 inches
The rear road wheel serves as an adjustable idler
Tracks: Center guide
Type: Double pin , 15 inch width, steel w/detachable rubber pad
Pitch: 6 inches
Ground Contact Length: 163 inches

ELECTRICAL SYSTEM

Nominal Voltage: 24 volts DC
Main Generator: (1) (alternator) 28 volts, 100 amperes, driven by main engine
Auxiliary Generator: (1) 28 volts, 100 amperes, driven by auxiliary engine
Battery: (4) 12 volts, 2 sets of 2 in series connected in parallel

COMMUNICATIONS

Radio: AN/VRC-47
Interphone: 3 stations plus external phone and loudspeaker

FIRE AND NBC PROTECTION

(1) 5 pound carbon dioxide, fixed
(1) 5 pound carbon dioxide, portable
E51 collective protector

PERFORMANCE

Maximum Speed: Level road		40 miles/hour
Water		3.8 miles/hour
Maximum Grade:		60 per cent
Maximum Trench:		8 feet
Maximum Vertical Wall:		36 inches
Maximum Fording Depth:		floats
Minimum Turning Circle: (diameter)		pivot
Cruising Range: Roads	approx.	350 miles

GENERAL DATA

Crew:	(12 optional)	11	men
Length: Overall		245	inches
Width: Overall		126	inches
Height: Overall		109	inches
Tread:		96	inches
Ground Clearance:		19	inches
Weight, Combat Loaded:		43,000	pounds
Weight, Unstowed:		39,000	pounds
Weight, Air Transport:		35,000	pounds
Power to Weight ratio: Gross hp		20.9	hp/ton
Ground Pressure: Zero penetration		6.8	psi

ARMOR

The turret is protected by a combination of steel, 5083 aluminum, and 7039 aluminum armor.

The hull is assembled with spaced laminate armor on the vertical sides and rear. The side slopes are 7039 aluminum armor. The top, bottom and front are 5083 aluminum armor. Steel applique armor is installed on the bottom front for mine protection.

ARMAMENT

Primary: 20mm automatic gun M139 or XM236 in turret mount

Traverse: Electrohydraulic and manual	360 degrees
Traverse Rate: (max)	6 seconds/360 degrees
Elevation: Electrohydraulic and manual	+60 to -10 degrees
Elevation Rate: (max)	60 degrees/second
Firing Rate:	850 to 1050 rounds/minute
Stabilizer System	azimuth and elevation

Secondary:

(1) 7.62mm M219 or XM238 MG coaxial w/20mm gun in turret

(6) .45 caliber SMG M3A1 ball mounted in firing ports

AMMUNITION

600 rounds 20mm

3400 rounds 7.62mm

4000 rounds .45 caliber

FIRE CONTROL AND VISION EQUIPMENT

Primary Weapon Sights: Periscope M36E2 (day/night)

External ring sight

Vision Devices	Direct	Indirect
Driver	Hatch	Periscope (4)
Commander	Hatch	Periscope (5)
Gunner	Hatch	Periscope (7)
Troop Compartment	Roof Hatch	Periscope (6)

Total Periscopes: (22)

ENGINE

Make and Model: Cummins VTA-903T

Type: 8 cylinder, 4 cycle, vee, turbosupercharged

Cooling System: Liquid	Ignition: Compression
Displacement:	903 cubic inches
Bore and Stroke:	5.5 x 4.75 inches
Compression Ratio:	15.5:1
Gross Horsepower: (max)	500 hp at 2600 rpm
Gross torque: (max)	1025 ft-lbs at 2350 rpm
Weight:	2450 pounds, dry
Fuel: diesel oil MIL-VV-F-800	197 gallons
Engine Oil:	22 quarts

POWER TRAIN

Transmission: General Electric HMPT-500

Hydromechanical, automatic range selection

Steering: Hydrostatic, steering yoke

Brakes: Multiple plate, oil cooled

Final Drive: Spur gear   Gear Ratio: 4.959:1

Drive Sprockets: At front of vehicle with 11 teeth

Pitch Diameter: 21 inches

RUNNING GEAR

Suspension: Torsion tube over bar

12 individually sprung dual road wheels (6/track)

Tire Size: 24 x 4 inches

2 single and 1 dual track return rollers per track

Dual adjustable idler at rear of each track

Idler Size: Outer, 20.0 x 3.7 inches

Inner, 19.9 x 5.7 inches

Shock absorbers on first 2 and last road wheels on each side

Tracks: Center guide

Type: Single pin, 21 inch width, steel w/detachable rubber pads

Pitch: 6 inches

Shoes per Vehicle: 166 (83/track)

Ground Contact Length: 150 inches

ELECTRICAL SYSTEM

Nominal Voltage: 24volts DC

Main Generator: 28 volts, 220 amperes, driven by main engine

Auxiliary Generator: None

Battery: (4) 12 volts, 2 sets of 2 in series connected in parallel

COMMUNICATIONS

Radio: AN/VRC-47, AN/VRC-64

Interphone: AN/VIC-1, 5 stations

FIRE PROTECTION

(1) 7 pound Halon, fixed in engine compartment

(1) 5 pound Halon, fixed in crew compartment

(1) 5 pound carbon dioxide, portable

PERFORMANCE

Maximum Speed: Level road		45 miles/hour
Water		5 miles/hour
Maximum Grade:		60 per cent
Maximum Trench:		8.3 feet
Maximum Vertical Wall:		36 inches
Maximum Fording Depth:		floats
Minimum Turning Circle: (diameter)		pivots
Cruising Range: Roads	approx.	300 miles

GENERAL DATA

Crew: M2	9	men
M2A1	10	men
Length: Overall	254	inches
Width: overall	126	inches
Height: Over commander's hatch	117	inches
Tread:	96	inches
Ground Clearance:	18	inches
Weight, Combat Loaded: M2	50,259	pounds
M2A1	50,261	pounds
Weight, Unstowed: M2	42,289	pounds
M2A1	43,500	pounds
Weight, Air Transport:	40,775	pounds
Power to Weight Ratio: Gross hp , M2	19.9	hp/ton
M2A1	19.9	hp/ton
Ground Pressure: Zero penetration, M2	7.8	psi
M2A1	7.8	psi

ARMOR

The turret is armored by a combination of steel, 5083 aluminum, and 7039 aluminum armor.

The hull is assembled with spaced laminate armor on the vertical sides and rear. This consists of two 0.25 inch thick steel plates one inch apart spaced 3 ½ inches from the 1 inch thick aluminum armor. The side slopes are 7039 aluminum armor. The top, bottom and front are 5083 aluminum armor. Steel applique armor 0.375 inches thick is installed on the front third of the bottom for mine protection.

ARMAMENT

Primary: 25mm automatic gun M242 in turret mount

Traverse: Electric and manual		360 degrees
Traverse Rate: (max)		6 seconds/360 degrees
Elevation: M2, Electric and manual		+59 to -9 degrees
M2A1, Electric and manual		+57 to -9 degrees
Elevation Rate: (max)		60 degrees/second
Firing Rate:	Single shot, 100, 200 rounds/minute	
Stabilizer System:	Electric, azimuth and elevation	

Primary: TOW missile system dual launcher on turret

Traverse: w/turret	360 degrees
Elevation:	+29 to -19 degrees
Elevation Rate: (max)	15 degrees/second

Secondary:

(1) 7.62mm M240C MG coaxial w/25mm gun in turret
(6) 5.56 M231 firing port weapons
(1) 7.62mm M60 MG stowed
(9) 5.56mm M16A1 rifles, M2 only
(10) 5.56mm M16A1 rifles, M2A1 only

AMMUNITION

900 rounds 25mm
2200 rounds 7.62mm (M240C)
2200 rounds 7.62mm (M60)
4200 rounds 5.56mm (M231)
2520 rounds 5.56mm (M16A1), M2 only
2800 rounds 5.56mm (M16A1), M2A1 only
3 66mm M72A2 light antitank weapons (LAW)
5 TOW or Dragon missiles, M2 only + 2 TOW in launcher
5 TOW 2 or Dragon missiles, M2A1 only + 2 TOW 2 in launcher

FIRE CONTROL AND VISION EQUIPMENT

Primary Weapon Sights:	Integrated sight unit (day/night)	
	5x auxiliary sight, late vehicles	
	External ring sight	
Vision Devices	Direct	Indirect
Driver	Hatch	Periscopes (4)
Commander	Hatch	Periscope M17 (7) early
		Periscope M17 (6) late
		Periscope M27 (1)
Gunner	Hatch	Periscope (2)
Troop compartment	Roof Hatch	Periscope (7)

Total Periscopes: (21)

ENGINE

Make and Model: Cummins VTA-903T	
Type: 8 cylinder, 4 cycle, vee, turbosupercharged	
Cooling System: Liquid	Ignition: Compression
Displacement:	903 cubic inches
Bore and Stroke:	5.5 x 4.75 inches
Compression Ratio:	15.5:1
Gross Horsepower: (max)	500 hp at 2600 rpm
Gross Torque: (max)	1025 ft-lbs at 2350 rpm
Weight:	2450 pounds, dry
Fuel: diesel oil MIL-VV-F-800	197 gallons
Engine Oil:	22 quarts

POWER TRAIN

Transmission: General Electric HMPT-500
Hydromechanical, automatic range selection
Steering: Hydrostatic, steering yoke
Brakes: Multiple plate, oil cooled

Final Drive: Spur gear	Gear Ratio: M2,	4.959:1
	M2A1,	5.561:1

Drive Sprockets: at front of Vehicle with 11 teeth
Pitch Diameter: 21 inches

RUNNING GEAR

Suspension Torsion bar
12 individually sprung dual road wheels (6/track)
Tire Size: 24 x 4 inches
2 single and 1 dual track return rollers per track
Dual adjustable idler at rear of each track
Idler Size: Outer, 20.0 x 3.7 inches
Inner, 19.9 x 5.7 inches
Shock absorbers on first 3 and last road wheels on each side
Tracks: Center guide
Type: Single pin, 21 inch width, steel w/detachable rubber pad
Pitch: 6 inches
Shoes per Vehicle: 166 (84 left, 82 right)
Ground Contact Length: 154 inches

ELECTRICAL SYSTEM

Nominal Voltage: 24 volts DC
Main Generator: 28 volts, 220 amperes, driven by main engine
Auxiliary Generator: None
Battery: (4) 12 volts, 2 sets of 2 in series connected in parallel

COMMUNICATIONS

Radio: AN/VRC-46, AN/GRC-160, M2 and M2A1
Interphone:

FIRE PROTECTION AND NBC PROTECTION

(1) 7 pound Halon, fixed in engine compartment
(2) 5 pound Halon, fixed in crew compartment
(2) 2.75 pound Halon, portable
Gas particulate filter unit, M2A1 only

PERFORMANCE

Maximum Speed: Level road		41 miles/hour
Water		4.5 miles /hour
Maximum Grade:		60 per cent
Maximum Trench:		8.3 feet
Maximum Vertical Wall:		36 inches
Maximum Fording Depth:		floats
Minimum Turning Circle: (diameter)		pivot
Cruising Range: Roads	approx.	300 miles

GENERAL DATA

Crew:	5	men
Length: Overall:	254	inches
Width: Overall	126	inches
Height: Over commander's hatch	117	inches
Tread:	96	inches
Ground Clearance:	18	inches
Weight, Combat Loaded: M3	49,945	pounds
M3A1	50,510	pounds
Weight, Unstowed: M3	41,975	pounds
M3A1	43,600	pounds
Weight, Air Transport:	40,775	pounds
Power to Weight Ratio: Gross hp, M3	20.0	hp/ton
M3A1	19.8	hp/ton
Ground Pressure: Zero penetration, M3	7.7	psi
M3A1	7.8	psi

ARMOR

The turret is armored by a combination of steel, 5083 aluminum, and 7039 aluminum armor.

The hull is assembled with spaced laminate armor on the vertical sides and rear. This consists of two 0.25 inch thick steel plates one inch apart spaced 3 ½ inches from the 1 inch thick aluminum armor. The side slopes are 7039 aluminum armor. The top, bottom and front are 5083 aluminum armor. Steel applique armor 0.375 inches thick is installed on the front third of the bottom for mine protection.

ARMAMENT

Primary: 25mm automatic gun M242 in turret mount

Traverse. Electric and manual	360 degrees
Traverse Rate: (max)	6 seconds/360 degrees
Elevation: M3, Electric and manual	+59 to -9 degrees
M3A1, Electric and manual	+57 to -9 degrees
Elevation Rate: (max)	60 degrees/second
Firing Rate:	Single shot, 100, 200 rounds/minute
Stabilizer System	Electric, azimuth and elevation

Primary: TOW missile system dual launcher on turret

Traverse: w/turret	360 degrees
Elevation:	+29 to -19 degrees
Elevation Rate: (max)	15 degrees/second

Secondary:

(1) 7.62mm M240C MG coaxial w/25mm gun in turret
(1) 7.62mm M60 MG stowed
(5) 5.56mm M16A1 rifles stowed

AMMUNITION

1500 rounds 25mm
4400 rounds 7.62mm (M240C)
3200 rounds 7.62mm (M60)
1680 rounds 5.56mm (M16A1)
3 66mm M72A2 light antitank weapons (LAW)
12 TOW missiles, M3
12 TOW 2 missiles M3A1

FIRE CONTROL AND VISION EQUIPMENT

Primary Weapon sights:	Integrated sight unit (day/night)
	5x auxiliary sight, late vehicles
	External ring sight

Vision Devices	Direct	Indirect
Driver	Hatch	Periscope (4)
Commander	Hatch	Periscope M17 (7) early
		(6) late
		Periscope M27 (1)
Gunner	Hatch	Periscope (2)
Troop Compartment	Roof Hatch	Periscope (7) M3
		Periscope (6) M3A1

Total Periscopes: (21) M3, (20) M3A1

ENGINE

Make and Model: Cummins VTA-903T	
Type: 8 cylinder, 4 cycle, vee, turbosuperchsrged	
Cooling System: Liquid	Ignition: Compression
Displacement:	903 cubic inches
Bore and Stroke:	5.5 x 4.75 inches
Compression Ratio:	15.5:1
Gross Horsepower: (max)	500 hp at 2600 rpm
Gross Torque: (max)	1025 ft-lbs at 2350 rpm
Weight:	2450 pounds, dry
Fuel: diesel oil MIL-VV-F-800	197 gallons
Engine Oil:	22 quarts

POWER TRAIN

Transmission: General Electric HMPT-500
      Hydromechanical, automatic range selection
Steering: Hydrostatic, steering yoke
Brakes: Multiple plate, oil cooled

Final Drive: Spur gear	Gear Ratio: M3, 4.959:1
	M3A1, 5.561:1

Drive Sprockets: At front of vehicle with 11 teeth
      Pitch Diameter: 21 inches

RUNNING GEAR

Suspension: Torsion bar
    12 individually sprung dual road wheels (6/track)
    Tire Size: 24 x 4 inches
    2 single and 1 dual track return rollers per track
    Dual adjustable idler at rear of each track
    Idler Size: Outer, 20.0 x 3.7 inches
             Inner, 19.0 x 5.7 inches
    Shock absorbers on first 3 and last road wheels on each side
Tracks: Center guide
    Type, Single pin, 21 inch width, steel w/detachable rubber pad
    Pitch: 6 inches
    Shoes per Vehicle: 166 (84 left, 82 right)
    Ground Contact Length: 154 inches

ELECTRICAL SYSTEM

Nominal Voltage: 24 volts DC
Main Generator: 28 volts, 220 amperes, driven by main engine
Auxiliary Generator: None
Battery: (4) 12 volts, 2 sets of 2 in series connected in parallel

COMMUNICATIONS

Radio: AN/VRC-12, AN/PRC-77
Interphone:

FIRE AND NBC PROTECTION

    (1) 7 pound Halon, fixed in engine compartment
    (2) 5 pound Halon, fixed in crew compartment
    (2) 2.75 pound Halon, portable
    Gas particulate filter unit, M3A1 only

PERFORMANCE

Maximum Speed: Level road		41 miles/hour
Water		4.5 miles/hour
Maximum Grade:		60 per cent
Maximum Trench:		8.3 feet
Maximum Vertical Wall:		36 inches
Maximum Fording Depth:		floats
Minimum Turning Circle: (diameter)		pivot
Cruising Range: Roads	approx.	300 miles

GENERAL DATA

Crew: M2A2	9	men
M2A2 Restow	10	men
Length: Overall	258	inches
Width: w/armor tile kit	142	inches
w/o armor tile kit	129	inches
Height: Over commander's hatch	117	inches
Tread:	96	inches
Ground Clearance:	18	inches
Weight, Combat Loaded: w/armor tile kit	approx. 66,000	pounds
w/o armor tile kit	approx. 60,000	pounds
Weight, Air Transport:	approx. 44,000	pounds
Power to Weight Ratio: w/armor tile kit	18.2	hp/ton
w/o armor tile kit	20.0	hp/ton
Ground Pressure: Zero penetration, w/armor tile kit	10.2	psi
w/o armor tile kit	9.3	psi

ARMOR

Turret: 5083 aluminum armor structure with steel applique armor. Provision for the installation of armor tiles on the front and right side. Hull: 5083 and 7039 aluminum armor structure with steel applique armor on the front, sides and bottom. Spaced laminate steel armor on rear and lower side skirts. Provision for the installation of armor tiles on the upper front and vertical sides.

ARMAMENT

Primary: 25mm automatic gun M242 in turret mount

Traverse: Electric and manual	360 degrees
Traverse Rate: (max)	6 seconds/360 degrees
Elevation: Electric and manual	+57 to -9 degrees
Elevation Rate: (max)	60 degrees/second
Firing Rate:	Single shot, 100, 200 rounds/minute
Stabilizer System:	Electric, azimuth and elevation

Primary: TOW missile system dual launcher on turret

Traverse: w/turret	360 degrees
Elevation:	+29 to -19 degrees
Elevation Rate: (max)	15 degrees/second

Secondary:

(1) 7.62mm M240C MG coaxial w/25mm gun in turret
(2) 5.56mm M231 firing port weapons
(1) 7.62mm M60 MG stowed
(9) 5.56mm M16A1 rifles stowed, M2A2 only
(10) 5.56mm M16A1 rifles stowed, M2A2 restow only

AMMUNITION

900 rounds 25mm
2200 rounds 7.62mm (M240C)
2200 rounds 7.62mm (M60)
2520 rounds 5.56mm (M231)
2520 rounds 5.56mm (M16A1), M2A2
2800 rounds 5.56mm (M16A1), M2A2 restow
3 66mm M72A2 light antitank weapons (LAW)
3 AT4 missiles, M2A2 restow only
6 Stinger missiles, M2A2 w/MAN/PADS only
5 TOW 2 or Dragon missiles + 2 TOW 2 in launcher
or
5 TOW 2 or 3 TOW 2 and 2 Javelin missiles, M2A2 restow only
+ 2 TOW 2 in launcher

FIRE CONTROL AND VISION EQUIPMENT

Primary Weapon Sights: Integrated sight unit (day/night)
                         5x auxiliary sight
                         External ring sight

Vision Devices	Direct	Indirect
Driver	Hatch	Periscope (4)
Commander	Hatch	Periscope M17 (6)
		Periscope M27 (1)
Gunner	Hatch	Periscope (2)
Troop Compartment	Roof Hatch	Periscope (6)

Total Periscopes (19)

ENGINE

Make and Model: Cummins VTA-903T	
Type: 8 cylinder, 4 cycle, vee, turbosuperchsrged	
Cooling System: Liquid	Ignition: Compression
Displacement:	903 cubic inches
Bore and Stroke:	5.5 x 4.75 inches
Compression Ratio:	14.5:1
Gross Horsepower: (max)	600 hp at 2600 rpm
Gross torque: (max)	1225 ft-lbs at 2300 rpm
Weight:	2650 pounds, dry
Fuel: diesel oil MIL-VV-F-800	175 gallons
Engine Oil:	26 quarts

POWER TRAIN

Transmission: General Electric HMPT-500-3 or HMPT-500-3EC
    Hydromechanical, automatic range selection
Steering: Hydrostatic, steering yoke
Brakes: Multiple plate, oil cooled
Final Drive: Spur gear   Gear Ratio: 5.561:1
Drive Sprockets: At front of vehicle with 11 teeth
    Pitch Diameter: 21 inches

RUNNING GEAR

Suspension: Torsion bar
    12 individually sprung dual road wheels (6/track)
    Tire Size: 24 x 4 inches
    2 single and 1 dual track return rollers per track
    Dual adjustable idler at rear of each track
    Idler Size: Outer, 20.0 x 3.7 inches
                Inner, 19.9 x 5.7 inches
    Shock absorbers on first 3 and last road wheels on each side
Tracks: Center guide
    Type: Single pin, 21 inch width, steel w/detachable rubber pad
    Pitch: 6 inches
    Shoes per Vehicle: 166 (84 left, 82 right)
    Ground Contact Length: 154 inches

ELECTRICAL SYSTEM

Nominal Voltage: 24 volts DC
Main Generator: 28 volts, 300 amperes, driven by main engine
Auxiliary Generator: None
Battery: (4) 12 volts, 2 sets of 2 in series connected in parallel

COMMUNICATIONS

Radio: AN/VRC-46, AN/GRC-160
Interphone:

FIRE AND NBC PROTECTION

(1) 7 pound Halon, fixed in engine compartment
(2) 5 pound Halon, fixed in crew compartment
(2) 2.75 pounds Halon, portable
Gas particulate filter unit

PERFORMANCE

Maximum Speed: Level road		35 miles/hour
Water		4.0 miles/hour
Maximum Grade:		60 per cent
Maximum Trench:		7 feet
Maximum Vertical Wall:		30 inches
Maximum Fording Depth:		floats
Minimum Turning Circle: (diameter)		pivot
Cruising Range: Roads	approx.	250 miles

GENERAL DATA

Crew:	5	men
Length: Overall	258	inches
Width: w/armor tile kit	142	inches
w/o armor tile kit	129	inches
Height: Over commander's hatch	117	inches
Tread:	96	inches
Ground Clearance:	18	inches
Weight, Combat Loaded: w/armor tile kit approx.	66,000	pounds
w/o armor tile kit approx.	60,000	pounds
Weight, Air Transport: approx.	44,000	pounds
Power to Weight Ratio: w/armor tile kit	18.2	hp/ton
w/o armor tile kit	20.0	hp/ton
Ground Pressure: Zero penetration, w/armor tile kit	10.2	psi
w/o armor tile kit	9.3	psi

ARMOR

Turret: 5083 aluminum armor structure with steel applique armor. Provision for the installation of armor tiles on the front and right side. Hull: 5083 and 7039 aluminum armor structure with steel applique armor on the front, sides and bottom. Spaced laminate steel armor on rear and lower side skirts. Provision for the installation of armor tiles on the upper front and vertical sides.

ARMAMENT

Primary: 25mm automatic gun M242 in turret mount

Traverse: Electric and manual	360 degrees
Traverse Rate: (max)	6 seconds/360 degrees
Elevation: Electric and manual	+57 to -9 degrees
Elevation Rate: (max)	60 degrees/second
Firing Rate:	Single shot, 100, 200 rounds/minute
Stabilizer System:	Electric, azimuth and elevation

Primary: TOW missile system dual launcher on turret

Traverse: w/turret	360 degrees
Elevation:	+29 to -19 degrees
Elevation Rate: (max)	15 degrees/second

Secondary:

(1) 7.62mm M240C MG coaxial w/25mm gun in turret
(1) 7.62mm MG M60 MG stowed
(5) 5.56 mm M16A1 rifles stowed

AMMUNITION

1500 rounds 25mm
4200 rounds 7.62mm (M240C)
3400 rounds 7.62mm (M60)
1680 rounds 5.56mm (M16A1)
3 66mm M72A2 light antitank weapons (LAW)
3 AT4 missiles, M3A2 restow only
12 TOW missiles

FIRE CONTROL AND VISION EQUIPMENT

Primary Weapon Sights:	Integrated sight unit (day/night)	
	5x auxiliary sight	
	External ring sight	
Vision Devices	Direct	Indirect
Driver	Hatch	Periscope (4)
Commander	Hatch	Periscope M17 (6)
		Periscope M27 (1)
Gunner	Hatch	Periscope (2)
Troop Compartment	Roof Hatch	Periscope (5)

Total Periscopes: (18)

ENGINE

Make and Model: Cummins VTA-903T
Type: 8 cylinder, 4 cycle, vee, turbosupercharged

Cooling System: Liquid	Ignition: Compression
Displacement:	903 cubic inches
Bore and Stroke:	5.5 x 4.75 inches
Compression Ratio:	14.5:1
Gross Horsepower: (max)	600 hp at 2600 rpm
Gross Torque: (max)	1225 ft-lbs at 2300 rpm
Weight:	2650 pounds, dry
Fuel: diesel oil MIL-VV-F-800	175 gallons
Engine Oil:	26 quarts

POWER TRAIN

Transmission: General Electric HMPT-500-3 or HMPT-500-3EC Hydromechanical, automatic range selection
Steering: Hydrostatic, steering yoke
Brakes: Multiple plate, oil cooled

Final Drive: Spur gear	Gear Ratio: 5.561:1

Drive Sprockets: At front of vehicle with 11 teeth
Pitch Diameter: 21 inches

RUNNING GEAR

Suspension: Torsion bar
12 individually sprung dual road wheels (6/track)
Tire Size: 24 x 4 inches
2 single and 1 dual track return rollers per track
Dual adjustable idler at rear of each track
Idler Size: Outer, 20.0 x 3.7 inches
Inner, 19.9 x 5.7 inches
Shock absorbers on first 3 and last road wheels on each side
Tracks: Center guide
Type: Single pin, 21 inch width, steel w/detachable rubber pad
Pitch: 6 inches
Shoes per Vehicle: 166 (84 left, 82 right)
Ground Contact Length: 154 inches

ELECTRICAL SYSTEM

Nominal Voltage: 24 volts DC
Main Generator: 28 volts, 300 amperes, driven by main engine
Auxiliary Generator: None
Battery: (4) 12 volts, 2 sets of 2 in series connected in parallel

COMMUNICATIONS

Radio: AN/VRC-46, AN/VRC-47, AN/PRC-77, AN/GRC-160
Interphone:

FIRE AND NBC PROTECTION

(1) 7 pound Halon, fixed in engine compartment
(2) 5 pound Halon, fixed in crew compartment
(2) 2.75 pound Halon, portable
Gas particulate filter unit

PERFORMANCE

Maximum Speed: Level road		35 miles/hour
Water		4.0 miles/hour
Maximum Grade:		60 per cent
Maximum Trench:		7 feet
Maximum Vertical Wall		30 inches
Maximum Fording Depth:		floats
Minimum Turning Circle: (diameter)		pivot
Cruising Range: Roads	approx.	250 miles

GENERAL DATA

Crew:	10	men
Length: Overall:	258	inches
Width: w/armor tile kit	142	inches
w/o armor tile kit	129	inches
Height: Over commander's hatch	117	inches
Tread:	96	inches
Ground Clearance:	18	inches
Weight, Combat Loaded: w/armor tile kit approx.	67,000	pounds
w/o armor tile kit approx.	61,000	pounds
Weight, Air Transport: approx.	45,000	pounds
Power to Weight Ratio: w/armor tile kit	17.9	hp/ton
w/o armor tile kit	19.7	hp/ton
Ground Pressure: Zero penetration, w/armor tile kit	10.4	ps
w/o armor tile kit	9.4	psi

ARMOR

Turret: 5083 aluminum armor structure with steel applique armor.
Provision for the installation of armor tiles on the front and right side.
Hull: 5083 and 7039 aluminum armor structure with steel applique
armor on the front, sides, and bottom. Spaced laminate steel armor
on rear and lower side skirts. Provision for the installation of armor
tiles on the upper front and vertical sides.

ARMAMENT

Primary: 25mm automatic gun M242 in turret mount

Traverse: Electric and manual	360 degrees
Traverse Rate: (max)	6 seconds/360 degrees
Elevation: Electric and manual	+57 to -9 degrees
Elevation Rate: (max)	60 degrees/second
Firing Rate:	Single shot, 100, 200 rounds/minute
Stabilizer System:	Electric, azimuth and elevation

Primary: TOW missile system dual launcher on turret

Traverse: w/turret	360 degrees
Elevation:	+29 to -19 degrees
Elevation Rate: (max)	15 degrees/second

Secondary:
  (1) 7.62mm M240C MG coaxial w/25mm gun in turret
  (2) 5.56mm M231 firing port weapons
  (1) 7.62mm M60 MG stowed
  (10) 5.56mm M16A1 rifles stowed

AMMUNITION

  900 rounds 25mm
  2200 rounds 7.62mm (M240C)
  2200 rounds 7.62mm (M60)
  2520 rounds 5.56mm (M231)
  2800 rounds 5.56mm (M16A1)
  3 66mm M72A2 light antitank weapons (LAW)
  3 AT4 missiles
  5 TOW2 or 3 TOW2 and 2 Javelin missiles + 2 TOW2 in launcher

FIRE CONTROL AND VISION EQUIPMENT

Primary Weapon Sights:	Integrated sight unit (day/night)
	5x auxiliary sight
	External ring sight
	Laser range finder
	Target acquisition system, M2A3 only

Vision Devices	Direct	Indirect
Driver	Hatch	Periscope (4)
		Thermal viewer
Commander	Hatch	Periscope M17 (6)
		Periscope M27 (1)
		Independent viewer, M2A3 only
		Forward looking infrared system
Gunner	Hatch	Periscope (2)
Troop Compartment	Roof Hatch	Periscope (6)

Total Periscopes: (19)

ENGINE

Make and Model: Cummins VTA-903T	
Type: 8 cylinder, 4 cycle, vee, turbosupercharged	
Cooling System: Liquid	Ignition: Compression
Displacement:	903 cubic inches
Bore and Stroke:	5.5 x 4.75 inches
Compression Ratio:	14.5:1
Gross Horsepower: (max)	600 hp at 2600 rpm
Gross Torque: (max)	1225 ft-lbs at 2300 rpm
Weight:	2650 pounds, dry
Fuel: diesel oil MIN-VV-F-800	175 gallons
Engine Oil:	26 quarts

POWER TRAIN

Transmission: General Electric HMPT-500-3EC
  Hydromechanical, automatic range selection
Steering: Hydrostatic, steering yoke
Brakes: Multiple plate, oil cooled
Final Drive: Spur gear   Gear Ratio: 5.561:1
Drive Sprockets: At front of vehicle with 11 teeth
  Pitch Diameter: 21 inches

RUNNING GEAR

Suspension: Torsion bar
  12 individually sprung dual road wheels (6/track)
  Tire Size: 24 x 4 inches
  2 single and 1 dual track return rollers per track
  Dual adjustable idler at rear of each track
  Idler Size: Outer, 20.0 x 3.7 inches
            Inner, 19.9 x 5.7 inches
  Shock absorbers on first 3 and last road wheels on each side
Tracks: Center guide
  Type: Single pin, 21 inch width, steel w/detachable rubber pad
  Pitch: 6 inches
  Shoes per Vehicle: 166 (84 left, 82 right)
  Ground Contact Length: 154 inches

ELECTRICAL SYSTEM

Nominal Voltage: 24 volts DC
Main Generator: 28 volts, 300 amperes, driven by main engine
Auxiliary Generator: None
Battery: (4) 12 volts, 2 sets of 2 in series connected in parallel

COMMUNICATIONS AND NAVIGATION

Radio: SINCGARS w/SIP/INC
Pos/Nav w/inertial GPS system
Commander's tactical display w/map graphics
Squad tactical display /FLIR monitor
Enhanced position location reporting system
Combat identification system
Vehicular intercommunication system (VIS)

FIRE AND NBC PROTECTION

  (1) 7 pounds Halon, fixed in engine compartment
  (2) 5 pound Halon, fixed in crew compartment
  (2) 2.75 pound Halon, portable

PERFORMANCE

Maximum Speed: Level road		35 miles/hour
Water		4.0 miles/hour
Maximum Grade:		60 per cent
Maximum Trench:		7 feet
Maximum Vertical Wall:		30 inches
Maximum Fording Depth:		floats
Minimum Turning Circle: (diameter)		pivot
Cruising Range: Roads	approx.	250 miles

ARMORED VEHICLE MOUNTED ROCKET LAUNCHER M270
(Rocket Launcher M269 on Carrier M993)

GENERAL DATA

Crew: (driver, commander, gunner)	3	men
Length: Overall:	274.5	inches
Width: Overall:	117	inches
Height: w/launcher stowed	102	inches
w/launcher fully elevated	232	inches
Tread:	96	inches
Ground Clearance:	17	inches
Weight, Combat Loaded: w/12 rockets	52,990	pounds
w/2 missiles	52,428	pounds
Weight, Empty:	42,800	pounds
Weight, M993 Carrier Only:	32,200	pounds
Power to Weight Ratio: w/12 rockets	18.9	hp/ton
Ground Pressure: Zero penetration, w/12 rockets	7.4	psi

ARMOR

Both the carrier and launcher are lightly armored.
The cab is 7039 aluminum armor and the hull is 5083 aluminum armor.
The cab windows are armor glass with louvers for splinter protection.

ARMAMENT

The launcher is armed with 2 rocket pods of 6 rockets each or 2 missile pods each carrying a single missile
>   (1) 5.56mm M16A1 rifle stowed

AMMUNITION
>   12 227mm M26 rockets
>   or
>   2 Army tactical missile systems (ATACMS) M39

FIRE CONTROL AND VISION EQUIPMENT

Computer controlled fire control system that can be quickly programmed to operate with several types of rockets and missiles.
Vision Devices:
Armor glass windows in cab
A vision block is installed on each side of the cab

ENGINE

Make and Model: Cummins VTA-903T
Type: 8 cylinder, 4 cycle, vee, turbosupercharged

Cooling System: Liquid	Ignition: Compression
Displacement:	903 cubic inches
Bore and Stroke:	5.5 x 4.75 inches
Compression Ratio:	15.5:1
Gross Horsepower: (max)	500 hp at 2600 rpm
Gross Torque: (max)	1025 ft-lbs at 2350 rpm
Weight:	2450 pounds, dry
Fuel: diesel oil MIL-V-V-F-800	163 gallons
Engine Oil:	22 quarts

POWER TRAIN

Transmission: General Electric HMPT-500
>   Hydromechanical, automatic range selection

Steering: Hydrostatic, steering yoke
Brakes: Multiple plate, oil cooled
Final Drive: Spur gear   Gear Ratio: 5.561:1
Drive Sprockets: At front of vehicle with 11 teeth
>   Pitch Diameter: 21 inches

RUNNING GEAR

Suspension: Torsion bar
>   12 individually sprung dual road wheels (6/track)
>   Tire Size: 24 x 4 inches
>   2 single and 2 dual track return rollers per track
>   Dual adjustable idler at rear of each track
>   Idler Size: Outer, 20.0 x 3.7 inches
>   Inner, 19.9 x 5.7 inches
>   Shock absorbers on first 2 and last road wheels on each side
>   Suspension lock-out to stabilize the vehicle during loading or launching operations.

Tracks: Center guide
>   Type: Single pin, 21 inch width, steel w/detachable rubber pad
>   Pitch: 6 inches
>   Shoes per Vehicle: 177 (89 left, 88 right)
>   Ground Contact Length: 170.5 inches

ELECTRICAL SYSTEM

Nominal Voltage: 24 volts DC
Main Generator: 28 volts, 300 amperes, driven by main engine
Auxiliary Generator: 28 volts, 300 amperes driven by power
>   take-off from transmission

Battery: (4) 12 volts, 2 sets of 2 in series connected in parallel

COMMUNICATIONS

Radio: RT-524A/VRC, R-442/VRC

FIRE AND NBC PROTECTION
>   (1) 7 pound Halon, fixed in engine compartment
>   (1) 2.75 pound Halon, portable
>   Gas particulate filter unit

PERFORMANCE

Maximum Speed: level road		40 miles/hour
Maximum Grade:		60 per cent
Maximum Trench:		7.6 feet
Maximum Vertical Wall:		30 inches
Maximum Fording Depth:		40 inches
Minimum Turning Circle:		pivot
Cruising Range: Roads	approx.	300 miles

GENERAL DATA
Crew: 2 to 4 men
Length: Overall, M29   125.75 inches
          M29C   192.1 inches
Width: Overall, M29   66 inches
          M29C   67.25 inches
Height: Over windshield   70.6 inches
Tread: 45.0 inches
Ground Clearance: M29   11.0 inches
          M29C   10.5 inches
Weight, Combat Loaded: M29   5277 pounds
          M29C   5971 pounds
Weight, Unstowed: M29   4077 pounds
          M29C   4771 pounds
Power to Weight Ratio: Net, M29   24.6 hp/ton
          M29C   21.8 hp/ton
Ground Pressure: Zero penetration: M29   1.7 psi
          M29C   1.9 psi

ARMOR
None
ARMAMENT
None
AMMUNITION
None
VISION EQUIPMENT
Open top vehicle

ENGINE
Make and Model: Studebaker 6-170
Type: 6 cylinder, 4 cycle, L-head, in-line
Cooling System: Liquid   Ignition: Battery
Displacement: 169.6 cubic inches
Bore and Stroke: 3 x 4 inches
Compression Ratio: 7:1
Net Horsepower: (max)   65 hp at 3600 rpm
Net Torque: (max)   130 ft-lbs at 1800 rpm
Weight: 500 pounds, dry
Fuel: 80 octane gasoline   35 gallons
Engine Oil: 5 quarts
POWER TRAIN
Clutch: Dry disc
Transmission: Synchromesh (2nd and 3rd), 3 speeds forward, 1 reverse

| Gear Ratios: | 1st | 2.66:1 | 3rd | 1.00:1 |
| | 2nd | 1.49:1 | reverse | 3.55:1 |

Transfer Case: Dual speed

| Gear Ratios: | High | 0.866:1 | Low | 2.74:1 |

Steering: Controlled differential

| Gear Ratio: 4.87:1 | Steering Ratio: 1.73:1 |

Brakes: Differential band
Drive Sprockets: At rear of vehicle with 9 teeth
RUNNING GEAR
Suspension: Transverse leaf spring
          16 dual wheels in 8 bogies (4 bogies/track)
          Wheel Size: 8 x 1.25 inches
          Dual Idler at front of each track
          Idler Size: 11.75 x 7 inches
          4 dual track return rollers (2/track)
Tracks: Center guide, T76E1
          Type: Rubber band, single piece, 20 inch width
          Pitch: Cross bar, 4.5 inches
          Shoes per Vehicle: Cross bars 112 (56/track)
          Ground Contact Length: 78.1 inches
ELECTRICAL SYSTEM
Nominal Voltage: 12 volts DC
Main Generator: (1) 12 volts, 40 amperes, driven by main engine
                                 or if radios are installed
                    (1) 12 volts, 55 amperes, driven by main engine
Auxiliary Generator: None
Battery: (2) 6 volts in series
COMMUNICATIONS
Radio: SCR 506, 508, 510, 528, 608, 610, 628, 694, 714 or
          British Number 19
Interphone: None
FIRE PROTECTION
          (1) 1 quart carbon tetrachloride, portable
PERFORMANCE
Maximum Speed: Level road   36 miles/hour
                    Water, M29C only   4 miles/hour
Maximum Tractive Effort: TE at stall   4200 pounds
          Per Cent of Vehicle Weight: TE/W, M29   80 per cent
                                          M29C   70 per cent
Maximum Grade: 100 per cent
Maximum Trench: 3 feet
Maximum Vertical Wall: 10 inches
Maximum Fording Depth: M29C   floats
Minimum Turning Circle: (diameter)   24 feet
Cruising Range: Roads   approx. 175 miles

GENERAL DATA

Crew:	2	men
Passengers:	8	men
Length: Overall	195.3	inches
Width: Over tracks	98	inches
Height: Overall	103.3	inches
Tread:	68	inches
Ground Clearance:	14.7	inches
Weight, Combat Loaded:	12,045	pounds
Weight, Unstowed:	8,329	pounds
Power to Weight Ratio: Net	18.3	hp/ton
Gross	22.4	hp/ton
Ground Pressure: Zero penetration	2.0	psi

ARMOR

None

ARMAMENT

(1) .50 caliber MG HB M2 over assistant driver's hatch

AMMUNITION

630 rounds .50 caliber

VISION EQUIPMENT

Windows in cab

WINCH

Located in cargo compartment under the rear seat

Capacity: 5000 pounds

Cable Size: 3/8 inches x 85 feet

ENGINE

Make and Model: Continental AOI-268-3A

Type: 4 cylinder, 4 cycle, opposed, fuel injection

Cooling System: Air	Ignition: Magneto
Displacement:	268.8 cubic inches
Bore and Stroke:	4.625 x 4.000 inches
Compression Ratio:	6.9:1
Net Horsepower: (max)	110 hp at 3200 rpm
Gross Horsepower: (max)	135 hp at 3200 rpm
Net Torque: (max)	208 ft-lbs at 2200 rpm
Gross Torque: (max)	222 ft-lbs at 2600 rpm
Weight:	620 pounds, dry
Fuel: 80 octane gasoline	70 gallons
Engine Oil:	12 quarts

POWER TRAIN

Transmission: Cross drive CD-150

Torque Converter: Hydraulic, single stage

Torque Converter Stall Ratio: 4.04:1

Overall Usable Ratios: Low 7.96:1   reverse 9.36:1
                        High 5.2:1

Steering: Land, mechanical steering bar
          Water, hydraulic, steering shaft

Brakes: Mechanical

Final Drive: Planetary gear   Gear Ratio: 4.105:1

Drive Sprockets: At front of vehicle with 12 teeth
          Pitch Diameter: 7.639 inches

RUNNING GEAR

Suspension: Torsion bar
    8 individually sprung dual road wheels (4/track)
    Tire Size: 6.60 x 15.00 pneumatic
    Dual idler at rear of each track
    Double acting shock absorbers on first and last road wheels

Tracks: Double center and outside guides
    Type: Band type, sectional, 30 inch width
    Section Length: 44 or 88 inches
    Pitch: Cross bar, 4 inches
    Shoes per Vehicle: Cross bars 176 (88/track)
    Ground Contact Length: 98.6 inches

ELECTRICAL SYSTEM

Nominal Voltage: 24 volts DC

Main Generator: (1) 28 volts, 25 amperes, driven by main engine

Auxiliary Generator: None

Battery: (2) 12 volts in series

COMMUNICATIONS

Radio: AN/GRC-3 thru 8, AN/VRQ-1 thru 3, AN/VRC-7,
       AN/VRC-16 thru 18, or AN/GRC-9

Interphone: None

FIRE PROTECTION

(1) 5 pound carbon dioxide, portable

PERFORMANCE

Maximum Speed: Level road		28 miles/hour
Water		4.5 miles/hour
Maximum Tractive Effort: TE at stall		8900 pounds
Per Cent of Vehicle Weight: TE/W		74 per cent
Maximum Grade:		60 per cent
Maximum Trench:		5 feet
Maximum Vertical Wall:		18 inches
Maximum Fording Depth:		floats
Minimum Turning Circle: (diameter)		pivot
Cruising Range: Roads	approx.	200 miles
Water	approx.	5.8 hours

CARGO CARRIER M116 "HUSKY"
AND
AMPHIBIOUS ASSAULT VEHICLE XM733

GENERAL DATA

Crew: M116	1	man
XM733 (driver, 5 gunners)	6	men
Passengers: M116 only, Summer gear	13	men
Winter gear	10	men
Length: Overall, M116	188.1	inches
XM733	195	inches
Width: Over track shrouds	82.1	inches
Height: M116, Over cab	79.1	inches
XM733, Overall	68	inches
Tread:	58.5	inches
Ground Clearance:	14	inches
Weight, Combat Loaded: M116	10,600	pounds
XM733	11,650	pounds
Weight, Unstowed: M116	6,700	pounds
XM733	10,000	pounds
Power to Weight Ratio: M116	30.2	hp/ton
XM733	27.5	hp/ton
Ground Pressure: Zero penetration, M116	2.6	psi
XM733	2.8	psi

ARMOR

M116: None

XM733: Light steel armor bolted to the aluminum hull

ARMAMENT

M116

    (1) 7.62mm rifle M14, stowed

XM733

    (4) 7.62mm MG M60D

    (1) 7.62mm rifle M14

    (1) 40mm automatic grenade launcher M175

        or

    (1) 7.62mm automatic gun (minigun)

AMMUNITION

M116

    180 rounds 7.62mm

XM733

    1200 rounds 7.62mm, M60D

    180 rounds 7.62mm, rifle

    100 rounds 40mm M384

        or

    500 rounds 7.62mm, linked (minigun)

VISION EQUIPMENT

Windows in cab (M116) or open top vehicle

WINCH

Located in the hull front

Capacity: 5000 to 6000 pounds

Cable Size: 3/8 inches x 100 feet

ENGINE

Make and Model: Chevrolet 283-V8 Military

Type: 8 cylinder, 4 cycle, vee

Cooling System: Liquid   Ignition: Delco

Displacement:	283 cubic inches
Bore and Stroke:	3.875 x 3 inches
Compression Ratio:	8:1
Net Horsepower: (max)	115 hp at 3600 rpm
Gross Horsepower: (max)	160 hp at 4200 rpm
Net Torque: (max)	210 ft-lbs at 2400 rpm
Gross Torque: (max)	240 ft-lbs at 2800 rpm
Fuel: 80 octane gasoline	65 gallon + 10 gallons stowed
Engine Oil:	7 quarts

POWER TRAIN

Transmission: Hydramatic Model 305MC

    Automatic: 4 speeds forward, 1 reverse

Gear Ratios:	1st	4.09:1	4th	1.00:
	2nd	2.63:1	reverse	4.54:1
	3rd	1.55:1		

Steering: Allison GS 100-3, geared, clutch-brake, steering wheel

Brakes: Multiple plate, oil cooled

Final Drive: Integral w/GS 100-3   Gear Ratio: 4.17:1

Drive Sprockets: At front of vehicle with 11 teeth

    Pitch Diameter: 14.00 inches

RUNNING GEAR

Suspension: Flat track, torsion bar

    10 individually sprung dual road wheels (5/track)

    Tire Size: 22 x 2 inches

    Dual adjustable idler at rear of each track

    Rotary shock absorbers on first and last road wheels on each side

Tracks: Double center guide

    Type: Band type, 20 inch width

    Pitch: Cross bar, 4 inches, 8 cross bars/32 inch section

    Sections per Vehicle: 22 (11/track)

    Ground Contact Length: 103 inches

ELECTRICAL SYSTEM

Nominal Voltage: 24 volts DC

Main Generator: (alternator) 24 volts, 100 amperes, driven by main engine

Auxiliary Generator: None

Battery: (4) 2 sets of 2 in series connected in parallel

COMMUNICATIONS

Radio: M116, AN/GRC-3 thru 9, AN/GRC-19, AN/GRR-5, AN/VRC-9,

        AN/VRC-10, AN/VRC-18, AN/VRC-24, AN/VRC-35,

        or AN/VRQ-1 thru 3

        XM733, AN/GRC-125, AN/VRC-46

Interphone: AN/UIC-1, 2 stations

FIRE PROTECTION

    (1) 5 pound carbon dioxide, portable

PERFORMANCE

Maximum Speed: Level road		37 miles/hour
Water		4.2 miles/hour
Maximum Tractive Effort: TE at Stall		9100 pounds
Per Cent of Vehicle Weight: TE/W, M116		86 per cent
XM733		78 per cent
Maximum Grade:		60 per cent
Maximum Trench:		4.8 feet
Maximum Vertical Wall:		18 inches
Maximum Fording Depth:		floats
Minimum Turning Circle: (diameter)		16 feet
Cruising Range: Roads	approx.	300 miles

GENERAL DATA

Crew: (driver, 3 passengers in cab)	4	men
Length: Overall	232	inches
Width: Over track shrouds	105.75	inches
Height: Over canopy bows	110.75	inches
Tread:	85.0	inches
Ground Clearance:	17.1	inches
Weight, Combat Loaded:	28,300	pounds
Weight, Unstowed:	16,300	pounds
Power to Weight ratio:	15.0	hp/ton
Ground Pressure: Zero penetration	8.5	psi

ARMOR
None
ARMAMENT
Optional, (1) .50 caliber MG HB M2 on M66 mount over cab
AMMUNITION
Optional, 600 rounds .50 caliber
VISION EQUIPMENT
Windows in cab or open top vehicle

ENGINE
Make and Model: General Motors 6V53
Type: 6 cylinder, 2 cycle, vee

Cooling System: Liquid	Ignition: Compression
Displacement:	318 cubic inches
Bore and Stroke:	3.875 x 4.5 inches
Compression Ratio:	17.0:1
Gross Horsepower: (max)	212 hp at 2800 rpm
Gross Torque: (max)	492 ft-lbs at 1300 rpm
Weight:	1310 pounds, dry
Fuel: diesel oil MIL-VV-F-800	105 gallons
Engine Oil:	22 quarts

POWER TRAIN
Transfer Case: Overall ratio 1.286:1 overdrive
Transmission: Allison TX-100
    Torque Converter: Hydraulic, single stage, multiphase w/lockup
    Torque Converter Stall ratio: 3.5:1

Transmission Ratios:	1st	3.81:1	3rd	1.00:1
	2nd	1.936:1	reverse	4.35:1

Steering: DS200 controlled differential, steering brake levers
    Input Ratio: 1.28:1   Steering Ratio: 1.1 to 1.786:1
Brakes: Differential band
Final Drive: Spur gear   Gear Ratio: 3.928:1
Drive Sprockets: At front of vehicle with 10 teeth
    Pitch Diameter: 19.618 inches
RUNNING GEAR
Suspension: Flat track, torsion bar
    10 individually sprung dual road wheels (5/track)
    Tire Size: 24 x 2.1 inches
    Dual adjustable idler at rear of each track
    Idler Size: 21 x 2.1 inches
    Shock absorbers on first 2 and last road wheels on each side
Tracks: Center guide, T130E1
    Type: Single pin, 15 inch width, steel w/detachable rubber pad
    Pitch: 6 inches
    Shoes per Vehicle: 130 (65/track)
    Ground Contact Length: 111 inches
ELECTRICAL SYSTEM
Nominal Voltage: 24 volts DC
Main Generator: (1) 28 volts, 100 amperes, driven by main engine
Auxiliary Generator: None
Battery: (2) 12 volts in series
COMMUNICATIONS
Radio: None
FIRE PROTECTION
    (1) 5 pound carbon dioxide, fixed
    (1) 5 pound carbon dioxide, portable
PERFORMANCE

Maximum Speed: Level road	35 miles/hour
Maximum Grade:	60 per cent
Maximum Trench:	5.5 feet
Maximum Vertical Wall:	24 inches
Maximum Fording Depth:	40 inches
Minimum Turning Circle: (diameter)	28 feet
Cruising Range: Roads approx.	300 miles

GENERAL DATA

Crew:	1	man
Passengers:	10	men
Length:	203	inches
Width: M4, M4C	97	inches
M4A1, M4A1C, M4A2	111	inches
Height:	99	inches
Tread: M4, M4C	80	inches
M4A1, M4A1C, M4A2	87.1	inches
Ground Clearance:	20	inches
Weight, Combat Loaded: M4, M4C	36,000	pounds
M4A1, M4A1C	36,500	pounds
M4A2	38,000	pounds
Power to Weight Ratio: M4, M4C	11.7	hp/ton
M4A1, M4A1C	11.5	hp/ton
M4A2	11.1	hp/ton
Ground Pressure: Zero penetration, M4, M4C	8.8	psi
M4A1, M4A1C	6.2	psi
M4A2	6.5	psi

ARMOR
None
ARMAMENT
   (1) .50 caliber MG HB M2 on mount M49C or M66
AMMUNITION
     500 rounds .50 caliber
     one of the following in M4 and M4A1
     54 rounds 90mm
     30 rounds 155mm gun
     20 rounds 8 inch howitzer
     12 rounds 240mm howitzer
     M4C and M4A1C carry additional rounds
VISION EQUIPMENT
Windows in cab
WINCH
Gar Wood Model 4M718 located in rear hull
Capacity: 30,000 pounds
Cable Size: 3/4 inches x 300 feet

ENGINE
Make and Model: Waukesha 145GZ
Type: 6 cylinder, 4 cycle, in-line

Cooling System: Liquid	Ignition: Battery
Displacement:	817 cubic inches
Bore and Stroke:	5.375 x 6 inches
Compression Ratio:	5.95:1
Net Horsepower: (max)	210 hp at 2100 rpm
Net Torque: (max)	585 ft-lbs at 1500 rpm
Weight:	2150 pounds, dry
Fuel: 70 octane gasoline	125 gallons
Engine Oil:	18 quarts

POWER TRAIN
Torque Converter: Twin-Disk Model T-10010
Torque Converter Gear Ratio: 1.372:1
Transmission: Selective

Gear Ratios:	1st	2.166:1	3rd	0.437:1
	2nd	1.555:1	reverse	1.822:1

Steering: Controlled Differential, steering levers
   Gear Ratio: 2.666:1   Steering Ratio: 1.747:1
Brakes: Mechanical on controlled differential
Final Drive: Spur gear   Gear Ratio: 2.764:1
Drive Sprockets: At front of vehicle with 13 teeth
   Pitch Diameter: 25.038 inches
RUNNING GEAR
Suspension: Horizontal volute spring
   8 wheels in 4 bogies (2 bogies/track)
   Tire Size: 20 x 9 inches
   4 track return rollers (2/track)
   Adjustable trailing idler at rear of each track
   Idler Size: 32 x 9 inches
Tracks: Outside guide, T48, T49, T54E1
   Type: T48, Double pin, 16.56 inch width, rubber block
      T49, T54E1, double pin 16.56 inch width, steel
      On M4A1, M4A1C and M4A2 extended end
      connectors increase track width to 23.69 inches.
   Pitch: 6 inches
   Shoes per Vehicle: 130 (65/track)
   Ground Contact Length: 124 inches
ELECTRICAL SYSTEM
Nominal Voltage: 12 volts DC
Main Generator: (1) 15 volts, 26 amperes, driven by main engine
Auxiliary generator: None
Battery: (1) 12 volts
COMMUNICATIONS
Radio: None
Interphone: None
FIRE AND GAS PROTECTION
   (2) 4 pound carbon dioxide, portable
   (1) 1 1/2 quart decontaminating apparatus M2
PERFORMANCE

Maximum Speed: Level road towing a 90mm gun	33 miles/hour
Maximum Grade:	60 per cent
Maximum Trench:	5 feet
Maximum Vertical Wall:	29 inches
Maximum Fording Depth:	41 inches
Minimum Turning Circle:	39 feet
Cruising Range: Roads approx.	180 miles

GENERAL DATA
Crew: 1 man
Passengers: M5, M5A2   8 men
    M5A1, M5A3   10 men
Length: M5, M5A2   191.1 inches
    M5A1, M5A3   196.4 inches
Width: M5, M5A1   100.0 inches
    M5A2, M5A3   114.5 inches
Height: M5, M5A2, w/o MG   104 inches
    M5A1, M5A3   114.5 inches
Tread: M5, M5A1   83 inches
    M5A2, M5A3   94.5 inches
Ground Clearance: 19.8 inches
Weight, Combat Loaded: M5   28,572 pounds
    M5A1   30,405 pounds
    M5A2   26,149 pounds
    M5A3   30,350 pounds
Weight, Unstowed: M5   23,308 pounds
    M5A1   24,438 pounds
    M5A2   20,885 pounds
    M5A3   23,025 pounds
Power to Weight Ratio: M5   16.7 hp/ton
    M5A1   15.5 hp/ton
    M5A2   18.0 hp/ton
    M5A3   15.5 hp/ton
Ground Pressure: Zero penetration, M5   11.4 psi
    M5A1   12.1 psi
    M5A2   5.7 psi
    M5A3   6.7 psi
ARMOR
None
ARMAMENT
   (1) .50 caliber MG HB M2 on mount M49C
   Field modification required on M5 and M5A2
AMMUNITION
   400 rounds .50 caliber
   one of the following
   56 rounds 105mm howitzer
   24 rounds 155mm howitzer
VISION EQUIPMENT
Windows in cab or open top vehicle

ENGINE
Make and Model: Continental R6572
Type: 6 cylinder, 4 cycle, in-line
Cooling System: Liquid   Ignition: Delco
Displacement: 571.7 cubic inches
Bore and Stroke: 4.75 x 5.375 inches
Compression Ratio: 6.5:1
Net Horsepower: (max)   235 hp at 2900 rpm
Net Torque: (max)   490 ft-lbs at 1600 rpm
Weight: 2026 pounds, dry
Fuel: 70 octane gasoline   100 gallons
Engine Oil: 22 quarts
POWER TRAIN
Clutch/Transfer Case:
   Gear ratio: Low   1.71:1   High   1.00:1
Transmission: Constant mesh
   Gear Ratio: 1st   5.43:1   4th   1.00:1
            2nd   3.20:1   reverse 5.36:1
            3rd   1.71:1
Steering: Controlled differential, steering brake levers
   Gear Ratio: 2.60:1   Steering Ratio: 1.844:1
Brakes: Mechanical on controlled differential
Final Drive: Spur gear   Gear Ratio: 2.35:1
Drive Sprockets: At front of vehicle with 14 teeth
   Pitch Diameter: 24.56 inches
RUNNING GEAR
Suspension: M5, M5A1, Vertical volute spring
     M5A2, M5A3, Horizontal volute spring
   M5, M5A1, 8 wheels in 4 bogies (2 bogies/track)
     Tire Size: 20 x 6 inches
     4 track return rollers (2/track)
     Adjustable trailing idler at rear of each track
     Idler Size: 28 x 6 inches
   M5A2, M5A3, 8 dual wheels in 4 bogies (2 bogies/track)
     4 dual track return rollers (2/track)
     Adjustable dual trailing idler at rear of each track
Tracks: M5, M5A1, Outside guide, M5A2, M5A3, Center guide
   Type: M5, M5A1, Double pin, 11.6 inch width, steel
     M5A2, M5A3, Single pin, 21 inch width, steel
   Pitch: 5.5 inches
   Shoes per Vehicle: M5, M5A1, 124 (62/track)
         M5A2, M5A3, 126 (63/track)
   Ground Contact Length: 108.5 inches
ELECTRICAL SYSTEM
Nominal Voltage: 12 volts DC
Main Generator: (1) 12 volts, driven by main engine
Auxiliary Generator: None
Battery: (1) 12 volts
COMMUNICATIONS
Radio: None
Interphone: None
FIRE PROTECTION
   (1) 4 pound carbon dioxide, portable
PERFORMANCE
Maximum Speed: Level roads   35 miles/hour
Maximum Grade: 72 per cent
Maximum Trench: 5.5 feet
Maximum Vertical Wall: 18 inches
Maximum Fording Depth: 53 inches
Minimum Turning Circle: (diameter) 36 feet
Cruising Range: Roads   approx. 125 miles

38 TON HIGH SPEED TRACTOR M6

GENERAL DATA
Crew: 1 man
Passengers: 10 men
Length: 257.8 inches
Width: 120.5 inches
Height: 104.1 inches
Tread: 98.5 inches
Ground Clearance: 20 inches
Weight, Combat Loaded: 75,000 pounds
Weight, Unstowed: 60,000 pounds
Power to Weight Ratio: Net 10.2 hp/ton
Ground Pressure: Zero penetration 9.9 psi
ARMOR
None
ARMAMENT
(1) 50 caliber MG HB M2 on mount M49C or M66
AMMUNITION
600 rounds .50 caliber
one of the following
24 rounds 120mm gun
24 rounds 8 inch gun
20 rounds 240mm howitzer
VISION EQUIPMENT
Windows in cab
WINCH
Gar Wood Model 6M823 located in rear hull
Capacity: 55,000 pounds
Cable Size: 1 inch x 300 feet

ENGINE
Make and Model: (2) Waukesha 145GZ
Type: 6 cylinder, 4 cycle, in-line
Cooling System: Liquid   Ignition: Battery
Displacement: 1634 cubic inches
(817 cubic inches/engine)
Bore and Stroke: 5.375 x 6 inches
Compression Ratio: 5.95:1
Net Horsepower: (max) 382 hp at 2100 rpm
(191 hp/engine)
Net Torque: (max) 1078 ft-lbs at 1500 rpm
(539 ft-lbs/engine)
Weight: 4300 pounds, dry
(2150 pounds/engine)
Fuel: 70 octane gasoline 250 gallons
Engine Oil: 18 quarts/engine
POWER TRAIN
Torque Converter: Twin Disk Model T-10010
Torque Converter Gear Ratio: 4.5:1
Transmission: Constant mesh
Gear ratios: 1st 2.12:1 reverse 2.76:1
2nd 1.05:1
Steering: Controlled differential, steering levers
Steering Ratio: 1.6:1
Brakes: Mechanical on controlled differential
Final Drive: Herringbone gear   Gear Ratio: 3.06:1
Drive Sprockets: At front of vehicle with 13 teeth
Pitch Diameter: 25.038 inches
RUNNING GEAR
Suspension: Horizontal volute spring
12 dual wheels in 6 bogies (3 bogies/track)
Tire Size: 20 x 6 inches
6 dual track return rollers (3/track)
Adjustable dual trailing idler at rear of each track
Idler Size: 32 x 6 inches
Tracks: Center guide
Type: Double pin, 21.56 inch width, steel chevron
Pitch: 6 inches
Shoes per Vehicle: 168 (84/track)
Ground Contact Length: 176.375 inches
ELECTRICAL SYSTEM
Nominal Voltage: 12 volts DC
Main Generator: (1) 15 volts, 25 amperes, driven by right engine
Auxiliary Generator: None
Battery: (2) 12 volts in parallel
COMMUNICATIONS
Radio: None
Interphone: None
FIRE PROTECTION
(2) 4 pound carbon dioxide, portable
PERFORMANCE
Maximum Speed: Level road 21 miles/hour
Maximum Grade: 60 per cent
Maximum Trench: 8 feet
Maximum Vertical Wall: 30 inches
Maximum Fording Depth: 54 inches
Minimum Turning Circle: (diameter) 53 feet
Cruising Range: Roads approx. 110 miles

460

GENERAL DATA

Crew:	2	men
Length:	265.1	inches
Width:	130.5	inches
Height:	117.25	inches
Tread:	102.5	inches
Ground Clearance:	18.8	inches
Weight, Combat Loaded:	63,000	pounds
Weight, Unstowed:	44,500	pounds
Power to Weight Ratio: Net	11.5	hp/ton
Gross	15.9	hp/ton
Ground Pressure: Zero penetration	9.5	psi

ARMOR

None

ARMAMENT

(1) .50 caliber MG HB M2 on ring mount
(2) .30 caliber carbine stowed
(1) 3.5 inch rocket launcher stowed

AMMUNITION

500 rounds .50 caliber
360 rounds .30 caliber, carbine
10 rounds 3.5 inch rockets

VISION EQUIPMENT

Windows in cab

ENGINE

Make and Model: M8A1, Continental AOS-895-3
M8A2, Continental AOSI-895-5
Type: AOS-895-3, 6 cylinder, 4 cycle, opposed, supercharged
AOSI-895-5, 6 cylinder, 4 cycle, opposed, supercharged, w/ fuel injection

Cooling System: Air	Ignition: Magneto
Displacement:	895.9 cubic inches
Bore and Stroke:	5.75 x 5.75 inches
Compression Ratio:	5.5:1
Net Horsepower: (max)	363 hp at 2800 rpm
Gross Horsepower: (max)	500 hp at 2800 rpm
Net Torque: (max)	760 ft-lbs at 2060 rpm
Gross Torque: (max)	985 ft-lbs at 2300 rpm
Weight:	approx. 1900 pounds, dry
Fuel: 80 octane gasoline	225 gallons
Engine Oil:	44 quarts

POWER TRAIN

Transmission: Allison CD-500-3, 2 speeds forward, 1 reverse
w/automatic lockup in high
Single stage hydraulic torque converter
Stall Multiplication: 4:1

Overall Usable Ratios:	Low	15.3:1	direct	1:1
	High	4.1:1	reverse	15.3:1

Steering: Mechanical, T-bar
Brakes: Multiple plate, oil cooled
Final Drive: Spur gear   Gear Ratio: 4.69:1
Drive Sprockets: At front of vehicle with 12 teeth
Pitch Diameter: 23.422 inches

RUNNING GEAR

Suspension: Torsion bar
12 dual road wheels (6/track)
Tire Size: 25.5 x 4.5 inches
8 dual track return rollers (4/track)
Adjustable idler at rear of each track
Tracks: Center guide T91E3
Type: Single pin, 21 inch width, steel chevron
Pitch: 6 inches
Shoes per Vehicle: 174 (87/track)
Ground Contact Length: 157.1 inches

ELECTRICAL SYSTEM

Nominal Voltage: 24 volts DC
Main Generator: (1) 24 volts, 150 amperes, driven by main engine
Auxiliary Generator: None
Battery: (2) 12 volts in series

COMMUNICATIONS

Radio: AN/VRC-8 to 10, AN/GRR-5
Interphone: None

FIRE PROTECTION

(3) 10 pound carbon dioxide, fixed
(1) 5 pound carbon dioxide, portable

PERFORMANCE

Maximum Speed: Level road	41 miles/hour
Maximum Tractive Effort: TE at stall	51,600 pounds
Per Cent of Vehicle Weight: TE/W	81 per cent
Maximum Grade:	60 per cent
Maximum Trench:	7 feet
Maximum Vertical Wall:	30 inches
Maximum Fording Depth:	42 inches
Minimum Turning Circle: (diameter)	pivot
Cruising Range: Roads	approx. 180 miles

GENERAL DATA

Crew:	3	men
Passengers:	34	men
Length: Overall	356	inches
Width: Overall	140.5	inches
Height: Over MG cupola	120.5	inches
Tread:	116.7	inches
Ground Clearance: Center	18	inches
Sides	11	inches
Weight, Combat Loaded: Land operation	87,780	pounds
Water operation	81,780	pounds
Weight: Unstowed:	64,200	pounds
Power to Weight Ratio: Land operation	18.5	hp/ton
Water operation	19.8	hp/ton
Ground Pressure: Zero penetration	9.2	psi

ARMOR

Steel armor 0.25 to 0.625 inches in thickness,
Welded assembly

ARMAMENT

(1) .30 caliber MG M1919A4 in cupola mount
Traverse: Manual   360 degrees
Elevation: Manual   +60 to -15 degrees

AMMUNITION

250 rounds .30 caliber

VISION EQUIPMENT

Vision Devices:	Direct	Indirect
Driver	Hatch	Periscope M17 (4)
		Periscope M17C (1)
Vehicle Commander	Hatch	Periscope M17 (4)
		Periscope M17C (1)
MG Cupola	Hatch and	None
	5 vision blocks	
Troop Commander	None	Periscopes M17 (2)

Total Periscopes: M17 (10), M17C (2)
Total Vision Blocks: (5) in MG cupola

ENGINE

Make and Model: Continental LV-1790-1
Type: 12 cylinder, 4 cycle, 90 degree vee

Cooling System: Liquid	Ignition: Magneto	
Displacement:	1791.7 cubic inches	
Bore and stroke:	5.75 x 5.75 inches	
Compression Ratio:	6.5:1	
Net Horsepower: (max)	704 hp at 2800 rpm	
Gross Horsepower: (max)	810 hp at 2800 rpm	
Net Torque: (max)	1440 ft-lbs at 2000 rpm	
Gross Torque: (max)	1610 ft-lbs at 2200 rpm	
Fuel: 80 octane gasoline	456 gallons	
Engine Oil:	64 quarts	

POWER TRAIN

Transmission: Allison CD-850-4A or CD-850-4B
    2 ranges forward, 1 reverse
    Single stage, multiphase, hydraulic torque converter
    Stall Multiplication: 4.3:1
    Overall Usable Ratios: low   13.0:1   reverse   17.8:1
                              high   4.5:1
Steering: Mechanical, wobble stick
Brakes: Multiple plate, oil cooled
Final Drive: ..LVTP5, Planetary gear
               LVTP5A1, Spur gear
Drive Sprockets: At rear of vehicle with 17 teeth
    Pitch Diameter: approx. 26.5 inches

RUNNING GEAR

Suspension: Torsilastic
    18 pairs of dual road wheels (9 pairs/track)
    Tire Size: 19.4 x 6 inches
    10 dual track return rollers (5/track)
    Dual compensating idler at front of each track
    Idler Size: 22 x 2.5 inches
Tracks: Center guide (inverted grouser)
    Type: Single pin, 20.75 inch width, steel
    Pitch: 5 inches
    Shoes per Vehicle: 268 (134/track)
    Ground Contact Length: 229.25 inches

ELECTRICAL SYSTEM

Nominal Voltage: 24 volts DC
Main Generator: (1) 24 volts, 300 amperes, driven by main engine
Auxiliary generator: (1) 28.5 volts, 175.5 amperes, driven by auxiliary engine
Battery: (4) 12 volts, 2 sets of 2 in series connected in parallel

COMMUNICATIONS

Radio: AN/GRC-5, AN/GRC-7 or AN/PRC-8 thru 10
Interphone: 3 stations

FIRE PROTECTION

(3) 5 pound carbon dioxide, fixed
(1) 5 pound carbon dioxide, portable

PERFORMANCE

Maximum Speed: Level Road	30 miles/hour	
Water	6.8 miles/hour	
Maximum Grade:	70 per cent	
Maximum Trench:	12 feet	
Maximum Vertical Wall:	36 inches	
Maximum Fording Depth:	floats	
Minimum Turning Circle: (diameter)	pivot	
Cruising Range: Roads	approx. 190 miles	
Water	approx. 57 miles	

GENERAL DATA

Crew:	7	men
Length: Overall	356	inches
Width: Overall	140.5	inches
Height: Over .50 caliber MG	160.5	inches
Tread:	116.7	inches
Turret Ring Diameter: (inside)	77.75	inches
Ground Clearance: Center	17	inches
Sides	11	inches
Weight, Combat Loaded: Land operation	86,600	pounds
Water operation	84,200	pounds
Weight, Unstowed:	74,210	pounds
Power to Weight Ratio: Land operation	18.7	hp/ton
Water operation	19.2	hp/ton
Ground Pressure: Zero penetration	9.1	psi

ARMOR

Type: Turret, rolled homogeneous steel; Hull rolled homogeneous steel; Welded assembly
Hull Thickness: 0.25 to 0.625 inches
Turret Thickness: Front 1 inch, Sides .075 inches, Top 0.25 inches

ARMAMENT

Primary: 105mm howitzer M49 in mount T172 in turret

Traverse: Hydraulic and manual	360 degrees
Traverse Rate: (max)	17 seconds/360 degrees
Elevation: Manual	+59 to -4.1 degrees
Firing Rate: (max)	
Loading System:	Manual
Stabilizer System:	Elevation only

Secondary:
(1) .50 caliber MG HB M2 on turret
(1) .30 caliber MG M1919A4E1 coaxial w/howitzer

AMMUNITION

151 rounds 105mm, land operation
or
100 rounds 105mm, water operation
1050 rounds .50 caliber
2000 rounds .30 caliber

FIRE CONTROL AND VISION EQUIPMENT

Primary Weapon:	Direct	Indirect
	Telescope T150E2	Telescope T149E2
	Periscope M23 (T38)	Azimuth Indicator T24
		Gunner's Quadrant M1A1

Vision Devices:

Driver	Hatch	Periscope M17 (4)
		Periscope M17C (1)
Asst. Driver	Hatch	Periscope M17 (4)
		Periscope M17C (1)
Commander	Hatch	Periscope M17 (4)
		Periscope M17C (1)
Gunner	Vision block (1)	Periscope M23 (T38)
		Telescope T149E2
Loader	Hatch and	None
	pistol port	
	Vision block (2)	

Total Periscopes: M17 (12), M17C (3), M23 (T28) (1),
Total Vision Blocks: (3) in turret side walls

ENGINE

Make and Model: Continental LV-1790-1	
Type: 12 cylinder, 4 cycle, 90 degree vee	
Cooling System: Liquid	Ignition: Magneto
Displacement:	1791.7 cubic inches
Bore and Stroke:	5.75 x 5.75 inches
Compression Ratio:	6.5:1
Net Horsepower: (max)	704 hp at 2800 rpm
Gross Horsepower: (max)	810 hp at 2800 rpm
Net Torque: (max)	1440 ft-lbs at 2000 rpm
Gross Torque: (max)	1610 ft-lbs at 2200 rpm
Fuel: 80 octane gasoline	456 gallons
Engine Oil:	64 quarts

POWER TRAIN

Transmission: Allison CD-850-4A or CD-850-4B
    2 ranges forward, 1 reverse
    Single stage, multiphase, hydraulic torque converter
    Stall Multiplication: 4.3:1

Overall Useable Ratios:	low	13.0:1	reverse 17.8:1
	high	4.5:1	

Steering: Mechanical, wobble stick
Brakes: Mechanical, oil cooled
Final Drive: LVTH6, Planetary gear
           LVTH6A1, Spur gear
Drive Sprockets: At rear of vehicle with 17 teeth
    Pitch Diameter: approx. 26.5 inches

RUNNING GEAR

Suspension: Torsilastic
    18 pairs of dual road wheels (9 pairs/track)
    Tire Size: 19.4 x 6 inches
    10 dual track return rollers (5/track)
    Dual compensating idler at front of each track
    Idler Size: 22 x 2.5 inches
Tracks: Center guide (inverted grouser)
    Type: Single pin, 20.75 inch width, steel
    Pitch: 5 inches
    Shoes per Vehicle: 268 (134/track)
    Ground Contact Length: 229.25 inches

ELECTRICAL SYSTEM

Nominal Voltage: 24 volts DC
Main Generator: (1) 24 volts, 300 amperes, driven by main engine
Auxiliary Generator: (1) 28.5 volts, 175.5 amperes, driven by auxiliary engine
Battery: (4) 12 volts, 2 sets of 2 in series connected in parallel

COMMUNICATIONS

Radio: AN/GRC-5
Interphone: AN/VIA-1, 5 stations

FIRE PROTECTION

    (3) 5 pound carbon dioxide, fixed
    (1) 5 pound carbon dioxide, portable

PERFORMANCE

Maximum Speed: Level road		30 miles/hour
Water		6.8 miles/hour
Maximum Grade:		60 per cent
Maximum Trench:		12 feet
Maximum Vertical Wall:		36 inches
Maximum Fording Depth:		floats
Minimum Turning Circle: (diameter)		pivot
Cruising Range: Roads	approx.	190 miles
Water	approx.	57 miles

GENERAL DATA

Crew:	3	men
Passengers:	25	men
Length:	312.75	inches
Width:	128.72	inches
Height:	128.5	inches
Tread:	123.8	inches
Ground Clearance:	16	inches
Weight, Combat Loaded:	50350	pounds
Weight, Unstowed:	38451	pounds
Power to Weight Ratio:	15.9	hp/ton
Ground Pressure: Zero penetration	7.7	psi

ARMOR
Type: 5083 aluminum alloy armor,
    Welded assembly

ARMAMENT
    (1) .50 caliber MG M85 in armament station
    Traverse: electrohydraulic and manual   360 degrees
    Traverse Rate: (max)   6 seconds/360 degrees
    Elevation: electrohydraulic and manual +60 to -15 degrees
    Elevation Rate: (max)   60 degrees/second

AMMUNITION
    400 rounds .50 caliber

FIRE CONTROL AND VISION EQUIPMENT
Armament Station: 8x optical sight and ring sight

Vision devices:	Direct	Indirect
Driver	Hatch and 7 vision blocks	Periscope M24 (infrared) (1)
Commander	Hatch and 7 vision blocks	Periscope M27 (1)
Gunner	Hatch and 9 vision blocks	None
Rear Ramp	1 vision block	None

ENGINE
Make and Model: General Motors 8V53T
Type: 8 cylinder, 2 cycle, vee, turbosupercharged

Cooling System: Liquid	Ignition: Compression
Displacement:	424 cubic inches
Bore and Stroke:	3.875 x 4.5 inches
Compression Ratio:	17.0:1
Gross Horsepower: (max)	400 hp at 2800 rpm
Gross Torque: (max)	825 ft-lbs at 2150 rpm
Weight:	2300 pounds, dry
Fuel: diesel oil MIL-VV-F-800	180 gallons
Engine Oil:	22 quarts

POWER TRAIN
Transmission: FMC HS-400-3
        Manually selected, full power shifting
        4 speeds forward, 2 reverse
Steering: Hydrostatic
Brakes: Mechanical
Final Drive: Spur gear   Gear Ratio: 3.06:1
Drive Sprockets: At front of vehicle with 11 teeth
        Pitch Diameter: 21 inches
Water Propulsion:
        Primary: Water jets
        Secondary: Tracks

RUNNING GEAR
Suspension: Flat track, torsion tube over bar
        12 dual road wheels (6/track)
        Tire Size: 26 x 4 inches
        Dual adjustable idler at rear of each track
        Idler Size: 20 x 3.25 inches
        Shock absorbers on first and last road wheels on each side
Tracks: Center guide
        Type: Single pin, 21 inch width, steel w/detachable rubber pad
        Pitch: 6 inches
        Shoes per Vehicle: 168 (84/track)
                        New Vehicle, 170 (85/track)
        Ground Contact Length: 155 inches

ELECTRICAL SYSTEM
Nominal Voltage: 24 volts DC
Main Generator: (1) 24 volts, 180 amperes, driven by main engine
Auxiliary Generator: None
Battery: (4) 12 volts, 2 sets of 2 in series connected in parallel

COMMUNICATIONS
Radio: AN/VRC-44, AN/VRC-46
Interphone: C-2298/VRC, 5 stations

FIRE PROTECTION
        (1) 17 pound Halon, fixed
        (1) 2.75 pound Halon, portable

PERFORMANCE

Maximum Speed: Level road		40 miles/hour
Water		8.4 miles/hour
Maximum Grade:		60 per cent
Maximum Trench:		8 feet
Maximum Vertical Wall:		36 inches
Maximum Fording Depth:		floats
Minimum Turning Circle: (diameter)		pivot
Cruising Range: Roads	approx.	300 miles
Water endurance		7 hours

GENERAL DATA

Crew:	3	men
Passengers:	25	men
Length:	321.3	inches
Width:	128.7	inches
Height:	130.5	inches
Tread:	123.8	inches
Ground Clearance:	16	inches
Weight, Combat Loaded: w/o applique armor	56,552	pounds
Weight, Unstowed:	42,108	pounds
Power to Weight ratio:	14.1	hp/ton
Ground Pressure: Zero penetration	8.7	psi

ARMOR

Type: 5083 aluminum alloy armor,
    Welded assembly
    Steel applique armor can be installed on sides of the vehicle.

ARMAMENT

(1) .50 caliber MG HB M2 in weapon station
(1) 40mm automatic grenade launcher Mk 19 in weapon station
    Traverse: Electric and manual   360 degrees
    Traverse Rate: (max)
    Elevation: Electric and manual   +60 to -15 degrees
    Elevation Rate: (max)
(2) 4 tube smoke grenade launchers on rear of weapon station

AMMUNITION

250 rounds .50 caliber, ready
48 rounds 40mm, ready

FIRE CONTROL AND VISION EQUIPMENT

Weapon Station: Periscope sight

Vision Devices	Direct	Indirect
Driver	Hatch and 7 vision blocks	Night vision viewer AN/VVS-2(V)1A
Commander	Hatch and 7 vision blocks	Periscope M27 (1)
Gunner	Hatch and vision blocks	Periscope sight
Rear ramp	1 vision block	None

ENGINE

Make and Model: Cummins VT400
Type: 8 cylinder, 4 cycle, vee, turbosupercharged

Cooling System: Liquid Ignition: Compression	
Displacement:	903 cubic inches
Bore and Stroke:	5.5 x 4.75 inches
Compression Ratio:	15.5:1
Gross Horsepower: (max)	400 hp at 2800 rpm
Gross torque: (max)	825 ft-lbs at 2050 rpm
Weight:	2450 pounds, dry
Fuel: diesel oil MIL-VV-F-800	171 gallons
Engine Oil:	26 quarts

POWER TRAIN

Transmission: FMC HS-400-3A1
    Manually selected, full power shifting
    4 speeds forward, 2 reverse
Steering: Hydrostatic
Brakes: Mechanical
Final Drive: Spur gear   Gear Ratio: 3.06:1
Drive Sprockets: At front of vehicle with 11 teeth
    Pitch Diameter: 21 inches
Water Propulsion:
    Primary: Water jets, 14,000 gallons/minute
    Secondary: Tracks

RUNNING GEAR

Suspension: Flat track, torsion tube over bar
    12 dual road wheels (6/track)
    Tire Size: 26 x 4 inches
    Dual adjustable idler at rear of each track
    Idler Size: 20 x 3.25 inches
    Shock absorbers on first and last road wheels on each side
Tracks: Center guide
    Type: Single pin, 21 inch width, steel w/detachable rubber pads
    Pitch: 6 inches
    Shoes per vehicle: 168 (84/track)
        New vehicle, 170 (85/track)
    Ground Contact Length: 155 inches

ELECTRICAL SYSTEM

Nominal Voltage: 24 volts DC
Main Generator: (1) 24 volts, 300 amperes, driven by main engine
Auxiliary Generator: None
Battery: (4) 12 volts, 2 sets of 2 in series connected in parallel

COMMUNICATIONS

Radio: AN/VRC-44, AN/VRC-46
Interphone: CC2298/VRC, 5 stations

FIRE PROTECTION

(2) 7 pound Halon, fixed
(1) 17 pound Halon, fixed
(1) 2.75 pound Halon, portable

PERFORMANCE

Maximum Speed: Level road		45 miles/hour
	Water	8.2 miles/hour
Maximum Grade:		60 per cent
Maximum Trench:		8 feet
Maximum Vertical Wall:		36 inches
Maximum Fording Depth:		floats
Minimum Turning Circle: (diameter)		pivot
Cruising Range: Roads	approx.	300 miles
	Water endurance	7 hours

REFERENCES AND SELECTED BIBLIOGRAPHY

Books and Published Articles

Flint, Captain Glenwood W. and Tixier, Captain Lewis B., "The M59", Armor, March-April 1954, p6

McLaughlin, Lieutenant John C., "The M75 Armored Personnel Carrier", Armor, January-February 1954, p6

Starry, General Donn A., "Mounted Combat in Vietnam", Department of the Army, Washington, D.C., 1978

Reports and Official Documents

"Fleet Marine Force Manual FMFM 9-2 Amphibian Vehicles", U.S. Marine Corps, Washington, D.C., 17 September 1964

"Notes on Materiel Armored Utility Vehicle T16 (M44)", Cadillac Motor Car Division, General Motors Corporation, Detroit, Michigan, 1 May 1947

"Notes on Development Type Materiel Infantry Vehicle, Armored, Tracked, T59 and Infantry Vehicle, Armored, Tracked, T59E1", Food Machinery and Chemical Corporation, San Jose, California, June 1952

"Notes on Development Type Materiel Armored Infantry Vehicle T73" International Harvester Company, Melrose Park, Illinois, 1 August 1952

"Research, Investigation & Experimentation in the Field of Amphibian Vehicles for the U.S. Marine Corps", Final Report, Ingersoll Kalamazoo Division, Borg-Warner Corporation, Kalamazoo, Michigan, December 1957

"Technical Manual TM9-772 Carrier, Cargo, M29 & Carrier, Cargo M29C", War Department, Washington, D.C., 5 July 1944

"Technical Manual TM9-785 18-ton High Speed Tractors M4, M4A1, M4C, and M4A1C", Department of the Army, Washington, D.C., April 1952

"Technical Manual TM9-786 13-ton High Speed Tractors M5, M5A1, M5A2, and M5A3" Department of the Army, Washington, D.C., April 1950

"Technical Manual TM9-788 38-ton High Speed Tractor M6", Department of the Army, Washington, D.C., May 1952

"Technical Manual TM9-1425-646-10-1 Operator's Manual Launcher, Rocket, Armored Vehicle Mounted M270, Volume 1", Department of the Army, Washington, D.C., 5 April 1990

"Technical Manual TM9-1450-646-10 Technical Manual Operator's Manual for Carrier, Multiple Launch Rocket System M993", Department of the Army, Washington, D.C., August 1984

"Technical Manual TM9-2300-203-12 Operation and Organizational Maintenance Full Tracked Armored Personnel Carrier M59 (T59) and 4.2 inch Full Tracked Self-propelled Mortar M84", Department of the Army, Washington, D.C., October 1958

"Technical Manual TM9-2300-224-10 Operators Manual for Carrier, Personnel, Full Tracked; Armored, M113" Department of the Army, Washington, D.C., 7 November 1961

"Technical Manual TM9-2300-224-20 Organizational Maintenance Manual for Carrier, Personnel, Full Tracked, Armored, M113" Department of the Army, Washington, D.C., 15 December 1961

"Technical Manual TM9-2300-257-10 Operator's Manual for Carrier, Personnel, Full Tracked, Armored M113A1, Carrier, Command Post, Light, Tracked M577A1, Carrier, Mortar, 107mm, Self-propelled M106A1, Carrier, Mortar, 81mm, Self-propelled M125A1, Carrier, Flame Thrower, Self-propelled M132A1" Department of the Army, Washington, D.C., November 1983

"Technical Manual TM9-2320-223-10 Operator's Manual for Carrier, Cargo, Amphibious: Tracked, M116 (T116E1)", Department of the Army, Washington, D.C., 9 July 1962

"Technical Manual TM9-2320-224-10 Operator's Manual for Carrier, Command and Reconnaissance: Armored M114 and M114A1", Department of the Army, Washington, D.C., 25 November 1964

"Technical Manual TM9-2350-247-10 Operator's Manual Carrier, Cargo, Tracked, 6-ton M548", Department of the Army, Washington, D.C., May 1974

"Technical Manual TM9-2350-252-10-1 Technical Manual Operator's Manual for Fighting Vehicle, Infantry M2, M2A1 and Fighting Vehicle Cavalry M3, M3A1, Hull", Department of the Army, Washington, D.C., November 1986

"Technical Manual TM9-2350-259-10 Operator's Manual for Combat Vehicle, Antitank, Improved TOW Vehicle M901 and M901A1" Department of the Army, Washington, D. C., 29 June 1979

"Technical Manual TM9-2350-261-10 Operator's Manual for Carrier, Personnel, Full Tracked, Armored M113A2, Carrier, Command Post, Light, , Tracked M577A2, Carrier, Mortar, 107mm M30, Self-propelled M106A2, Carrier, Mortar, 81mm M29A1, Self-propelled M125A2, Carrier, Smoke Generator, Full Tracked M1059, Carrier Mortar, 120mm Self-propelled M1064, Carrier, Standardized Integrated Command Post System M1068" Department of the Army, Washington, D.C., 19 August 1993

"Technical Manual TM9-2350-277-10 Operator's Manual for Carrier, Personnel, Full Tracked, Armored M113A3, Carrier Command Post, Light, Tracked M577A3, Carrier, Antitank (TOW), Full Tracked, Armored M901A3, Carrier, Fire Support Personnel, Full Tracked, Armored M981A3, Carrier, Smoke Generator, Full Tracked M1059A3, Carrier, Mortar, 120mm, Self-propelled M1064A3, Carrier, Standardized Integrated Command Post System M1068A3", Department of the Army, Washington, D.C., 25 July 1994

"Technical Manual TM9-2350-284-10-1 Technical Manual Operator's Manual High Survivability for Fighting Vehicle, Infantry M2A2 and Fighting Vehicle, Cavalry M3A2, Hull" Department of the Army, Washington, D.C., January 1996

"Technical Manual TM9-2350-284-10-2 Operator's Manual High Survivability Fighting Vehicle, Infantry M2A2 and Fighting Vehicle, Cavalry M3A2, Turret", Department f the Army, Washington, D.C., 16 July 1996

"Technical Manual TM9-2350-300-10 Operation and Maintenance Manual for Gun, Air Defense artillery, Self-propelled, 20mm M163A1" Department of the Army, Washington, D.C. , May 1976

"Technical Manual TM9-7418 Armored Full Tracked Personnel Carrier M75", Department of the Army , Washington, D.C., 30 January 1956

"Technical Manual TM9-7604 Amphibious Cargo Carrier M76", Department of the Army, Washington, D.C., March 1957

"U.S. Marine Corps Maintenance Manual USMC-ORD-MM-7000 for Landing Vehicle Tracked LVTP5", U.S. Marine Corps, Washington, D.C., May 1955

"U.S. Marine Corps Maintenance Manual USMC-ORD-MM-7001 for Landing Vehicle Tracked LVTH6", U.S. Marine Corps, Washington, D.C., May 1958

"U.S. Marine Corps Technical Manual TM 02642A-10/2 Landing Vehicle Tracked Engineer Model 1 LVTE1", U.S. Marine Corps, Washington, D.C. November 1966

"U.S. Marine Corps Technical Manual TM-00512C-10/1 for Landing Vehicle Tracked Howitzer Model H6A1 LVTH6A1" U.S. Marine Corps, Washington, D.C., January 1965

"U.S. Marine Corps Technical Manual TM-07007A-10A Operator's Manual for Landing Vehicle Tracked, Personnel, Model 7 LVTP7" U.S. Marine Corps, Washington, D.C., 1 August 1978

"U.S. Marine Corps Technical Manual TM-07007B-10 Operator's Manual for Landing Vehicle Tracked, Personnel, Model 7A1 LVTP7A1" U.S. Marine Corps, Washington, D.C., 1 June, 1983

"U.S. Marine Corps Technical Manual TM-07267A-10/1 Operator's Manual for Landing Vehicle Tracked, Recovery, Model 7 LVTR7" U.S. Marine Corps, Washington, D.C., August 1980

"U.S. Marine Corps Technical Manual TM-08600A-10 Operator's Manual for Electric Drive M85 Weapon Station Assault Amphibian Vehicle, Personnel, Model 7A1, AAVP7A1", U.S. Marine Corps, Washington, D.C., December 1984

468

www.ingramcontent.com/pod-product-compliance
Lightning Source LLC
Chambersburg PA
CBHW050237220326
41598CB00047B/7435